STRUCTURAL AND STRESS ANALYSIS

To my wife Margaret with love
and gratitude

STRUCTURAL AND STRESS ANALYSIS

Dr. T. H. G. MEGSON
Senior Lecturer in Civil Engineering
University of Leeds

OXFORD AUCKLAND BOSTON JOHANNESBURG MELBOURNE NEW DELHI

Butterworth-Heinemann
Linacre House, Jordan Hill, Oxford OX2 8DP
225 Wildwood Avenue, Woburn, MA 01801-2041
A division of Reed Educational and Professional Publishing Ltd

A member of the Reed Elsevier plc group

First published in Great Britain by Arnold 1996
Reprinted by Butterworth-Heinemann 2000

British Library Cataloguing in Publication Data
A catalogue record for this book is available from the British Library

Library of Congress Cataloguing in Publication Data
A catalogue record for this book is available from the Library of Congress

ISBN 0 340 63196 1

Typset in 10/12 pt Times by
Mathematical Composition Setters Ltd, Salisbury, UK.
Printed and bound in Great Britain by MPG Books Ltd, Bodmin, Cornwall

Contents

Chapter 18   Structural Instability

Preface

The purpose of this book is to provide, in a unified form, a text covering the associated topics of structural and stress analysis for students of civil engineering during the first two years of their degree course. The book is also intended for students studying for Higher National Diplomas, Higher National Certificates and related courses in civil engineering.

Frequently, textbooks on these topics concentrate on structural analysis or stress analysis and often they are lectured as two separate courses. There is, however, a degree of overlap between the two subjects and, moreover, they are closely related. In this book, therefore, they are presented in a unified form which illustrates their interdependence. This is particularly important at the first-year level where there is a tendency for students to 'compartmentalize' subjects so that an overall appreciation of the subject is lost.

The subject matter presented here is confined to the topics students would be expected to study in their first two years since third- and fourth-year courses in structural and/or stress analysis can be relatively highly specialized and are therefore best served by specialist texts. Furthermore, the topics are arranged in a logical manner so that one follows naturally on from another. Thus, for example, internal force systems in statically determinate structures are determined before their associated stresses and strains are considered, while complex stress and strain systems produced by the simultaneous application of different types of load follow the determination of stresses and strains due to the loads acting separately.

Although in practice modern methods of analysis are largely computer-based, the methods presented in this book form, in many cases, the basis for the establishment of the flexibility and stiffness matrices that are used in computer-based analysis. It is therefore advantageous for these methods to be studied since, otherwise, the student would not obtain an appreciation of structural behaviour, an essential part of the structural designer's background.

In recent years some students enrolling for degree courses in civil engineering, while being perfectly qualified from the point of view of pure mathematics, lack a knowledge of structural mechanics, an essential basis for the study of structural and stress analysis. Therefore a chapter devoted to those principles of statics that are a necessary preliminary has been included.

As stated above, the topics have been arranged in a logical sequence so that they form a coherent and progressive 'story'. Hence, in Chapter 1, structures are considered in terms of their function, their geometries in different roles, their methods of support and the differences between their statically determinate and indeterminate forms. Also considered is the role of analysis in the design process and methods of idealizing structures so that they become amenable to analysis. In Chapter 2 the necessary principles of statics are discussed and applied directly to the calculation of support reactions. Chapters 3–6 are concerned with the determination of internal force distributions in statically determinate beams, trusses, cables and arches, while in Chapter 7 stress and strain are discussed and stress–strain relationships established. The relationships between the elastic constants are then derived and the concept of strain energy in axial tension and compression introduced. This is then applied to the determination of the effects of impact loads, the calculation of displacements in axially loaded members and the deflection of a simple truss. Subsequently, some simple statically indeterminate systems are analysed and the compatibility of displacement condition introduced. Finally, expressions for the stresses in thin-walled pressure vessels are derived. The properties of the different materials used in civil engineering are investigated in Chapter 8 together with an introduction to the phenomena of strain-hardening, creep and relaxation and fatigue; a table of the properties of the more common civil engineering materials is given at the end of the chapter. Chapters 9, 10 and 11 are respectively concerned with the stresses produced by the bending, shear and torsion of beams while Chapter 12 investigates composite beams. Deflections due to bending and shear are determined in Chapter 13, which also includes the application of the theory to the analysis of some statically indeterminate beams. Having determined stress distributions produced by the separate actions of different types of load, we consider, in Chapter 14, the state of stress and strain at a point in a structural member when the loads act simultaneously. This leads directly to the experimental determination of surface strains and stresses and the theories of elastic failure for both ductile and brittle materials. Chapter 15 contains a detailed discussion of the principle of virtual work and the various energy methods. These are applied to the determination of the displacements of beams and trusses and to the determination of the effects of temperature gradients in beams. Finally, the reciprocal theorems are derived and their use illustrated. Chapter 16 is concerned solely with the analysis of statically indeterminate structures. Initially methods for determining the degree of statical and kinematic indeterminacy of a structure are described and then the methods presented in Chapter 15 are used to analyse statically indeterminate beams, trusses, braced beams, portal frames and two-pinned arches. Special methods of analysis, i.e. slope-deflection and moment distribution, are then applied to continuous beams and frames. The chapter is concluded by an introduction to matrix methods. Chapter 17 covers influence lines for beams, trusses and continuous beams while Chapter 18 investigates the stability of columns.

Numerous worked examples are presented in the text to illustrate the theory, while a selection of unworked problems with answers is given at the end of each chapter.

T.H.G. MEGSON

CHAPTER 1

Introduction

In the past it was common practice to teach structural analysis and stress analysis, or theory of structures and strength of materials as they were frequently known, as two separate subjects where, generally, structural analysis was concerned with the calculation of internal force systems and stress analysis involved the determination of the corresponding internal stresses and associated strains. Inevitably a degree of overlap occurred. For example, the calculation of shear force and bending moment distributions in beams would be presented in both structural and stress analysis courses, as would the determination of displacements. In fact, a knowledge of methods of determining displacements is essential in the analysis of some statically indeterminate structures. Clearly, therefore, it is logical to present a unified approach in which the 'story' can be told progressively with one topic following naturally on from another.

Initially we shall examine the functions and forms of structures together with support systems and the difference between statically determinate and statically indeterminate structures. We shall also discuss the role of analysis in the design process and the idealization of structures into forms amenable to analysis.

1.1 Function of a structure

The basic function of any structure is to support loads. These arise in a variety of ways and depend, generally, upon the purpose for which the structure has been built. Thus in a steel-framed multistorey building the steel frame supports the roof and floors, the external walls or cladding and also resists the action of wind loads. In turn, the external walls provide protection for the interior of the building and transmit wind loads through the floor slabs to the frame, while the roof carries snow and wind loads which are also transmitted to the frame. In addition, the floor slabs carry people, furniture, floor coverings, etc. Ultimately, of course, the steel frame is supported on the foundations of the building which comprise a structural system in their own right.

Other structures carry other types of load. A bridge structure supports a deck which allows the passage of pedestrians and vehicles, dams hold back large volumes of water, retaining walls prevent the slippage of embankments and offshore structures carry drilling rigs, accommodation for their crews, helicopter pads and resist the action of the sea and the elements. Harbour docks and jetties carry cranes for unloading cargo and must resist the impact of docking ships. Petroleum and gas storage tanks must be able to resist internal pressure and, at the same time, possess the strength and stability to carry wind and snow loads. Television transmitting masts are usually extremely tall and placed in elevated positions where wind and snow loads are major factors. Other structures, such as ships, aircraft, space vehicles, cars, etc., carry equally complex loading systems but fall outside the realm of structural engineering. However, no matter how simple or how complex a structure may be or whether the structure is intended to carry loads or merely act as a protective covering, there will be one load to which it will always be subjected, its own weight.

1.2   Structural Forms

The decision as to the form of a structure rests with the structural designer and is governed by the purpose for which the structure is required, the materials that are to be used and any aesthetic considerations that may apply. At the same time a designer may face a situation in which more than one structural form will satisfy the requirements of the problem so that the designer must then rely on skill and experience to select the best solution. On the other hand there may be scope for a new and novel structure which provides savings in cost and improvements in appearance. Structures, for construction and analysis purposes, are divided into a number of structural elements, although an element of one structure may, in another situation, form a complete structure in its own right. Thus, for example, a beam may support a footpath across a stream (Fig. 1.1) or form part of a large framework (Fig. 1.2). Beams are the most common structural elements and carry loads by developing shear forces and bending moments along their length, as we shall see in Chapter 3.

As spans increase, the use of beams to support bridge decks becomes uneconomical. For moderately large spans, trusses may be used. Trusses carry loads by developing axial forces in their members, and their depth, for the same span and load, is larger than that of a beam but, because of the skeletal nature of their construction, will be lighter. The Warren truss shown in Fig. 1.3 is typical of those used in bridge construction; other geometries form roof supports and also carry bridge decks. Portal frames (Fig. 1.4) are commonly used in building construction

Fig. 1.1 Beam as a simple bridge

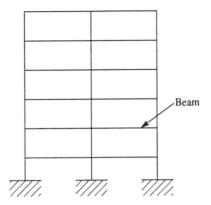

Fig. 1.2 Beam as a structural element

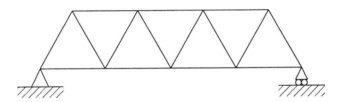

Fig. 1.3 Warren truss

and generally comprise arrangements of beams and columns. The frames derive their stability under load from their rigid joints; the frames would, of course, be stable if their feet were on pinned supports (see Section 1.3). The arrangement shown in Fig. 1.4(a) frequently forms the basic unit in a multistorey, multibay building such as that shown in Fig. 1.2, whereas the frame shown in Fig. 1.4(b) is often used in single storey multibay buildings such as warehouses and factories (Fig. 1.5). Frames are comparatively easy to erect; the Empire State Building in New York, for example, was completed in eighteen months. However, frames frequently need to be reinforced by bracing or shear walls against large lateral forces produced by wind or earthquake loads.

The use of trusses to support bridge decks becomes impracticable for longer than moderate spans. In this situation arches are often used. Figure 1.6(a) shows an arch

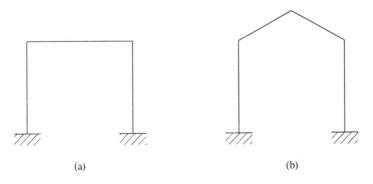

(a)   (b)

Fig. 1.4 Portal frames

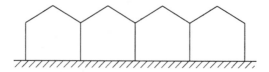

Fig. 1.5   Multibay single storey building

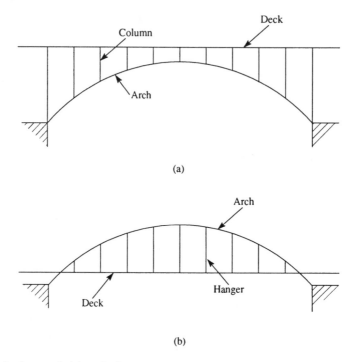

Fig. 1.6   Arches as bridge deck supports

in which the bridge deck is carried by columns supported, in turn, by the arch. Alternatively the bridge deck may be suspended from the arch by hangers, as shown in Fig. 1.6(b). Arches carry most of their loads by developing compressive stresses within the arch itself and therefore in the past were frequently constructed using materials of high compressive strength and low tensile strength such as masonry. In addition to bridges, arches are used to support roofs. They may be constructed in a variety of geometries; they may be semicircular, parabolic or even linear where the members comprising the arch are straight.

For exceptionally long-span bridges, and sometimes for short spans, cables are used to support the bridge deck. Generally, the cables pass over saddles on the tops of towers and are fixed at each end within the ground by massive anchor blocks. The cables carry hangers from which the bridge deck is suspended; a typical arrangement is shown in Fig. 1.7.

Other structural forms include slabs, which are used as floors in buildings, as raft foundations and as bridge decks, and continuum structures which include shells, folded plate roofs, arch dams, etc.; generally, continuum structures require computer-based methods of analysis.

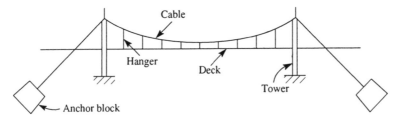

Fig. 1.7   Suspension bridge

1.3   Support systems

The loads applied to a structure are transferred to its foundations by its supports. In practice supports may be complex, in which case they are idealized into a form that may readily be analysed. Thus a support that allows rotation but prevents translation in practice would be as shown in Fig. 1.8(a), but is represented for analysis purposes by the idealized form shown in Fig. 1.8(b); this type of support is called a *pinned support*.

A beam that is supported at one end by a pinned support would not necessarily be supported in the same way at the other. One support of this type is sufficient to maintain the horizontal equilibrium of a beam and it may be advantageous to allow horizontal movement of the other end so that, for example, expansion and contraction caused by temperature variations do not induce additional stresses. Such a support may take the form of a composite steel and rubber bearing as shown in Fig. 1.9(a) or consist of a roller sandwiched between steel plates. In an idealized form, this type of support is represented as shown in Fig. 1.9(b) and is called a *roller support*. It is assumed that such a support allows horizontal movement and rotation but prevents movement vertically, up or down.

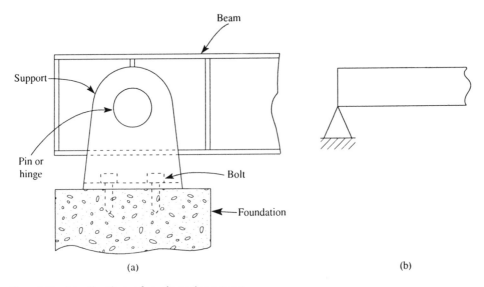

Fig. 1.8   Idealization of a pinned support

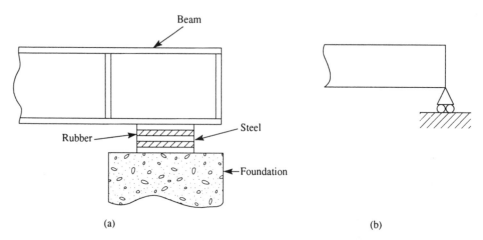

(a)   (b)

Fig. 1.9 Idealization of a sliding or roller support

It is worth noting that a horizontal beam on two pinned supports would be statically indeterminate for other than purely vertical loads since, as we shall see in Section 2.5, there would be two vertical and two horizontal components of support reaction but only three independent equations of statical equilibrium.

In some instances beams are supported in such a way that both translation and rotation are prevented. In Fig. 1.10(a) the steel I-beam is connected through brackets to the flanges of a steel column and therefore cannot rotate or move in any direction; the idealized form of this support is shown in Fig. 1.10(b) and is called a *fixed, built-in* or *encastré support*. A beam that is supported by a pinned support and a roller support as shown in Fig. 1.11(a) is called a *simply supported beam*; note that the supports will not necessarily be positioned at the ends of a beam. A beam supported by combinations of more than two pinned and roller supports (Fig. 1.11(b)) is known as a *continuous beam*. A beam that is built-in at one end and free at the other (Fig. 1.12(a)) is a *cantilever beam* while a beam that is built-in at both ends (Fig. 1.12(b)) is a *fixed, built-in* or *encastré beam*.

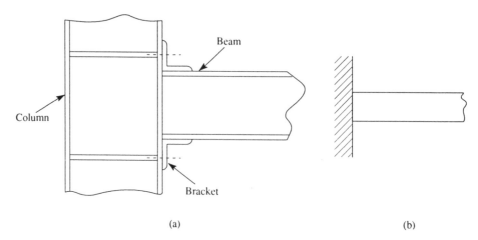

(a)   (b)

Fig. 1.10 Idealization of a built-in support

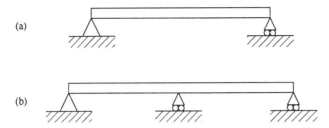

(a)

(b)

Fig. 1.11 (a) Simply supported beam; (b) continuous beam

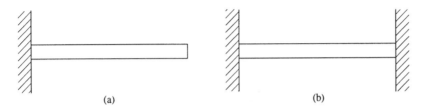

(a)

(b)

Fig. 1.12 (a) Cantilever beam; (b) fixed or built-in beam

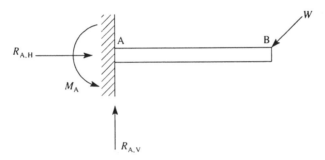

Fig. 1.13 Support reactions in a cantilever beam subjected to an inclined load at its free end

When loads are applied to a structure, reactions are generated in the supports and in many structural analysis problems the first step is to calculate their values. It is important, therefore, to identify correctly the type of reaction associated with a particular support. Thus, supports that prevent translation in a particular direction produce a force reaction in that direction while supports that prevent rotation induce moment reactions. For example, in the cantilever beam of Fig. 1.13, the applied load W has horizontal and vertical components which induce horizontal ($R_{A,H}$) and vertical ($R_{A,V}$) reactions of force at the built-in end A, while the rotational effect of W is balanced by the moment reaction M_A. We shall consider the calculation of support reactions in detail in Section 2.5.

1.4 Statically determinate and indeterminate structures

In many structural systems the principles of statical equilibrium (Section 2.4) may be used to determine support reactions and internal force distributions; such systems

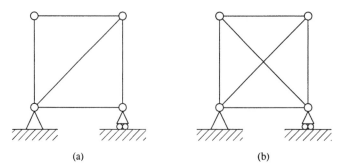

Fig. 1.14   (a) Statically determinate truss; (b) statically indeterminate truss

are called *statically determinate*. Systems for which the principles of statical equilibrium are insufficient to determine support reactions and/or internal force distributions, i.e. there are a greater number of unknowns than the number of equations of statical equilibrium, are known as *statically indeterminate* or *hyperstatic* systems. However, it is possible that even though the support reactions are statically determinate, the internal forces are not, and vice versa. Thus, for example, the truss in Fig. 1.14(a) is, as we shall see in Chapter 4, statically determinate both for support reactions and forces in the members whereas the truss shown in Fig. 1.14(b) is statically determinate only as far as the calculation of support reactions is concerned.

Another type of indeterminacy, *kinematic indeterminacy*, is associated with the ability to deform, or the degrees of freedom of, a structure and is discussed in detail in Section 16.3. A degree of freedom is a possible displacement of a joint (or node as it is often called) in a structure. Thus a joint in a plane truss has three possible modes of displacement or degrees of freedom, two of translation in two mutually perpendicular directions and one of rotation, all in the plane of the truss. On the other hand a joint in a three-dimensional space truss or frame possesses six degrees of freedom, three of translation in three mutually perpendicular directions and three of rotation about three mutually perpendicular axes.

1.5   Analysis and design

Some students in the early stages of their studies have only a vague idea of the difference between an analytical problem and a design problem. It will be instructive, therefore, to examine the various steps in the design procedure and to consider the role of analysis in that procedure.

Initially the structural designer is faced with a requirement for a structure to fulfil a particular role. This may be a bridge of a specific span, a multistorey building of a given floor area, a retaining wall having a required height, and so on. At this stage the designer will decide on a possible form for the structure. In the case of a bridge, for example, the designer must decide whether to use beams, trusses, arches or cables to support the bridge deck. To some extent, as we have seen, the choice is governed by the span required, although other factors may influence the decision. Thus, in Scotland, the Firth of Tay is crossed by a multispan bridge supported on columns, whereas the road bridge crossing the Firth of Forth is a suspension bridge.

In the latter case a large height clearance is required to accommodate shipping. In addition it is possible that the designer may consider different schemes for the same requirement. Further decisions are required as to the materials to be used: steel, reinforced concrete, timber, etc.

Having decided on a form for the structure, the loads on the structure are calculated. These arise in different ways. *Dead loads* are loads that are permanently present, such as the structure's self-weight, fixtures, cladding, etc. *Live* or *imposed loads* are movable or actually moving loads, such as temporary partitions, people, vehicles on a bridge, snow, etc. *Wind loads* are live loads but require special consideration since they are affected by the location, size and shape of the structure. Other live loads may include soil or hydrostatic pressure and dynamic effects produced, for example, by vibrating machinery, wind gusts, wave action or, in some parts of the world, earthquake action.

In some instances values of the above loads are given in Codes of Practice. Thus, for floors in office buildings designed for general use, CP3: Chapter V: Part I specifies a distributed load of $2 \cdot 5$ kN/m^2 together with a concentrated load of $2 \cdot 7$ kN applied over any square of side 300 mm, while CP3: Chapter V: Part 2 gives details of how wind loads should be calculated.

When the loads have been determined, the structure is *analysed*, i.e. the external and internal forces and moments are calculated, from which are obtained the internal stress distributions and also the strains and displacements. The structure is then checked for *safety*, i.e. that it possesses sufficient strength to resist loads without danger of collapse, and for *serviceability*, which determines its ability to carry loads without excessive deformation or local distress; Codes of Practice are used in this procedure. It is possible that this check may show that the structure is underdesigned (unsafe and/or unserviceable) or overdesigned (uneconomic) so that adjustments must be made to the arrangement and/or the sizes of the members; the analysis and design check are then repeated.

Analysis, as can be seen from the above discussion, forms only part of the complete design process and is concerned with a given structure subjected to given loads. Thus, generally, there is a unique solution to an analytical problem whereas there may be one, two or more perfectly acceptable solutions to a design problem.

1.6 Structural idealization

Generally, structures are complex and must be *idealized* or simplified into a form that can be analysed. This idealization depends upon factors such as the degree of accuracy required from the analysis because, usually, the more sophisticated the method of analysis employed the more time consuming, and therefore more costly, it is. Thus a preliminary evaluation of two or more possible design solutions would not require the same degree of accuracy as the check on the finalized design. Other factors affecting the idealization include the type of load being applied, since it is possible that a structure will require different idealizations under different loads.

We have seen in Section 1.3 how actual supports are idealized. An example of structural idealization is shown in Fig. 1.15 where the simple roof truss of Fig. 1.15(a) is supported on columns and forms one of a series comprising a roof structure. The roof cladding is attached to the truss through purlins which connect

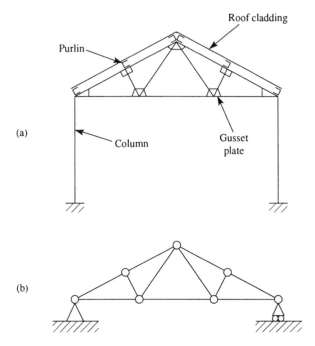

Fig. 1.15   (a) Actual truss; (b) idealized truss

each truss, and the truss members are connected to each other by gusset plates which may be riveted or welded to the members forming rigid joints. This structure possesses a high degree of statical indeterminacy and its analysis would probably require a computer-based approach. However, the assumption of a simple support system, the replacement of the rigid joints by pinned or hinged joints and the assumption that the forces in the members are purely axial, result, as we shall see in Chapter 4, in a statically determinate structure (Fig. 1.15(b)). Such an idealization might appear extreme but, so long as the loads are applied at the joints and the truss is supported at joints, the forces in the members are predominantly axial and bending moments and shear forces are negligibly small.

At the other extreme a continuum structure, such as a folded plate roof, would be idealized into a large number of *finite elements* connected at *nodes* and analysed using a computer; the *finite element method* is, in fact, an exclusively computer-based technique. A large range of elements is available in finite element packages including simple beam elements, plate elements, which can model both in-plane and out-of-plane effects, and three-dimensional 'brick' elements for the idealization of solid three-dimensional structures. A wide range of literature devoted to finite element analysis is available but will not be considered here as the method is outside the scope of this book.

CHAPTER 2

Principles of Statics

Statics, as the name implies, is concerned with the study of bodies at rest or, in other words, in equilibrium, under the action of a force system. Actually, a moving body is in equilibrium if the forces acting on it are producing neither acceleration nor deceleration. However, in structural analysis, structural members are generally at rest and therefore in a state of *statical equilibrium*.

In this chapter we shall discuss those principles of statics that are essential to structural and stress analysis; an elementary knowledge of vectors is assumed.

2.1 Force

The definition of a force is derived from Newton's First Law of Motion which states that a body will remain in its state of rest or in its state of uniform motion in a straight line unless compelled by an external force to change that state. Force is therefore associated with a *change* in motion, i.e. it causes acceleration or deceleration.

We all have direct experience of force systems. The force of the earth's gravitational pull acts vertically downwards on our bodies giving us weight; wind forces, which can vary in magnitude, tend to push us horizontally. Forces therefore possess magnitude and direction. At the same time the effect of a force depends upon its position. For example, a door may be opened or closed by pushing horizontally at its free edge, but if the same force is applied at any point on the vertical line through its hinges the door will neither open nor close. Thus we see that a force is specified by its magnitude, direction and position and is therefore a *vector* quantity. As such it must obey the laws of vector addition, which is a fundamental concept that may be verified experimentally.

Since a force is a vector it may be represented graphically as shown in Fig. 2.1, where the force F is considered to be acting on an infinitesimally small particle at the point A and in a direction from left to right. The magnitude of F is represented, to a suitable scale, by the length of the line AB and its direction by the direction of the arrow. In vector notation the force F is written as **F**.

Consider now a cube of material placed on a horizontal surface and acted upon by a force F_1 as shown in Fig. 2.2(a). If F_1 is greater than the frictional force between the surface and the cube, the cube will move in the direction of F_1. Similarly if a force F_2 is applied as shown in Fig. 2.2(b) the cube will move in the direction of F_2.

Fig. 2.1   Representation of a force by a vector

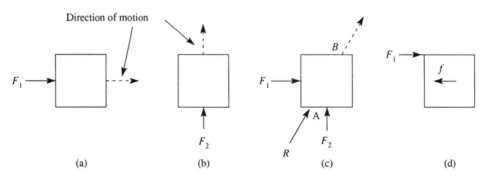

Fig. 2.2   Action of forces on a cube

It follows that if F_1 and F_2 were applied simultaneously, the cube would move in some inclined direction as though it were acted on by a single inclined force R (Fig. 2.2(c)); clearly R is the *resultant* of F_1 and F_2.

Note that F_1 and F_2 (and R) are in a horizontal plane and that their lines of action pass through the centre of gravity of the cube, otherwise rotation as well as translation would occur since, if F_1, say, were applied at one corner of the cube as shown in Fig. 2.2(d), the frictional force f, which would act at the centre of the bottom face of the cube would, with F_1, form a couple (see Section 2.2).

The effect of the force R on the cube would be the same whether it was applied at the point A or at the point B (so long as the cube is rigid). Thus a force may be considered to be applied at any point on its line of action, a principle known as the *transmissibility of a force*.

Parallelogram of forces

The resultant of two concurrent and coplanar forces, whose lines of action pass through a single point and lie in the same plane (Fig. 2.3(a)), may be found using the theorem of the parallelogram of forces which states that:

If two forces acting at a point are represented by two adjacent sides of a parallelogram drawn from that point their resultant is represented in magnitude and direction by the diagonal of the parallelogram drawn through the point.

Thus in Fig. 2.3(b) R is the resultant of F_1 and F_2. This result may be verified experimentally or, alternatively, demonstrated to be true using the laws of vector addition. Thus in Fig. 2.3(b) the side BC of the parallelogram is equal in magnitude and direction to the force F_1 represented by the side OA. Therefore, in vector notation

$$\mathbf{R} = \mathbf{F}_2 + \mathbf{F}_1$$

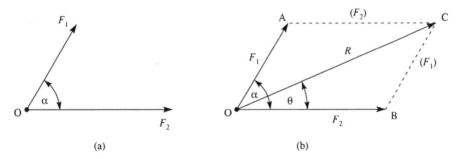

Fig. 2.3 Resultant of two concurrent forces

The same result would be obtained by considering the side AC of the parallelogram which is equal in magnitude and direction to the force F_2. Thus

$$\mathbf{R} = \mathbf{F}_1 + \mathbf{F}_2$$

Note that vectors obey the *commutative law*, i.e.

$$\mathbf{F}_2 + \mathbf{F}_1 = \mathbf{F}_1 + \mathbf{F}_2$$

The determination of the actual magnitude and direction of R may be carried out graphically by drawing the vectors representing F_1 and F_2 to the *same scale* (i.e. OB and BC) and then completing the triangle OBC by drawing in the vector, along OC, representing R. Alternatively R and θ may be calculated using the trigonometry of triangles. Hence

$$R^2 = F_1^2 + F_2^2 + 2F_1F_2 \cos \alpha \tag{2.1}$$

and

$$\tan \theta = \frac{F_1 \sin \alpha}{F_2 + F_1 \cos \alpha} \tag{2.2}$$

In Fig. 2.3(a) both F_1 and F_2 are 'pulling away' from the particle at O. In Fig. 2.4(a) F_1 is a 'thrust' whereas F_2 remains a 'pull'. To use the parallelogram of forces the system must be reduced to either two 'pulls' as shown in Fig. 2.4(b) or two 'thrusts' as shown in Fig. 2.4(c). In all three systems we see that the effect on the particle at O is the same.

As we have seen, the combined effect of the two forces F_1 and F_2 acting simultaneously is the same as if they had been replaced by the single force R. Conversely, if R were to be replaced by F_1 and F_2 the effect would again be the same. F_1 and F_2 may therefore be regarded as the *components* of R in the directions OA and OB; R is then said to have been *resolved* into two components, F_1 and F_2.

Of particular interest in structural analysis is the resolution of a force into two components at right angles to each other. In this case the parallelogram of Fig. 2.3(b) becomes a rectangle in which $\alpha = 90°$ (Fig. 2.5) and, clearly,

$$F_2 = R \cos \theta, \qquad F_1 = R \sin \theta \tag{2.3}$$

It follows from Fig. 2.5, or from Eqs (2.1) and (2.2), that

$$R^2 = F_1^2 + F_2^2, \qquad \tan \theta = F_1/F_2 \tag{2.4}$$

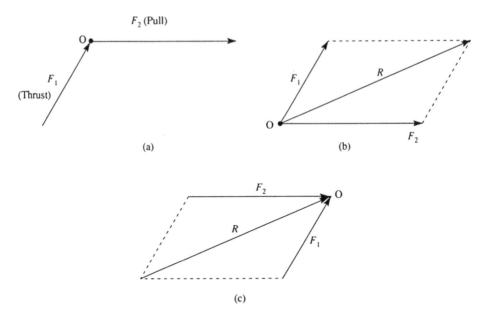

(a)

(b)

(c)

Fig. 2.4   Reduction of a force system

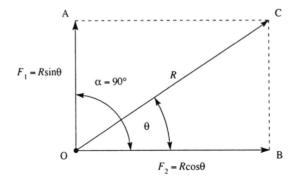

Fig. 2.5   Resolution of a force into two components at right angles

We note, by reference to Figs 2.2(a) and (b), that a force does not induce motion in a direction perpendicular to its line of action; in other words a force has no effect in a direction perpendicular to itself. This may also be seen by setting $\theta = 90°$ in Eqs (2.3), then

$$F_1 = R, \qquad F_2 = 0$$

and the component of R in a direction perpendicular to its line of action is zero.

The resultant of a system of concurrent forces

So far we have considered the resultant of just two concurrent forces. The method used for that case may be extended to determine the resultant of a system of any number of concurrent coplanar forces such as that shown in Fig. 2.6(a). Thus in the vector diagram of Fig. 2.6(b)

$$\mathbf{R}_{12} = \mathbf{F}_1 + \mathbf{F}_2$$

where $\mathbf{R}_{12}$ is the resultant of $\mathbf{F}_1$ and $\mathbf{F}_2$. Further

$$\mathbf{R}_{123} = \mathbf{R}_{12} + \mathbf{F}_3 = \mathbf{F}_1 + \mathbf{F}_2 + \mathbf{F}_3$$

so that $\mathbf{R}_{123}$ is the resultant of $\mathbf{F}_1$, $\mathbf{F}_2$ and $\mathbf{F}_3$. Finally

$$\mathbf{R} = \mathbf{R}_{123} + \mathbf{F}_4 = \mathbf{F}_1 + \mathbf{F}_2 + \mathbf{F}_3 + \mathbf{F}_4$$

where $\mathbf{R}$ is the resultant of $\mathbf{F}_1$, $\mathbf{F}_2$, $\mathbf{F}_3$ and $\mathbf{F}_4$.

The actual value and direction of $\mathbf{R}$ may be found graphically by constructing the vector diagram of Fig. 2.6(b) to scale or by resolving each force into components parallel to two directions at right angles, say the x and y directions shown in Fig. 2.6(a). Thus

$$F_x = F_1 + F_2 \cos \alpha - F_3 \cos \beta - F_4 \cos \gamma$$
$$F_y = F_2 \sin \alpha + F_3 \sin \beta - F_4 \sin \gamma$$

Then

$$R = \sqrt{F_x^2 + F_y^2}$$

and·

$$\tan \theta = \frac{F_y}{F_x}$$

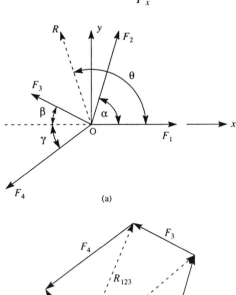

(a)

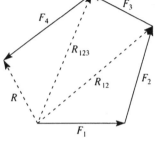

(b)

Fig. 2.6 Resultant of a system of concurrent forces

The forces F_1, F_2, F_3 and F_4 in Fig. 2.6(a) do not have to be taken in any particular order when constructing the vector diagram of Fig. 2.6(b). Identical results for the magnitude and direction of R are obtained if the forces in the vector diagram are taken in the order F_1, F_4, F_3, F_2 as shown in Fig. 2.7 or, in fact, are taken in any order so long as the directions of the forces are adhered to and one force vector is drawn from the end of the previous force vector.

Equilibrant of a system of concurrent forces

In Fig. 2.3(b) the resultant R of the forces F_1 and F_2 represents the combined effect of F_1 and F_2 on the particle at O. It follows that this effect may be eliminated by introducing a force R_E which is equal in magnitude but opposite in direction to R at O, as shown in Fig. 2.8(a). R_E is known at the *equilibrant* of F_1 and F_2 and the

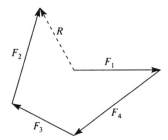

Fig. 2.7   Alternative construction of force diagram for system of Fig. 2.6(a)

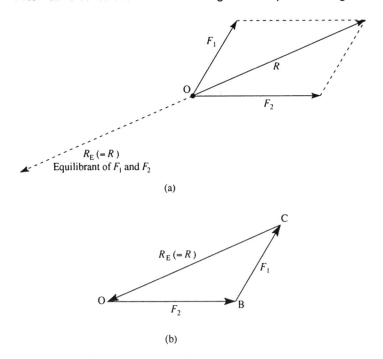

Fig. 2.8   Equilibrant of two concurrent forces

particle at O will then be in *equilibrium* and remain stationary. In other words the forces F_1, F_2 and R_E are in equilibrium and, by reference to Fig. 2.3(b), we see that these three forces may be represented by the triangle of vectors OBC as shown in Fig. 2.8(b). This result leads directly to the law of the *triangle of forces* which states that:

If three forces acting at a point are in equilibrium they may be represented in magnitude and direction by the sides of a triangle taken in order.

The law of the triangle of forces may be used in the analysis of a plane, pin-jointed truss in which, say, one of three concurrent forces is known in magnitude and direction but only the lines of action of the other two. The law enables us to find the magnitudes of the other two forces and also the direction of their lines of action.

The above arguments may be extended to a system comprising any number of concurrent forces. Thus, for the force system of Fig. 2.6(a), R_E, shown in Fig. 2.9(a), is the equilibrant of the forces F_1, F_2, F_3 and F_4. Then F_1, F_2, F_3, F_4 and R_E may be represented by the force polygon OBCDE as shown in Fig. 2.9(b). The law of the *polygon of forces* follows:

If a number of forces acting at a point are in equilibrium they may be represented in magnitude and direction by the sides of a closed polygon taken in order.

Again, the law of the polygon of forces may be used in the analysis of plane, pin-jointed trusses where several members meet at a joint but where no more than two forces are unknown in magnitude.

The resultant of a system of non-concurrent forces

In most structural problems the lines of action of the different forces acting on the structure do not meet at a single point; such a force system is non-concurrent.

Consider the system of non-concurrent forces shown in Fig. 2.10(a); their resultant may be found graphically using the parallelogram of forces as demonstrated in Fig. 2.10(b). Produce the lines of action of F_1 and F_2 to their point

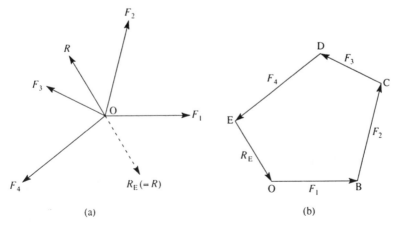

(a)　　　　　　　　　　　　　　　(b)

Fig. 2.9 Equilibrant of a number of concurrent forces

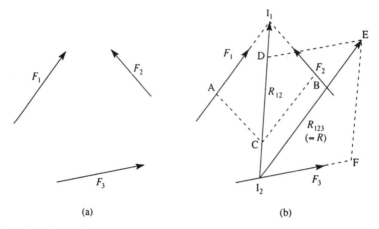

Fig. 2.10   Resultant of a system of non-concurrent forces

of intersection, I_1. Measure $I_1A = F_1$ and $I_1B = F_2$ to the same scale, then complete the parallelogram I_1ACB; the diagonal CI_1 represents the resultant, R_{12}, of F_1 and F_2. Now produce the line of action of R_{12} backwards to intersect the line of action of F_3 at I_2. Measure $I_2D = R_{12}$ and $I_2 F = F_3$ to the same scale as before, then complete the parallelogram I_2DEF; the diagonal $I_2E = R_{123}$, the resultant of R_{12} and F_3. It follows that $R_{123} = R$, the resultant of F_1, F_2 and F_3. Note that only the line of action and the magnitude of R can be found, not its point of action, since the vectors F_1, F_2 and F_3 in Fig. 2.10(a) define the lines of action of the forces, not their points of action.

If the points of action of the forces are known, defined, say, by coordinates referred to a convenient xy axis system, the magnitude, direction and point of action of their resultant may be found by resolving each force into components parallel to the x and y axes and then finding the magnitude and position of the resultants R_x and R_y of each set of components using the method described in Section 2.3 for a system of parallel forces. The resultant R of the force system is then given by

$$R = \sqrt{R_x^2 + R_y^2}$$

and its point of action is the point of intersection of R_x and R_y; finally, its inclination θ to the x axis, say, is

$$\theta = \tan^{-1} \frac{R_y}{R_x}$$

2.2   Moment of a force

So far we have been concerned with the translational effect of a force, i.e. the tendency of a force to move a body in a straight line from one position to another. A force may, however, exert a rotational effect on a body so that the body tends to turn about some given point or axis.

In Fig. 2.11(a) the bar AB is attached to a hinge which allows it to rotate in a horizontal plane (Fig. 2.11(a) is a plan view). A force F whose line of action passes through the hinge will have no rotational effect on the bar but, when acting at some point along the bar as in Fig. 2.11(b), will cause the bar to rotate about the hinge. Further, it is common experience that the greater the distance of F from the hinge, the greater will be its effect. (Thus a greater force is required to close a door when the force is applied near to the vertical line through its hinges than if the force were applied close to the free edge, the usual position for a door knob or handle.) At the same time, the force F exerts its greatest effect when it acts at right angles to the bar. If it were inclined, as shown in Fig. 2.11(c), such that its line of action passed through the hinge it would exert no rotational effect on the bar at all.

In Fig. 2.11(b) F is said to exert a *moment* on the bar about the hinge, which is usually referred to as the *fulcrum*. Clearly the rotational effect of F depends upon its magnitude and also on its distance from the hinge. We therefore define the moment of a force, F, about a given point O (Fig. 2.12) as the product of the force and the perpendicular distance of its line of action from the point. Thus, in Fig. 2.12, the moment, M, of F about O is given by

$$M = Fa \qquad (2.5)$$

where 'a' is known as the *lever arm* or *moment arm* of F about O; note that the units of a moment are the units of force × distance.

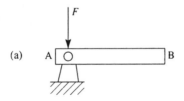

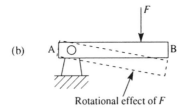

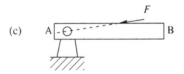

Fig. 2.11 Rotational effect of a force

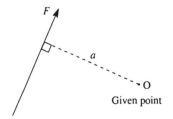

Fig. 2.12   Moment of a force about a given point

It can be seen from the above that a moment possesses both magnitude and a rotational sense; in Fig. 2.12, for example, F exerts a clockwise moment about O. A moment is therefore a vector (an alternative argument is that the product of a vector, F, and a scalar, a, is a vector). It is conventional to represent a moment vector graphically by a double-headed arrow, where the direction of the arrow designates a clockwise moment when looking in the direction of the arrow. Therefore, in Fig. 2.12, the moment $M(=Fa)$ would be represented by a double headed arrow through O with its direction into the plane of the paper.

Moments, being vectors, may be resolved into components in the same way as forces. Consider the moment, M, (Fig. 2.13(a)) in a plane inclined at an angle θ to the xz plane. The component of M in the xz plane, M_{xz}, may be imagined to be produced by rotating the plane containing M through the angle θ into the xz plane. Similarly, the component of M in the yz plane, M_{yz}, is obtained by rotating the plane containing M through the angle $90 - \theta$. Vectorially, the situation is that shown in Fig. 2.13(b), where the directions of the arrows represent clockwise moments when viewed in the directions of the arrows. Then

$$M_{xz} = M \cos \theta, \qquad M_{yz} = M \sin \theta$$

The action of a moment on a structural member depends upon the plane in which it acts. In Fig. 2.14(a), for example, the moment, M, which is applied in the

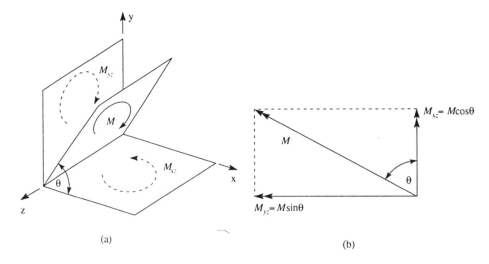

(a)   (b)

Fig. 2.13   Resolution of a moment

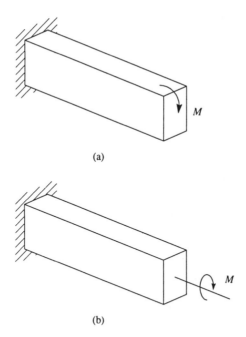

(a)

(b)

Fig. 2.14   Action of a moment in different planes

longitudinal vertical plane of symmetry, will cause the beam to bend in a vertical plane. In Fig. 2.14(b) the moment, M, is applied in the plane of the cross-section of the beam and will therefore produce twisting; in this case M is called a *torque*.

Couples

Consider the two coplanar, equal and parallel forces F which act in opposite directions as shown in Fig. 2.15. The sum of their moments, M_O, about *any* point O in their plane is

$$M_O = F \times BO - F \times AO$$

where OAB is perpendicular to both forces. Then

$$M_O = F(BO - AO) = F \times AB$$

and we see that the sum of the moments of the two forces F about any point in their plane is equal to the product of one of the forces and the perpendicular distance between their lines of action; this system is termed a *couple* and the distance AB is the *arm* or *lever arm* of the couple.

Since a couple is, in effect, a pure moment (not to be confused with the moment of a force about a specific point which varies with the position of the point) it may be resolved into components in the same way as the moment M in Fig. 2.13.

Equivalent force systems

In structural analysis it is often convenient to replace a force system acting at one

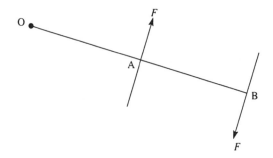

Fig. 2.15 Moment of a couple

point by an equivalent force system acting at another. In Fig. 2.16(a), for example, the effect on the cylinder of the force F acting at A on the arm AB may be determined as follows.

If we apply equal and opposite forces F at B as shown in Fig. 2.16(b), the overall effect on the cylinder is unchanged. However, the force F at A and the equal and opposite force F at B form a couple which, as we have seen, has the same moment (Fa) about any point in its plane. Thus the single force F at A may be replaced by a

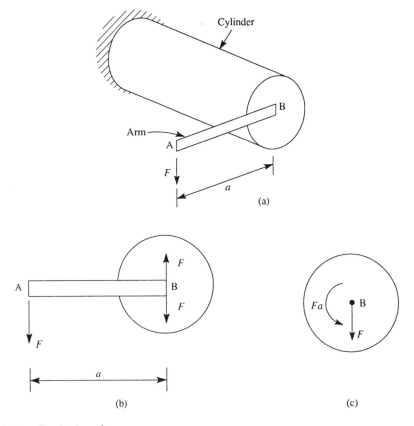

Fig. 2.16 Equivalent force system

single force F at B together with a moment equal to Fa as shown in Fig. 2.16(c). The effects of the force F at B and the moment (actually a torque) Fa may be calculated separately and then combined using the principle of superposition (see Section 3.8).

2.3   The resultant of a system of parallel forces

Since, as we have seen, a system of forces may be replaced by their resultant, it follows that a particular action of a force system, say the combined moments of the forces about a point, must be identical to the same action of their resultant. This principle may be used to determine the magnitude and line of action of a system of parallel forces such as that shown in Fig. 2.17(a).

The point of intersection of the lines of action of F_1 and F_2 is at infinity so that the parallelogram of forces (Fig. 2.3(b)) degenerates into a straight line as shown in Fig. 2.17(b) where, clearly

$$R = F_1 + F_2 \tag{2.6}$$

The position of the line of action of R may be found using the principle stated above, i.e. the sum of the moments of F_1 and F_2 about any point must be equivalent to the moment of R about the same point. Thus from Fig. 2.17(a) and taking moments about, say, the line of action of F_1 we have

$$F_2 a = Rx = (F_1 + F_2)x$$

Hence $$x = \frac{F_2}{F_1 + F_2} a \tag{2.7}$$

Note that the action of R is *equivalent* to that of F_1 and F_2, so that, in this case, we equate clockwise to clockwise moments.

The principle of equivalence may be extended to any number of parallel forces irrespective of their directions and is of particular use in the calculation of the position of centroids of area, as we shall see in Section 9.6.

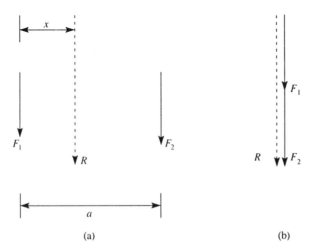

(a)   (b)

Fig. 2.17   Resultant of a system of parallel forces

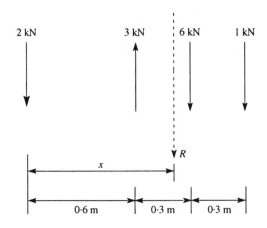

Fig. 2.18 Force system of Ex. 2.1

Example 2.1 Find the magnitude and position of the line of action of the resultant of the force system shown in Fig. 2.18.

In this case the polygon of forces (Fig. 2.6(b)) degenerates into a straight line and

$$R = 2 - 3 + 6 + 1 = 6 \text{ kN} \tag{i}$$

Suppose that the line of action of R is at a distance x from the 2 kN force, then, taking moments about the 2 kN force,

$$Rx = -3 \times 0 \cdot 6 + 6 \times 0 \cdot 9 + 1 \times 1 \cdot 2$$

Substituting for R from Eq. (i) we have

$$6x = -1 \cdot 8 + 5 \cdot 4 + 1 \cdot 2$$

which gives $\qquad\qquad x = 0 \cdot 8 \text{ m}$

We could, in fact, take moments about any point, say now the 6 kN force. Then

$$R(0 \cdot 9 - x) = 2 \times 0 \cdot 9 - 3 \times 0 \cdot 3 - 1 \times 0 \cdot 3$$

so that $\qquad\qquad x = 0 \cdot 8 \text{ m as before.}$

Note that in the second solution, anticlockwise moments have been selected as positive.

2.4 Equilibrium of force systems

We have seen in Section 2.1 that, for a particle or a body to remain stationary, that is in statical equilibrium, the resultant force on the particle or body must be zero. Thus, if a body (generally in structural analysis we are concerned with bodies, i.e. structural members, not particles) is not to move in a particular direction, the resultant force in that direction must be zero. Furthermore, the prevention of the movement of a body in two directions at right angles ensures that the body will not move in any direction. It follows that, for such a body to be in equilibrium, the sum of the

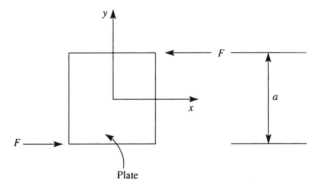

Fig. 2.19 Couple produced by out-of-line forces

components of all the forces acting on the body in any two mutually perpendicular directions must be zero. In mathematical terms and choosing, say, the x and y directions as the mutually perpendicular directions, the condition may be written

$$\Sigma F_x = 0, \qquad \Sigma F_y = 0 \tag{2.8}$$

However, the condition specified by Eqs (2.8) is not sufficient to guarantee the equilibrium of a body acted on by a system of coplanar forces. In Fig. 2.19, for example, the forces F acting on a plate resting on a horizontal surface satisfy the condition $\Sigma F_x = 0$ (there are no forces in the y direction so that $\Sigma F_y = 0$ is automatically satisfied), but form a couple Fa which will cause the plate to rotate in an anticlockwise sense so long as its magnitude is sufficient to overcome the frictional resistance between the plate and the surface. We have also seen that a couple exerts the same moment about any point in its plane so that we may deduce a further condition for the statical equilibrium of a body acted upon by a system of coplanar forces, namely, that the sum of the moments of all the forces acting on the body about *any* point in their plane must be zero. Therefore, designating a moment in the xy plane about the z axis as M_z, we have

$$\Sigma M_z = 0 \tag{2.9}$$

Combining Eqs (2.8) and (2.9) we obtain the necessary conditions for a system of coplanar forces to be in equilibrium. Thus

$$\Sigma F_x = 0, \qquad \Sigma F_y = 0, \qquad \Sigma M_z = 0 \tag{2.10}$$

The above arguments may be extended to a three-dimensional force system which is, again, referred to an xyz axis system. Thus for equilibrium

$$\Sigma F_x = 0, \qquad \Sigma F_y = 0, \qquad \Sigma F_z = 0 \tag{2.11}$$

and

$$\Sigma M_x = 0, \qquad \Sigma M_y = 0, \qquad \Sigma M_z = 0 \tag{2.12}$$

2.5 Calculation of support reactions

The conditions of statical equilibrium, Eqs (2.10), are used to calculate reactions at supports in structures so long as the support system is statically determinate (see

Section 1.4). Generally the calculation of support reactions is a necessary preliminary to the determination of internal force and stress distributions and displacements.

Example 2.2 Calculate the support reactions in the simply supported beam ABCD shown in Fig. 2.20.

The different types of support have been discussed in Section 1.3. In Fig. 2.20 the support at A is a pinned support which allows rotation but no translation in any direction, while the support at D allows rotation and translation in a horizontal direction but not in a vertical direction. Therefore there will be no moment reactions at A or D and only a vertical reaction at D, R_D. It follows that the horizontal component of the 5 kN load can only be resisted at A, $R_{A.H}$, which, in addition, will provide a vertical reaction, $R_{A.V}$.

Since the forces acting on the beam are coplanar, Eqs (2.10) are used. From the first of these, i.e. $\Sigma F_x = 0$, we have

$$R_{A.H} - 5 \cos 60° = 0$$

which gives

$$R_{A.H} = 2·5 \text{ kN}$$

The use of the second equation, $\Sigma F_y = 0$, at this stage would not lead directly to either $R_{A.V}$ or R_D since both would be included in the single equation. A better approach is to use the moment equation, $\Sigma M_z = 0$, and take moments about either A or D (it is immaterial which), thereby eliminating one of the vertical reactions. Taking moments, say, about D, we have

$$R_{A.V} \times 1·2 - 3 \times 0·9 - (5 \sin 60°) \times 0·4 = 0 \qquad \text{(i)}$$

Note that in Eq. (i) the moment of the 5 kN force about D may be obtained either by calculating the perpendicular distance of its line of action from D ($0·4 \sin 60°$) or by resolving it into vertical and horizontal components ($5 \sin 60°$ and $5 \cos 60°$, respectively) where only the vertical component exerts a moment about D. From Eq. (i)

$$R_{A.V} = 3·7 \text{ kN}$$

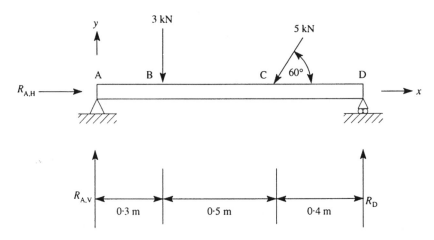

Fig. 2.20 Beam of Ex. 2.2

The vertical reaction at D may now be found using $\Sigma F_y = 0$ or by taking moments about A, which would be slightly lengthier. Thus

$$R_D + R_{A,V} - 3 - 5 \sin 60° = 0$$

so that

$$R_D = 3.6 \text{ kN}$$

Example 2.3 Calculate the reactions at the support in the cantilever beam shown in Fig. 2.21.

The beam has a fixed support at A which prevents translation in any direction and also rotation. The loads applied to the beam will therefore induce a horizontal reaction, $R_{A,H}$, at A and a vertical reaction, $R_{A,V}$, together with a moment reaction M_A. Using the first of Eqs (2.10), $\Sigma F_x = 0$, we obtain

$$R_{A,H} - 2 \cos 45° = 0$$

whence

$$R_{A,H} = 1.4 \text{ kN}$$

From the second of Eqs (2.10), $\Sigma F_y = 0$

$$R_{A,V} - 5 - 2 \sin 45° = 0$$

which gives

$$R_{A,V} = 6.4 \text{ kN}$$

Finally from the third of Eqs (2.10), $\Sigma M_z = 0$, and taking moments about A, thereby eliminating $R_{A,H}$ and $R_{A,V}$,

$$M_A - 5 \times 0.4 - (2 \sin 45°) \times 1.0 = 0$$

from which

$$M_A = 3.4 \text{ kN m}$$

In Exs 2.2 and 2.3, the directions or sense of the support reactions is reasonably obvious. However, where this is not the case, a direction or sense is assumed which, if incorrect, will result in a negative value.

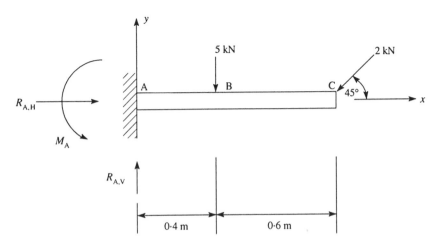

Fig. 2.21 Beam of Ex. 2.3

Occasionally the resultant reaction at a support is of interest. In Ex. 2.2 the resultant reaction at A is found using the first of Eqs (2.4), i.e.

$$R_A^2 = R_{A,H}^2 + R_{A,V}^2$$

which gives

$$R_A^2 = 2 \cdot 5^2 + 3 \cdot 7^2$$

whence

$$R_A = 4 \cdot 5 \text{ kN}$$

The inclination of R_A to, say, the vertical is found from the second of Eqs (2.4). Thus

$$\tan \theta = \frac{R_{A,H}}{R_{A,V}} = \frac{2 \cdot 5}{3 \cdot 7} = 0 \cdot 676$$

from which

$$\theta = 34 \cdot 0°$$

Example 2.4   Calculate the reactions at the supports in the plane truss shown in Fig. 2.22.

The truss is supported in the same manner as the beam in Ex. 2.2 so that there will be horizontal and vertical reactions at A and only a vertical reaction at B.
   The angle of the truss, α, is given by

$$\alpha = \tan^{-1} \frac{2 \cdot 4}{3} = 38 \cdot 7°$$

From the first of Eqs (2.10) we have

$$R_{A,H} - 5 \sin 38 \cdot 7° - 10 \sin 38 \cdot 7° = 0$$

from which

$$R_{A,H} = 9 \cdot 4 \text{ kN}$$

Now taking moments about B, say, $(\Sigma M_B = 0)$

$$R_{A,V} \times 6 - (5 \cos 38 \cdot 7°) \times 4 \cdot 5 + (5 \sin 38 \cdot 7°) \times 1 \cdot 2 + (10 \cos 38 \cdot 7°) \times 1 \cdot 5$$
$$+ (10 \sin 38 \cdot 7°) \times 1 \cdot 2 - 3 \times 4 - 2 \times 2 = 0$$

which gives

$$R_{A,V} = 1 \cdot 8 \text{ kN}$$

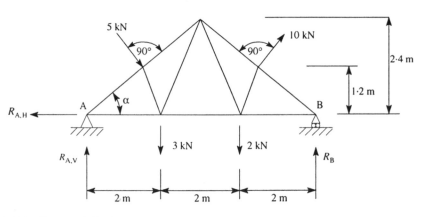

Fig. 2.22   Truss of Ex. 2.4

Note that in the moment equation it is simpler to resolve the 5 kN and 10 kN loads into horizontal and vertical components at their points of application and then take moments rather than calculate the perpendicular distance of each of their lines of action from B.

The reaction at B, R_B, is now most easily found by resolving vertically ($\Sigma F_y = 0$). Thus

$$R_B + R_{A,V} - 5 \cos 38 \cdot 7° + 10 \cos 38 \cdot 7° - 3 - 2 = 0$$

which gives $\qquad\qquad R_B = -0 \cdot 7 \text{ kN}$

In this case the negative sign of R_B indicates that the reaction is downward, not upward, as initially assumed.

Problems

P.2.1 Determine the magnitude and inclination of the resultant of the two forces acting at the point O in Fig. P.2.1, (a) by a graphical method, (b) by calculation.

Ans. 21·8 kN, 23.4° to the direction of the 15 kN load.

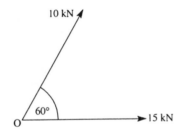

Fig. P.2.1

P.2.2 Determine the magnitude and inclination of the resultant of the system of concurrent forces shown in Fig. P.2.2, (a) by a graphical method, (b) by calculation.

Ans. 8·6 kN, 23.9° down and to the left.

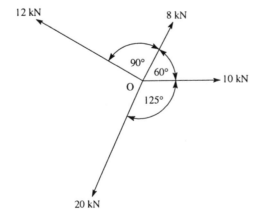

Fig. P.2.2

P.2.3   Calculate the magnitude, inclination and point of action of the resultant of the system of non-concurrent forces shown in Fig. P.2.3. The coordinates of the points of action are given in m.

Ans.   130·5 kN, 49·6° to the x direction at the point (0·86, 1·22).

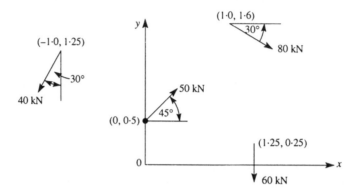

Fig. P.2.3

P.2.4   Calculate the support reactions in the beams shown in Figs. P.2.4(a)–(d).

Ans. (a) $R_{A,H} = 9·2$ kN to left, $R_{A,V} = 6·9$ kN upwards, $R_B = 7·9$ kN upwards.
       (b) $R_A = 65$ kN, $M_A = 400$ kN m anticlockwise.
       (c) $R_{A,H} = 20$ kN to right, $R_{A,V} = 22·5$ kN upwards, $R_B = 12·5$ kN upwards.
       (d) $R_A = 41·8$ kN upwards, $R_B = 54·2$ kN upwards.

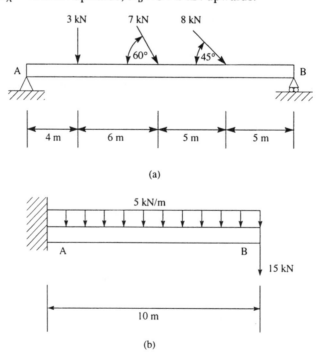

Fig. P.2.4

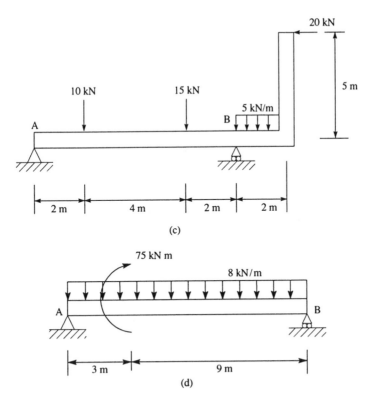

(c)

(d)

Fig. P.2.4 (*continued*)

P.2.5 Calculate the support reactions in the plane trusses shown in Figs P.2.5(a) and (b).

Ans. (a) $R_A = 57$ kN upwards, $R_B = 2$ kN downwards.

(b) $R_{A,H} = 3710$ N to left, $R_{A,V} = 835$ N downwards, $R_B = 4733$ N downwards.

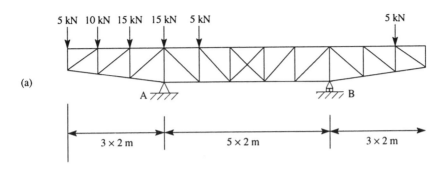

Fig. P.2.5

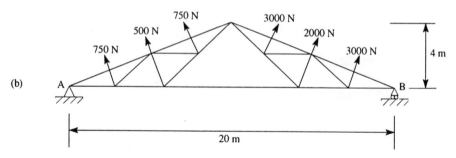

(b)

Fig. P.2.5   (*continued*)

CHAPTER 3

Normal Force, Shear Force, Bending Moment and Torsion

The purpose of a structure is to support the loads for which it has been designed. To accomplish this it must be able to transmit a load from one point to another, i.e. from the loading point to the supports. In Fig. 2.21, for example, the beam transmits the effects of the loads at B and C to the built-in end A. This it achieves by developing an *internal force* system and it is the distribution of these internal forces which must be determined before corresponding stress distributions and displacements can be found.

A knowledge of stress is essential in structural design where the cross-sectional area of a member must be such that stresses do not exceed values that would cause breakdown in the crystalline structure of the material of the member; in other words, a structural failure. In addition to stresses, strains, and thereby displacements, must be calculated to ensure that as well as strength a structural member possesses sufficient stiffness to prevent excessive distortions damaging surrounding portions of a complete structure.

In this chapter we shall examine the different types of load to which a structural member may be subjected and then determine corresponding internal force distributions.

3.1 Types of load

Structural members may be subjected to complex loading systems apparently comprised of several different types of load. However, no matter how complex such systems appear to be they consist of a maximum of four basic load types: axial loads, shear loads, bending moments and torsion.

Axial load

Axial loads are applied along the longitudinal or centroidal axis of a structural member. If the action of the load is to increase the length of the member, the member is said to be in *tension* (Fig. 3.1(a)) and the applied load is *tensile*. A load that tends to shorten a member places the member in *compression* and is known as a *compressive* load (Fig. 3.1(b)). Members such as those shown in Figs 3.1(a) and (b) are commonly found in pin-jointed frameworks where a member in tension is called a *tie* and one in compression a *strut* or *column*. More frequently, however, the name

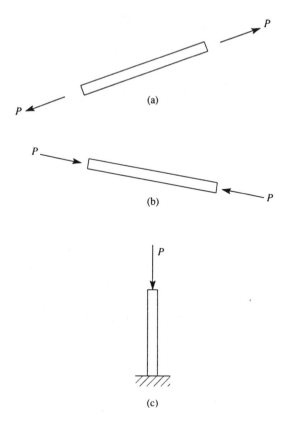

Fig. 3.1 Axially loaded members

'column' is associated with a vertical member carrying a compressive load, as illustrated in Fig. 3.1(c).

Shear load

Shear loads act perpendicularly to the axis of a structural member and have one of the forms shown in Fig. 3.2; in this case the members are *beams*. Fig. 3.2(a) shows a *concentrated* shear load, W, applied to a cantilever beam. The shear load in Fig. 3.2(b) is *distributed* over a length of the beam and is of *intensity w* (force units) per unit length.

A concentrated load on a beam is one which, theoretically, can be regarded as acting wholly at one point. In reality such a situation could not arise since this would imply that the bearing pressure between the load and the beam was infinitely large. Thus, in practice, all loads must be distributed over a finite length of beam. It is when this length of beam is small that we can consider, for the purposes of calculation, the load to be concentrated at one point.

Practical examples of loads that may be regarded as concentrated arise when a beam supports other transverse beams. Distributed loads occur in situations where a girder, for example, supports a wall or floor slab; other distributed loads result from wind forces. All beams in fact support a distributed load, their self-weight.

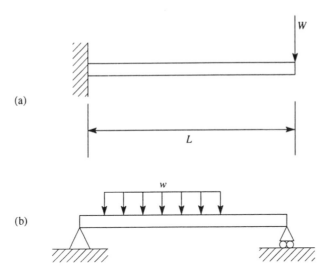

Fig. 3.2 Shear loads applied to beams

Bending moment

In practice it is difficult to apply a pure bending moment such as that shown in Fig. 3.3(a) to a beam. Generally, pure bending moments arise through the application of other types of load to adjacent structural members. For example, in Fig. 3.3(b), a vertical member BC is attached to the cantilever AB and carries a horizontal shear load, P (as far as BC is concerned). AB is therefore subjected to a pure moment, $M = Ph$, at B together with an axial load, P.

Torsion

A similar situation arises in the application of a pure torque, T (Fig. 3.4(a)), to a beam. A practical example of a torque applied to a cantilever beam is given in Fig. 3.4(b) where the horizontal member BC supports a vertical shear load at C. The cantilever AB is then subjected to a pure torque, $T = Wh$, plus a shear load, W.

All the loads illustrated in Figs 3.1–3.4 are applied to the various members by some external agency and are therefore *externally applied loads*. Each of these loads induces reactions in the support systems of the different beams; examples of the calculation of support reactions are given in Section 2.5. Since structures are in equilibrium under a force system of externally applied loads and support reactions, it follows that the support reactions are themselves externally applied loads.

Now consider the cantilever beam of Fig. 3.2(a). If we were to physically cut through the beam at some section mm (Fig. 3.5(a)) the portion BC would no longer be able to support the load, W. The portion AB of the beam therefore performs the same function for the portion BC as does the wall for the complete beam. Thus at the section mm the portion AB applies a force W and a moment M to the portion BC at B, thereby maintaining its equilibrium (Fig. 3.5(b)); by the law of action and reaction, BC exerts an equal force system on AB, but opposite in direction. The complete force systems acting on the two faces of the section mm are shown in Fig. 3.5(b).

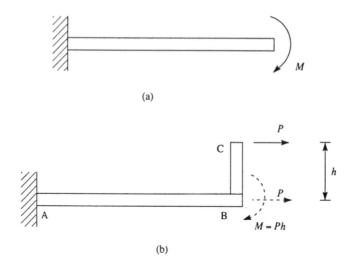

(a)

(b)

Fig. 3.3   Moments applied to beams

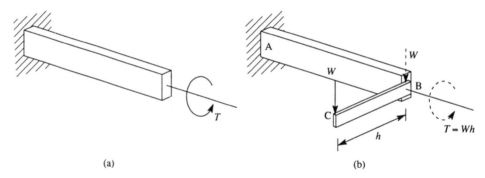

(a)   (b)

Fig. 3.4   Torques applied to a beam

Systems of forces such as those at the section mm are known as *internal forces*. Generally, they vary throughout the length of a structural member as can be seen from Fig. 3.5(b) where the internal moment, M, increases in magnitude as the built-in end is approached due to the increasing rotational effect of W. We note that applied loads of one type can induce internal forces of another. Thus in Fig. 3.5(b) the external shear load, W, produces both shear and bending at the section mm.

Internal forces are distributed throughout beam sections in the form of stresses. It follows that the resultant of each individual stress distribution must be the corresponding internal force; internal forces are therefore often known as *stress resultants*. However, before an individual stress distribution can be found it is necessary to determine the corresponding internal force. Also, in design problems, it is necessary to determine the position and value of maximum stress and displacement. Thus, usually, the first step in the analysis of a structure is to calculate the distribution of each of the four basic internal force types throughout the component structural members. We shall therefore determine the distributions of the four internal force systems in a variety of structural members. First, however, we shall establish a notation and sign convention for each type of force.

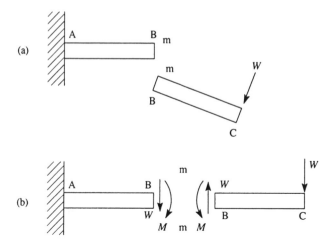

Fig. 3.5 Internal force system generated by an external shear load

3.2 Notation and sign convention

We shall be concerned initially with structural members having at least one longitudinal plane of symmetry. Normally this will be a vertical plane and will contain the externally applied loads. Later, however, we shall investigate the bending and shear of beams having unsymmetrical sections so that as far as possible the notation and sign convention we adopt now will be consistent with that required later. Some modifications will, however, be necessary.

The coordinates of all points in a structural member will be referred to a right-handed system of axes $Oxyz$ as shown in Fig. 3.6. Oz forms the longitudinal axis of the beam and Oy is vertically downwards. Vertical externally applied loads W(concentrated) and w(distributed) are positive in the positive direction of the y axis as are vertical displacements of the member, which are given the symbol v. An axial load P is positive when tensile while a torque T is positive if applied in an anticlockwise sense when viewed in the direction zO.

Positive internal force systems are shown acting on a length of a structural member in Fig. 3.7. For equilibrium of the length of beam the internal forces must

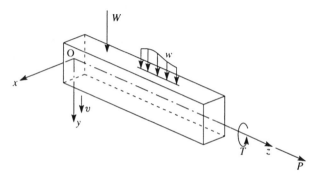

Fig. 3.6 Notation and sign conventions for displacements and externally applied loads

Fig. 3.7   Positive internal force systems

act in opposite directions on the opposite faces of the length of beam. Thus when deciding the sign of an internal force, attention must be paid to the face of the section on which it acts. For example, a positive shear force, S, acts downwards on the right-hand face of the element but upwards on the left-hand face.

A positive bending moment, M, will cause the upper surface of a beam to become concave and is therefore known as a *sagging* bending moment. Negative bending moments produce convex upper surfaces on beams and are termed *hogging* bending moments.

The axial force, N, acts normally to the cross-section of a member and is therefore referred to as a normal force; N is positive when tensile.

A positive internal torque is anticlockwise when viewed in the direction zO.

We see from the above that positive externally applied loads are associated with positive corresponding internal forces.

Generally it is advantageous to represent the distribution of internal forces, moments and torques by *internal force diagrams*; the methods of construction of these diagrams will be illustrated by examples.

3.3   Normal force

Example 3.1   Construct a normal force diagram for the beam AB shown in Fig. 3.8(a).

The first step is to calculate the support reactions using the methods described in Section 2.5. In this case, since the beam is on a roller support at B, the horizontal load at B is reacted at A; clearly $R_{A.H} = 10$ kN acting to the left.

Generally the distribution of an internal force will change at a loading discontinuity. In this case there is no loading discontinuity at any section of the beam so that we can determine the complete distribution of the normal force by calculating the normal force at any section Z, a distance z from A.

Consider the length AZ of the beam as shown in Fig. 3.8(b) (equally we could consider the length ZB). The internal normal force acting at Z is N_{AB} which is shown acting in a positive (tensile) direction. The length AZ of the beam is in equilibrium under the action of $R_{A.H}$ ($=10$ kN) and N_{AB}. Thus, from Section 2.4, for equilibrium in the z direction,

$$N_{AB} - R_{A.H} = N_{AB} - 10 = 0$$

whence

$$N_{AB} = +10 \text{ kN}$$

N_{AB} is positive and therefore acts in the assumed positive direction; the normal force diagram for the complete beam is then as shown in Fig. 3.8(c).

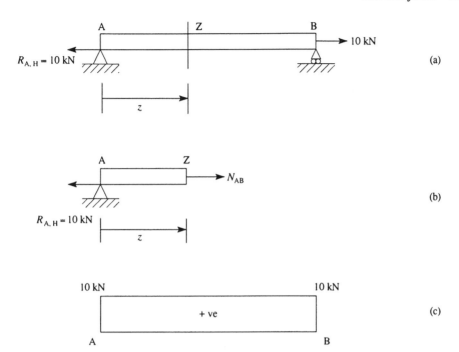

Fig. 3.8 Normal force diagram for the beam of Ex. 3.1

When the equilibrium of a portion of a structure is considered as in Fig. 3.8(b) we are using what is termed a *free body diagram*.

Example 3.2 Draw a normal force diagram for the beam ABC shown in Fig. 3.9(a).

Again by considering the overall equilibrium of the beam we see that $R_{A,H} = 10$ kN acting to the left (C is the roller support).

In this example there is a loading discontinuity at B so that the distribution of the normal force in AB will be different to that in BC. We must therefore determine the normal force at an arbitrary section Z_1 between A and B and then at an arbitrary section Z_2 between B and C.

The free body diagram for the portion of the beam AZ_1 is shown in Fig. 3.9(b). (Alternatively we could consider the portion Z_1C.) As before, we draw in a positive normal force, N_{AB}. Then, for equilibrium of AZ_1 in the z direction.

$$N_{AB} - 10 = 0$$

so that $\qquad\qquad N_{AB} = +10$ kN (tension)

Now consider the length ABZ_2 of the beam; again we draw in a positive normal force, N_{BC}. Then for equilibrium of ABZ_2 in the z direction.

$$N_{BC} + 10 - 10 = 0$$

which gives $\qquad\qquad N_{BC} = 0$

Note that we would have obtained the same result by considering the portion Z_2C of the beam.

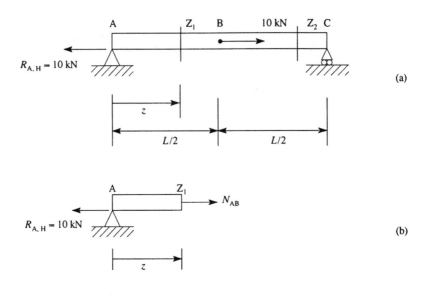

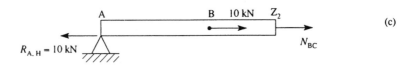

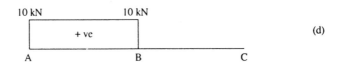

Fig. 3.9 Normal force diagram for the beam of Ex. 3.2

Finally the complete normal force diagram for the beam is drawn as shown in Fig. 3.9(d).

Example 3.3 Fig. 3.10(a) shows a beam ABCD supporting three concentrated loads, two of which are inclined to the longitudinal axis of the beam. Construct the normal force diagram for the beam and determine the maximum value.

In this example we are only concerned with determining the normal force distribution in the beam, so that it is unnecessary to calculate the vertical reactions at the supports. Further, the horizontal components of the inclined loads can only be resisted at A since D is a roller support. Thus, considering the horizontal equilibrium of the beam,

$$R_{A,H} + 6 \cos 60° - 4 \cos 60° = 0$$

whence
$$R_{A,H} = -1 \text{ kN}$$

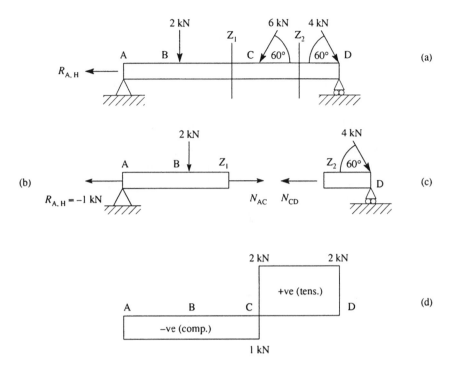

Fig. 3.10 Normal force diagram for the beam of Ex. 3.3

The negative sign of $R_{A.H}$ indicates that the reaction acts to the right and not to the left as originally assumed. However, rather than change the direction of $R_{A.H}$ in the diagram, it is simpler to retain the assumed direction and then insert the negative value as required.

Although there is an apparent loading discontinuity at B, the 2 kN load acts perpendicularly to the longitudinal axis of the beam and will therefore not affect the normal force. We may therefore consider the normal force at any section Z_1 between A and C. The free body diagram for the portion AZ_1 of the beam is shown in Fig. 3.10(b); again we draw in a positive normal force N_{AC}. For equilibrium of AZ_1

$$N_{AC} - R_{A.H} = 0$$

so that $\qquad N_{AC} = R_{A.H} = -1$ kN (compression)

The horizontal component of the inclined load at C produces a loading discontinuity so that we now consider the normal force at any section Z_2 between C and D. Here it is slightly simpler to consider the equilibrium of the length Z_2D of the beam rather than the length AZ_2. Thus, from Fig. 3.10(c)

$$N_{CD} - 4 \cos 60° = 0$$

which gives $\qquad N_{CD} = +2$ kN (tension)

From the completed normal force diagram in Fig. 3.10(d) we see that the maximum normal force in the beam is 2 kN (tension) acting at all sections between C and D.

3.4 Shear force and bending moment

It is convenient to consider shear force and bending moment distributions in beams simultaneously since, as we shall see in Section 3.5, they are directly related. Again the method of construction of shear force and bending moment diagrams will be illustrated by examples.

Example 3.4   Cantilever beam with a concentrated load at the free end (Fig. 3.11).

Generally, as in the case of normal force distributions, we require the variation in shear force and bending moment along the length of a beam. Again, loading discontinuities, such as concentrated loads and/or a sudden change in the intensity of a distributed load, cause discontinuities in the distribution of shear force and bending moment so that it is necessary to consider a series of sections, one between

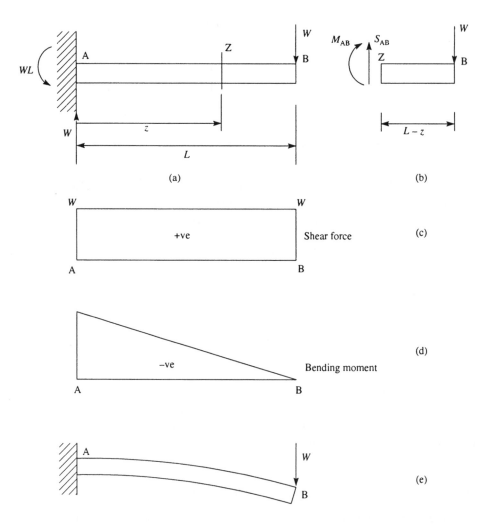

Fig. 3.11   Shear force and bending moment diagrams for the beam of Ex. 3.4

each loading discontinuity. In this example, however, there are no loading discontinuities between the built-in end A and the free end B so that we may consider a section Z at any point between A and B.

For many beams the value of each support reaction must be calculated before the shear force and bending moment distributions can be obtained. In Fig. 3.11(a) a consideration of the overall equilibrium of the beam (see Section 2.5) gives a vertical reaction, W, and a moment reaction, WL, at the built-in end. However, if we consider the equilibrium of the length ZB of the beam as shown in the free body diagram in Fig. 3.11(b), this calculation is unnecessary.

As in the case of normal force distributions we assign positive directions to the shear force, S_{AB}, and bending moment, M_{AB}, at the section Z. Then, for vertical equilibrium of the length ZB of the beam we have

$$S_{AB} - W = 0$$

which gives

$$S_{AB} = +W$$

Thus the shear force is constant along the length of the beam and the shear force diagram is rectangular in shape, as shown in Fig. 3.11(c).

The bending moment, M_{AB}, is now found by considering the moment equilibrium of the length ZB of the beam about the section Z. Alternatively we could take moments about B, but this would involve the moment of the shear force, S_{AB}, about B. This approach, although valid, is not good practice since it includes a previously calculated quantity; in some cases, however, this is unavoidable. Thus, taking moments about the section Z we have

$$M_{AB} + W(L - z) = 0$$

whence

$$M_{AB} = -W(L - z) \tag{i}$$

Eq. (i) shows that M_{AB} varies linearly along the length of the beam, is negative, i.e. hogging, at all sections and increases from zero at the free end ($z = L$) to $-WL$ at the built-in end where $z = 0$.

It is usual to draw the bending moment diagram on the tension side of a beam. This procedure is particularly useful in the design of reinforced concrete beams since it shows directly the surface of the beam near which the major steel reinforcement should be provided. Also, drawing the bending moment diagram on the tension side of a beam can give an indication of the deflected shape as illustrated in Exs 3.4–3.7. This is not always the case, however, as we shall see in Exs 3.8 and 3.9.

In this case the beam will bend as shown in Fig. 3.11(e) so that the upper surface of the beam is in tension and the lower one in compression; the bending moment diagram is therefore drawn on the upper surface as shown in Fig. 3.11(d). Note that negative (hogging) bending moments applied in a vertical plane will always result in the upper surface of a beam being in tension.

Example 3.5 Cantilever beam carrying a uniformly distributed load of intensity w.

Again it is unnecessary to calculate the reactions at the built-in end of the cantilever; their values are, however, shown in Fig. 3.12(a). Note that for the purpose of calculating the moment reaction the uniformly distributed load may be replaced by a concentrated load ($= wL$) acting at a distance $L/2$ from A.

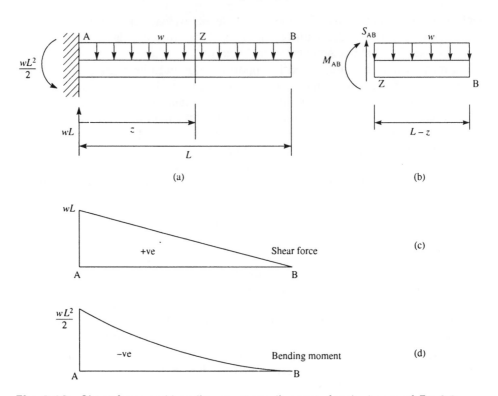

Fig. 3.12   Shear force and bending moment diagrams for the beam of Ex. 3.5

There is no loading discontinuity between A and B so that we may consider the shear force and bending moment at any section Z between A and B. As before, we insert positive directions for the shear force, S_{AB}, and bending moment, M_{AB}, in the free body diagram of Fig. 3.12(b). Then, for vertical equilibrium of the length ZB of the beam,

$$S_{AB} - w(L - z) = 0$$

so that $$S_{AB} = +w(L - z)$$ (i)

Therefore S_{AB} varies linearly with z and increases from zero at B to $+wL$ at A (Fig. 3.12(c)).

Now consider the moment equilibrium of the length AB of the beam and take moments about Z:

$$M_{AB} + \frac{w}{2}(L - z)^2 = 0$$

which gives $$M_{AB} = -\frac{w}{2}(L - z)^2$$ (ii)

Note that the total load on the length ZB of the beam is $w(L - z)$, which we may consider acting as a concentrated load at a distance $(L - z)/2$ from Z. From Eq. (ii) we see that the bending moment, M_{AB}, is negative at all sections of the beam and varies parabolically as shown in Fig. 3.12(d) where the bending moment diagram is again drawn on the tension side of the beam. The actual shape of the bending moment diagram may be found by plotting values or, more conveniently, by examining Eq. (ii). Differentiating with respect to z we obtain

$$\frac{\mathrm{d}M_{AB}}{\mathrm{d}z} = w(L - z) \tag{iii}$$

so that when $z = L$, $\mathrm{d}M_{AB}/\mathrm{d}z = 0$ and the bending moment diagram is tangential to the datum line AB at B. Furthermore it can be seen from Eq. (iii) that the gradient $(\mathrm{d}M_{AB}/\mathrm{d}z)$ of the bending moment diagram decreases as z increases, so that its shape is as shown in Fig. 3.12(d).

Example 3.6   Simply supported beam carrying a central concentrated load.

In this example it is necessary to calculate the value of the support reactions, both of which are seen, from symmetry, to be $W/2$ (Fig. 3.13(a)). Also, there is a loading discontinuity at B, so that we must consider the shear force and bending moment first at an arbitrary section Z_1 say, between A and B and then at an arbitrary section Z_2 between B and C.

From the free body diagram in Fig. 3.13(b) in which both S_{AB} and M_{AB} are in positive directions we see, by considering the vertical equilibrium of the length AZ_1 of the beam, that

$$S_{AB} - \frac{W}{2} = 0$$

whence
$$S_{AB} = +\frac{W}{2}$$

S_{AB} is therefore constant at all sections of the beam between A and B, in other words, from a section immediately to the right of A to a section immediately to the left of B.

Now consider the free body diagram of the length Z_2C of the beam in Fig. 3.13(c). Note that, equally, we could have considered the length ABZ_2, but this would have been slightly more complicated in terms of the number of loads acting. For vertical equilibrium of Z_2C

$$S_{AB} + \frac{W}{2} = 0$$

from which
$$S_{AB} = -\frac{W}{2}$$

and we see that S_{BC} is constant at all sections of the beam between B and C so that the complete shear force diagram has the form shown in Fig. 3.13(d). Note that the *change* in shear force from that at a section immediately to the left of B to that at a section immediately to the right of B is $-W$. We shall consider the implications of this later in the chapter.

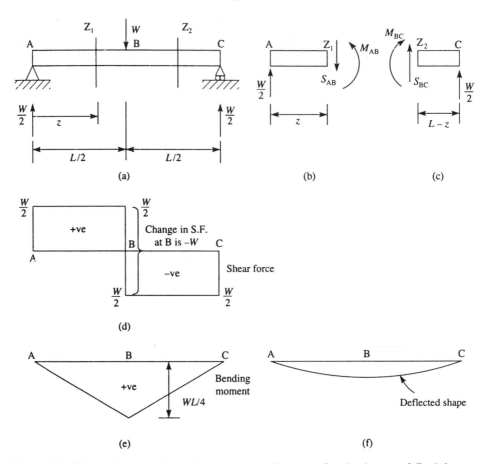

Fig. 3.13   Shear force and bending moment diagrams for the beam of Ex 3.6

It would also appear from Fig. 3.13(d) that there are two different values of shear force at the same section B of the beam. This results from the assumption that W is concentrated at a point which, practically, is impossible since there would then be an infinite bearing pressure on the surface of the beam. In practice, the load W and the support reactions would be distributed over a small length of beam (Fig. 3.14(a)) so that the actual shear force distribution would be that shown in Fig. 3.14(b).

The distribution of the bending moment in AB is now found by considering the moment equilibrium about Z_1 of the length AZ_1 of the beam in Fig. 3.13(b). Thus

$$M_{AB} - \frac{W}{2}z = 0$$

or
$$M_{AB} = \frac{W}{2}z \qquad\qquad\text{(i)}$$

Therefore M_{AB} varies linearly from zero at A $(z = 0)$ to $+WL/4$ at B$(z = L/2)$.
Now considering the length Z_2C of the beam in Fig 3.13(c) and taking moments about Z_2

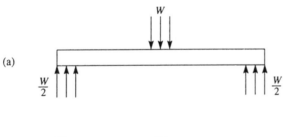

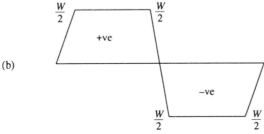

Fig. 3.14 Shear force diagram in a practical situation

$$M_{BC} - \frac{W}{2}(L - z) = 0$$

which gives

$$M_{BC} = + \frac{W}{2}(L - z) \qquad \text{(ii)}$$

From Eq. (ii) we see that M_{BC} varies linearly from $+WL/4$ at B $(z = L/2)$ to zero at C $(z = L)$.

The complete bending moment diagram is shown in Fig. 3.13(e). Note that the bending moment is positive (sagging) at all sections of the beam so that the lower surface of the beam is in tension. In this example the deflected shape of the beam would be that shown in Fig. 3.13(f).

Example 3.7 Simply supported beam carrying a uniformly distributed load.

The symmetry of the beam and its load may again be used to determine the support reactions which are each $wL/2$. Furthermore, there is no loading discontinuity between the ends A and B of the beam so that it is sufficient to consider the shear force and bending moment at just one section Z, a distance z, say, from A; again we draw in positive directions for the shear force and bending moment at the section Z in the free body diagram shown in Fig. 3.15(b).

Considering the vertical equilibrium of the length AZ of the beam gives

$$S_{AB} + wz - w\frac{L}{2} = 0$$

whence

$$S_{AB} = + w\left(\frac{L}{2} - z\right) \qquad \text{(i)}$$

S_{AB} therefore varies linearly along the length of the beam from $+wL/2$ at A $(z = 0)$ to $-wL/2$ at B $(z = L)$. Note that $S_{AB} = 0$ at mid-span $(z = L/2)$.

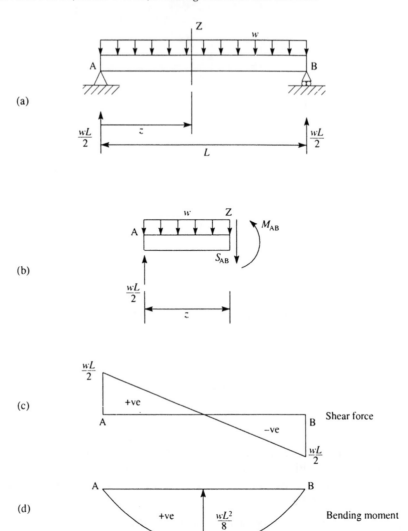

Fig. 3.15   Shear force and bending moment diagrams for the beam of Ex. 3.7

Now taking moments about Z for the length AZ of the beam in Fig. 3.15(b) we have

$$M_{AB} + \frac{wz^2}{2} - \frac{wL}{2} z = 0$$

from which

$$M_{AB} = +\frac{wz}{2} (L - z) \qquad\qquad (ii)$$

Thus M_{AB} varies parabolically along the length of the beam and is positive (sagging) at all sections of the beam except at the supports ($z = 0$ and $z = L$) where it is zero.

Also, differentiating Eq. (ii) with respect to z gives

$$\frac{dM_{AB}}{dz} = w\left(\frac{L}{2} - z\right) \tag{iii}$$

From Eq. (iii) we see that $dM_{AB}/dz = 0$ at mid-span where $z = L/2$, so that the bending moment diagram has a turning value or mathematical maximum at this section. In this case this mathematical maximum is the maximum value of the bending moment in the beam and is, from Eq. (ii), $+wL^2/8$.

The bending moment diagram for the beam is shown in Fig. 3.15(d) where it is again drawn on the tension side of the beam; the deflected shape of the beam will be identical in form to the bending moment diagram.

Examples 3.4–3.7 may be regarded as 'standard' cases and it is useful to memorize the form that the shear force and bending moment diagrams take including the principal values.

Example 3.8　Simply supported beam with cantilever overhang (Fig. 3.16(a)).

The support reactions are calculated using the methods described in Section 2.5. Thus, taking moments about B in Fig. 3.16(a) we have

$$R_A \times 2 - 2 \times 3 \times 0.5 + 1 \times 1 = 0$$

which gives

$$R_A = 1 \text{ kN}$$

From vertical equilibrium

$$R_B + R_A - 2 \times 3 - 1 = 0$$

so that

$$R_B = 6 \text{ kN}$$

The support reaction at B produces a loading discontinuity at B so that we must consider the shear force and bending moment at two arbitrary sections of the beam, Z_1 in AB and Z_2 in BC. Free body diagrams are therefore drawn for the lengths AZ_1 and Z_2C of the beam and positive directions for the shear force and bending moment drawn in as shown in Figs 3.16(b) and (c). Alternatively, we could have considered the lengths Z_1BC and ABZ_2, but this approach would have involved slightly more complicated solutions in terms of the number of loads applied.

Now from the vertical equilibrium of the length AZ_1 of the beam in Fig. 3.16(b) we have

$$S_{AB} + 2z - 1 = 0$$

or

$$S_{AB} = 1 - 2z \tag{i}$$

The shear force therefore varies linearly in AB from $+1$ kN at A ($z = 0$) to -3 kN at B ($z = 2$ m). Note that $S_{AB} = 0$ at $z = 0 \cdot 5$ m.

Consideration of the vertical equilibrium of the length Z_2C of the beam in Fig. 3.16(c) gives

$$S_{BC} - 2(3 - z) - 1 = 0$$

from which

$$S_{BC} = 7 - 2z \tag{ii}$$

Eq. (ii) shows that S_{BC} varies linearly in BC from $+3$ kN at B ($z = 2$ m) to $+1$ kN at C ($z = 3$ m).

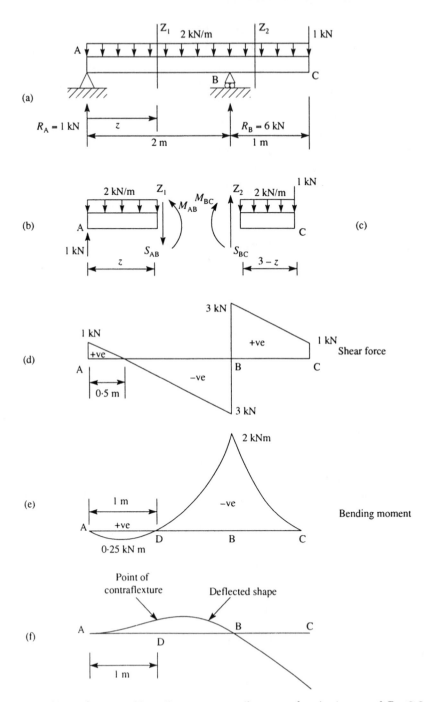

Fig. 3.16   Shear force and bending moment diagrams for the beam of Ex. 3.8

The complete shear force diagram for the beam is shown in Fig. 3.16(d).

The bending moment, M_{AB}, is now obtained by considering the moment equilibrium of the length AZ_1 of the beam about Z_1 in Fig. 3.16(b). Hence

$$M_{AB} + 2z\frac{z}{2} - 1z = 0$$

so that
$$M_{AB} = z - z^2 \qquad\qquad (iii)$$

which is a parabolic function of z. The distribution may be plotted by selecting a series of values of z and calculating the corresponding values of M_{AB}. However, this would not necessarily produce accurate estimates of either the magnitudes and positions of the maximum values of M_{AB} or, say, the positions of the zero values of M_{AB} which, as we shall see later, are important in beam design. A better approach is to examine Eq. (iii) as follows. Clearly when $z = 0$, $M_{AB} = 0$ as would be expected at the simple support at A. Also at B, where $z = 2$ m, $M_{AB} = -2$ kN so that although the support at B is a simple support and allows rotation of the beam, there is a moment at B; this is produced by the loads on the cantilever overhang BC. Rewriting Eq. (iii) in the form

$$M_{AB} = z(1 - z) \qquad\qquad (iv)$$

we see immediately that $M_{AB} = 0$ at $z = 0$ (as demonstrated above) and that $M_{AB} = 0$ at $z = 1$ m, the point D in Fig. 3.16(e). We shall see later in Chapter 9 that at the point in the beam where the bending moment changes sign the curvature of the beam is zero; this point is known as a *point of contraflexure* or *point of inflection*. Now differentiating Eq. (iii) with respect to z we obtain

$$\frac{dM_{AB}}{dz} = 1 - 2z \qquad\qquad (v)$$

and we see that $dM_{AB}/dz = 0$ at $z = 0.5$ m. In other words M_{AB} has a turning value or mathematical maximum at $z = 0.5$ m at which point $M_{AB} = 0.25$ kN m. Note that this is not the greatest value of bending moment in the span AB. Also it can be seen that for $0 < z < 0.5$ m, dM_{AB}/dz decreases with z while for 0.5 m $< z < 2$ m, dM_{AB}/dz increases negatively with z.

Now we consider the moment equilibrium of the length Z_2C of the beam in Fig. 3.16(c) about Z_2.

$$M_{BC} + \frac{2}{2}(3 - z)^2 + 1(3 - z) = 0$$

so that
$$M_{BC} = -12 + 7z - z^2 \qquad\qquad (vi)$$

from which we see that dM_{BC}/dz is not zero at any point in BC and that as z increases dM_{BC}/dz decreases.

The complete bending moment diagram is therefore as shown in Fig. 3.16(e). Note that the value of zero shear force in AB coincides with the turning value of the bending moment.

In this particular example it is not possible to deduce the displaced shape of the beam from the bending moment diagram. Only three facts relating to the displaced shape can be stated with certainty; these are, the deflections at A and B are zero and there is a point of contraflexure at D, 1 m from A. However, using the method described in Section 13.2 gives the displaced shape shown in Fig. 3.16(f). Note that, although the beam is subjected to a sagging bending moment over the length AD, the actual deflection is upwards; clearly this could not have been deduced from the bending moment diagram.

Example 3.9   Simply supported beam carrying a point moment.

From a consideration of the overall equilibrium of the beam (Fig. 3.17(a)) the support reactions are $R_A = M_0/L$ acting vertically upward and $R_C = M_0/L$ acting vertically downward. Note that R_A and R_C are independent of the point of application of M_0.

   Although there is a loading discontinuity at B it is a point moment and will not affect the distribution of shear force. Thus, by considering the vertical equilibrium of either AZ_1 in Fig. 3.17(b) or Z_2C in Fig. 3.17(c) we see that

$$S_{AB} = S_{BC} = + \frac{M_0}{L} \tag{i}$$

The shear force is therefore constant along the length of the beam as shown in Fig. 3.17(d).

   Now considering the moment equilibrium about Z_1 of the length AZ_1 of the beam in Fig. 3.17(b),

$$M_{AB} - \frac{M_0}{L} z = 0$$

or

$$M_{AB} = \frac{M_0}{L} z \tag{ii}$$

M_{AB} therefore increases linearly from zero at A $(z = 0)$ to $+3M_0/4$ at B $(z = 3L/4)$. From Fig. 3.17(c) and taking moments about Z_2 we have

$$M_{BC} + \frac{M_0}{L} (L - z) = 0$$

or

$$M_{BC} = \frac{M_0}{L} (z - L) \tag{iii}$$

M_{BC} therefore decreases linearly from $-M_0/4$ at B $(z = 3L/4)$ to zero at C $(z = L)$; the complete distribution of bending moment is shown in Fig. 3.17(e). The deflected form of the beam is shown in Fig. 3.17(f) where a point of contraflexure occurs at B, the section at which the bending moment changes sign.

   In this example, as in Ex. 3.8, the exact form of the deflected shape cannot be deduced from the bending moment diagram without analysis. However, using the method of singularities described in Section 13.2, it may be shown that the deflection at B is positive and that the slope of the beam at C is negative, giving the displaced shape shown in Fig. 3.17(f).

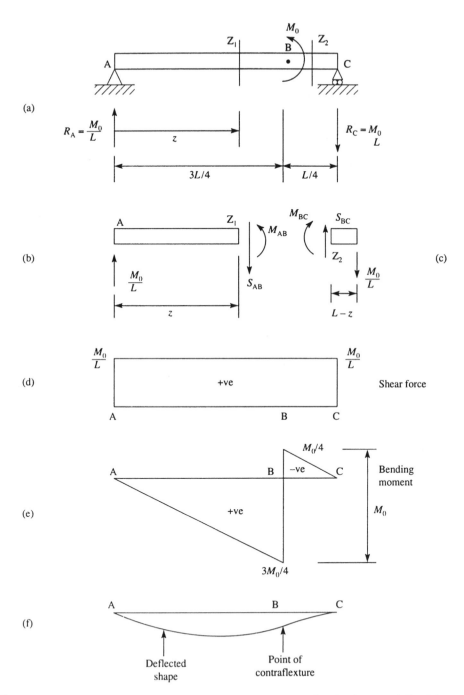

Fig. 3.17 Shear force and bending moment diagrams for the beam of Ex. 3.9

3.5   Load, shear force and bending moment relationships

It is clear from Exs 3.4–3.9 that load, shear force and bending moment are related. Thus, for example, uniformly distributed loads produce linearly varying shear forces and maximum values of bending moment coincide with zero shear force. We shall now examine these relationships mathematically.

The length of beam shown in Fig. 3.18(a) carries a general system of loading comprising concentrated loads and a distributed load $w(z)$. An elemental length δz of the beam is subjected to the force and moment system shown in Fig. 3.18(b); since δz is very small the distributed load may be regarded as constant over the length δz. For vertical equilibrium of the element

$$S - w(z)\delta z - (S + \delta S) = 0$$

so that

$$-w(z)\delta z - \delta S = 0$$

Thus, in the limit as $\delta z \to 0$

$$\frac{dS}{dz} = -w(z) \tag{3.1}$$

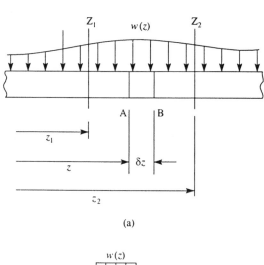

(a)

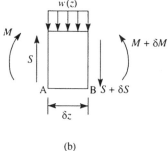

(b)

Fig. 3.18   Load, shear force and bending moment relationships

From Eq. (3.1) we see that the rate of change of shear force at a section of a beam, in other words the gradient of the shear force diagram, is equal to minus the value of the load intensity at that section. In Fig. 3.12(c), for example, the shear force changes linearly from wL at A to zero at B so that the gradient of the shear force diagram at any section of the beam is $-wL/L = -w$ where w is the load intensity. Equation (3.1) also applies at beam sections subjected to concentrated loads. In Fig. 3.13(a) the load intensity at B, theoretically, is infinite, as is the gradient of the shear force diagram at B (Fig. 3.13(d)). In practice the shear force diagram would have a finite gradient at this section as illustrated in Fig. 3.14.

Now integrating Eq. (3.1) with respect to z we obtain

$$S = -\int w(z)\,dz + C_1 \tag{3.2}$$

in which C_1 is a constant of integration which may be determined in a particular case from the loading boundary conditions.

If, for example, $w(z)$ is a uniformly distributed load of intensity w, that is it is not a function of z, Eq. (3.2) becomes

$$S = -wz + C_1$$

which is the equation of a straight line of gradient $-w$ as demonstrated for the cantilever beam of Fig. 3.12 in the previous paragraph. Furthermore, for this particular example, $S = 0$ at $z = L$ so that $C_1 = wL$ and $S = w(L - z)$ as before.

In the case of a beam carrying only concentrated loads then, in the bays between the loads, $w(z) = 0$ and Eq. (3.2) reduces to

$$S = C_1$$

so that the shear force is constant over the unloaded length of beam (see Figs 3.11 and 3.13).

Suppose now that Eq. (3.1) is integrated over the length of beam between the sections Z_1 and Z_2. Then

$$\int_{z_1}^{z_2} \frac{dS}{dz}\,dz = - \int_{z_1}^{z_2} w(z)\,dz$$

which gives

$$S_2 - S_1 = \int_{z_1}^{z_2} w(z)\,dz \tag{3.3}$$

where S_1 and S_2 are the shear forces at the sections Z_1 and Z_2 respectively. Equation (3.3) shows that the *change* in shear force between two sections of a beam is equal to minus the area under the load distribution curve over that length of beam.

The argument may be applied to the case of a concentrated load W which may be regarded as a uniformly distributed load acting over an extremely small elemental length of beam, say δz. The area under the load distribution curve would then be $w\delta z (=W)$ and the change in shear force from the section z to the section $z + \delta z$ would be $-W$. In other words, the change in shear force from a section immediately to the left of a concentrated load to a section immediately to the right is equal to minus the value of the load, as noted in Ex. 3.6.

Now consider the rotational equilibrium of the element δz in Fig. 3.18(b) about B. Thus

$$M + S\,\delta z - w(z)\,\delta z\,\frac{\delta z}{2} - (M + \delta M) = 0$$

The term involving the square of δz is a second-order term and may be neglected. Hence

$$S\delta z - \delta M = 0$$

or, in the limit as $\delta z \to 0$

$$\frac{dM}{dz} = S \qquad\qquad (3.4)$$

Equation (3.4) establishes for the general case what may be observed in particular in the shear force and bending moment diagrams of Exs 3.4–3.9, i.e. the gradient of the bending moment diagram at a beam section is equal to the value of the shear force at that section. For example, in Fig. 3.16(e) the bending moment in AB is a mathematical maximum at the section where the shear force is zero.

Integrating Eq. (3.4) with respect to z we have

$$M = \int S\,dz + C_2 \qquad\qquad (3.5)$$

in which C_2 is a constant of integration. Substituting for S in Eq. (3.5) from Eq. (3.2) gives

$$M = \int \left[-\int w(z)\,dz + C_1 \right] dz + C_2$$

or

$$M = -\iint w(z)\,dz + C_1 z + C_2 \qquad\qquad (3.6)$$

If $w(z)$ is a uniformly distributed load of intensity w, Eq. (3.6) becomes

$$M = -w\,\frac{z^2}{2} + C_1 z + C_2$$

which shows that the equation of the bending moment diagram on a length of beam carrying a uniformly distributed load is parabolic.

In the case of a beam carrying concentrated loads only then, between the loads, $w(z) = 0$ and Eq. (3.6) reduces to

$$M = C_1 z + C_2$$

which shows that the bending moment varies linearly between the loads and has a gradient C_1.

The constants C_1 and C_2 in Eq. (3.6) may be found, for a given beam, from the loading boundary conditions. Thus, for the cantilever beam of Fig. 3.12, we have already shown that $C_1 = wL$ so that $M = -wz^2/2 + wLz + C_2$. Also, when $z = L$, $M = 0$ which gives $C_2 = -wL^2/2$ and hence $M = -wz^2/2 + wLz - wL^2/2$ as before.

Now integrating Eq. (3.4) over the length of beam between the sections Z_1 and Z_2 (Fig. 3.18(a)),

$$\int_{z_1}^{z_2} \frac{dM}{dz}\, dz = \int_{z_1}^{z_2} S\, dz$$

which gives

$$M_2 - M_1 = \int_{z_1}^{z_2} S\, dz \qquad (3.7)$$

where M_1 and M_2 are the bending moments at the sections Z_1 and Z_2, respectively. Equation (3.7) shows that the *change* in bending moment between two sections of a beam is equal to the area of the shear force diagram between those sections. Again, using the cantilever beam of Fig. 3.12 as an example, we see that the change in bending moment from A to B is $wL^2/2$ and that the area of the shear force diagram between A and B is $wL^2/2$.

Finally, from Eqs (3.1) and (3.4)

$$\frac{d^2 M}{dz^2} = \frac{dS}{dz} = -w(z) \qquad (3.8)$$

The relationships established above may be used to construct shear force and bending moment diagrams for some beams more readily than when the methods illustrated in Exs 3.4–3.9 are employed. In addition they may be used to provide simpler solutions in some beam problems.

Example 3.10   Construct shear force and bending moment diagrams for the beam shown in Fig. 3.19(a).

Initially the support reactions are calculated using the methods described in Section 2.5. Hence, for moment equilibrium of the beam about E

$$R_A \times 4 - 2 \times 3 - 5 \times 2 - 4 \times 1 \times 0 \cdot 5 = 0$$

from which
$$R_A = 4 \cdot 5 \text{ kN}$$

Now considering the vertical equilibrium of the beam

$$R_E + R_A - 2 - 5 - 4 \times 1 = 0$$

so that
$$R_E = 6 \cdot 5 \text{ kN}$$

In constructing the shear force diagram we can make use of the facts that, as established above, the shear force is constant over unloaded bays of the beam, varies linearly when the loading is uniformly distributed and changes negatively as a vertically downward concentrated load is crossed in the positive z direction by the value of the load. Thus in Fig. 3.19(b) the shear force increases positively by 4·5 kN as we move from the left of A to the right of A, is constant between A and B, changes negatively by 2 kN as we move from the left of B to the right of B, and so on. In effect the shear force diagram is constructed by following the loading pattern. Note that between D and E the shear force changes linearly from −2·5 kN at D to −6·5 kN at a section immediately to the left of E, in other words it changes by −4 kN, the total value of the downward-acting uniformly distributed load.

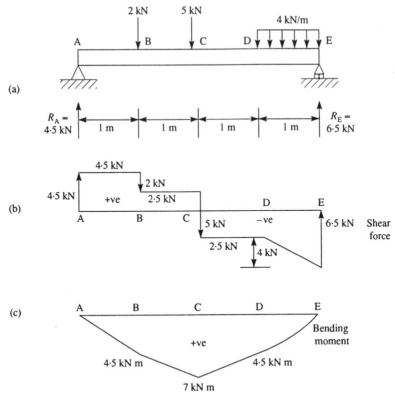

Fig. 3.19 Shear force and bending moment diagrams for the beam of Ex. 3.10

The bending moment diagram may also be constructed using the above relationships, namely, the bending moment varies linearly over unloaded lengths of beam and parabolically over lengths of beam carrying a uniformly distributed load. Also, the change in bending moment between two sections of a beam is equal to the area of the shear force diagram between those sections. Thus in Fig. 3.19(a) we know that the bending moment at the pinned support at A is zero and that it varies linearly in the bay AB. The bending moment at B is then equal to the area of the shear force diagram between A and B, which is +4·5 kN m. This represents, in fact, the change in bending moment from the zero value at A to the value at B. At C the area of the shear force diagram to the right or left of C is 7 kN m (note that the bending moment at E is also zero), and so on. In the bay DE the shape of the parabolic curve representing the distribution of bending moment over the length of the uniformly distributed load may be found using part of Eq. (3.8), i.e.

$$\frac{\mathrm{d}^2 M}{\mathrm{d}z^2} = -w(z)$$

For a vertically downward uniformly distributed load this expression becomes

$$\frac{\mathrm{d}^2 M}{\mathrm{d}z^2} = -w$$

which from mathematical theory shows that the curve representing the variation in bending moment is convex in the positive direction of bending moment. This may be observed in the bending moment diagrams in Figs 3.12(d), 3.15(d) and 3.16(e). In this example the bending moment diagram for the complete beam is shown in Fig. 3.19(c) and is again drawn on the tension side of the beam.

Example 3.11   A precast concrete beam of length L is to be lifted from the casting bed and transported so that the maximum bending moment is as small as possible. If the beam is lifted by two slings placed symmetrically, show that each sling should be $0.21L$ from the adjacent end.

The external load on the beam is comprised solely of its own weight, which is uniformly distributed along its length. The problem is therefore resolved into that of a simply supported beam carrying a uniformly distributed load in which the supports are positioned at some distance a from each end (Fig. 3.20(a)).

The shear force and bending moment diagrams may be constructed in terms of a using the methods described above and would take the forms shown in Figs 3.20(b) and (c). Examination of the bending moment diagram shows that there are two possible positions for the maximum bending moment. First at B and C where the bending moment is hogging and has equal values from symmetry; second at the mid-span point where the bending moment has a turning value and is sagging if the supports at B and C are spaced a sufficient distance apart. Suppose that B and C are positioned such that the value of the hogging bending moment at B and C is

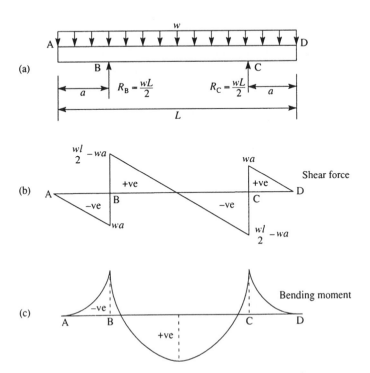

Fig. 3.20 Determination of optimum position for supports in a precast concrete beam (Ex. 3.11)

numerically equal to the sagging bending moment at the mid-span point. If now B and C are moved further apart the mid-span moment will increase while the moment at B and C decreases. Conversely, if B and C are brought closer together, the hogging moment at B and C increases while the mid-span moment decreases. It follows that the maximum bending moment will be as small as possible when the hogging moment at B and C is numerically equal to the sagging moment at mid-span.

The solution will be simplified if use is made of the relationship in Eq. (3.7). Thus, when the supports are in the optimum position, the change in bending moment from A to B (negative) is equal to minus half the change in the bending moment from B to the mid-span point (positive). It follows that the area of the shear force diagram between A and B is equal to minus half of that between B and the mid-span point. Then

$$-\frac{1}{2}awa = -\frac{1}{2}\left[\frac{1}{2}\left(\frac{L}{2} - a\right)w\left(\frac{L}{2} - a\right)\right]$$

which reduces to

$$a^2 + La - L^2/4 = 0$$

the solution of which gives

$$a = 0 \cdot 21L \text{(the negative solution has no practical significance)}$$

3.6 Torsion

The distribution of torque along a structural member may be obtained by considering the equilibrium in free body diagrams of lengths of member in a similar manner to that used for the determination of shear force distributions in Exs 3.4–3.9.

Example 3.12   Construct a torsion diagram for the beam shown in Fig. 3.21(a).

There is a loading discontinuity at B so that we must consider the torque at separate sections Z_1 and Z_2 in AB and BC, respectively. Thus, in the free body diagrams shown in Figs 3.21(b) and (c) we insert positive internal torques.
From Fig. 3.21(b)

$$T_{AB} - 10 + 8 = 0$$

so that
$$T_{AB} = +2 \text{ kN m}$$

From Fig. 3.21(c)

$$T_{BC} + 8 = 0$$

from which
$$T_{BC} = -8 \text{ kN m}$$

The complete torsion diagram is shown in Fig. 3.21(d).

Example 3.13   The structural member ABC shown in Fig. 3.22 carries a distributed torque of 2 kN m/m together with a concentrated torque of 10 kN m at mid-span. The supports at A and C prevent rotation of the member in planes perpendicular to its axis. Construct a torsion diagram for the member and determine the maximum value of torque.

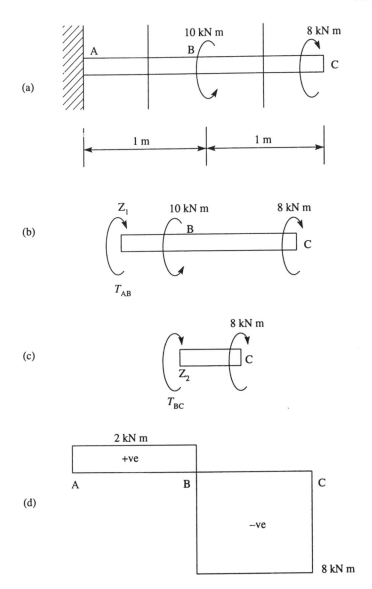

Fig. 3.21 Torsion diagram for a cantilever beam

From the rotational equilibrium of the member about its longitudinal axis and its symmetry about the mid-span section at B, we see that the reactive torques T_A and T_C are each -9 kN m, i.e. clockwise when viewed in the direction CBA. In general, as we shall see in Chapter 11, reaction torques at supports form a statically indeterminate system.

In this particular problem there is a loading discontinuity at B so that we must consider the internal torques at two arbitrary sections Z_1 and Z_2 as shown in Fig. 3.23(a).

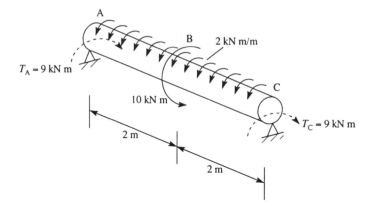

Fig. 3.22   Beam of Ex. 3.13

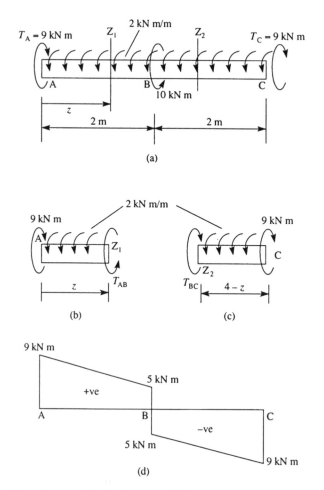

Fig. 3.23   Torsion diagram for the beam of Ex. 3.13

From the free body diagram in Fig. 3.23(b)

$$T_{AB} + 2z - 9 = 0$$

which gives

$$T_{AB} = 9 - 2z \qquad \text{(i)}$$

From Eq. (i) we see that T_{AB} varies linearly from $+9$ kN m at A ($z = 0$) to $+5$ kN m at a section immediately to the left of B ($z = 2$ m). Furthermore, from Fig. 3.23(c)

$$T_{BC} - 2(4 - z) + 9 = 0$$

so that

$$T_{BC} = -2z - 1 \qquad \text{(ii)}$$

from which we see that T_{BC} varies linearly from -5 kN m at a section immediately

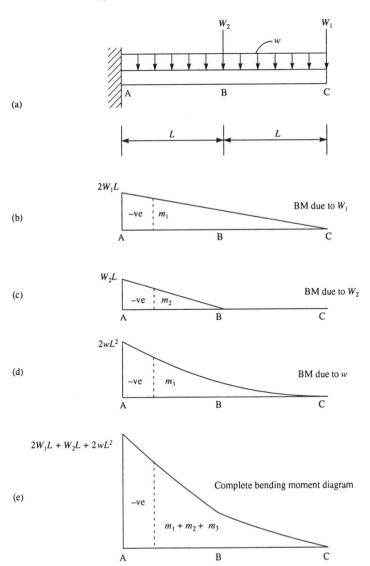

Fig. 3.24 Bending moment (BM) diagram using the principle of superposition

to the right of B ($z = 2$ m) to -9 kN m at C ($z = 4$ m). The resulting torsion diagram is shown in Fig. 3.23(d).

3.7 Principle of superposition

An extremely useful principle in the analysis of linearly elastic structures (see Chapter 8) is that of superposition. The principle states that if the displacements at all points in an elastic body are proportional to the forces producing them, that is the body is linearly elastic, the effect (i.e. stresses and displacements) on such a body of a number of forces acting simultaneously is the sum of the effects of the forces applied separately.

This principle can sometimes simplify the construction of shear force and bending moment diagrams.

Example 3.14   Construct the bending moment diagram for the beam shown in Fig. 3.24(a).

Figures 3.24(b), (c) and (d) show the bending moment diagrams for the cantilever when each of the three loading systems acts separately. The bending moment diagram for the beam when the loads act simultaneously is obtained by adding the ordinates of the separate diagrams and is shown in Fig. 3.24(e).

Problems

P.3.1   A transmitting mast of height 40 m and weight 4·5 kN/m length is stayed by three groups of four cables attached to the mast at heights of 15, 25 and 35 m. If each cable is anchored to the ground at a distance of 20 m from the base of the mast and tensioned to a force of 15 kN, draw a diagram of the compressive force in the mast.

Ans.   Max. force = 315 kN.

P.3.2   Construct the normal force, shear force and bending moment diagrams for the beam shown in Fig. P.3.2.

Ans. $N_{AB} = 9\cdot2$ kN, $N_{BC} = 9\cdot2$ kN, $N_{CD} = 5\cdot7$ kN, $N_{DE} = 0$.
   $S_{AB} = 6\cdot8$ kN, $S_{BC} = 3\cdot8$ kN, $S_{CD} = -2\cdot3$ kN, $S_{DE} = -7\cdot9$ kN.
   $M_B = 27\cdot2$ kN m, $M_C = 50$ kN m, $M_D = 39\cdot5$ kN m.

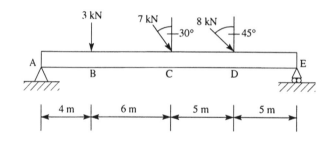

Fig. P.3.2

P.3.3 Draw dimensioned sketches of the diagrams of normal force, shear force and bending moment for the beam shown in Fig. P.3.3.

Ans. $N_{AB} = N_{BC} = N_{CD} = 0$, $N_{DE} = -6$ kN.

$S_A = 0$, S_B (in AB) $= -10$ kN, S_B (in BC) $= 10$ kN,

$S_C = 4$ kN, S_D (in CD) $= 4$ kN, $S_{DE} = -4$ kN.

$M_B = -25$ kN m, $M_C = -4$ kN m, $M_D = 12$ kN m.

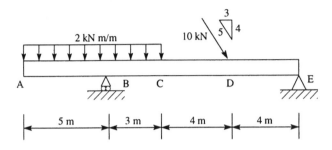

Fig. P.3.3

P.3.4 Draw shear force and bending moment diagrams for the beam shown in Fig. P.3.4.

Ans. $S_{AB} = W$, $S_{BC} = 0$, $S_{CD} = -W$.

$M_B = M_C = WL/4$.

Note zero shear and constant bending moment in central span.

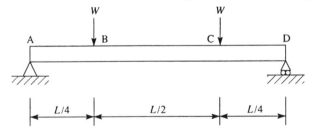

Fig. P.3.4

P.3.5 The cantilever AB shown in Fig. P.3.5 carries a uniformly distributed load of 5 kN/m and a concentrated load of 15 kN at its free end. Construct the shear force and bending moment diagrams for the beam.

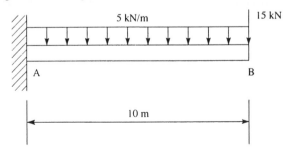

Fig. P.3.5

Ans. $S_B = 15$ kN, $S_C = 65$ kN.

   $M_B = 0$, $M_A = -400$ kN m.

P.3.6   Sketch the bending moment and shear force diagrams for the simply supported beam shown in Fig. P.3.6 and insert the principal values.

Ans. S_B (in AB) = -5 kN, S_B (in BC) = $3 \cdot 75$ kN, S_C (in BC) = $-6 \cdot 25$ kN,

   $S_{CD} = 5$ kN, $M_B = -12 \cdot 5$ kN m, $M_C = -25$ kN m.

   Turning value of bending moment of $-5 \cdot 5$ kN m in BC, $3 \cdot 75$ m from B.

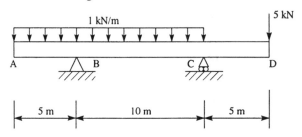

Fig. P.3.6

P.3.7   Draw the shear force and bending moment diagrams for the beam shown in Fig. P.3.7 indicating the principal values.

Ans. $S_{AB} = 5 \cdot 6$ kN, S_B (in BC) = $-4 \cdot 4$ kN, S_C (in BC) = $-7 \cdot 4$ kN, S_C

   (in CD) = $1 \cdot 5$ kN.

   $M_B = 16 \cdot 69$ kN m, $M_C = -1 \cdot 13$ kN m.

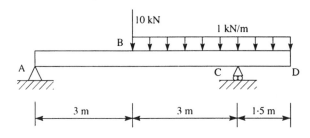

Fig. P.3.7

P.3.8   Find the value of w in the beam shown in Fig. P.3.8 for which the maximum sagging bending moment occurs at a point $10/3$ m from the left-hand support and determine the value of this moment.

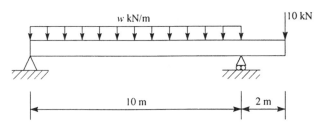

Fig. P.3.8

Ans. $w = 1 \cdot 2$ kN/m, $6 \cdot 7$ kN m.

P.3.9 Find the value of n for the beam shown in Fig. P.3.9 such that the maximum sagging bending moment occurs at $L/3$ from the right-hand support. Using this value of n determine the position of the point of contraflexure in the beam.

Ans. $n = 4/3$, $L/3$ from left-hand support.

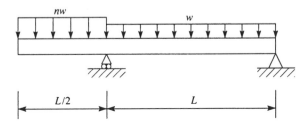

Fig. P.3.9

P.3.10 Sketch the shear force and bending moment diagrams for the simply supported beam shown in Fig. P.3.10 and determine the positions of maximum bending moment and point of contraflexure. Calculate the value of the maximum moment.

Ans. $S_A = 45$ kN, S_B (in AB) $= -55$ kN, $S_{BC} = 20$ kN.

$M_{max} = 202 \cdot 5$ kN m at 9 m from A, $M_B = -100$ kN m.

Point of contraflexure is 18 m from A.

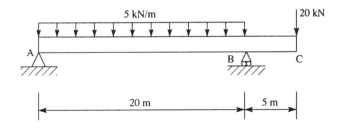

Fig. P.3.10

P.3.11 Determine the position of maximum bending moment in a simply supported beam, 8 m span, which carries a load of 100 kN uniformly distributed over its complete length and, in addition, a load of 120 kN uniformly distributed over $2 \cdot 5$ m to the right from a point 2 m from the left support. Calculate the value of maximum bending moment and the value of bending moment at mid-span.

Ans. $M_{max} = 294$ kN m at $3 \cdot 6$ m from left-hand support.

M (mid-span) $= 289$ kN m.

P.3.12 A simply supported beam AB has a span of 6 m and carries a distributed load which varies linearly in intensity from zero at A to 2 kN/m at B. Sketch the shear force and bending moment diagrams for the beam and insert the principal values.

Ans. $S_{AB} = 2 - z^2/6$, $S_A = 2$ kN, $S_B = -4$ kN.

$M_{AB} = 2z - z^3/18$, $M_{max} = 4 \cdot 62$ kN m at $3 \cdot 46$ m from A.

P.3.13   A precast concrete beam of length L is to be lifted by a single sling and has one end resting on the ground. Show that the optimum position for the sling is $0 \cdot 29$ m from the nearest end.

P.3.14   Construct shear force and bending moment diagrams for the framework shown in Fig. P.3.14.

Ans. $S_{AB} = 60$ kN, $S_{BC} = 10$ kN, $S_{CD} = -140$ kN.

$M_B = 480$ kN m, $M_C = 560$ kN m.

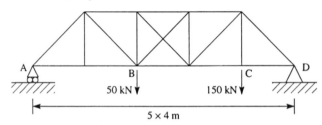

Fig. P.3.14

P3.15   Draw shear force and bending moment diagrams for the framework shown in Fig. P.3.15.

Ans. $S_{AB} = -5$ kN, $S_{BC} = -15$ kN, $S_{CD} = -30$ kN, $S_{DE} = 12$ kN, $S_{EF} = 7$ kN,

$S_{FG} = 5$ kN, $S_{GH} = 0$.

$M_B = -10$ kN m, $M_C = -40$ kN m, $M_D = -100$ kN m, $M_E = -76$ kN m,

$M_F = -20$ kN m, $M_G = M_H = 0$.

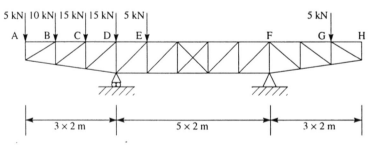

Fig. P.3.15

P.3.16   The cranked cantilever ABC shown in Fig. P.3.16 carries a load of 3 kN at its free end. Draw shear force, bending moment and torsion diagrams for the complete beam.

Ans. $S_{CB} = 3$ kN, $S_{BA} = 3$ kN.

$M_C = 0$, M_B (in CB) $= -6$ kN m, M_B (in BA) $= 0$, $M_A = -9$ kN m.

$T_{CB} = 0$, $T_{BA} = 6$ kN m.

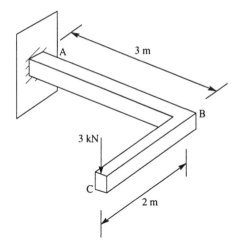

Fig. P.3.16

P.3.17 Construct a torsion diagram for the beam shown in Fig. P.3.17.

Ans. $T_{CB} = -300$ N m, $T_{BA} = -400$ N m.

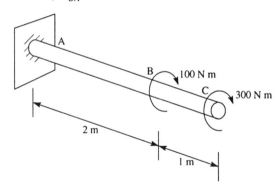

Fig. P.3.17

P.3.18 The beam ABC shown in Fig. P.3.18 carries a distributed torque of 1 N m/mm over its outer half BC and a concentrated torque of 500 N m at B. Sketch the torsion diagram for the beam inserting the principal values.

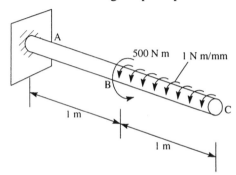

Fig. P.3.18

Ans. $T_C = 0$, T_B (in BC) $= 1000$ N m, T_B (in AB) $= 1500$ N m.

P.3.19   The cylindrical bar ABCD shown in Fig. P.3.19 is supported symmetrically at B and C by supports that prevent rotation of the bar about its longitudinal axis. The bar carries a uniformly distributed torque of 2 N m/mm together with concentrated torques of 400 N m at each end. Draw the torsion diagram for the bar and determine the maximum value of torque.

Ans. $T_{DC} = 400 + 2z$, $T_{CB} = 2z - 2000$, $T_{BA} = 2z - 4400$ (T in N m when z is in mm).
   $T_{max} = 1400$ N m at C and B.

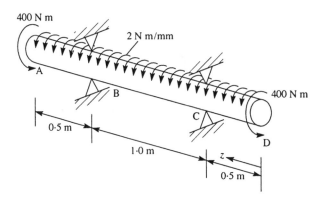

Fig. P.3.19

CHAPTER 4

Analysis of Pin-jointed Trusses

In Chapter 1 we discussed various structural forms and saw that for moderately large spans, simple beams become uneconomical and may be replaced by trusses. These structures comprise members connected at their ends and are constructed in a variety of arrangements. In general, trusses are lighter, stronger and stiffer than solid beams of the same span; they do, however, take up more room and are more expensive to fabricate.

Initially in this chapter we shall discuss types of truss, their function and the idealization of a truss into a form amenable to analysis. Subsequently, we shall investigate the criterion which indicates the degree of their statical determinacy, examine the action of the members of a truss in supporting loads and, finally, examine methods of analysis of both trusses and space frames.

4.1 Types of truss

Generally the form selected for a truss depends upon the purpose for which it is required. Examples of different types of truss are shown in Figs 4.1(a)–(f); some are named after the railway engineers who invented them.

The Pratt, Howe, Warren and K trusses would, for example, be used to support bridge decks and large-span roofing systems (the Howe truss is no longer used for reasons we shall discuss in Section 4.5) whereas the Fink truss would be used to support gable-ended roofs. The Bowstring truss is somewhat of a special case in that if the upper chord members are arranged such that the joints lie on a parabola and the loads, all of equal magnitude, are applied at the upper joints, the internal members carry no load. This result derives from arch theory (Chapter 6) but is rarely of practical significance since, generally, the loads would be applied to the lower chord joints as in the case of the truss being used to support a bridge deck.

Frequently, plane trusses are connected together to form a three-dimensional structure. For example, in the overhead crane shown in Fig. 4.2, the tower would usually comprise four plane trusses joined together to form a 'box' while the jibs would be constructed by connecting three plane trusses together to form a triangular cross-section.

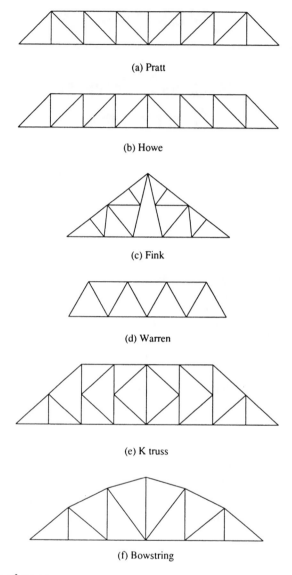

(a) Pratt

(b) Howe

(c) Fink

(d) Warren

(e) K truss

(f) Bowstring

Fig. 4.1   Types of truss

4.2   Assumptions in truss analysis

It can be seen from Fig. 4.1 that trusses consist of a series of triangular units. The triangle, even when its members are connected together by hinges or pins as in Fig. 4.3(a), is an inherently stable structure, i.e. it will not collapse under any arrangement of loads applied in its own plane. On the other hand, the rectangular structure shown in Fig. 4.3(b) would be unstable if vertical loads were applied at the joints and would collapse under the loading system shown; in other words it is a mechanism.

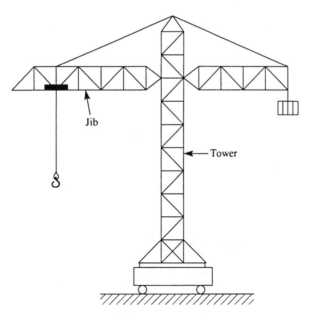

Fig. 4.2 Overhead crane structure

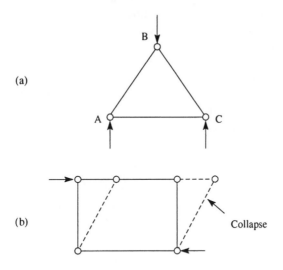

Fig. 4.3 Basic unit of a truss

Further properties of a pin-jointed triangular structure are that the forces in the members are purely axial and that it is statically determinate (see Section 4.4) so long as the structure is loaded and supported at the joints. Thus the forces in the members can be found using the equations of statical equilibrium (Eqs (2.10)). It follows that a truss comprising pin-jointed triangular units is also statically determinate if the above loading and support conditions are satisfied. In Section 4.4 we shall derive a simple test for determining whether or not a pin-jointed truss is statically determinate; this test, although applicable in most cases is not, as we shall see, foolproof.

The assumptions on which the analysis of trusses is based are as follows:

(1) The members of the truss are connected at their ends by frictionless pins or hinges.
(2) The truss is loaded and supported only at its joints.
(3) The forces in the members of the truss are purely axial.

Assumptions (2) and (3) are interdependent since the application of a load at some point along a truss member would, in effect, convert the member into a simply supported beam and, as we have seen in Chapter 3, generate, in addition to axial loads, shear forces and bending moments; the truss would then become statically indeterminate.

4.3 Idealization of a truss

In practice trusses are not pin-jointed but are constructed, in the case of steel trusses, by bolting, riveting or welding the ends of the members to gusset plates as shown in Fig. 4.4. In a timber roof truss the members are connected using spiked plates driven into their vertical surfaces on each side of a joint. The joints in trusses are therefore semi-rigid and can transmit moments, unlike a frictionless pinned joint. Furthermore, if the loads are applied at points on a member away from its ends, that member behaves as a fixed or built-in beam with unknown moments and shear forces as well as axial loads at its ends. Such a truss would possess a high degree of statical indeterminacy and would require a computer-based analysis.

However, if such a truss is built up using the basic triangular unit and the loads and support points coincide with the member joints then, even assuming rigid joints, a computer-based analysis would show that the shear forces and bending moments in the members are extremely small compared to the axial forces which, themselves, would be very close in magnitude to those obtained from an analysis based on the assumption of pinned joints.

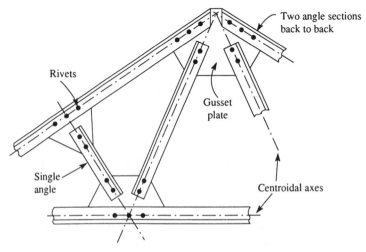

Fig. 4.4 Actual truss construction

A further condition in employing a pin-jointed idealization of an actual truss is that the centroidal axes of the members in the actual truss are concurrent, as shown in Fig. 4.4. We shall see in Section 9.2 that a load parallel to, but offset from, the centroidal axis of a member induces a bending moment in the cross-section of the member; this situation is minimized in an actual truss if the centroidal axes of all members meeting at a joint are concurrent.

4.4 Statical determinacy

It was stated in Section 4.2 that the basic triangular pin-jointed unit is statically determinate and the forces in the members are purely axial so long as the loads and support points coincide with the joints. The justification for this is as follows. Consider the joint B in the triangle in Fig. 4.3(a). The forces acting on the actual pin or hinge are the externally applied load and the axial forces in the members AB and BC; the system is shown in the free body diagram in Fig. 4.5. The internal axial forces in the members BA and BC, F_{BA} and F_{BC}, are drawn to show them pulling away from the joint B; this indicates that the members are in tension. Actually, we can see by inspection that both members will be in compression since their combined vertical components are required to equilibrate the applied vertical load. The assumption of tension, however, would only result in negative values in the calculation of F_{BA} and F_{BC} and is therefore a valid approach. In fact we shall adopt the method of initially assuming tension in all members of a truss when we consider methods of analysis, since a negative value for a member force will then always signify compression and will be in agreement with the sign convention adopted in Section 3.2.

Since the pin or hinge at the joint B is in equilibrium and the forces acting on the pin are coplanar, Eqs (2.10) apply. Thus the sum of the components of all the forces acting on the pin in any two directions at right angles must be zero. The moment equation, $\sum M = 0$, is automatically satisfied since the pin cannot transmit a moment and the lines of action of all the forces acting on the pin must therefore be concurrent. Thus, for the joint B, we can write down two equations of force equilibrium which are sufficient to solve for the unknown member forces F_{BA} and F_{BC}. The same argument may then be applied to either joint A or joint C to solve for the remaining unknown internal force $F_{AC}(=F_{CA})$. Thus we see that the basic triangular unit is statically determinate.

Now let us consider the construction of a simple pin-jointed truss. Initially we start with a single triangular unit ABC as shown in Fig. 4.6. A further triangle BCD

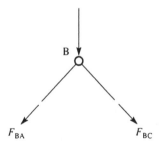

Fig. 4.5 Joint equilibrium in a triangular structure

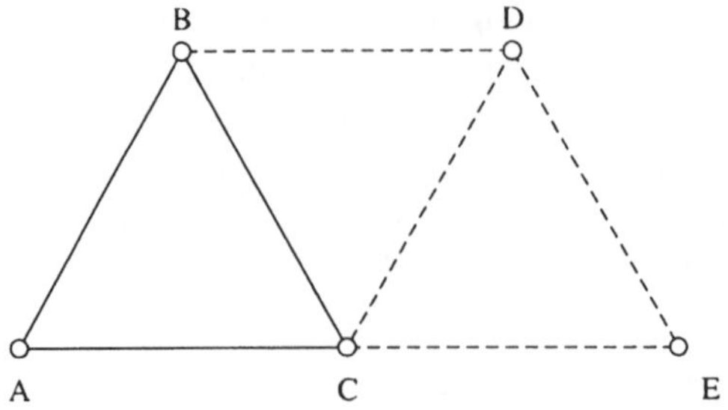

Fig. 4.6 Construction of a single Warren truss

is created by adding the *two* members BD and CD and the *single* joint D. The third triangle CDE is then formed by the addition of the *two* members CE and DE and the *single* joint E and so on for as many triangular units as required. Thus, after the initial triangle is formed, each additional triangle requires *two* members and a *single* joint. In other words the number of additional members is equal to twice the number of additional joints. This relationship may be expressed qualitatively as follows.

Suppose that m is the total number of members in a truss and j the total number of joints. Then, noting that initially there are three members and three joints, the above relationship may be written

$$m - 3 = 2(j - 3)$$

so that
$$m = 2j - 3 \qquad\qquad (4.1)$$

If Eq. (4.1) is satisfied, the truss is constructed from a series of statically determinate triangles and the truss itself is statically determinate. Furthermore, if $m < 2j - 3$ the structure is unstable (see Fig. 4.3(b)) or if $m > 2j - 3$, the structure is statically indeterminate. Note that Eq. (4.1) applies only to the internal forces in a truss; the support system must also be statically determinate to enable the analysis to be carried out.

Example 4.1 Test the statical determinacy of the pin-jointed trusses shown in Figs 4.7(a), (b) and (c).

In Fig. 4.7(a) the truss has five members and four joints. Thus $m = 5$ and $j = 4$ so that

$$2j - 3 = 5 = m$$

and Eq. (4.1) is satisfied.

The truss in Fig. 4.7(b) has an additional member so that $m = 6$ and $j = 4$. Therefore

$$m > 2j - 3$$

and the truss is statically indeterminate.

The truss in Fig. 4.7(c) comprises a series of triangular units which suggests that it is statically determinate. However, in this case, $m = 8$ and $j = 5$. Thus $2j - 3 = 7$ so that $m > 2j - 3$ and the truss is statically indeterminate. In fact any single member may be removed and the truss would retain its stability under any loading system in its own plane.

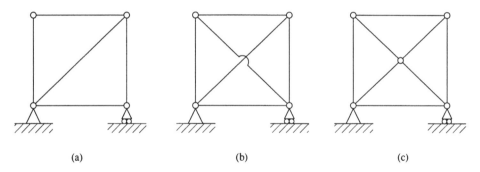

(a)   (b)   (c)

Fig. 4.7   Statical determinacy of trusses

Unfortunately, in some cases, Eq. (4.1) is satisfied but the truss may be statically indeterminate or a mechanism. The truss in Fig. 4.8, for example, has nine members and six joints so that Eq. (4.1) is satisfied. However, clearly the left-hand half is a mechanism and the right-hand half is statically indeterminate. Theoretically, assuming that the truss members are weightless, the truss could support vertical loads applied to the left- and/or right-hand vertical members; this would, of course, be an unstable condition. Any other form of loading would cause a collapse of the left hand half of the truss and consequently of the truss itself.

The presence of a rectangular region in a truss such as that in the truss in Fig. 4.8 does not necessarily result in collapse. The truss in Fig. 4.9 has nine members and six joints so that Eq. (4.1) is satisfied. This does not, as we have seen, guarantee either a stable or statically determinate truss. If, therefore, there is some doubt we can return to the procedure of building up a truss from a single triangular unit as demonstrated in Fig. 4.6. Thus, remembering that each additional triangle is created by adding two members and one joint and that the resulting truss is stable and statically determinate, we can examine the truss in Fig. 4.9 as follows.

Suppose that ACD is the initial triangle. The additional triangle ACB is formed by adding the two members AB and BC and the single joint B. The triangle DCE follows by adding the two members CE and DE and the joint E. Finally, the two members BF and EF and the joint F are added to form the rectangular portion CBFE. We therefore conclude that the truss in Fig. 4.9 is stable and statically determinate. Compare the construction of this truss with that of the statically indeterminate truss in Fig. 4.7(c).

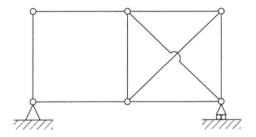

Fig. 4.8   Applicability of test for statical determinacy

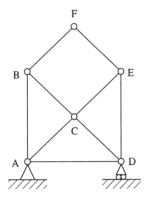

Fig. 4.9 Investigation into truss stability

A condition, similar to Eq. (4.1), applies to space frames; the result for a space frame having m members and j pinned joints is

$$m = 3j - 6 \tag{4.2}$$

4.5 Resistance of a truss to shear force and bending moment

Although the members of a truss carry only axial loads, the truss itself acts as a beam and is subjected to shear forces and bending moments. Therefore, before we consider methods of analysis of trusses, it will be instructive to examine the manner in which a truss resists shear forces and bending moments.

The Pratt truss shown in Fig. 4.10(a) carries a concentrated load W applied at a joint on the bottom chord at mid-span. Using the methods described in Section 3.4, the shear force and bending moment diagrams for the truss are constructed as shown in Figs 4.10(b) and (c), respectively.

First we shall consider the shear force. In the bay ABCD the shear force is $W/2$ and is positive. Thus at any section mm between A and B (Fig. 4.11) we see that the internal shear force is $W/2$. Since the horizontal members AB and DC are unable to resist shear forces, the internal shear force can only be equilibrated by the vertical component of the force F_{AC} in the member AC. Fig. 4.11 shows the direction of the internal shear force applied at the section mm so that F_{AC} is tensile. Hence

$$F_{AC} \cos 45° = \frac{W}{2}$$

The same result applies to all the internal diagonals whether to the right or left of the mid-span point since the shear force is constant, although reversed in sign, either side of the load. The two outer diagonals are in compression since their vertical components must be in equilibrium with the vertically upward support reactions. Alternatively, we arrive at the same result by considering the internal shear force at a section just to the right of the left-hand support and just to the left of the right-hand support.

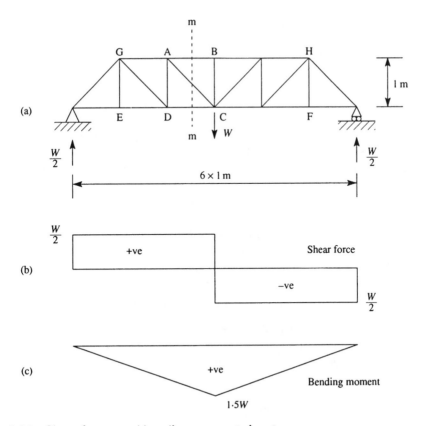

Fig. 4.10 Shear forces and bending moments in a truss

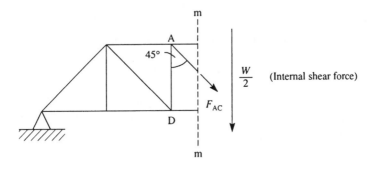

Fig. 4.11 Internal shear force in a truss

If the diagonal AC was repositioned to span between D and B it would be subjected to an axial compressive load. This situation would be undesirable since the longer a compression member, the smaller the load required to cause buckling (see Chapter 18). Therefore, the aim of truss design is to ensure that the forces in the longest members, the diagonals in this case, are predominantly tensile. Hence the Howe truss (Fig. 4.1(b)), whose diagonals for downward loads would be in compression, is no longer in use.

In some situations the loading on a truss could be reversed so that a diagonal that is usually in tension would be in compression. To counter this an extra diagonal inclined in the opposite direction is included (spanning, say, from D to B in Fig. 4.12). This, as we have seen, would result in the truss becoming statically indeterminate. However, if it is assumed that the original diagonal (AC in Fig. 4.12) has buckled under the compressive load and therefore carries no load, the truss is once again statically determinate.

We shall now consider the manner in which a truss resists bending moments. The bending moment at a section immediately to the left of the mid-span vertical BC in the truss in Fig. 4.10(a) is, from Fig. 4.10(c), $1.5W$ and is positive, as shown in Fig. 4.12. This bending moment is equivalent to the moment resultant, about any point in their plane, of the member forces at this section. In Fig. 4.12, analysis by the method of sections (Section 4.7) gives $F_{BA} = 1.5W$ (compression), $F_{AC} = 0.707W$ (tension) and $F_{DC} = 1.0W$ (tension). Therefore at C, F_{DC} plus the horizontal component of F_{AC} is equal to $1.5W$ which, together with F_{BA}, produces a couple of magnitude $1.5W \times 1$ which is equal to the applied bending moment. Alternatively, we could take moments of the internal forces about B (or C). Hence

$$M_B = F_{DC} \times 1 + F_{AC} \times 1 \sin 45° = 1.0W \times 1 + 0.707W \times 1 \sin 45° = 1.5W$$

as before. Note that in Fig. 4.12 the moment resultant of the internal force system is *equivalent* to the applied moment, i.e. it is in the same sense as the applied moment.

Now let us consider the bending moment at, say, the mid-point of the bay AB, where its magnitude is, from Fig. 4.10(c), $1.25W$. The internal force system is shown in Fig. 4.13 in which F_{BA}, F_{AC} and F_{DC} have the same values as before. Thus, taking moments about, say, the mid-point of the top chord member AB, we have

$$M = F_{DC} \times 1 + F_{AC} \times 0.5 \sin 45° = 1.0W \times 1 + 0.707W \times 0.5 \sin 45° = 1.25W$$

the value of the applied moment.

From the discussion above it is clear that, in trusses, shear loads are resisted by inclined members, while all members combine to resist bending moments. Furthermore, positive (sagging) bending moments induce compression in upper chord members and tension in lower chord members.

Finally, note that in the truss in Fig. 4.10 the forces in the members GE, BC and HF are all zero, as can be seen by considering the vertical equilibrium of joints E, B and F. Forces would only be induced in these members if external loads were applied directly at the joints E, B and F. Generally, if three coplanar members meet

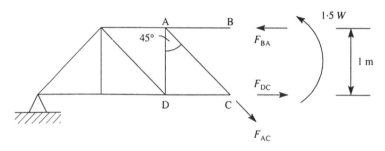

Fig. 4.12   Internal bending moment in a truss

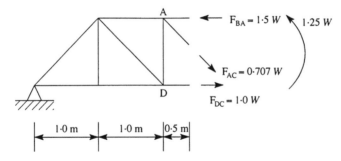

Fig. 4.13 Resistance of a bending moment at a mid-bay point

at a joint and two of them are collinear, the force in the third member is zero if no external force is applied at the joint.

4.6 Method of joints

We have seen in Section 4.4 that the axial forces in the members of a simple pin-jointed triangular structure may be found by examining the equilibrium of their connecting pins or hinges in two directions at right angles (Eqs (2.10)). This approach may be extended to plane trusses to determine the axial forces in all their members; the method is known as the *method of joints* and will be illustrated by an example.

Example 4.2   Determine the forces in the members of the Warren truss shown in Fig. 4.14; all members are 1 m long.

Generally, although not always, the support reactions must be calculated first. Hence, taking moments about D for the truss in Fig. 4.14 we obtain

$$R_A \times 2 - 2 \times 1 \cdot 5 - 1 \times 1 - 3 \times 0 \cdot 5 = 0$$

which gives $\qquad R_A = 2 \cdot 75$ kN

Then, resolving vertically

$$R_D + R_A - 2 - 1 - 3 = 0$$

so that $\qquad R_D = 3 \cdot 25$ kN

Note that there will be no horizontal reaction at A (D is a roller support) since no horizontal loads are applied.

The next step is to assign directions to the forces *acting on each joint*. In one approach the truss is examined to determine whether the force in a member is tensile or compressive. For some members this is straightforward. For example, in Fig. 4.14, the vertical reaction at A, R_A, can only be equilibrated by the vertical component of the force in AB which must therefore act downwards, indicating that the member is in compression (a compressive force in a member will push towards a joint whereas a tensile force will pull away from a joint). In some cases, where several members meet at a joint, the nature of the force in a particular member is difficult, if not impossible, to determine by inspection. Then a direction must be

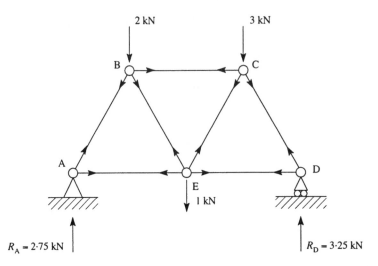

Fig. 4.14 Analysis of a Warren truss

assumed which, if incorrect, will result in a negative value for the member force. It follows that, in the same truss, both positive and negative values may be obtained for tensile forces and also for compressive forces, a situation leading to possible confusion. Therefore, if every member in a truss is initially assumed to be in tension, negative values will always indicate compression and the solution will then agree with the sign convention adopted in Section 3.2.

We now assign tensile forces to the members of the truss in Fig. 4.14 using arrows to indicate the *action of the force in the member on the joint*; thus all arrows are shown to pull away from the adjacent joint.

The analysis, as we have seen, is based on a consideration of the equilibrium of each pin or hinge under the action of *all* the forces at the joint. Thus for each pin or hinge we can write down two equations of equilibrium. It follows that a solution can only be obtained if there are no more than two unknown forces acting at the joint. In Fig. 4.14, therefore, we can only begin the analysis at the joint A or at the joint D, since at each of the joints B and C there are three unknown forces while at E there are four.

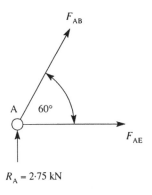

Fig. 4.15 Equilibrium of forces at joint A

Consider joint A. The forces acting on the pin at A are shown in the free body diagram in Fig. 4.15. F_{AB} may be determined directly by resolving forces vertically. Hence

$$F_{AB} \sin 60° + 2{\cdot}75 = 0 \tag{i}$$

so that
$$F_{AB} = -3{\cdot}18 \text{ kN}$$

the negative sign indicating that AB is in compression as expected.

Referring again to Fig. 4.15 and resolving forces horizontally

$$F_{AE} + F_{AB} \cos 60° = 0 \tag{ii}$$

Substituting the *negative* value of F_{AB} in Eq. (ii) we obtain

$$F_{AE} - 3{\cdot}18 \cos 60° = 0$$

which gives
$$F_{AE} = +1{\cdot}59 \text{ kN}$$

the positive sign indicating that F_{AB} is a tensile force.

We now inspect the truss to determine the next joint at which there are no more than two unknown forces. At joint E there remain three unknowns since only $F_{EA}(=F_{AE})$ has yet been determined. At joint B there are now two unknowns since $F_{BA}(=F_{AB})$ has been determined; we can therefore proceed to joint B. The forces acting at B are shown in Fig. 4.16. Since F_{BA} is now known we can resolve forces vertically and therefore obtain F_{BE} directly. Thus

$$F_{BE} \cos 30° + F_{BA} \cos 30° + 2 = 0 \tag{iii}$$

Substituting the negative value of F_{BA} in Eq. (iii) gives

$$F_{BE} = +0{\cdot}87 \text{ kN}$$

which is positive and therefore tensile.

Resolving forces horizontally at the joint B we have

$$F_{BC} + F_{BE} \cos 60° - F_{BA} \cos 60° = 0 \tag{iv}$$

Substituting the positive value of F_{BE} and the negative value of F_{BA} in Eq. (iv) yields

$$F_{BC} = -2{\cdot}03 \text{ kN}$$

the negative sign indicating that the member BC is in compression.

We have now calculated four of the seven unknown member forces. There are in fact just two unknown forces at each of the remaining joints C, D and E so that, theoretically, it is immaterial which joint we consider next. From a solution viewpoint there are three forces at D, four at C and five at E so that the arithmetic will be slightly simpler if we next consider D to obtain F_{DC} and F_{DE} and then C to obtain F_{CE}. At C, F_{CE} could be determined by resolving forces in the direction CE rather than horizontally or vertically. Carrying out this procedure gives

$$F_{DC} = -3{\cdot}75 \text{ kN (compression)}, \quad F_{DE} = +1{\cdot}88 \text{ kN (tension)},$$
$$F_{CE} = +0{\cdot}29 \text{ kN (tension)}$$

The reader should verify these values using the method suggested above.

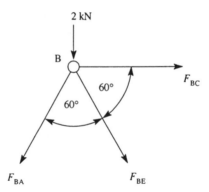

Fig. 4.16 Equilibrium of forces at joint B

It may be noted that in this example we could write down ten equations of equilibrium, two for each of the five joints, and yet there are only seven unknown member forces. The apparently extra three equations result from the use of overall equilibrium to calculate the support reactions. Thus an alternative approach would be to write down the ten equilibrium equations which would include the three unknown support reactions (there would be a horizontal reaction at A if horizontal as well as vertical loads were applied) and solve the resulting ten equations simultaneously. Overall equilibrium could then be examined to check the accuracy of the solution. Generally, however, the method adopted above produces a quicker solution.

4.7 Method of sections

It will be appreciated from Section 4.5 that in many trusses the maximum member forces, particularly in horizontal members, will occur in the central region where the applied bending moment would possibly have its maximum value. It will also be appreciated from Ex. 4.2 that the calculation of member forces in the central region of a multibay truss such as the Pratt truss shown in Fig. 4.1(a) would be extremely tedious since the calculation must begin at an outside support and then proceed inwards joint by joint. This approach may be circumvented by using the *method of sections*.

The method is based on the premise that if a structure is in equilibrium, any portion or component of the structure will also be in equilibrium under the action of any external forces and the internal forces acting between the portion or component and the remainder of the structure. We shall illustrate the method by an example.

Example 4.3 Calculate the forces in the members CD, CF and EF in the Pratt truss shown in Fig. 4.17.

Initially the support reactions are calculated and are readily shown to be

$$R_{A,V} = 4 \cdot 5 \text{ kN}, \ R_{A,H} = 2 \text{ kN}, \ R_B = 5 \cdot 5 \text{ kN}$$

We now 'cut' the members CD, CF and EF by a section mm, thereby dividing the truss into two separate parts. Consider the left-hand part shown in Fig. 4.18 (equally

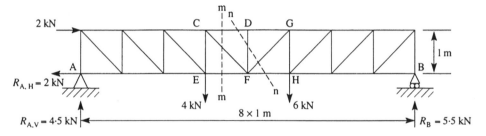

Fig. 4.17 Calculation of member forces using the the method of sections

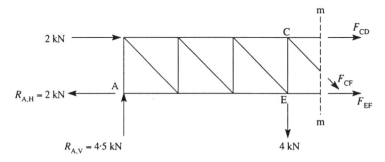

Fig. 4.18 Equilibrium of a portion of a truss

we could consider the right-hand part). Clearly, if we actually cut the members CD, CF and EF, both the left-hand and right-hand parts would collapse. However, the equilibrium of the left-hand part, say, could be maintained by applying the forces F_{CD}, F_{CF} and F_{EF} to the cut ends of the members. Therefore, in Fig. 4.18, the left-hand part of the truss is in equilibrium under the action of the externally applied loads, the support reactions and the forces F_{CD}, F_{CF} and F_{EF} which are, as in the method of joints, initially assumed to be tensile; Eqs (2.10) are then used to calculate the three unknown forces.

Thus, resolving vertically gives

$$F_{CF} \cos 45° + 4 - 4 \cdot 5 = 0 \tag{i}$$

whence

$$F_{CF} = +0 \cdot 71 \text{ kN}$$

and is tensile.

Now taking moments about the point of intersection of F_{CF} and F_{EF} we have

$$F_{CD} \times 1 + 2 \times 1 + 4 \cdot 5 \times 4 - 4 \times 1 = 0 \tag{ii}$$

so that

$$F_{CD} = -16 \text{ kN}$$

and is compressive.

Finally F_{EF} is obtained by taking moments about C, thereby eliminating F_{CF} and F_{CD} from the equation. Alternatively, we could resolve forces horizontally since F_{CF} and F_{CD} are now known; however, this approach would involve a slightly lengthier calculation. Hence

$$F_{EF} \times 1 - 4 \cdot 5 \times 3 - 2 \times 1 = 0 \qquad \text{(iii)}$$

which gives $F_{EF} = +15 \cdot 5 \text{ kN}$

the positive sign indicating tension.

Note that Eqs (i), (ii) and (iii) each include just one of the unknown member forces so that it is immaterial which is calculated first. In some problems, however, a preliminary examination is worthwhile to determine the optimum order of solution.

In Ex. 4.3 we see that there are just three possible equations of equilibrium so that we cannot solve for more than three unknown forces. It follows that a section such as mm which *must divide the frame into two separate parts* must also *not cut through more than three members in which the forces are unknown*. If, for example, we wished to determine the forces in CD, DF, FG and FH we would first calculate F_{CD} using the section mm as above and then determine F_{DF}, F_{FG} and F_{FH} using the section nn. Actually, in this particular example F_{DF} may be seen to be zero by inspection (see Section 4.5) but the principle holds.

4.8 Method of tension coefficients

An alternative form of the method of joints which is particularly useful in the analysis of pin-jointed space frames is the *method of tension coefficients*.

Consider the member AB, shown in Fig. 4.19, which connects two pinned joints A and B whose coordinates, referred to arbitrary xy axes, are (x_A, y_A) and (x_B, y_B) respectively; the member carries a *tensile* force, T_{AB}, is of length L_{AB} and is inclined at an angle α to the x axis. The component of T_{AB} parallel to the x axis at A is given by

$$T_{AB} \cos \alpha = T_{AB} \frac{(x_B - x_A)}{L_{AB}} = \frac{T_{AB}}{L_{AB}} (x_B - x_A)$$

Similarly the component of T_{AB} at A parallel to the y axis is

$$T_{AB} \sin \alpha = \frac{T_{AB}}{L_{AB}} (y_B - y_A)$$

We now define a *tension coefficient* $t_{AB} = T_{AB}/L_{AB}$ so that the above components of T_{AB} become:

$$\text{parallel to the } x \text{ axis: } t_{AB}(x_B - x_A) \qquad \text{(4.3)}$$

$$\text{parallel to the } y \text{ axis: } t_{AB}(y_B - y_A) \qquad \text{(4.4)}$$

Equilibrium equations may be written down for each joint in turn in terms of tension coefficients and joint coordinates referred to some convenient axis system. The solution of these equations gives t_{AB}, etc, whence $T_{AB} = t_{AB} L_{AB}$ in which L_{AB}, unless given, may be calculated using Pythagoras' theorem, i.e. $L_{AB} = \sqrt{(x_B - x_A)^2 + (y_B - y_A)^2}$. Again the initial assumption of tension in a member results in negative values corresponding to compression. Note the order of suffixes in Eqs (4.3) and (4.4).

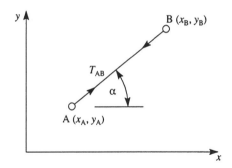

Fig. 4.19 Method of tension coefficients

Example 4.4 Determine the forces in the members of the pin-jointed truss shown in Fig. 4.20.

The support reactions are first calculated and are as shown in Fig. 4.20.

The next step is to choose an *xy* axis system and then insert the joint coordinates in the diagram. In Fig. 4.20 we shall choose the support point A as the origin of axes although, in fact, any joint would suffice; the joint coordinates are then as shown.

Again, as in the method of joints, the solution can only begin at a joint where there are no more than two unknown member forces, in this case joints A and E. Theoretically it is immaterial at which of these joints the analysis begins but since A is the origin of axes we shall start at A. Note that it is unnecessary to insert arrows to indicate the directions of the member forces since the members are assumed to be in tension and the directions of the components of the member forces are automatically specified when written in terms of tension coefficients and joint coordinates (Eqs (4.3) and (4.4)).

The equations of equilibrium at joint A are

$$x \text{ direction: } t_{AB}(x_B - x_A) + t_{AC}(x_C - x_A) - R_{A,H} = 0 \tag{i}$$

$$y \text{ direction: } t_{AB}(y_B - y_A) + t_{AC}(y_C - y_A) + R_{A,V} = 0 \tag{ii}$$

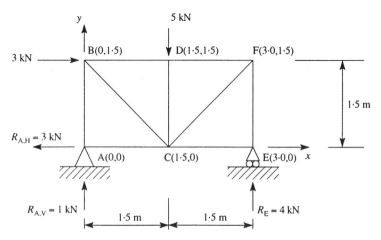

Fig. 4.20 Analysis of a truss using tension coefficients (Ex. 4.4)

Substituting the values of $R_{A,H}$, $R_{A,V}$ and the joint coordinates in Eqs (i) and (ii) we obtain, from Eq. (i),

$$t_{AB}(0-0) + t_{AC}(1\cdot5-0) - 3 = 0$$

whence
$$t_{AC} = +2\cdot0$$

and from Eq. (ii)

$$t_{AB}(1\cdot5-0) + t_{AC}(0-0) + 1 = 0$$

so that
$$t_{AB} = -0\cdot67$$

We see from the derivation of Eqs (4.3) and (4.4) that the units of a tension coefficient are force/unit length, in this case kN/m. Generally, however, we shall omit the units.

We can now proceed to joint B at which, since t_{BA} ($=t_{AB}$) has been calculated, there are two unknowns.

$$x \text{ direction: } t_{BA}(x_A - x_B) + t_{BC}(x_C - x_B) + t_{BD}(x_D - x_B) + 3 = 0 \qquad \text{(iii)}$$

$$y \text{ direction: } t_{BA}(y_A - y_B) + t_{BC}(y_C - y_B) + t_{BD}(y_D - y_B) = 0 \qquad \text{(iv)}$$

Substituting the values of the joint coordinates and t_{BA} in Eqs (iii) and (iv) we have, from Eq. (iii),

$$-0\cdot67(0-0) + t_{BC}(1\cdot5-0) + t_{BD}(1\cdot5-0) + 3 = 0$$

which simplifies to

$$1\cdot5\,t_{BC} + 1\cdot5\,t_{BD} + 3 = 0 \qquad \text{(v)}$$

and from Eq. (iv)

$$-0\cdot67(0-1\cdot5) + t_{BC}(0-1\cdot5) + t_{BD}(1\cdot5-1\cdot5) = 0$$

whence
$$t_{BC} = +0\cdot67$$

Hence, from Eq. (v)

$$t_{BD} = -2\cdot67$$

There are now just two unknown member forces at joint D. Hence, at D

$$x \text{ direction: } t_{DB}(x_B - x_D) + t_{DF}(x_F - x_D) + t_{DC}(x_C - x_D) = 0 \qquad \text{(vi)}$$

$$y \text{ direction: } t_{DB}(y_B - y_D) + t_{DF}(y_F - y_D) + t_{DC}(y_C - y_D) - 5 = 0 \qquad \text{(vii)}$$

Substituting values of joint coordinates and the previously calculated value of t_{DB} ($=t_{BD}$) in Eqs (vi) and (vii) we obtain, from Eq. (vi),

$$-2\cdot67(0-1\cdot5) + t_{DF}(3\cdot0-1\cdot5) + t_{DC}(1\cdot5-1\cdot5) = 0$$

so that
$$t_{DF} = -2\cdot67$$

and from Eq. (vii)

$$-2\cdot67(1\cdot5-1\cdot5) + t_{DF}(1\cdot5-1\cdot5) + t_{DC}(0-1\cdot5) - 5 = 0$$

from which
$$t_{DC} = -3\cdot33$$

The solution then proceeds to joint C to obtain t_{CF} and t_{CE} or to joint F to determine t_{FC} and t_{FE}; joint F would be preferable since fewer members meet at F than at C. Finally, the remaining unknown tension coefficient (t_{EC} or t_{EF}) is found by considering the equilibrium of joint E. Thus

$$t_{FC} = +2\cdot67, \quad t_{FE} = -2\cdot67, \quad t_{EC} = 0$$

which the reader should verify.

The forces in the truss members are now calculated by multiplying the tension coefficients by the member lengths. Thus

$$T_{AB} = t_{AB}L_{AB} = -0\cdot67 \times 1\cdot5 = -1\cdot0 \text{ kN (compression)}$$

$$T_{AC} = t_{AC}L_{AC} = +2\cdot0 \times 1\cdot5 = +3\cdot0 \text{ kN (tension)}$$

$$T_{BC} = t_{BC}L_{BC}$$

in which $\qquad L_{BC} = \sqrt{(x_B - x_C)^2 + (y_B - y_C)^2} = \sqrt{(0 - 1\cdot5)^2 + (1\cdot5 - 0)^2} = 2\cdot12 \text{ m}$

whence $\qquad\qquad T_{BC} = +0\cdot67 \times 2\cdot12 = +1\cdot42 \text{ kN (tension)}$

Note that in the calculation of member lengths it is immaterial in which order the joint coordinates occur in the brackets since the brackets are squared. Also

$$T_{BD} = t_{BD}L_{BD} = -2\cdot67 \times 1\cdot5 = -4\cdot0 \text{ kN (compression)}$$

Similarly $\qquad\qquad T_{DF} = -4\cdot0 \text{ kN (compression)},$

$$T_{DC} = -5\cdot0 \text{ kN (compression)},$$

$$T_{FC} = +5\cdot67 \text{ kN (tension)},$$

$$T_{FE} = -4\cdot0 \text{ kN (compression)}, \ T_{EC} = 0.$$

4.9 Graphical method of solution

In some instances, particularly when a rapid solution is required, the member forces in a truss may be found using a graphical method.

The method is based upon the condition that each joint in a truss is in equilibrium so that the forces acting at a joint may be represented in magnitude and direction by the sides of a closed polygon (see Section 2.1). The directions of the forces must be drawn in the same directions as the corresponding members and there must be no more than two unknown forces at a particular joint otherwise a polygon of forces cannot be constructed. The method will be illustrated by applying it to the truss in Ex. 4.2.

Example 4.5 Determine the forces in the members of the Warren truss shown in Fig. 4.21; all members are 1 m long.

It is convenient in this approach to designate forces in members in terms of the areas between them rather than referring to the joints at their ends. Thus, in Fig. 4.21, we number the areas between all forces, both internal and external; the reason for this will become clear when the force diagram for the complete structure is constructed.

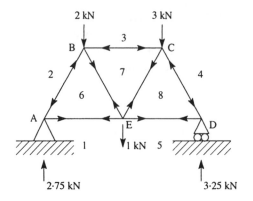

Fig. 4.21 Analysis of a truss by a graphical method

The support reactions were calculated in Ex. 4.2 and are shown in Fig. 4.21. We must start at a joint where there are no more than two unknown forces, in this example either A or D; here we select A. The force polygon for joint A is constructed by going round A in, say, a clockwise sense. We must then go round every joint in the same sense.

First we draw a vector 12 to represent the support reaction at A of 2·75 kN to a convenient scale (see Fig. 4.22). Note that we are moving clockwise from the region 1 to the region 2 so that the vector 12 is vertically upwards, the direction of the reaction at A (if we had decided to move round A in an anticlockwise sense the vector would be drawn as 21 vertically upwards). The force in the member AB at A will be represented by a vector 26 in the direction AB or BA, depending on whether it is tensile or compressive, while the force in the member AE at A is represented by the vector 61 in the direction AE or EA depending, again, on whether it is tensile or compressive. The point 6 in the force polygon is therefore located by drawing a line through the point 2 parallel to the member AB to intersect, at 6, a line drawn through

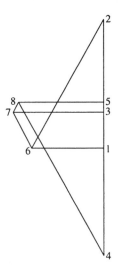

Fig. 4.22 Force polygon for the truss of Ex. 4.5

the point 1 parallel to the member AE. Thus we see from the force polygon that the direction of the vector 26 is towards A so that the member AB is in compression while the direction of the vector 61 is away from A indicating that the member AE is in tension. We now insert arrows on the members AB and AE in Fig. 4.21 to indicate compression and tension, respectively.

We next consider joint B where there are now just two unknown member forces since we have previously determined the force in the member AB; note that, moving clockwise round B, this force is represented by the vector 62, which means that it is acting towards B as it must since we have already established that AB is in compression. Rather than construct a separate force polygon for the joint B we shall superimpose the force polygon on that constructed for joint A since the vector 26 (or 62) is common to both; we thereby avoid repetition. Thus, through the point 2, we draw a vector 23 vertically downwards to represent the 2 kN load to the same scale as before. The force in the member BC is represented by the vector 37 parallel to BC (or CB) while the force in the member BE is represented by the vector 76 drawn in the direction of BE (or EB); thus we locate the point 7 in the force polygon. Hence we see that the force in BC (vector 37) acts towards B indicating compression, while the force in BE (vector 76) acts away from B indicating tension; again, arrows are inserted in Fig. 4.21 to show the action of the forces.

Now we consider joint C where the unknown member forces are in CD and CE. The force in the member CB at C is represented in magnitude and direction by the vector 73 in the force polygon. From the point 3 we draw a vector 34 vertically downwards to represent the 3 kN load. The vectors 48 and 87 are then drawn parallel to the members CD and CE and represent the forces in the members CD and CE respectively. Thus we see that the force in CD (vector 48) acts towards C, i.e. CD is in compression, while the force in CE (vector 87) acts away from C indicating tension; again we insert corresponding arrows on the members in Fig. 4.21.

Finally the vector 45 is drawn vertically upwards to represent the vertical reaction (=3·25 kN) at D and the vector 58, which must be parallel to the member DE, inserted (since the points 5 and 8 are already located in the force polygon this is a useful check on the accuracy of construction). From the direction of the vector 58 we deduce that the member DE is in tension.

Note that in the force polygon the vectors may be read in both directions. Thus the vector 26 represents the force in the member AB acting at A, while the vector 62 represents the force in AB acting at B. It should also be clear why there must be consistency in the sense in which we move round each joint; for example, the vector 26 represents the direction of the force at A in the member AB when we move in a clockwise sense round A. However, if we then moved in an anticlockwise sense round the joint B the vector 26 would represent the magnitude and direction of the force in AB at B and would indicate that AB is in tension, which clearly it is not.

4.10 Compound trusses

In some situations simple trusses are connected together to form a compound truss, in which case it is generally not possible to calculate the forces in all the members by the method of joints even though the truss is statically determinate.

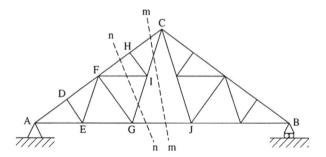

Fig. 4.23 Compound truss

Figure 4.23 shows a compound truss comprising two simple trusses AGC and BJC connected at the apex C and by the linking bar GJ; all the joints are pinned and we shall suppose that the truss carries loads at all its joints. We note that the truss has twenty-seven members and fifteen joints so that Eq. (4.1) is satisfied and the truss is statically determinate.

Initially we would calculate the support reactions at A and B and commence a method of joints solution at the joint A (or at the joint B) where there are no more than two unknown member forces. Thus the magnitudes of F_{AD} and F_{AE} would be obtained. Then, by considering the equilibrium of joint D, we would calculate F_{DE} and F_{DF} and then F_{EF} and F_{EG} by considering the equilibrium of joint E. At this stage, however, the analysis can proceed no further, since at each of the next joints to be considered, F and G, there are three unknown member forces: F_{FG}, F_{FI} and F_{FH} at F and F_{GF}, F_{GI} and F_{GJ} at G. An identical situation would have arisen if the analysis had commenced in the right-hand half of the truss at B. This difficulty is overcome by taking a section mm to cut the three members HC, IC and GJ and using the method of sections to calculate the corresponding member forces. Having obtained F_{GJ} we can consider the equilibrium of joint G to calculate F_{GI} and F_{GF}. Hence F_{FI} and F_{FH} follow by considering the equilibrium of joint F; the remaining unknown member forces follow. Note that obtaining F_{GJ} by taking the section mm allows all the member forces in the right-hand half of the truss to be found by the method of joints.

The method of sections could be used to solve for all the member forces. First we could obtain F_{HC}, F_{IC} and F_{GJ} by taking the section mm and then F_{FH}, F_{FI} and F_{GI} by taking the section nn where F_{GJ} is known, and so on.

4.11   Pin-jointed space frames

The most convenient method of analysing statically determinate stable space frames (see Eq. (4.2)) is that of tension coefficients. In the case of space frames, however, there are three possible equations of equilibrium for each joint (Eqs (2.11)); the moment equations (Eqs (2.12)) are automatically satisfied since, as in the case of plane trusses, the lines of action of all the forces in the members meeting at a joint pass through the joint and the pin cannot transmit moments. Therefore the analysis must begin at a joint where there are no more than three unknown forces.

The calculation of the reactions at supports in space frames can be complex. If a space frame has a statically determinate support system, a maximum of six reaction components can exist since there are a maximum of six equations of overall equilibrium (Eqs (2.11) and (2.12)). However, for the frame to be stable the reactions must be orientated in such a way that they can resist the components of the forces and moments about each of the three coordinate axes. Fortunately, in many problems, it is unnecessary to calculate support reactions since there is usually one joint at which there are no more than three unknown member forces.

Example 4.6 Calculate the forces in the members of the space frame whose elevations and plan are shown in Figs 4.24(a), (b) and (c), respectively.

In this particular problem the exact nature of the support points is not specified so that the support reactions cannot be calculated. However, we note that at joint F there are just three unknown member forces so that the analysis may begin at F.

The first step is to choose an axis system and an origin of axes. Any system may be chosen so long as care is taken to ensure that there is agreement between the axis directions in each of the three views. Also, any point may be chosen as the origin of axes and need not necessarily coincide with a joint. In this problem it would appear logical to choose F, since the analysis will begin at F. Furthermore, it will be helpful to sketch the axis directions on each of the three views as shown and to insert the joint coordinates on the plan view (Fig. 4.24(c)).

At joint F

$$x \text{ direction: } t_{FD}(x_D - x_F) + t_{FB}(x_B - x_F) + t_{FE}(x_E - x_F) - 40 = 0 \tag{i}$$

$$y \text{ direction: } t_{FD}(y_D - y_F) + t_{FB}(y_B - y_F) + t_{FE}(y_E - y_F) = 0 \tag{ii}$$

$$z \text{ direction: } t_{FD}(z_D - z_F) + t_{FB}(z_B - z_F) + t_{FE}(z_E - z_F) = 0 \tag{iii}$$

Substituting the values of the joint coordinates in Eqs (i), (ii) and (iii) in turn we obtain, from Eq. (i),

$$t_{FD}(2 - 0) + t_{FB}(-2 - 0) + t_{FE}(0 - 0) - 40 = 0$$

whence

$$t_{FD} - t_{FB} - 20 = 0 \tag{iv}$$

from Eq. (ii)

$$t_{FD}(-2 - 0) + t_{FB}(-2 - 0) + t_{FE}(0 - 0) = 0$$

which gives

$$t_{FD} + t_{FB} = 0 \tag{v}$$

and from Eq. (iii)

$$t_{FD}(2 - 0) + t_{FB}(2 - 0) + t_{FE}(-2 - 0) = 0$$

so that

$$t_{FD} + t_{FB} - t_{FE} = 0 \tag{vi}$$

From Eqs (v) and (vi) we see by inspection that

$$t_{FE} = 0$$

Now adding Eqs (iv) and (v),

$$2t_{FD} - 20 = 0$$

whence

$$t_{FD} = 10$$

Therefore, from Eq. (v)

$$t_{FB} = -10$$

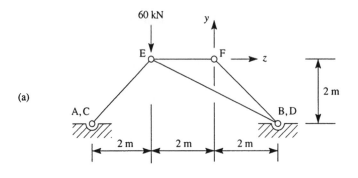

(a)

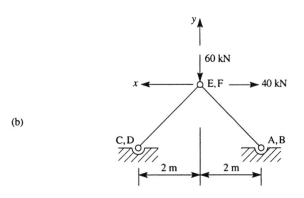

(b)

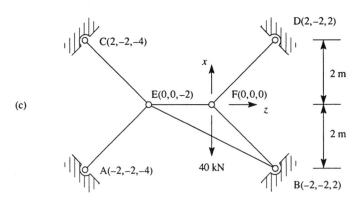

(c)

Fig. 4.24 Elevations and plan of space frame of Ex. 4.6

We now proceed to joint E where, since $t_{EF} = t_{FE}$, there are just three unknown member forces.

x direction: $t_{EB}(x_B - x_E) + t_{EC}(x_C - x_E) + t_{EA}(x_A - x_E) + t_{EF}(x_F - x_E) = 0$ (vii)

y direction: $t_{EB}(y_B - y_E) + t_{EC}(y_C - y_E) + t_{EA}(y_A - y_E) + t_{EF}(y_F - y_E) - 60 = 0$ (viii)

z direction: $t_{EB}(z_B - z_E) + t_{EC}(z_C - z_E) + t_{EA}(z_A - z_E) + t_{EF}(z_F - z_E) = 0$ (ix)

Substituting the values of the coordinates and $t_{EF}(=0)$ in Eqs (vii)–(ix) in turn gives, from Eq. (vii),

$$t_{EB}(-2-0) + t_{EC}(2-0) + t_{EA}(-2-0) = 0$$

so that
$$t_{EB} - t_{EC} + t_{EA} = 0 \qquad \text{(x)}$$

from Eq. (viii)

$$t_{EB}(-2-0) + t_{EC}(-2-0) + t_{EA}(-2-0) - 60 = 0$$

whence
$$t_{EB} + t_{EC} + t_{EA} + 30 = 0 \qquad \text{(xi)}$$

and from Eq. (ix)

$$t_{EB}(2+2) + t_{EC}(-4+2) + t_{EA}(-4+2) = 0$$

which gives
$$t_{EB} - 0.5t_{EC} - 0.5t_{EA} = 0 \qquad \text{(xii)}$$

Subtracting Eq. (xi) from Eq. (x) we have

$$-2t_{EC} - 30 = 0$$

so that
$$t_{EC} = -15$$

Now subtracting Eq. (xii) from Eq. (xi) (or Eq. (x)) yields

$$1.5t_{EC} + 1.5t_{EA} + 30 = 0$$

which gives
$$t_{EA} = -5$$

Finally, from any of Eqs (x)–(xii),

$$t_{EB} = -10$$

The length of each of the members is now calculated, except that of EF which is given (=2 m). Thus, using Pythagoras' theorem,

$$L_{FB} = \sqrt{(x_B - x_F)^2 + (y_B - y_F)^2 + (z_B - z_F)^2}$$

whence
$$L_{FB} = \sqrt{(-2-0)^2 + (-2-0)^2 + (2-0)^2} = 3.46 \text{ m}$$

Similarly
$$L_{FD} = L_{EC} = L_{EA} = 3.46 \text{ m}, \; L_{EB} = 4.90 \text{ m}$$

The forces in the members follow. Thus

$$T_{FB} = t_{FB}L_{FB} = -10 \times 3.46 \text{ kN} = -34.6 \text{ kN (compression)}$$

Similarly

$$T_{FD} = +34.6 \text{ kN (tension)}, \; T_{FE} = 0, \; T_{EC} = -51.9 \text{ kN (compression)}$$
$$T_{EA} = -17.3 \text{ kN (compression)}, \; T_{EB} = -49.0 \text{ kN (compression)}$$

The solution of Eqs (iv)–(vi) and (x)–(xii) in Ex. 4.6 was relatively straightforward in that many of the coefficients of the tension coefficients could be reduced to unity. This is not always the case, so that it is possible that the solution of three simultaneous equations must be carried out. In this situation an elimination method, described in standard mathematical texts, may be used.

Problems

P.4.1 Determine the forces in the members of the truss shown in Fig. P.4.1 using the method of joints and check the forces in the members JK, JD and DE by the method of sections.

Ans. AG = +37·5, AB = −22·5, GB = −20.0, BC = −22·5, GC = −12·5,

GH = +30·0, HC = 0, HJ = +30·0, CJ = +12·5, CD = −37·5, JD = −10·0,

JK = +37·5, DK = +12·5, DE = −45·0, KE = −70·0, EF = −45·0, KF = +75·0.

All in kN.

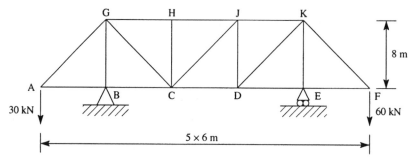

Fig. P.4.1

P.4.2 Calculate the forces in the members of the truss shown in Fig. P.4.2.

Ans. AC = −30·0, AB = +26·0, CP = −8·7, CE = −25·0, EP = +8·7,

PF = +17·3, EF = −17·3, EG = −20·0, EH = +8·7, FH = +17·3,

GH = −8·7, GJ = −15·0, HJ = +26·0, FB = 0.

All in kN.

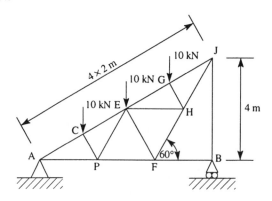

Fig. P.4.2

P.4.3 Calculate the forces in the members EF, EG, EH and FH of the truss shown in Fig. P.4.3. Note that the horizontal load of 4 kN is applied at the joint C.

Ans. EF = −20·0, EG = −80·0, EH = −33·3, FH = +106·6 kN.

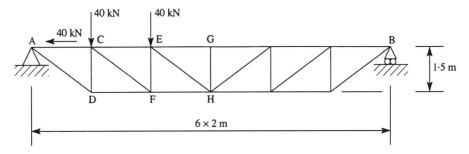

Fig. P.4.3

P.4.4 The roof truss shown in Fig. P.4.4 is comprised entirely of equilateral triangles; the wind loads of 6 kN at J and B act perpendicularly to the member JB. Calculate the forces in the members DF, EF, EG and EK.

Ans. DF = +106·5, EF = +1·7, EG = −107·4, EK = −20·8 kN.

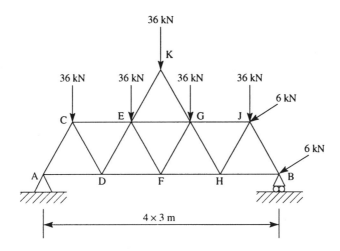

Fig. P.4.4

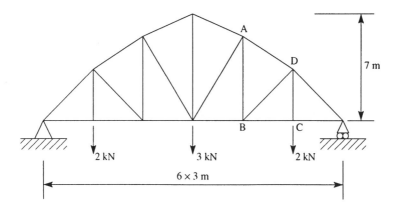

Fig. P.4.5

P.4.5   The upper chord joints of the bowstring truss shown in Fig. P.4.5 lie on a parabola whose equation has the form $y = kx^2$ referred to axes whose origin coincides with the uppermost joint. Calculate the forces in the members AD, BD and BC.

Ans.   AD $= -3 \cdot 1$, BD $= -0 \cdot 5$, BC $= +2 \cdot 7$ kN.

P.4.6   The truss shown in Fig. P.4.6 is supported by a hinge at A and a cable at D which is inclined at an angle of $45°$ to the horizontal members. Calculate the tension, T, in the cable and hence the forces in all the members by the method of tension coefficients.

Ans. $T = 13 \cdot 55$ kN. AB $= -9 \cdot 2$, BC $= -9 \cdot 4$, CD $= -4 \cdot 7$, DE $= +7 \cdot 1$,
EF $= -5 \cdot 0$, FG $= -0 \cdot 3$, GH $= -3 \cdot 1$, AH $= -4 \cdot 4$, BH $= +3 \cdot 1$,
BG $= +4 \cdot 0$, CF $= -6 \cdot 6$, GC $= +4 \cdot 7$, FD $= +4 \cdot 7$.
All in kN.

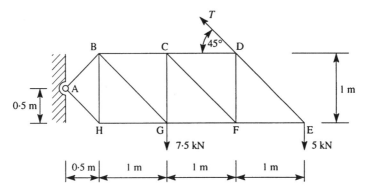

Fig. P.4.6

P.4.7   Check your answers to problems P.4.1, P.4.2 and P.4.6 using a graphical method.

P.4.8   Find the forces in the members of the space frame shown in Fig. P.4.8.

Ans.   OA $= +24 \cdot 2$, OB $= +11 \cdot 9$, OC $= -40 \cdot 2$ kN.

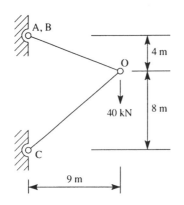
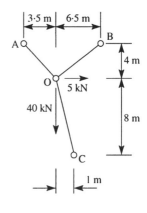

Fig. P.4.8

P.4.9 Use the method of tension coefficients to calculate the forces in the members of the space frame shown in Fig. P.4.9. Note that the loads P_2 and P_3 act in a horizontal plane and at angles of 45° to the vertical plane BAD.

Ans. AB = +13.1, AD = +13·1, AC = −59·0 kN.

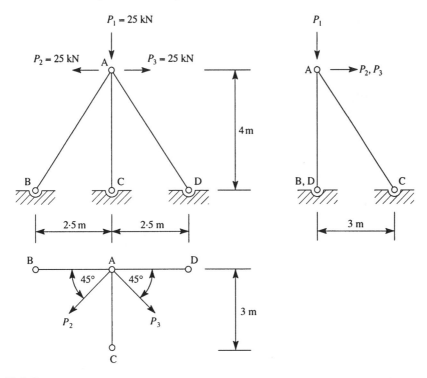

Fig. P.4.9

P.4.10 The pin-jointed frame shown in Fig. P.4.10 is attached to a vertical wall at the points A, B, C and D; the members BE, BF, EF and AF are in the same horizontal plane. The frame supports vertically downward loads of 9 kN and 6 kN at E and F, respectively, and a horizontal load of 3 kN at E in the direction EF.

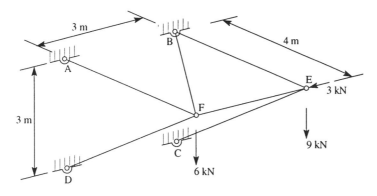

Fig. P.4.10

Calculate the forces in the members of the frame using the method of tension coefficients.

Ans.   EF = −3·0, EC = −15·0, EB = +12·0, FB = +5·0, FA = +4·0,
FD = −10·0.
All in kN.

P.4.11   Fig. P.4.11 shows the plan of a space frame which consists of six pin-jointed members. The member DE is horizontal and 4 m above the horizontal plane containing A, B and C while the loads applied at D and E act in a horizontal plane. Calculate the forces in the members.

Ans.   AD = 0, DC = 0, DE = +40·0, AE = 0, CE = −60·0, BE = +60·0 kN.

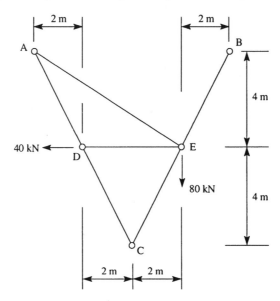

Fig. P.4.11

CHAPTER 5

Cables

Flexible cables have been used to form structural systems for many centuries. Some of the earliest man-made structures of any size were hanging bridges constructed from jungle vines and creepers and spanning ravines and rivers. In European literature the earliest description of an iron suspension bridge was published by Verantius in 1607, while ropes have been used in military bridging from at least 1600. In modern times, cables formed by binding a large number of steel wires together are employed in bridge construction where the bridge deck is suspended on hangers from the cables themselves. The cables in turn pass over the tops of towers and are fixed to anchor blocks embedded in the ground; in this manner large, clear spans are achieved. Cables are also used in cable-stayed bridges, as part of roof support systems, for prestressing in concrete beams and for guyed structures such as pylons and television masts.

Structurally, cables are extremely efficient because they make the most effective use of structural material in that their loads are carried solely through tension. There is, therefore, no tendency for buckling to occur either from bending or from compressive axial loads (see Chapter 18). However, many of the structures mentioned above are statically indeterminate to a high degree. In other situations, particularly in guyed towers and cable-stayed bridges, the extension of the cables affects the internal force system and the analysis becomes non-linear. Such considerations are outside the scope of this book so that we shall concentrate on cables in which loads are suspended directly from the cable.

Two categories of cable arise; the first is relatively lightweight and carries a limited number of concentrated loads, while the second is heavier with a more uniform distribution of load. We shall also examine, in the case of suspension bridges, the effects of different forms of cable support at the towers.

5.1 Lightweight cables carrying concentrated loads

In the analysis of this type of cable we shall assume that the self-weight of the cable is negligible, that it can only carry tensile forces and that the extension of the cable does not affect the geometry of the system. We shall illustrate the method by examples.

Example 5.1 The cable shown in Fig. 5.1 is pinned to supports at A and B and carries a concentrated load of 10 kN at a point C. Calculate the tension in each part of the cable and the reactions at the supports.

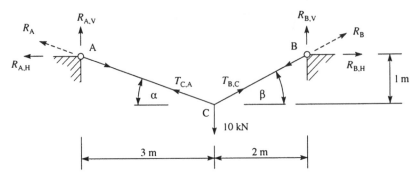

Fig. 5.1   Lightweight cable carrying a concentrated load

Since the cable is weightless the lengths AC and CB are straight. The tensions T_{AC} and T_{CB} in the parts AC and CB, respectively, may be found by considering the equilibrium of the forces acting at C where, from Fig. 5.1, we see that

$$\alpha = \tan^{-1} 1/3 = 18{\cdot}4°, \ \beta = \tan^{-1} 1/2 = 26{\cdot}6°$$

Thus, resolving forces in a direction *perpendicular* to CB (thereby eliminating T_{CB}) we have

$$T_{CA} \cos 45° - 10 \cos 26{\cdot}6° = 0$$

whence

$$T_{CA} = 12{\cdot}6 \text{ kN}$$

Now resolving forces horizontally (or alternatively vertically or perpendicular to CA) gives

$$T_{CB} \cos 26{\cdot}6° - T_{CA} \cos 18{\cdot}4° = 0$$

so that

$$T_{CB} = 13{\cdot}4 \text{ kN}$$

Since the bending moment in the cable is everywhere zero we can take moments about B (or A) to find the vertical component of the reaction at A, $R_{A.V}$, (or $R_{B.V}$) directly. Hence

$$R_{A.V} \times 5 - 10 \times 2 = 0 \tag{i}$$

whence

$$R_{A.V} = 4 \text{ kN}$$

Now resolving forces vertically for the complete cable

$$R_{B.V} + R_{A.V} - 10 = 0 \tag{ii}$$

which gives

$$R_{B.V} = 6 \text{ kN}$$

From the horizontal equilibrium of the cable the horizontal components of the reactions at A and B are equal, i.e. $R_{A.H} = R_{B.H}$. Thus, taking moments about C for the forces to the left of C,

$$R_{A.H} \times 1 - R_{A.V} \times 3 = 0 \tag{iii}$$

from which

$$R_{A.H} = 12 \text{ kN } (= R_{B.H})$$

Note that the horizontal component of the reaction at A, $R_{A,H}$, would be included in the moment equation (Eq. (i)) if the support points A and B were on different levels. In this case Eqs (i) and (iii) could be solved simultaneously for $R_{A,V}$ and $R_{A,H}$. Note also that the tensions T_{CA} and T_{CB} could be found from the components of the support reactions since the resultant reaction at each support, R_A at A and R_B at B, must be equal and opposite in direction to the tension in the cable otherwise the cable would be subjected to shear forces, which we have assumed is not possible. Hence

$$T_{CA} = R_A = \sqrt{4^2 + 12^2} = 12\cdot6 \text{ kN}$$

$$T_{CB} = R_B = \sqrt{6^2 + 12^2} = 13\cdot4 \text{ kN}$$

as before.

In Ex. 5.1 the geometry of the loaded cable was specified. We shall now consider the case of a cable carrying more than one load. Thus, in the cable in Fig. 5.2(a), the loads W_1 and W_2 at the points C and D produce a different deflected shape to the loads W_3 and W_4 at C and D in Fig. 5.2(b). The analysis is then affected by the change in geometry as well as the change in loading, a different situation to that in beam and truss analysis. The cable becomes, in effect, a mechanism and changes shape to maintain its equilibrium; the analysis then becomes non-linear and therefore statically indeterminate. However, if the geometry of the deflected cable is partially specified, say the maximum deflection or sag is given, the system becomes statically determinate.

Example 5.2   Calculate the tension in each of the parts AC, CD and DB of the cable shown in Fig. 5.3.

There are different possible approaches to the solution of this problem. For example, we could investigate the equilibrium of the forces acting at the point C and resolve horizontally and vertically. We would then obtain two equations in which the unknowns would be T_{CA}, T_{CD}, α and β. From the geometry of the cable $\alpha = \tan^{-1} (0\cdot5/1\cdot5) = 18\cdot4°$ so that there would be three unknowns remaining. A third equation could be obtained by examining the moment equilibrium of the length AC of the cable about A, where the moment is zero since the cable is flexible. The solution of these three simultaneous equations would be rather tedious so that a simpler approach is preferable.

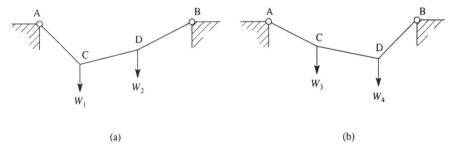

(a)   (b)

Fig. 5.2   Effect on cable geometry of load variation

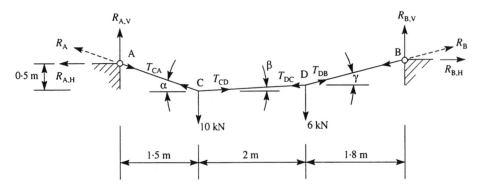

Fig. 5.3 Cable of Ex. 5.2

In Ex. 5.1 we saw that the resultant reaction at the supports is equal and opposite to the tension in the cable at the supports. Thus, by determining $R_{A,V}$ and $R_{A,H}$ we can obtain T_{CA} directly. Hence, taking moments about B we have

$$R_{A,V} \times 5 \cdot 3 - 10 \times 3 \cdot 8 - 6 \times 1 \cdot 8 = 0$$

from which $\qquad\qquad R_{A,V} = 9 \cdot 2 \text{ kN}$

Since the cable is perfectly flexible the internal moment at any point is zero. Therefore, taking moments of forces to the left of C about C gives

$$R_{A,H} \times 0 \cdot 5 - R_{A,V} \times 1 \cdot 5 = 0$$

so that $\qquad\qquad R_{A,H} = 27 \cdot 6 \text{ kN}$

Alternatively we could have obtained $R_{A,H}$ by using the fact that the resultant reaction, R_A, at A is in line with the cable at A, i.e. $R_{A,V}/R_{A,H} = \tan \alpha = \tan 18 \cdot 4°$, whence $R_{A,H} = 27 \cdot 6 \text{ kN}$ as before. Having obtained $R_{A,V}$ and $R_{A,H}$, T_{CA} follows. Thus

$$T_{CA} = R_A = \sqrt{R_{A,H}^2 + R_{A,V}^2} = \sqrt{27 \cdot 6^2 + 9 \cdot 2^2}$$

whence $\qquad\qquad T_{CA} = 29 \cdot 1 \text{ kN}$

From a consideration of the vertical equilibrium of the forces acting at C we have

$$T_{CD} \sin \beta + T_{CA} \sin \alpha - 10 = T_{CD} \sin \beta + 29 \cdot 1 \sin 18 \cdot 4° - 10 = 0$$

which gives $\qquad\qquad T_{CD} \sin \beta = 0 \cdot 815 \qquad\qquad\qquad\text{(i)}$

From the horizontal equilibrium of the forces at C

$$T_{CD} \cos \beta - T_{CA} \cos \alpha = T_{CD} \cos \beta - 29 \cdot 1 \cos 18 \cdot 4° = 0$$

so that $\qquad\qquad T_{CD} \cos \beta = 27 \cdot 612 \qquad\qquad\qquad\text{(ii)}$

Dividing Eq. (i) by Eq. (ii) yields

$$\tan \beta = 0 \cdot 0295$$

whence $\qquad\qquad \beta = 1 \cdot 69°$

Hence, from either of Eqs (i) or (ii)

$$T_{CD} = 27 \cdot 6 \text{ kN}$$

We can obtain the tension in DB in a similar manner. Thus, from the vertical equilibrium of the forces at D, we have

$$T_{DB} \sin \gamma - T_{CD} \sin \beta - 6 = T_{DB} \sin \gamma - 27 \cdot 6 \sin 1 \cdot 69° - 6 = 0$$

from which $\qquad\qquad T_{DB} \sin \gamma = 6 \cdot 815$ \hfill (iii)

From the horizontal equilibrium of the forces at D we see that

$$T_{DB} \cos \gamma - T_{CB} \cos \beta = T_{DB} \cos \gamma - 27 \cdot 6 \cos 1 \cdot 69° = 0$$

from which $\qquad\qquad T_{DB} \cos \gamma = 27 \cdot 618$ \hfill (iv)

Dividing Eq. (iii) by Eq. (iv) we obtain

$$\tan \gamma = 0 \cdot 2468$$

so that $\qquad\qquad\qquad \gamma = 13 \cdot 86°$

T_{DB} follows from either of Eqs (iii) or (iv) and is

$$T_{DB} = 28 \cdot 4 \text{ kN}$$

Alternatively we could have calculated T_{DB} by determining $R_{B,H}$ $(=R_{A,H})$ and $R_{B,V}$. Then

$$T_{DB} = R_B = \sqrt{R_{B,H}^2 + R_{B,V}^2}$$

and $\qquad\qquad\qquad \gamma = \tan^{-1} (R_{B,V}/R_{B,H})$

This approach would, in fact, be a little shorter than the one given above. However, in the case where the cable carries more than two loads, the above method must be used at loading points adjacent to the support points.

5.2   Heavy cables

We shall now consider the more practical case of cables having a significant self-weight.

Governing equation for deflected shape

The cable AB shown in Fig. 5.4(a) carries a distributed load $w(z)$ per unit of its horizontally projected length. An element of the cable, whose horizontal projection if δz, is shown, together with the forces acting on it, in Fig. 5.4(b). Since δz is infinitesimally small, the load intensity may be regarded as constant over the length of the element. Suppose that T is the tension in the cable at the point z and that $T + \delta T$ is the tension at the point $z + \delta z$; the vertical and horizontal components of T are V and H, respectively. In the absence of any externally applied horizontal loads we see that

$$H = \text{constant}$$

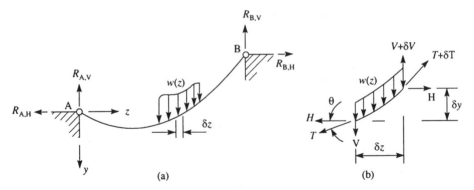

Fig. 5.4   Cable subjected to a distributed load

and from the vertical equilibrium of the element we have

$$V + \delta V - w(z)\delta z - V = 0$$

so that, in the limit as $\delta z \rightarrow 0$

$$\frac{dV}{dz} = w(z) \tag{5.1}$$

From Fig. 5.4(b)

$$\frac{V}{H} = \tan\theta = -\frac{dy}{dz}$$

where y is the vertical deflection of the cable at any point referred to the z axis.

Hence
$$V = -H \frac{dy}{dz}$$

so that
$$\frac{dV}{dz} = -H \frac{d^2y}{dz^2} \tag{5.2}$$

Substituting for dV/dz from Eq. (5.1) into Eq. (5.2) we obtain the *governing equation* for the deflected shape of the cable. Thus

$$H \frac{d^2y}{dz^2} = -w(z) \tag{5.3}$$

We are now in a position to investigate cables subjected to different load applications.

Cable under its own weight

In this case let us suppose that the weight per actual unit length of the cable is w_s. Then, by referring to Fig. 5.5, we see that the weight per unit of the horizontally

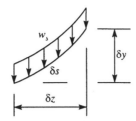

Fig. 5.5 Elemental length of cable under its own weight

projected length of the cable, $w(z)$, is given by

$$w(z)\delta z = w_s \delta s \tag{5.4}$$

Now, in the limit as $\delta s \to 0$, $ds = (dz^2 + dy^2)^{1/2}$
Whence, from Eq. (5.4),

$$w(z) = w_s \left[1 + \left(\frac{dy}{dz} \right)^2 \right]^{1/2} \tag{5.5}$$

Substituting for $w(z)$ from Eq. (5.5) in Eq. (5.3) gives

$$H \frac{d^2 y}{dz^2} = -w_s \left[1 + \left(\frac{dy}{dz} \right)^2 \right]^{1/2} \tag{5.6}$$

Let $dy/dz = p$. Then Eq. (5.6) may be written

$$H \frac{dp}{dz} = -w_s (1 + p^2)^{1/2}$$

or, rearranging and integrating

$$\int \frac{dp}{(1 + p^2)^{1/2}} = -\int \frac{w_s}{H} dz \tag{5.7}$$

The term on the left-hand side of Eq. (5.7) is a standard integral. Thus

$$\sinh^{-1} p = -\frac{w_s}{H} z + C_1$$

in which C_1 is a constant of integration. Thus

$$p = \sinh \left(-\frac{w_s}{H} z + C_1 \right)$$

Now substituting for p $(=dy/dz)$ we obtain

$$\frac{dy}{dz} = \sinh \left(-\frac{w_s}{H} z + C_1 \right)$$

which, when integrated, becomes

$$y = -\frac{H}{w_s} \cosh \left(-\frac{w_s}{H} z + C_1 \right) + C_2 \tag{5.8}$$

in which C_2 is a second constant of integration.

The deflected shape defined by Eq. (5.8) is known as a *catenary*; the constants C_1 and C_2 may be found using the boundary conditions of a particular problem.

Example 5.3   Determine the equation of the deflected shape of the symmetrically supported cable shown in Fig. 5.6 if its self-weight is w_s per unit of its actual length.

The equation of its deflected shape is given by Eq. (5.8), i.e.

$$y = -\frac{H}{w_s} \cosh\left(-\frac{w_s}{H} z + C_1\right) + C_2 \tag{i}$$

Differentiating Eq. (i) with respect to z we have

$$\frac{dy}{dz} = \sinh\left(-\frac{w_s}{H} z + C_1\right) \tag{ii}$$

From symmetry, the slope of the cable at mid-span is zero, i.e. $dy/dz = 0$ when $z = L/2$. Thus, from Eq. (ii)

$$0 = \sinh\left(-\frac{w_s}{H} \frac{L}{2} + C_1\right)$$

from which

$$C_1 = \frac{w_s}{H} \frac{L}{2}$$

Eq. (i) then becomes

$$y = -\frac{H}{w_s} \cosh\left[-\frac{w_s}{H}\left(z - \frac{L}{2}\right)\right] + C_2 \tag{iii}$$

The deflection of the cable at its supports is zero, i.e. $y = 0$ when $z = 0$ and $z = L$. From the first of these conditions

$$0 = -\frac{H}{w_s} \cosh\left(\frac{w_s}{H} \frac{L}{2}\right) + C_2$$

so that

$$C_2 = \frac{H}{w_s} \cosh\left(\frac{w_s}{H} \frac{L}{2}\right)$$

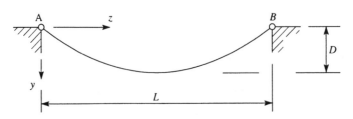

Fig. 5.6   Deflected shape of a symmetrically supported cable

Eq. (ii) is then written as

$$y = -\frac{H}{w_s}\left\{\cosh\left[-\frac{w_s}{H}\left(z-\frac{L}{2}\right)\right] - \cosh\frac{w_s}{H}\frac{L}{2}\right\}$$ (iv)

Equation (iv) gives the deflected shape of the cable in terms of its self-weight, its length and the horizontal component, H, of the tension in the cable. In a particular case where, say, w_s, L and H are specified, the sag, D, of the cable is obtained directly from Eq. (iv). Alternatively if, instead of H, the sag D is fixed, H is obtained from Eq. (iv) which then becomes a transcendental equation; this may be solved graphically.

Since H is constant the maximum tension in the cable will occur at the point where the vertical component of the tension in the cable is greatest. In the above example this will occur at the support points where the vertical component of the tension in the cable is equal to half its total weight. For a cable having supports at different heights, the maximum tension will occur at the highest support since the length of cable from its lowest point to this support is greater than that on the opposite side of the lowest point. Furthermore, the slope of the cable at the highest support is a maximum (see Fig. 5.4(a)).

Cable subjected to a uniform horizontally distributed load

This loading condition is, as we shall see when we consider suspension bridges, more representative of that in actual suspension structures than the previous case.

For the cable shown in Fig. 5.7, Eq. (5.3) becomes

$$H\frac{d^2y}{dz^2} = -w$$ (5.9)

Integrating Eq. (5.9) with respect to z we have

$$H\frac{dy}{dz} = -wz + C_1$$ (5.10)

and

$$Hy = -w\frac{z^2}{2} + C_1z + C_2$$ (5.11)

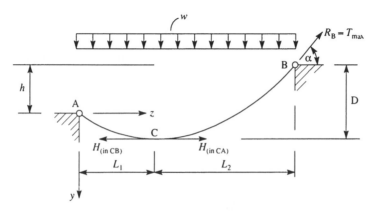

Fig. 5.7 Cable carrying a uniform horizontally distributed load

The boundary conditions are $y = 0$ at $z = 0$ and $y = -h$ at $z = L$. The first of these gives $C_2 = 0$ while from the second we have

$$H(-h) = -w\frac{L^2}{2} + C_1 L$$

so that

$$C_1 = \frac{wL}{2} - H\frac{h}{L}$$

Thus Eqs. (5.10) and (5.11) become, respectively,

$$\frac{dy}{dz} = -\frac{w}{H}z + \frac{wL}{2H} - \frac{h}{L} \tag{5.12}$$

and

$$y = -\frac{w}{2H}z^2 + \left(\frac{wL}{2H} - \frac{h}{L}\right)z \tag{5.13}$$

Thus the cable in this case takes up a parabolic shape.

Equations (5.12) and (5.13) are expressed in terms of the horizontal component, H, of the tension in the cable, the applied load and the cable geometry. If, however, the maximum sag, D, of the cable is known, H may be eliminated as follows.

The position of maximum sag coincides with the point of zero slope. Thus from Eq. (5.12)

$$0 = -\frac{w}{H}z + \frac{wL}{2H} - \frac{h}{L}$$

whence

$$z = \frac{L}{2} - \frac{Hh}{wL} = L_1 \qquad \text{(see Fig. 5.7)}$$

Then the horizontal distance, L_2, from the lowest point of the cable to the support at B is given by

$$L_2 = L - L_1 = \frac{L}{2} + \frac{Hh}{wL}$$

Now considering the moment equilibrium of the length CB of the cable about B we have, from Fig. 5.7,

$$HD - w\frac{L_2^2}{2} = 0$$

so that

$$HD - \frac{w}{2}\left(\frac{L}{2} + \frac{Hh}{wL}\right)^2 = 0 \tag{5.14}$$

Equation (5.14) is a quadratic equation in H and may be solved for a specific case using the formula.

Alternatively, H may be determined by considering the moment equilibrium of the lengths AC and CB about A and C, respectively. Thus, for AC

$$H(D - h) - w\frac{L_1^2}{2} = 0$$

whence
$$H = \frac{wL_1^2}{2(D - h)} \qquad (5.15)$$

For CB

$$HD - \frac{wL_2^2}{2} = 0$$

so that
$$H = \frac{wL_2^2}{2D} \qquad (5.16)$$

Hence, equating Eqs (5.15) and (5.16),

$$\frac{wL_1^2}{2(D - h)} = \frac{wL_2^2}{2D}$$

which gives
$$L_1 = \sqrt{\frac{D - h}{D}}\, L_2$$

But
$$L_1 + L_2 = L$$

therefore
$$L_2\left[\sqrt{\frac{D - h}{D}} + 1\right] = L$$

from which
$$L_2 = \frac{L}{\left(\sqrt{\dfrac{D - h}{D}} + 1\right)} \qquad (5.17)$$

Thus, from Eq. (5.16)

$$H = \frac{wL^2}{2D\left[\sqrt{\dfrac{D - h}{D}} + 1\right]^2} \qquad (5.18)$$

As in the case of the catenary the maximum tension will occur, since H = constant, at the point where the vertical component of the tension is greatest. Thus, in the cable of Fig. 5.7, the maximum tension occurs at B where, as $L_2 > L_1$, the vertical component of the tension ($= wL_2$) is greatest. Hence

$$T_{max} = \sqrt{(wL_2)^2 + H^2} \qquad (5.19)$$

in which L_2 is obtained from Eq. (5.17) and H from one of Eqs. (5.14), (5.16) or (5.18).

At B the slope of the cable is given by

$$\alpha = \tan^{-1}\frac{wL^2}{H} \tag{5.20}$$

or, alternatively, from Eq. (5.12)

$$\left(\frac{dy}{dz}\right)_{z=L} = -\frac{w}{H}L + \frac{wL}{2H} - \frac{h}{L} = -\frac{wL}{2H} - \frac{h}{L} \tag{5.21}$$

For a cable in which the supports are on the same horizontal level, i.e. $h=0$, Eqs. (5.12), (5.13), (5.14) and (5.19) reduce, respectively, to

$$\frac{dy}{dz} = \frac{w}{H}\left(\frac{L}{2} - z\right) \tag{5.22}$$

$$y = \frac{w}{2H}(Lz - z^2) \tag{5.23}$$

$$H = \frac{wL^2}{8D} \tag{5.24}$$

$$T_{max} = \frac{wL}{2}\sqrt{1 + \left(\frac{L}{4D}\right)^2} \tag{5.25}$$

We observe from the above that the analysis of a cable under its own weight, that is a catenary, yields a more complex solution than that in which the load is assumed to be uniformly distributed horizontally. However, if the sag in the cable is small relative to its length, this assumption gives results that differ only slightly from the more accurate but more complex catenary approach. Thus, in practice, the loading is generally assumed to be uniformly distributed horizontally.

Example 5.4 Determine the maximum tension and the maximum slope in the cable shown in Fig. 5.8 if it carries a uniform horizontally distributed load of intensity 10 kN/m.

From Eq. (5.17)

$$L_2 = \frac{200}{\left(\sqrt{\frac{18-6}{18}} + 1\right)} = 110 \cdot 1\,\text{m}$$

Then, from Eq. (5.16)

$$H = \frac{10 \times 110 \cdot 1^2}{2 \times 18} = 3367 \cdot 2\,\text{kN}$$

The maximum tension follows from Eq. (5.19), i.e.

$$T_{max} = \sqrt{(10 \times 110 \cdot 1)^2 + 3367 \cdot 2^2} = 3542 \cdot 6\,\text{kN}$$

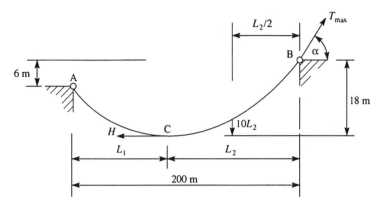

Fig. 5.8 Suspension cable of Ex. 5.4

Then, from Eq. (5.20)

$$\alpha_{max} = \tan^{-1} \frac{10 \times 110 \cdot 1}{3367 \cdot 2} = 18 \cdot 1° \text{ at B}$$

Suspension bridges

A typical arrangement for a suspension bridge is shown diagrammatically in Fig. 5.9. The bridge deck is suspended by hangers from the cables which pass over the tops of the towers and are secured by massive anchor blocks embedded in the ground. The advantage of this form of bridge construction is its ability to provide large clear spans so that sea-going ships, say, can pass unimpeded. Typical examples in the UK are the suspension bridges over the rivers Humber and Severn, the Forth road bridge and the Menai Straits bridge in which the suspension cables comprise chain links rather than tightly bound wires. Suspension bridges are also used for much smaller spans such as pedestrian footbridges and for light vehicular traffic over narrow rivers.

The major portion of the load carried by the cables in a suspension bridge is due to the weight of the deck, its associated stiffening girder and the weight of the vehicles crossing the bridge. By comparison, the self-weight of the cables is negligible. Thus we may assume that the cables carry a uniform horizontally

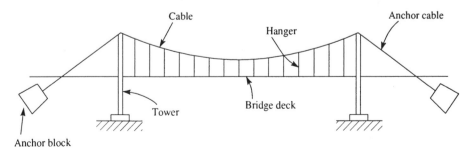

Fig. 5.9 Diagrammatic representation of a suspension bridge

distributed load and therefore take up a parabolic shape; the analysis described in the preceding section therefore applies.

The cables, as can be seen from Fig. 5.9, are continuous over the tops of the towers. In practice they slide in grooves in saddles located on the tops of the towers. For convenience we shall idealize this method of support into two forms, the actual method lying somewhere between the two. In Fig. 5.10(a) the cable passes over a frictionless pulley, which means that the tension, T_A, in the anchor cable is equal to T_C, the tension at the tower in the suspension cable. Generally the inclination, β, of the anchor cable is a fixed value and will not be equal to the inclination, α, of the suspension cable at the tower, There will therefore be a resultant horizontal force, H_T, on the top of the tower given by

$$H_T = T_C \cos \alpha - T_A \cos \beta$$

or, since $T_A = T_C$

$$H_T = T_C (\cos \alpha - \cos \beta) \tag{5.26}$$

H_T, in turn, produces a bending moment, M_T, in the tower which is a maximum at the tower base. Hence

$$M_{T(max)} = H_T h_T = T_C (\cos \alpha - \cos \beta) h_T \tag{5.27}$$

Also, the vertical compressive load, V_T, on the tower is

$$V_T = T_C (\sin \alpha + \sin \beta) \tag{5.28}$$

In the arrangement shown in Fig. 5.10(b) the cable passes over a saddle which is supported on rollers on the top of the tower. The saddle therefore cannot resist a horizontal force and adjusts its position until

$$T_A \cos \beta = T_C \cos \alpha \tag{5.29}$$

For a given value of β, Eq. (5.29) determines the necessary value of T_A. Clearly, since there is no resultant horizontal force on the top of the tower, the bending moment in the tower is everywhere zero. Finally, the vertical compressive load on the tower is given by

$$V_T = T_C \sin \alpha + T_A \sin \beta \tag{5.30}$$

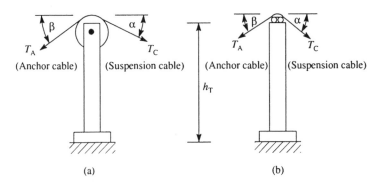

(a)   (b)

Fig. 5.10   Idealization of cable supports

Example 5.5 The cable of a suspension bridge, shown in Fig. 5.11, runs over a frictionless pulley on the top of each of the towers at A and B and is fixed to anchor blocks at D and E. If the cable carries a uniform horizontally distributed load of 120 kN/m determine the diameter required if the permissible working stress on the gross area of the cable, including voids, is 600 N/mm². Also calculate the bending moment and direct load at the base of a tower and the required weight of the anchor blocks.

The tops of the towers are on the same horizontal level, so that the tension in the cable at these points is the same and will be the maximum tension in the cable. The maximum tension is found directly from Eq. (5.25) and is

$$T_{max} = \frac{120 \times 300}{2} \sqrt{1 + \left(\frac{300}{4 \times 30}\right)^2} = 48\,466 \cdot 5 \text{ kN}$$

The maximum direct stress, σ_{max}, is given by

$$\sigma_{max} = \frac{T_{max}}{\pi d^2/4} \text{ (see Section 7.1)}$$

in which d is the cable diameter. Hence

$$600 = \frac{48\,466 \cdot 5 \times 10}{\pi d^2/4}$$

which gives $d = 320 \cdot 7$ mm

The angle of inclination of the suspension cable to the horizontal at the top of the tower is obtained using Eq. (5.20) in which $L_2 = L/2$. Hence

$$\alpha = \tan^{-1} \frac{wL}{2H} = \tan^{-1} \frac{120 \times 300}{2H}$$

where H is given by Eq. (5.24). Thus

$$H = \frac{120 \times 300^2}{8 \times 30} = 45\,000 \text{ kN}$$

so that $$\alpha = \tan^{-1} \frac{120 \times 300}{2 \times 45\,000} = 21 \cdot 8°$$

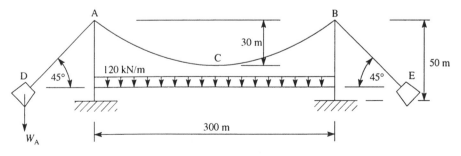

Fig. 5.11 Suspension bridge of Ex. 5.5

Therefore, from Eq. (5.27), the bending moment at the base of the tower is

$$M_T = 48\ 466 \cdot 5\ (\cos 21.8° - \cos 45°) \times 50$$

whence $M_T = 536\ 000\ \text{kN m}$

The direct load at the base of the tower is found using Eq. (5.28), i.e.

$$V_T = 48\ 466 \cdot 5\ (\sin 21 \cdot 8° + \sin 45°)$$

which gives $V_T = 52\ 269 \cdot 9\ \text{kN}$

Finally the weight, W_A, of an anchor block must resist the vertical component of the tension in the anchor cable. Thus

$$W_A = T_A \cos 45° = 48\ 466 \cdot 5 \cos 45°$$

from which $W_A = 34\ 271 \cdot 0\ \text{kN}.$

Problems

P.5.1 Calculate the tension in each segment of the cable known in Fig. P.5.1 and also the vertical distance of the points B and E below the support points A and F.

Ans. $T_{AB} = T_{FE} = 26 \cdot 9$ kN, $T_{BC} = T_{ED} = 25 \cdot 5$ kN, $T_{CD} = 25 \cdot 0$ kN, 1·0 m.

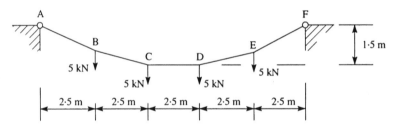

Fig. P.5.1

P.5.2   Calculate the sag at the point B in the cable shown in Fig. P.5.2 and the tension in each of its segments.

Ans.   0·81 m relative to A. $T_{AB} = 4 \cdot 9$ kN, $T_{BC} = 4 \cdot 6$ kN, $T_{CD} = 4 \cdot 7$ kN.

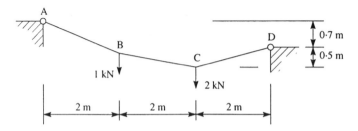

Fig. P.5.2

P.5.3   Calculate the sag, relative to A, of the points C and D in the cable shown in Fig. P.5.3. Determine also the tension in each of its segments.

Ans.   C = 4·2 m, D = 3·1 m. $T_{AB} = 11 \cdot 98$ kN, $T_{BC} = 9 \cdot 68$ kN, $T_{CD} = 9 \cdot 43$ kN.

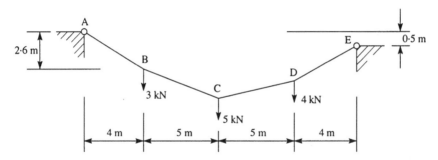

Fig. P.5.3

P.5.4 A cable that carries a uniform horizontally distributed load of 10 kN/m is suspended between two points that are at the same level and 80 m apart. Determine the minimum sag that may be allowed at mid-span if the maximum tension in the cable is limited to 1000 kN.

Ans. 8·73 m.

P.5.5 A suspension cable is suspended from two points 102 m apart and at the same horizontal level. The self-weight of the cable can be considered to be equivalent to 36 N/m of horizontal length. If the cable carries two concentrated loads each of 10 kN at 34 m and 68 m horizontally from the left-hand support and the maximum sag in the cable is 3 m, determine the maximum tension in the cable and the vertical distance between the concentrated loads and the supports.

Ans. 129·5 kN, 2·96 m.

P.5.6 A cable of a suspension bridge has a span of 80 m, a sag of 8 m and carries a uniform horizontally distributed load of 24 kN/m over the complete span. The cable passes over frictionless pulleys at the top of each tower which are of the same height. If the anchor cables are to be arranged such that there is no bending moment in the towers calculate the inclination of the anchor cables to the horizontal. Calculate also the maximum tension in the cable and the vertical force on a tower.

Ans. 21·8°, 2584·9 kN, 1919·9 kN.

P.5.7 A suspension cable passes over saddles supported by roller bearings on the top of two towers 120 m apart and differing in height by 2·5 m. The maximum sag in the cable is 10 m and each anchor cable is inclined at 55° to the horizontal. If the cable carries a uniform horizontally distributed load of 25 kN/m and is to be made of steel having an allowable tensile stress of 240 N/mm², determine its minimum diameter. Calculate also the vertical load on the tallest tower.

Ans. 218·3 mm, 8962·9 kN.

P.5.8 A suspension cable has a sag of 40 m and is fixed to two towers of the same height and 400 m apart; the effective cross-sectional area of the cable is 0·08 m². However, due to corrosion, the effective cross-sectional area of the central half of the cable is reduced by 20%. If the stress in the cable is limited to

500 N/mm^2 calculate the maximum allowable distributed load the cable can support. Calculate also the inclination of the cable to the horizontal at the top of the towers.

Ans. 62.8 kN/m, $21.8°$.

P.5.9   A suspension bridge with two main cables has a span of 250 m and a sag of 25 m. It carries a uniform horizontally distributed load of 25 kN/m and the allowable stress in the cables is 800 N/mm^2. If each anchor cable makes an angle of $45°$ with the towers calculate:

(a) the required cross-sectional area of the cables,
(b) the load in an anchor cable and the overturning force on a tower when
    (i) the cables run over a pulley device,
    (ii) the cables are attached to a saddle resting on rollers.

Ans.   (a) 5270 mm^2, (b) (i) 4210 kN, 930 kN (ii) 5530 kN, 0.

P.5.10   A suspension cable passes over two towers 80 m apart and carries a load of 5 kN per metre of span. If the top of the left-hand tower is 4 m below the top of the right-hand tower, calculate the maximum tension in the cables. Also, if the cable passes over saddles on rollers on the tops of the towers with the anchor cable at $45°$ to the horizontal, calculate the vertical thrust on the right-hand tower.

Ans.   360 kN, 502 kN.

CHAPTER 6

Arches

The Romans were the first to use arches as major structural elements, employing them, mainly in semicircular form, in bridge and aqueduct construction and for roof supports, particularly the barrel vault. Their choice of the semicircular shape was due to the ease with which such an arch could be set out. Generally these arches, as we shall see, carried mainly compressive loads and were therefore constructed from stone blocks, or *voussoirs*, where the joints were either dry or used weak mortar.

During the Middle Ages, Gothic arches, distinguished by their pointed apex, were used to a large extent in the construction of the great European cathedrals. The horizontal thrust developed at the supports, or *springings*, and caused by the tendency of an arch to 'flatten' under load was frequently resisted by *flying buttresses*. This type of arch was also used extensively in the 19th century.

In the 18th century masonry arches were used to support bridges over the large number of canals that were built in that period. Many of these bridges survive to the present day and carry loads unimagined by their designers.

Today arches are usually made of steel or of reinforced or prestressed concrete and can support both tensile as well as compressive loads. They are used to support bridge decks and roofs and vary in span from a few metres in a roof support system to several hundred metres in bridges. A fine example of a steel arch bridge is the Sydney harbour bridge in which the deck is supported by hangers suspended from the arch (see Figs 1.6(a) and (b) for examples of bridge decks supported by arches).

Arches are constructed in a variety of forms. Their components may be straight or curved, but generally fall into two categories. The first, which we shall consider in this chapter, is the three-pinned arch which is statically determinate, whereas the second, the two-pinned arch, is statically indeterminate and will be considered in Chapter 16.

Initially we shall examine the manner in which arches carry loads.

6.1 The linear arch

There is a direct relationship between the action of a flexible cable in carrying loads and the action of an arch. In Section 5.1 we determined the tensile forces in the segments of lightweight cables carrying concentrated loads and saw that the geometry of a cable changed under different loading systems; hence, for example, the two geometries of the same cable in Figs 5.2(a) and (b).

Let us suppose that the cable in Fig. 5.2(a) is made up of three bars or links AC, CD and DB hinged together at C and D and pinned to the supports at A and B. If the loading remains unchanged the deflected shape of the three-link structure will be identical to that of the cable in Fig. 5.2(a) and is shown in Fig. 6.1(a). Furthermore the tension in a link will be exactly the same as the tension in the corresponding segment of the cable. Now suppose that the three-link structure of Fig. 6.1(a) is inverted as shown in Fig. 6.1(b) and that the loads W_1 and W_2 are applied as before. In this situation the forces in the links will be identical in magnitude to those in Fig. 6.1(a) but will now be compressive as opposed to tensile; the structure shown in Fig. 6.1(b) is patently an arch.

The same argument can be applied to any cable and loading system so that the internal forces in an arch may be deduced by analysing a cable having exactly the same shape and carrying identical loads, a fact first realized by Robert Hooke in the 17th century. As in the example in Fig. 6.1 the internal forces in the arch will have the same magnitude as the corresponding cable forces but will be compressive, not tensile.

It is obvious from the above that the internal forces in the arch act along the axes of the different components and that the arch is therefore not subjected to internal shear forces and bending moments; an arch in which the internal forces are purely axial is called a *linear arch*. We also deduce, from Section 5.2, that the internal forces in an arch whose shape is that of a parabola and which carries a uniform horizontally distributed load are purely axial. Further, it will now have become clear why the internal members of a bowstring truss (Section 4.1) carrying loads of equal magnitude along its upper chord joints carry zero force.

There is, however, a major difference between the behaviour of the two structures in Figs 6.1(a) and (b). A change in the values of the loads W_1 and W_2 will merely result in a change in the geometry of the structure in Fig. 6.1(a), whereas the slightest changes in the values of W_1 and W_2 in Fig. 6.1(b) will result in the collapse of the arch as a mechanism. In this particular case collapse could be prevented by replacing the pinned joint at C (or D) by a rigid joint as shown in Fig. 6.2. The forces in the members remain unchanged since the geometry of the structure is unchanged, but the arch is now stable and has become a *three-pinned arch* which, as we shall see, is statically determinate.

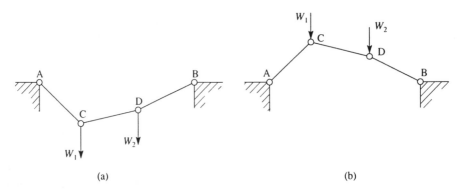

(a)   (b)

Fig. 6.1   Equivalence of cable and arch structures

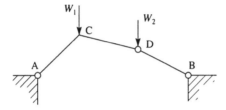

Fig. 6.2 Linear three-pinned arch

If now the pinned joint at D was replaced by a rigid joint, the forces in the members would remain the same, but the arch has become a *two-pinned arch*. In this case, because of the tension cable equivalence, the arch is statically determinate. It is important to realize, however, that the above arguments only apply for the set of loads W_1 and W_2 which produce the particular shape of cable shown in Fig. 6.1(a). If the loads were repositioned or changed in magnitude, the two-pinned arch would become statically indeterminate and would probably cease to be a linear arch so that bending moments and shear forces would be induced. The three-pinned arch of Fig. 6.2 would also become non-linear if the loads were repositioned or changed in magnitude.

In the above we have ignored the effect on the geometry of the arch caused by the shortening of the members. The effect of this on the three-pinned arch is negligible since the pins can accommodate the small changes in angle between the members which this causes. This is not the case in a two-pinned arch or in an arch with no pins at all (in effect a portal frame) so that bending moments and shear forces are induced. However, so long as the loads (W_1 and W_2 in this case) remain unchanged in magnitude and position, the corresponding stresses are 'secondary' and will have little effect on the axial forces.

The linear arch, in which the internal forces are purely axial, is important for the structural designer since the linear arch shape gives the smallest stresses. If, however, the thrust line is not axial, bending stresses are induced and these can cause tension on the inner or outer faces (the *intrados* and *extrados*) of the arch. In a masonry arch in which the joints are either dry or made using a weak mortar, this can lead to cracking and possible failure. Furthermore, if the thrust line lies outside the faces of the arch, instability leading to collapse can also occur. We shall deduce in Section 9.2 that for no tension to be developed in a rectangular cross-section, the compressive force on the section must lie within the middle third of the section.

In small-span arch bridges, these factors are not of great importance since the greatest loads on the arch come from vehicular traffic. These loads vary with the size of the vehicle and its position on the bridge, so that it is generally impossible for the designer to achieve a linear arch. On the other hand, in large-span arch bridges, the self-weight of the arch forms the major portion of the load the arch has to carry. In Section 5.2 we saw that a cable under its own weight takes up the shape of a catenary. It follows that the ideal shape for an arch of constant thickness is an inverted catenary. However, in the analysis of the three-pinned arch we shall assume a general case in which shear forces and bending moments, as well as axial forces, are present.

6.2   The three-pinned arch

A three-pinned arch would be used in situations where there is a possibility of support displacement; this, in a two-pinned arch, would induce additional stresses. In the analysis of a three-pinned arch the first step, generally, is to determine the support reactions.

Support reactions – supports on same horizontal level

Consider the arch shown in Fig. 6.3. It carries an inclined concentrated load, W, at a given point D, a horizontal distance a from the support point A. The equation of the shape of the arch will generally be known so that the position of specified points on the arch, say D, can be obtained. We shall suppose that the third pin is positioned at the crown, C, of the arch, although this need not necessarily be the case; the height or *rise* of the arch is h.

The supports at A and B are pinned but neither can be a roller support or the arch would collapse. Therefore, in addition to the two vertical components of the reactions at A and B, there will be horizontal components $R_{A,H}$ and $R_{B,H}$. Thus there are four unknown components of reaction but only three equations of overall equilibrium (Eqs (2.10)) so that an additional equation is required. This is obtained from the fact that the third pin at C is unable to transmit bending moments although, obviously, it is able to transmit shear forces.

Thus, from the overall vertical equilibrium of the arch in Fig. 6.3, we have

$$R_{A,V} + R_{B,V} - W \cos \alpha = 0 \qquad (6.1)$$

and from the horizontal equilibrium

$$R_{A,H} - R_{B,H} - W \sin \alpha = 0 \qquad (6.2)$$

Now taking moments about, say, B,

$$R_{A,V} L - W \cos \alpha (L - a) - W \sin \alpha\, h_D = 0 \qquad (6.3)$$

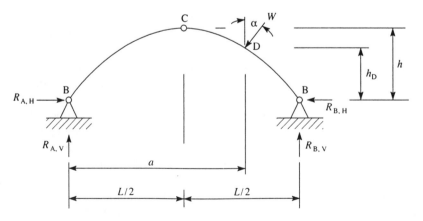

Fig. 6.3   Three-pinned arch

The internal moment at C is zero so that we can take moments about C of forces to the left or right of C. A slightly simpler expression results by considering forces to the left of C; thus

$$R_{A,V} \frac{L}{2} - R_{A,H} h = 0 \qquad (6.4)$$

Equations (6.1)–(6.4) enable the four components of reaction to be found; the normal force, shear force and bending moment at any point in the arch follow.

Example 6.1 Calculate the normal force, shear force and bending moment at the point X in the semicircular arch shown in Fig. 6.4.

In this example we can find either vertical component of reaction directly by taking moments about one of the support points. Hence, taking moments about B, say,

$$R_{A,V} \times 12 - 60 \ (6 \cos 30° + 6) - 100 \ (6 \sin 30° + 6) = 0$$

which gives
$$R_{A,V} = 131 \cdot 0 \text{ kN}$$

Now resolving forces vertically: $R_{B,V} + R_{A,V} - 60 - 100 = 0$

which, on substituting for $R_{A,V}$, gives

$$R_{B,V} = 29 \cdot 0 \text{ kN}$$

Since no horizontal loads are present, we see by inspection that

$$R_{A,H} = R_{B,H}$$

Finally, taking moments of forces to the right of C about C (this is a little simpler than considering forces to the left of C) we have

$$R_{B,H} \times 6 - R_{B,V} \times 6 = 0$$

from which
$$R_{B,H} = 29 \cdot 0 \text{ kN} = R_{A,H}$$

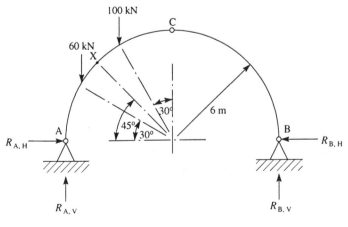

Fig. 6.4 Three-pinned arch of Ex. 6.1

The normal force at the point X is obtained by resolving the forces to one side of X in a direction tangential to the arch at X. Thus, considering forces to the left of X and taking tensile forces as positive,

$$N_X = -R_{A,V} \cos 45° - R_{A,H} \sin 45° + 60 \cos 45°$$

so that

$$N_X = -70 \cdot 7 \text{ kN}$$

and is compressive.

The shear force at X is found by resolving the forces to one side of X in a direction perpendicular to the tangent at X. We shall take a positive shear force as acting radially outwards when it is to the left of a section. Thus, considering forces to the left of X

$$S_X = R_{A,V} \sin 45° - R_{A,H} \cos 45° - 60 \sin 45°$$

which gives

$$S_X = +29 \cdot 7 \text{ kN}$$

Now taking moments about X for forces to the left of X and regarding a positive moment as causing tension on the underside of the arch, we have

$$M_X = R_{A,V} (6 - 6 \cos 45°) - R_{A,H} \times 6 \sin 45° - 60 (6 \cos 30° - 6 \cos 45°)$$

Whence

$$M_X = +50 \cdot 0 \text{ kN m}$$

Note that in Ex. 6.1 the sign conventions adopted for normal force, shear force and bending moment are the same as those specified in Chapter 3.

Support reactions – supports on different levels

In the three-pinned arch shown in Fig. 6.5 the support at B is a known height, h_B, above A. Let us suppose that the equation of the shape of the arch is known so that all dimensions may be calculated. Now, resolving forces vertically gives

$$R_{A,V} + R_{B,V} - W \cos \alpha = 0 \tag{6.5}$$

and horizontally we have

$$R_{A,H} - R_{B,H} - W \sin \alpha = 0 \tag{6.6}$$

Also, taking moments about B, say,

$$R_{A,V} L - R_{A,H} h_B - W \cos \alpha (L - a) - W \sin \alpha (h_D - h_B) = 0 \tag{6.7}$$

Note that, unlike the previous case, the horizontal component of the reaction at A is included in the overall moment equation (Eq. (6.7)).

Finally we can take moments of all the forces to the left or right of C about C since the internal moment at C is zero. In this case the overall moment equation (Eq. (6.7)) includes both components, $R_{A,V}$ and $R_{A,H}$, of the support reaction at A. Thus, if we now consider moments about C of forces to the left of C, we shall obtain a moment equation in terms of $R_{A,V}$ and $R_{A,H}$. This equation, with Eq. (6.7), provides two simultaneous equations which may be solved for $R_{A,V}$ and $R_{A,H}$. Alternatively if, when we were considering the overall moment equilibrium of the arch, we had taken moments about A, Eq. (6.7) would have been expressed in terms

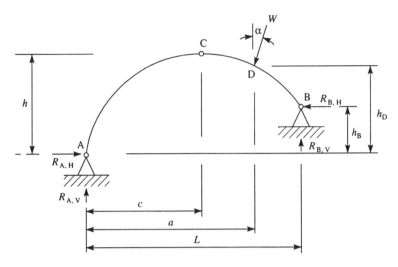

Fig. 6.5 Three-pinned arch with supports at different levels

of $R_{B,V}$ and $R_{B,H}$. Then we would obtain the fourth equation by taking moments about C of the forces to the right of C and the two simultaneous equations would be in terms of $R_{B,V}$ and $R_{B,H}$. Theoretically this approach is not necessary but it leads to a simpler solution. Thus, referring to Fig. 6.5

$$R_{A,V}c - R_{A,H}h = 0 \tag{6.8}$$

The solution of Eqs (6.7) and (6.8) gives $R_{A,V}$ and $R_{A,H}$, then $R_{B,V}$ and $R_{B,H}$ follow from Eqs (6.5) and (6.6), respectively.

Example 6.2 The parabolic arch shown in Fig. 6.6 carries a uniform horizontally distributed load of intensity 10 kN/m over the portion AC of its span. Calculate the values of the normal force, shear force and bending moment at the point D.

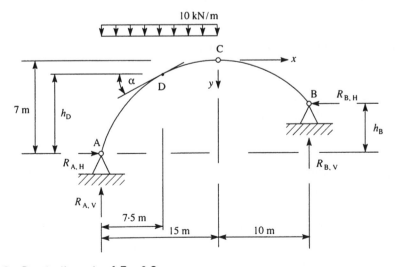

Fig. 6.6 Parabolic arch of Ex. 6.2

Initially we must determine the equation of the arch so that the heights of B and D may be calculated. The simplest approach is to choose the origin of axes at C so that the equation of the parabola may be written in the form

$$y = kx^2 \tag{i}$$

in which k is a constant. At A, $y = 7$ m when $x = -15$ m. Hence, from Eq. (i)

$$7 = k \times (-15)^2$$

whence $$k = 0\cdot0311$$

and Eq. (i) becomes

$$y = 0\cdot0311 x^2 \tag{ii}$$

Then $$y_B = 0\cdot0311 \times (10)^2 = 3\cdot11 \text{ m}$$

Hence $$h_B = 7 - 3\cdot11 = 3\cdot89 \text{ m}$$

Also $$y_D = 0\cdot0311 \times (-7\cdot5)^2 = 1\cdot75 \text{ m}$$

so that $$h_D = 7 - 1\cdot75 = 5\cdot25 \text{ m}$$

Taking moments about A for the overall equilibrium of the arch we have

$$R_{B,V} \times 25 + R_{B,H} \times 3\cdot89 - 10 \times 15 \times 7\cdot5 = 0$$

which simplifies to

$$R_{B,V} + 0\cdot16 R_{B,H} - 45\cdot0 = 0 \tag{iii}$$

Now taking moments about C for the forces to the right of C we obtain

$$R_{B,V} \times 10 - R_{B,H} \times 3\cdot11 = 0$$

Whence $$R_{B,V} - 0\cdot311 R_{B,H} = 0 \tag{iv}$$

The simultaneous solution of Eqs (iii) and (iv) gives

$$R_{B,V} = 29\cdot7 \text{ kN}, \ R_{B,H} = 95\cdot5 \text{ kN}$$

From the horizontal equilibrium of the arch we have

$$R_{A,H} = R_{B,H} = 95\cdot5 \text{ kN}$$

and from the vertical equilibrium

$$R_{A,V} + R_{B,V} - 10 \times 15 = 0$$

which gives $$R_{A,V} = 120\cdot3 \text{ kN}$$

To calculate the normal force and shear force at the point D we require the slope of the arch at D. From Eq. (ii)

$$\left(\frac{dy}{dx}\right)_D = 2 \times 0\cdot0311 \times (-7\cdot5) = -0\cdot4665 = -\tan\alpha$$

Hence $$\alpha = 25\cdot0°$$

Now resolving forces to the left (or right) of D in a direction parallel to the tangent at D we obtain the normal force at D. Hence

$$N_D = -R_{A,V} \sin 25 \cdot 0° - R_{A,H} \cos 25 \cdot 0° + 10 \times 7 \cdot 5 \sin 25 \cdot 0°$$

which gives $\qquad N_D = -105 \cdot 7 \text{ kN (compression)}$

The shear force at D is then

$$S_D = R_{A,V} \cos 25 \cdot 0° - R_{A,H} \sin 25 \cdot 0° - 10 \times 7 \cdot 5 \cos 25.0°$$

so that $\qquad S_D = +0 \cdot 7 \text{ kN}$

Finally the bending moment at D is

$$M_D = R_{A,V} \times 7 \cdot 5 - R_{A,H} \times 5 \cdot 25 - 10 \times 7 \cdot 5 \times \frac{7 \cdot 5}{2}$$

from which $\qquad M_D = +119 \cdot 6 \text{ kN m}$

6.3 A three-pinned parabolic arch carrying a uniform horizontally distributed load

In Section 5.2 we saw that a flexible cable carrying a uniform horizontally distributed load took up the shape of a parabola. It follows that a three-pinned parabolic arch carrying the same loading would experience zero shear force and bending moment at all sections. We shall now investigate the bending moment in the symmetrical three-pinned arch shown in Fig. 6.7.

The vertical components of the support reactions are, from symmetry,

$$R_{A,V} = R_{B,V} = \frac{wL}{2}$$

Also, in the absence of any horizontal loads

$$R_{A,H} = R_{B,H}$$

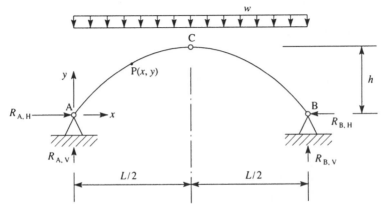

Fig. 6.7 Parabolic arch carrying a uniform horizontally distributed load

Now taking moments of forces to the left of C about C,

$$R_{A,H}h - R_{A,V}\frac{L}{2} + \frac{wL}{2}\frac{L}{4} = 0$$

which gives

$$R_{A,H} = \frac{wL^2}{8h}$$

With the origin of axes at A, the equation of the parabolic shape of the parabola may be shown to be

$$y = \frac{4h}{L^2}(Lx - x^2)$$

The bending moment at any point $P(x, y)$ in the arch is given by

$$M_P = R_{A,V}x - R_{A,H}y - \frac{wx^2}{2}$$

or, substituting for $R_{A,V}$ and $R_{A,H}$ and for y in terms of x,

$$M_P = \frac{wL}{2}x - \frac{wL^2}{8h}\frac{4h}{L^2}(Lx - x^2) - \frac{wx^2}{2}$$

Simplifying this expression

$$M_P = \frac{wL}{2}x - \frac{wL}{2}x - \frac{wx^2}{2} - \frac{wx^2}{2} = 0$$

as expected.

The shear force may also be shown to be zero at all sections of the arch.

6.4   Bending moment diagram for a three-pinned arch

Consider the arch shown in Fig. 6.8; we shall suppose that the equation of the arch referred to the xy axes is known. The load W is applied at a given point D (x_D, y_D) and the support reactions may be calculated by the methods previously described. The bending moment, M_{P1}, at any point P_1 (x, y) between A and D is given by

$$M_{P1} = R_{A,V}x - R_{A,H}y \tag{6.9}$$

and the bending moment, M_{P2}, at the point P_2 (x, y) between D and B is

$$M_{P2} = R_{A,V}x - W(x - x_D) - R_{A,H}y \tag{6.10}$$

Now let us consider a simply supported beam AB having the same span as the arch and carrying a load, W, at the same horizontal distance, x_D, from the left-hand support (Fig. 6.9(a)). The vertical reactions, R_A and R_B will have the same magnitude as the vertical components of the support reactions in the arch. Thus the bending moment at any point between A and D and a distance x from A is

$$M_{AD} = R_A x = R_{A,V}x \tag{6.11}$$

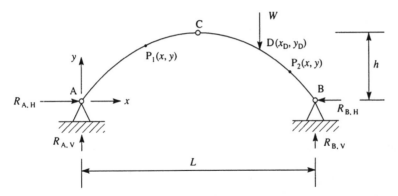

Fig. 6.8 Determination of the bending moment diagram for a three-pinned arch

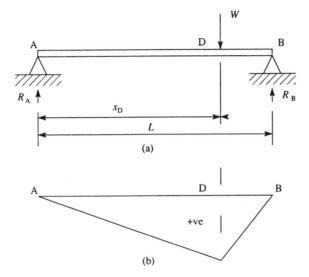

Fig. 6.9 Bending moment diagram for a simply supported beam (tension on undersurface of beam)

Also the bending moment at any point between D and B a distance x from A is

$$M_{DB} = R_A x - W(x - x_D) = R_{A,V} x - W(x - x_D) \qquad (6.12)$$

giving the bending moment diagram shown in Fig. 6.9(b). Comparing Eqs (6.11) and (6.12) with Eqs (6.9) and (6.10), respectively, we see that Eq. (6.9) may be written

$$M_{P1} = M_{AD} - R_{A,H} y \qquad (6.13)$$

and Eq. (6.10) may be written

$$M_{P2} = M_{DB} - R_{A,H} y \qquad (6.14)$$

Thus the complete bending moment diagram for the arch may be regarded as the sum of a 'simply supported beam' bending moment diagram and an 'arch' bending moment diagram in which the 'arch' diagram has the same shape as the arch itself, since its ordinates are equal to a constant multiplied by y. The two bending moment diagrams may be superimposed as shown in Fig. 6.10 to give the complete bending moment diagram for the arch. Note that the curve of the arch forms the baseline of the bending moment diagram and that the bending moment at the crown of the arch where the third pin is located is zero.

In the above it was assumed that the mathematical equation of the curve of the arch is known. However, in a situation where, say, only a scale drawing of the curve of the arch is available, a semigraphical procedure may be adopted if the loads are vertical. The 'arch' bending moment at the crown C of the arch is $R_{AH}h$ as shown in Fig. 6.10. The magnitude of this bending moment may be calculated so that the scale of the bending moment diagram is then fixed by the rise (at C) of the arch in the scale drawing. Also this bending moment is equal in magnitude but opposite in sign to the 'simply supported beam' bending moment at this point. Other values of 'simply supported beam' bending moment may be calculated at, say, load positions and plotted on the complete bending moment diagram to the already determined scale. The diagram is then completed, enabling values of bending moment to be scaled off as required.

In the arch of Fig. 6.8 a simple construction may be used to produce the complete bending moment diagram. In this case the arch shape is drawn as in Fig. 6.10 and this, as we have seen, fixes the scale of the bending moment diagram. Then, since the final bending moment at C is zero and is also zero at A and B, a line drawn from A through C to meet the vertical through the point of application of the load at E represents the 'simply supported beam' bending

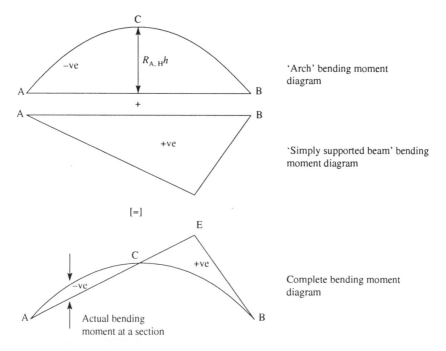

Fig. 6.10 Complete bending moment diagram for a three-pinned arch

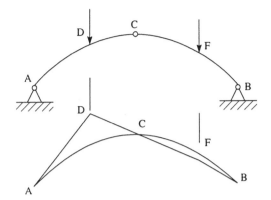

Fig. 6.11 Bending moment diagram for a three-pinned arch carrying two loads

moment diagram between A and D. The bending moment diagram is then completed by drawing in the line EB.

This construction is only possible when the arch carries a single load. In the case of an arch carrying two or more loads as in Fig. 6.11, the 'simply supported beam' bending moments must be calculated at D and F and their values plotted to the same scale as the 'arch' bending moment diagram. Clearly the bending moment at C remains zero.

We shall consider the statically indeterminate two-pinned arch in Chapter 16.

Problems

P.6.1 Determine the position and calculate the value of the maximum bending moment in the loaded half of the semicircular three-pinned arch shown in Fig. P.6.1.

Ans. 6·59 m from A, 84·2 kN m (sagging).

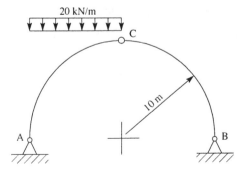

Fig. P.6.1

P.6.2 Figure P.6.2 shows a three-pinned arch of radius 12 m. Calculate the normal force, shear force and bending moment at the point D.

Ans. 14·4 kN (compression), 5·5 kN, 21·9 kN m (hogging).

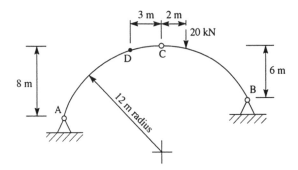

Fig. P.6.2

P.6.3   The three-pinned arch shown in Fig. P.6.3 is parabolic in shape. If the arch carries a uniform horizontally distributed load of intensity 40 kN/m over the part CB, calculate the bending moment at D.

Ans.   140·5 kN m (sagging).

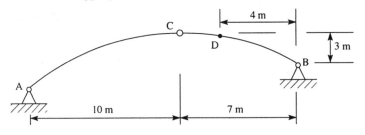

Fig. P.6.3

P.6.4   In the three-pinned arch ACB shown in Fig. P.6.4 the portion AC has the shape of a parabola with its origin at C, while CB is straight. The portion AC carries a uniform horizontally distributed load of intensity 30 kN/m, while the portion CB carries a uniform horizontally distributed load of intensity 18 kN/m. Calculate the normal force, shear force and bending moment at the point D.

Ans.   91·2 kN (compression), 8·9 kN, 210·0 kN m (sagging).

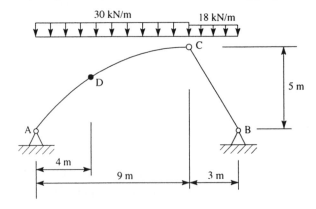

Fig. P.6.4

P.6.5 Draw normal force, shear force and bending moment diagrams for the loaded half of the three-pinned arch shown in Fig. P.6.5.

Ans. $N_{BD} = 26 \cdot 5$ kN, $N_{DE} = 19 \cdot 5$ kN, $N_{EF} = N_{FC} = 15$ kN (all compression).

$S_{BD} = -5 \cdot 3$ kN, $S_{DE} = +1.8$ kN, $S_{EF} = -2 \cdot 5$ kN, $S_{FC} = +7 \cdot 5$ kN.

$M_D = 11 \cdot 3$ kN m, $M_E = 7 \cdot 5$ kN m, $M_F = 11 \cdot 3$ kN m (sagging).

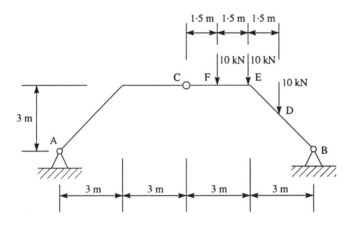

Fig. P.6.5

P.6.6 Calculate the components of the support reactions at A and D in the three-pinned arch shown in Fig. P.6.6 and hence draw the bending moment diagram for the member DC; draw the diagram on the tension side of the member. All members are 1·5 m long.

Ans. $R_{A.V} = 6 \cdot 3$ kN, $R_{A.H} = 11.12$ kN, $R_{D.V} = 21 \cdot 43$ kN, $R_{D.H} = 3 \cdot 88$ kN.

$M_D = 0$, $M_C = 5 \cdot 82$ kN m (tension on left of CD).

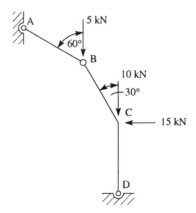

Fig. P.6.6

CHAPTER 7

Stress and Strain

We are now in a position to calculate internal force distributions in a variety of structural forms, i.e. normal forces, shear forces and bending moments in beams and arches, axial forces in truss and space frame members and in suspension cables and torque distributions in beams. These internal force systems are distributed throughout the cross-section of a structural member in the form of stresses. However, although there are four basic types of internal force, there are only two types of stress: one which acts perpendicularly to the cross-section of a member and one which acts tangentially. The former is known as a *direct stress*, the latter as a *shear stress*.

The distribution of these stresses over the cross-section of a structural member depends upon the internal force system at the section and also upon the geometry of the cross-section. In some cases, as we shall see later, these distributions are complex, particularly those produced by the bending and shear of unsymmetrical sections. We can, however, examine the nature of each of these stresses by considering simple loading systems acting on structural members whose cross-sections have some degree of symmetry. At the same time we shall define the corresponding strains and investigate the relationships between the two.

7.1 Direct stress in tension and compression

The simplest form of direct stress system is that produced by an axial load. Suppose that a structural member has a uniform 'I' cross-section of area A and is subjected to an axial tensile load, P, as shown in Fig. 7.1(a). At any section mm the internal force is a normal force which, from the arguments presented in Chapter 3, is equal to P (Fig. 7.1(b)). It is clear that this normal force is not resisted at just one point on each face of the section as Fig. 7.1(b) indicates but at every point as shown in Fig. 7.2. We assume in fact that P is distributed uniformly over the complete face of the section so that at any point in the cross-section there is an intensity of force, i.e. stress, to which we give the symbol σ and which we define as

$$\sigma = \frac{P}{A} \tag{7.1}$$

This direct stress acts in the direction shown in Fig. 7.2 when P is tensile and in the reverse direction when P is compressive. The sign convention for direct stress is

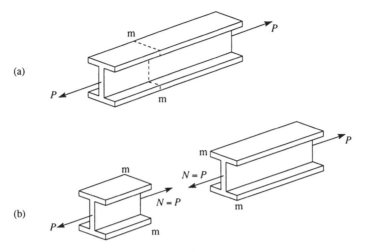

(a)

(b)

Fig. 7.1 Structural member with axial load

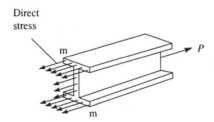

Fig. 7.2 Internal force distribution in a beam section

identical to that for normal force; a tensile stress is therefore positive while a compressive stress is negative.

In Fig. 7.1 the section mm is some distance from the point of application of the load. At sections in the proximity of the applied load the distribution of direct stress will depend upon the method of application of the load, and only in the case where the applied load is distributed uniformly over the cross-section will the direct stress be uniform over sections in this region. In other cases *stress concentrations* arise which require specialized analysis; this topic is covered in more advanced texts on strength of materials and stress analysis.

We shall see in Chapter 8 that it is the level of stress that governs the behaviour of structural materials. For a given material, failure, or breakdown of the crystalline structure of the material under load, occurs at a constant value of stress. For example, in the case of steel subjected to simple tension failure begins at a stress of about 300 N/mm², although variations occur in steels manufactured to different specifications. This stress is independent of size or shape and may therefore be used as the basis for the design of structures fabricated from steel. Failure stress varies considerably from material to material and in some cases depends upon whether the material is subjected to tension or compression.

A knowledge of the failure stress of a material is essential in structural design where, generally, a designer wishes to determine a minimum size for a structural

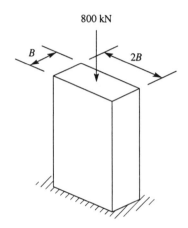

Fig. 7.3 Column of Ex. 7.1

member carrying a given load. Thus, for a member fabricated from a given material and subjected to axial load, we would use Eq. (7.1) either to determine a minimum area of cross-section for a given load or to check the stress level in a given member carrying a given load.

Example 7.1 A short column has a rectangular cross-section with sides in the ratio 1:2 (Fig. 7.3). Determine the minimum dimensions of the column section if the column carries an axial load of 800 kN and the failure stress of the material of the column is 400 N/mm^2.

From Eq. (7.1) the minimum area of the cross-section is given by

$$A_{min} = \frac{P}{\sigma_{max}} = \frac{800 \times 10^3}{400} = 2000 \text{ mm}^2$$

But

$$A_{min} = 2B^2 = 2000 \text{ mm}^2$$

from which

$$B = 31 \cdot 6 \text{ mm}$$

Therefore the minimum dimensions of the column cross-section are 31·6 mm × 63·2 mm. In practice these dimensions would be rounded up to 32 mm × 64 mm or, if the column were of some standard section, the next section having a cross-sectional area greater than 2000 mm^2 would be chosen. Also the column would not be designed to the limit of its failure stress but to a working or design stress which would incorporate some safety factor (see Section 8.7).

7.2 Shear stress in shear and torsion

An externally applied shear load induces an internal shear force which is tangential to the faces of a beam cross-section. Fig. 7.4(a) illustrates such a situation for a cantilever beam carrying a shear load W at its free end. We have seen in Chapter 3 that the action of W is to cause sliding of one face of the cross-section relative to the other; W also induces internal bending moments which produce internal direct

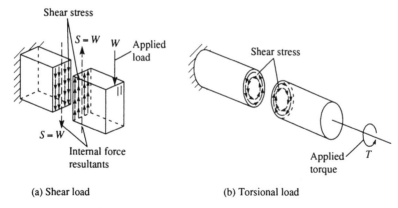

(a) Shear load   (b) Torsional load

Fig. 7.4 Generation of shear stresses in beam sections

stress systems; these are considered in a later chapter. The internal shear force S ($=W$) required to maintain the vertical equilibrium of the portions of the beam is distributed over each face of the cross-section. Thus at any point in the cross-section there is a tangential intensity of force which is termed *shear stress*. This shear stress is not distributed uniformly over the faces of the cross-section as we shall see in Chapter 10. For the moment, however, we shall define the average shear stress over the faces of the cross-section as

$$\tau_{av} = \frac{W}{A} \qquad (7.2)$$

where A is the cross-sectional area of the beam.

A system of shear stresses is induced in a different way in the circular-section bar shown in Fig. 7.4(b) where the internal torque (T) tends to produce a relative rotational sliding of the two faces of the cross-section. The shear stresses are tangential to concentric circular paths in the faces of the cross-section. We shall examine the shear stress due to torsion in various cross-sections in Chapter 11.

7.3 Complementary shear stress

Consider the cantilever beam shown in Fig. 7.5(a). Let us suppose that the beam is of rectangular cross-section having a depth h and unit thickness; it carries a vertical shear load W at its free end. The internal shear forces on the opposite faces mm and nn of an elemental length δz of the beam are distributed as shear stresses in some manner over each face as shown in Fig. 7.5(b). Suppose now that we isolate a small rectangular element ABCD of depth δh of this elemental length of beam (Fig. 7.5(c)) and consider its equilibrium. Since the element is small, the shear stresses τ on the faces AD and BC may be regarded as constant. The shear force resultants of these shear stresses clearly satisfy vertical equilibrium of the element but rotationally produce a clockwise couple. This must be equilibrated by an anticlockwise couple which can only be produced by shear forces on the horizontal faces AB and CD of the element. Let τ' be the shear

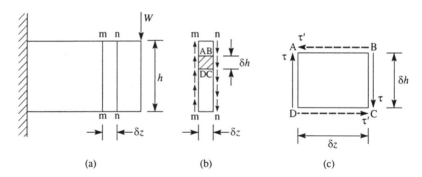

Fig. 7.5 Complementary shear stress

stresses induced by these shear forces. Then for rotational equilibrium of the element about the corner D

$$\tau' \times \delta z \times 1 \times \delta h = \tau \times \delta h \times 1 \times \delta z$$

which gives $$\tau' = \tau \qquad (7.3)$$

We see, therefore, that a shear stress acting on a given plane is always accompanied by an equal *complementary shear stress* acting on planes perpendicular to the given plane and in the opposite sense.

7.4　Direct strain

Since no material is completely rigid, the application of loads produces distortion. Thus, as we observed in Chapter 3, an axial tensile load will cause a structural member to increase in length, whereas a compressive load would shorten its length.

Suppose that δ is the change in length produced by either a tensile or compressive axial load. We now define the *direct strain*, ε, in the member in non-dimensional form as the change in length per unit length of the member. Hence

$$\varepsilon = \frac{\delta}{L_0} \qquad (7.4)$$

where L_0 is the length of the member in its unloaded state. Clearly ε may be either a tensile (positive) strain or a compressive (negative) strain. Equation (7.4) is applicable only when distortions are relatively small and can be used for values of strain up to and around 0.001, which is adequate for most structural problems. For larger values, load–displacement relationships become complex and are therefore left for more advanced texts.

We shall see in Section 7.7 that it is convenient to measure distortion in this non-dimensional form since there is a direct relationship between the stress in a member and the accompanying strain. The strain in an axially loaded member therefore depends solely upon the level of stress in the member and is independent of its length or cross-sectional geometry.

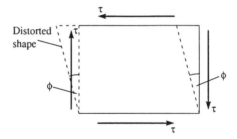

Fig. 7.6 Shear strain in an element

7.5 Shear strain

In Section 7.3 we established that shear loads applied to a structural member induce a system of shear and complementary shear stresses on any small rectangular element. The distortion in such an element due to these shear stresses does not involve a change in length but a change in shape as shown in Fig. 7.6. We define the *shear strain*, γ, in the element as the change in angle between two originally mutually perpendicular edges. Thus in Fig. 7.6

$$\gamma = \phi \text{ radians} \tag{7.5}$$

7.6 Volumetric strain due to hydrostatic pressure

A rather special case of strain which we shall find useful later occurs when a cube of material is subjected to equal compressive stresses, σ, on all six faces as shown in Fig. 7.7. This state of stress is that which would be experienced by the cube if it were immersed at some depth in a fluid, hence the term hydrostatic pressure. The analysis would, in fact, be equally valid if σ were a tensile stress.

Suppose that the original length of each side of the cube is L_0 and that δ is the decrease in length of each side due to the stress. Then, defining the *volumetric strain* as the change in volume per unit volume, we have

$$\text{volumetric strain} = \frac{L_0^3 - (L_0 - \delta)^3}{L_0^3}$$

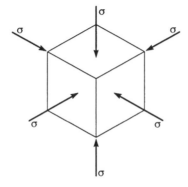

Fig. 7.7 Cube subjected to hydrostatic pressure

Expanding the bracketed term and neglecting second- and higher-order powers of δ gives

$$\text{volumetric strain} = 3L_0^2\delta/L_0^3$$

from which $$\text{volumetric strain} = \frac{3\delta}{L_0} \tag{7.6}$$

Thus we see that for this case the volumetric strain is three times the linear strain in any of the three stress directions.

7.7 Stress–strain relationships

Hooke's law and Young's modulus

The relationship between direct stress and strain for a particular material may be determined experimentally by a *tensile test* which is described in detail in Chapter 8. A tensile test consists basically of applying an axial tensile load in known increments to a specimen of material of a given length and cross-sectional area and measuring the corresponding increases in length. The stress produced by each value of load may be calculated from Eq. (7.1) and the corresponding strain from Eq. (7.4). A stress–strain curve is then drawn which, for some materials, would have a shape similar to that shown in Fig. 7.8. Stress–strain curves for other materials differ in detail but, generally, all have a linear portion such as ab in Fig. 7.8. In this region stress is directly proportional to strain, a relationship that was discovered in 1678 by Robert Hooke and which is known as *Hooke's law*. It may be expressed mathematically as

$$\sigma = E\varepsilon \tag{7.7}$$

where E is the constant of proportionality. E is known as *Young's modulus* or the *elastic modulus* of the material and has the same units as stress. For mild steel E is of the order of 200 kN/mm². Equation (7.7) may be written in alternative form as

$$\frac{\sigma}{\varepsilon} = E \tag{7.8}$$

For many materials E has the same value in tension and compression.

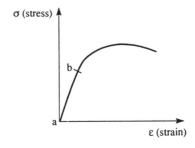

Fig. 7.8 Typical stress–strain curve

Shear modulus

By comparison with Eq. (7.8) we can define the *shear modulus* or *modulus of rigidity*, G, of a material as the ratio of shear stress to shear strain; thus

$$\frac{\tau}{\gamma} = G \tag{7.9}$$

Volume or bulk modulus

Again, the *volume modulus* or *bulk modulus*, K, of a material is defined in a similar manner as the ratio of volumetric stress to volumetric strain, i.e.

$$\frac{\text{volumetric stress}}{\text{volumetric strain}} = K \tag{7.10}$$

It is not usual to assign separate symbols to volumetric stress and strain since they may, respectively, be expressed in terms of direct stress and linear strain. Thus in the case of hydrostatic pressure (Section 7.6),

$$K = \frac{\sigma}{3\varepsilon} \tag{7.11}$$

Example 7.2 A mild steel column is hollow and circular in cross-section with an external diameter of 350 mm and an internal diameter of 300 mm. It carries a compressive axial load of 2000 kN. Determine the direct stress in the column and also the shortening of the column if its initial height is 5 m. Take $E = 200\ 000\ \text{N/mm}^2$.

The cross-sectional area A of the column is given by

$$A = \frac{\pi}{4}(350^2 - 300^2) = 25\ 525 \cdot 4\ \text{mm}^2$$

The direct stress σ in the column is, therefore, from Eq. (7.1)

$$\sigma = -\frac{2000 \times 10^3}{25\ 525 \cdot 4} = -78 \cdot 4\ \text{N/mm}^2 \text{ (compression)}$$

The corresponding strain is obtained from either Eq. (7.7) or Eq. (7.8) and is

$$\varepsilon = \frac{-78 \cdot 4}{200\ 000} = -0 \cdot 000\ 39$$

Finally the shortening, δ, of the column follows from Eq. (7.4), i.e.

$$\delta = 0 \cdot 000\ 39 \times 5 \times 10^3 = 1 \cdot 95\ \text{mm}$$

Example 7.3 A short, deep cantilever beam is 500 mm long by 200 mm deep and is 2 mm thick. It carries a vertically downward load of 10 kN at its free end. Assuming that the shear stress is uniformly distributed over the cross-section of the beam, calculate the deflection due to shear at the free end. Take $G = 25\ 000\ \text{N/mm}^2$.

The internal shear force is constant along the length of the beam and equal to 10 kN. Since the shear stress is uniform over the cross-section of the beam, we may use Eq. (7.2) to determine its value, i.e.

$$\tau_{av} = \frac{W}{A} = \frac{10 \times 10^3}{200 \times 2} = 25 \ N/mm^2$$

This shear stress is constant along the length of the beam; it follows from Eq. (7.9) that the shear strain is also constant along the length of the beam and is given by

$$\gamma = \frac{\tau_{av}}{G} = \frac{25}{25\,000} = 0.001 \ rad$$

This value is in fact the angle that the beam makes with the horizontal. The deflection, Δ_s, due to shear at the free end is therefore

$$\Delta_s = 0.001 \times 500 = 0.5 \ mm$$

In practice, the solution of this particular problem would be a great deal more complex than this since the shear stress distribution is not uniform. Deflections due to shear are investigated in Chapter 13.

7.8 Poisson effect

It is common experience that a material such as rubber suffers a reduction in cross-sectional area when stretched under a tensile load. This effect, known as the *Poisson effect*, also occurs in structural materials subjected to tensile and compressive loads, although in the latter case the cross-sectional area increases. In the region where the stress–strain curve of a material is linear, the ratio of lateral strain to longitudinal strain is a constant which is known as *Poisson's ratio* and is given the symbol v. The effect is illustrated in Fig. 7.9.

Consider now the action of different direct stress systems acting on an elemental cube of material (Fig. 7.10). The stresses are all tensile stresses and are given suffixes which designate their directions in relation to the system of axes specified in Section 3.2. In Fig. 7.10(a) the direct strain, ε_z, in the z direction is obtained directly from either Eq. (7.7) or Eq. (7.8) and is

$$\varepsilon_z = \frac{\sigma_z}{E}$$

Due to the Poisson effect there are accompanying strains in the x and y directions given by

$$\varepsilon_x = -v\varepsilon_z, \quad \varepsilon_y = -v\varepsilon_z$$

or, substituting for ε_z in terms of σ_z,

$$\varepsilon_z = -v\,\frac{\sigma_z}{E}, \qquad \varepsilon_y = -v\,\frac{\sigma_z}{E} \qquad\qquad (7.12)$$

These strains are negative since they are associated with contractions as opposed to positive strains produced by extensions.

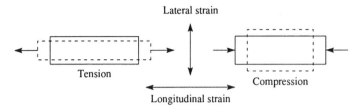

Fig. 7.9 The Poisson effect

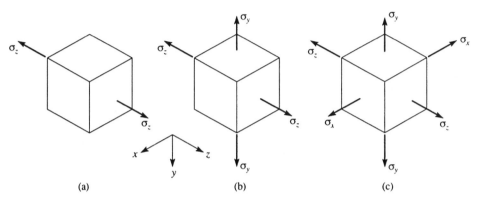

Fig. 7.10 The Poisson effect in a cube of material

In Fig. 7.10(b) the direct stress σ_y has an effect on the direct strain ε_z as does σ_z on ε_y. Thus

$$\varepsilon_z = \frac{\sigma_z}{E} - \frac{v\sigma_y}{E}, \qquad \varepsilon_y = \frac{\sigma_y}{E} - \frac{v\sigma_z}{E}, \qquad \varepsilon_x = -\frac{v\sigma_z}{E} - \frac{v\sigma_y}{E} \qquad (7.13)$$

By a similar argument, the strains in the z, y and x directions for the cube of Fig. 7.10(c) are

$$\varepsilon_z = \frac{\sigma_z}{E} - \frac{v\sigma_y}{E} - \frac{v\sigma_x}{E}, \qquad \varepsilon_y = \frac{\sigma_y}{E} - \frac{v\sigma_z}{E} - \frac{v\sigma_x}{E}, \qquad \varepsilon_x = \frac{\sigma_x}{E} - \frac{v\sigma_z}{E} - \frac{v\sigma_y}{E}$$

$$(7.14)$$

Let us now suppose that the cube of material in Fig. 7.10(c) is subjected to a uniform stress on each face such that $\sigma_z = \sigma_y = \sigma_x = \sigma$. The strain in each of the axial directions is therefore the same and is, from any one of Eqs (7.14)

$$\varepsilon = \frac{\sigma}{E}(1 - 2v)$$

In Section 7.6 we showed that the volumetric strain in a cube of material subjected to equal stresses on all faces is three times the linear strain. Thus in this case

$$\text{Volumetric strain} = \frac{3\sigma}{E}(1 - 2v) \qquad (7.15)$$

It would be unreasonable to suppose that the volume of a cube of material subjected to tensile stresses on all faces could decrease. It follows that Eq. (7.15) cannot have a negative value. We conclude, therefore, that v must always be less than 0·5. For most metals v has a value in the region of 0·3 while for concrete v can be as low as 0·1.

Collectively E, G, K and v are known as the *elastic constants* of a material.

7.9   Relationships between the elastic constants

There are different methods for determining the relationships between the elastic constants. The one presented here is relatively simple in approach and does not require a knowledge of topics other than those already covered.

In Fig. 7.11(a), ABCD is a square element of material of unit thickness and is in equilibrium under a shear and complementary shear stress system τ. Imagine now that the element is 'cut' along the diagonal AC as shown in Fig. 7.11(b). In order to maintain the equilibrium of the triangular portion ABC it is possible that a direct force and a shear force are required on the face AC. These forces, if they exist, will be distributed over the face of the element in the form of direct and shear stress systems, respectively. Since the element is small, these stresses may be assumed to be constant along the face AC. Let the direct stress on AC in the direction BD be σ_{BD} and the shear stress on AC be τ_{AC}. Then resolving forces on the element in the direction BD we have

$$\sigma_{BD} AC \times 1 = -\tau AB \times 1 \times \cos 45° - \tau BC \times 1 \times \cos 45°$$

Dividing through by AC

$$\sigma_{BD} = -\tau \frac{AB}{AC} \cos 45° - \tau \frac{BC}{AC} \cos 45°$$

or

$$\sigma_{BD} = -\tau \cos^2 45° - \tau \cos^2 45°$$

from which

$$\sigma_{BD} = -\tau \tag{7.16}$$

The negative sign indicates that σ_{BD} is a compressive stress. Similarly, resolving forces in the direction AC

$$\tau_{AC} AC \times 1 = \tau AB \times 1 \times \cos 45° - \tau BC \times 1 \times \cos 45°$$

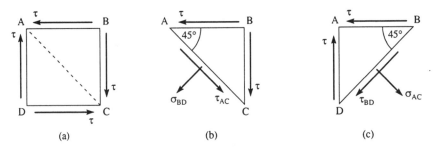

(a)   (b)   (c)

Fig. 7.11   Determination of the relationships between the elastic constants

Again dividing through by AC we obtain

$$\tau_{AC} = \tau \cos^2 45° - \tau \cos^2 45° = 0$$

A similar analysis of the triangular element ABD in Fig. 7.11(c) shows that

$$\sigma_{AC} = \tau \qquad (7.17)$$

and

$$\tau_{BD} = 0$$

Hence we see that on planes parallel to the diagonals of the element there are direct stresses σ_{BD} (compressive) and σ_{AC} (tensile) both numerically equal to τ as shown in Fig. 7.12. It follows from Section 7.8 that the direct strain in the direction AC is given by

$$\varepsilon_{AC} = \frac{\sigma_{AC}}{E} + \frac{\nu \sigma_{BD}}{E} = \frac{\tau}{E}(1 + \nu) \qquad (7.18)$$

Note that the compressive stress σ_{BD} makes a positive contribution to the strain ε_{AC}.

In Section 7.5 we defined shear strain and saw that under pure shear, only a change of shape is involved. Thus the element ABCD of Fig. 7.11(a) distorts into the shape A'B'CD shown in Fig. 7.13. The shear strain γ produced by the shear stress τ is then given by

$$\gamma = \phi \text{ radians} = \frac{B'B}{BC} \qquad (7.19)$$

since ϕ is a small angle. The increase in length of the diagonal AC to A'C is approximately equal to A'F where AF is perpendicular to A'C. Thus

$$\varepsilon_{AC} = \frac{A'C - AC}{AC} = \frac{A'F}{AC}$$

Again, since ϕ is a small angle, $A\hat{A}'F \approx 45°$ so that

$$A'F = A'A \cos 45°$$

Also

$$AC = BC/\cos 45°$$

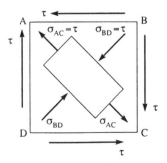

Fig. 7.12 Stresses on diagonal planes in element

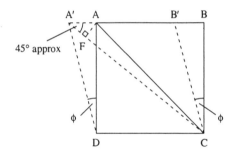

Fig. 7.13   Distortion due to shear in element

Hence $$\varepsilon_{AC} = \frac{A'A \cos^2 45°}{BC} = \frac{B'B \cos^2 45°}{BC} = \frac{1}{2}\frac{B'B}{BC}$$

Therefore, from Eq. (7.19)

$$\varepsilon_{AC} = \tfrac{1}{2}\gamma \qquad (7.20)$$

Substituting for ε_{AC} in Eq. (7.18) we obtain

$$\frac{1}{2}\gamma = \frac{\tau}{E}(1+v)$$

or, since $\tau/\gamma = G$ from Eq. (7.9),

$$G = \frac{E}{2(1+v)} \qquad \text{or} \quad E = 2G(1+v) \qquad (7.21)$$

The relationship between Young's modulus E and bulk modulus K is obtained directly from Eqs (7.10) and (7.15). Thus, from Eq. (7.10)

$$\text{Volumetric strain} = \frac{\sigma}{K}$$

where σ is the volumetric stress. Substituting in Eq. (7.15)

$$\frac{\sigma}{K} = \frac{3\sigma}{E}(1-2v)$$

from which $$K = \frac{E}{3(1-2v)} \qquad (7.22)$$

Eliminating E from Eqs (7.21) and (7.22) gives

$$K = \frac{2G(1+v)}{3(1-2v)} \qquad (7.23)$$

Example 7.4   A cube of material is subjected to a compressive stress σ on each of its faces. If $v = 0\cdot3$ and $E = 200\,000$ N/mm^2, calculate the value of this stress if the volume of the cube is reduced by 0·1%. Calculate also the percentage reduction in length of one of the sides.

From Eq. (7.22)

$$K = \frac{200\,000}{3(1 - 2 \times 0·3)} = 167\,000 \text{ N/mm}^2$$

The volumetric strain is 0·001 since the volume of the block is reduced by 0·1%. Therefore, from Eq. (7.10),

$$0·001 = \frac{\sigma}{K}$$

or
$$\sigma = 0·001 \times 167\,000 = 167 \text{ N/mm}^2$$

In Section 7.6 we established that the volumetric strain in a cube subjected to a uniform stress on all six faces is three times the linear strain. Thus in this case

$$\text{linear strain} = \tfrac{1}{3} \times 0·001 = 0·00033$$

The length of one side of the cube is therefore reduced by 0·033%.

7.10 Strain energy in simple tension or compression

An important concept in the analysis of structures is that of *strain energy*. The total strain energy of a structural member may comprise the separate strain energies due to axial load, bending moment, shear and torsion. In this section we shall concentrate on the strain energy due to tensile or compressive loads; the strain energy produced by each of the other loading systems is considered in the relevant, later chapters.

A structural member subjected to a gradually increasing tensile load P gradually increases in length (Fig. 7.14(a)). The load–extension curve for the member is linear until the limit of proportionality is exceeded, as shown in Fig. 7.14(b). The geometry of the non-linear portion of the curve depends upon the properties of the material of the member (see Chapter 8). Clearly the load P moves through small displacements Δ and therefore does work on the member. This work, which causes the member to extend, is stored in the member as strain energy. If the value of P is

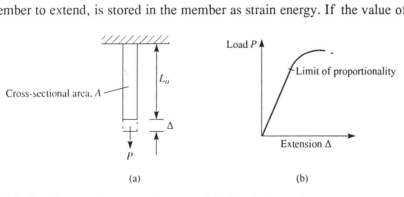

Fig. 7.14 Load–extension curve for an axially loaded member

restricted so that the limit of proportionality is not exceeded, the gradual removal of P results in the member returning to its original length and the strain energy stored in the member may be recovered in the form of work. When the limit of proportionality is exceeded, not all of the work done by P is recoverable; some is used in producing a permanent distortion of the member (see Chapter 8), the related energy appearing largely as heat.

Suppose the structural member of Fig. 7.14(a) is gradually loaded to some value of P within the limit of proportionality of the material of the member, the corresponding elongation being Δ. Let the elongation corresponding to some intermediate value of load, say P_1, be Δ_1 (Fig. 7.15). Then a small increase in load of δP_1 will produce a small increase, $\delta \Delta_1$, in elongation. The incremental work done in producing this increment in elongation may be taken as equal to the average load between P_1 and $P_1 + \delta P_1$ multiplied by $\delta \Delta_1$. Thus

$$\text{Incremental work done} = \left[\frac{P_1 + (P_1 + \delta P_1)}{2} \right] \delta \Delta_1$$

which, neglecting second-order terms, becomes

$$\text{Incremental work done} = P_1 \, \delta \Delta_1$$

The total work done on the member by the load P in producing the elongation Δ is therefore given by

$$\text{Total work done} = \int_0^\Delta P_1 \, d\Delta_1 \tag{7.24}$$

Since the load–extension relationship is linear, then

$$P_1 = K \Delta_1 \tag{7.25}$$

where K is some constant whose value depends upon the material properties of the member. Substituting the particular values of P and Δ in Eq. (7.25), we obtain

$$K = \frac{P}{\Delta}$$

whence Eq. (7.25) becomes

$$P_1 = \frac{P}{\Delta} \Delta_1$$

Now substituting for P_1 in Eq. (7.24) we have

$$\text{Total work done} = \int_0^\Delta \frac{P}{\Delta} \Delta_1 \, d\Delta_1$$

Integration of this equation yields

$$\text{Total work done} = \tfrac{1}{2} P \Delta \tag{7.26}$$

Alternatively, we see that the right-hand side of Eq. (7.24) represents the area under the load–extension curve, so that again we obtain

$$\text{Total work done} = \tfrac{1}{2} P \Delta$$

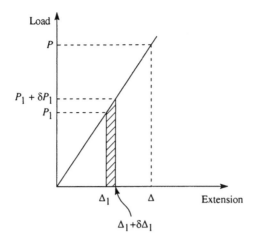

Fig. 7.15 Work done by a gradually applied load

By the law of conservation of energy, the total work done is equal to the strain energy, U, stored in the member. Thus

$$U = \tfrac{1}{2} P\Delta \tag{7.27}$$

The direct stress, σ, in the member of Fig. 7.14(a) corresponding to the load P is given by Eq. (7.1), i.e.

$$\sigma = \frac{P}{A}$$

Also the direct strain, ε, corresponding to the elongation Δ is, from Eq. (7.4),

$$\varepsilon = \frac{\Delta}{L_0}$$

Furthermore, since the load–extension curve is linear, the direct stress and strain are related by Eq. (7.7), so that

$$\frac{P}{A} = E\,\frac{\Delta}{L_0}$$

from which
$$\Delta = \frac{PL_0}{AE} \tag{7.28}$$

In Eq. (7.28) the quantity L_0/AE determines the magnitude of the displacement produced by a given load; it is therefore known as the *flexibility* of the member. Conversely, by transposing Eq. (7.28) we see that

$$P = \frac{AE}{L_0}\,\Delta$$

in which the quantity AE/L_0 determines the magnitude of the load required to produce a given displacement. Thus AE/L_0 is the *stiffness* of the member.

Substituting for Δ in Eq. (7.27) gives

$$U = \frac{P^2 L_0}{2AE} \qquad (7.29)$$

It is often convenient to express strain energy in terms of the direct stress σ. Thus, rewriting Eq. (7.29) in the form

$$U = \frac{1}{2} \frac{P^2}{A^2} \frac{AL_0}{E}$$

we obtain

$$U = \frac{\sigma^2}{2E} \times AL_0 \qquad (7.30)$$

in which we see that AL_0 is the volume of the member. The strain energy per unit volume of the member is then

$$\frac{\sigma^2}{2E}$$

The greatest amount of strain energy per unit volume that can be stored in a member without exceeding the limit of proportionality is known as the *modulus of resilience* and is attained when the direct stress in the member is equal to the direct stress corresponding to the elastic limit of the material of the member.

The strain energy, U, may also be expressed in terms of the elongation, Δ, or the direct strain, ε. Thus, substituting for P in Eq. (7.29)

$$U = \frac{EA\Delta^2}{2L_0} \qquad (7.31)$$

or, substituting for σ in Eq. (7.30)

$$U = \tfrac{1}{2} E\varepsilon^2 \times AL_0 \qquad (7.32)$$

The above expressions for strain energy also apply to structural members subjected to compressive loads since the work done by P in Fig. 7.14(a) is independent of the direction of movement of P. It follows that strain energy is always a positive quantity.

Example 7.5   A concrete column of height 3 m has a square cross-section of side 200 mm. It is designed to support an axial load of 100 kN. At mid-height a recess is cut in one face of the column to receive a floor beam (Fig. 7.16). Calculate the strain energy of the column produced by the axial load before and after the recess is cut. Take Young's modulus $E = 20\,000$ N/mm^2.

The strain energy, U_1, of the column before the recess is cut is obtained directly from Eq. (7.29). Thus

$$U_1 = \frac{(100 \times 10^3)^2 \times 3 \times 10^6}{2 \times 200^2 \times 20\,000 \times 10^6} = 18\text{·}75 \text{ N m}$$

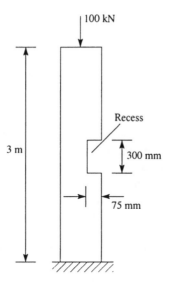

Fig. 7.16   Column of Ex. 7.5

After the recess has been cut, the stress in the reduced cross-section will be greater than that in the remainder of the column. The total strain energy, U_2, may then be found using Eq. (7.29) but will comprise two parts. Hence

$$U_2 = \frac{(100 \times 10^3)^2}{2 \times 20\,000 \times 10^6}\left(\frac{2{\cdot}7 \times 10^6}{200^2} + \frac{0{\cdot}3 \times 10^6}{200 \times 125}\right) = 19{\cdot}88 \text{ N m}$$

Alternatively we could calculate the direct stress in the different sections of the column and use Eq. (7.30). In the complete section

$$\sigma = \frac{100 \times 10^3}{200^2} = 2{\cdot}5 \text{ N/mm}^2$$

whereas in the recessed section

$$\sigma = \frac{100 \times 10^3}{200 \times 125} = 4{\cdot}0 \text{ N/mm}^2$$

Thus $U_2 = \dfrac{1}{2 \times 20\,000 \times 10^6}$

$$\times \left[(2{\cdot}5 \times 10^6)^2 \times \frac{200^2}{10^6} \times 2{\cdot}7 + (4{\cdot}0 \times 10^6)^2 \times \frac{200 \times 125}{10^6} \times 0{\cdot}3\right]$$

i.e. $U_2 = 19{\cdot}88 \text{ N m}$ as before.

A comparison of U_1 and U_2 shows that the strain energy of the column increases when the volume decreases. Hence we see from Eq. (7.30) that such a change could increase the value of stress (which depends upon the ratio of strain energy to volume) by a comparatively large amount. The ability to absorb energy is of primary importance in dynamic loading situations where the presence of a recess or cut-out can lead to high values of stress.

7.11   Impact loads on structural members

Possibly the most common controlled form of impact loading in civil engineering occurs when a pile is driven into the ground to form part of a foundation system. A given weight is allowed to fall through a predetermined height on to the head of the pile. Obviously at the instant of impact the stress generated in the pile is very much greater than that which would occur in the static case where the weight just rests on the head of the pile. The concept of strain energy may be used to determine this maximum stress and the accompanying deformation.

Suppose a weight, P, falls through a height, h, onto a column of original length, L_0, and causes a maximum deformation, δ_{max}, as shown in Fig. 7.17(a). We can obtain an approximate solution by neglecting energy losses during impact such as those producing deformation of the weight, noise and heat. We shall further assume that the maximum direct stress, σ_{max}, produced in the column is below the limit of proportionality so that no energy is dissipated in causing plastic deformations. Thus all the work done by the falling weight is transformed into strain energy of the column. Since the column is elastic, the stress and deflection in the column follow oscillations which decrease in amplitude with time from their maximum values to values corresponding to the static case as shown in Figs 7.17(b) and (c).

The weight, P, falls through a height $(h + \delta_{max})$ and therefore loses energy equal to $P(h + \delta_{max})$. This is converted into strain energy of the column which, in terms of σ_{max}, is given by Eq. (7.30). Thus

$$P(h + \delta_{max}) = \frac{\sigma_{max}^2}{2E} \times AL_0 \qquad (7.33)$$

The maximum strain, ε_{max}, in the column is related to σ_{max} by Eq. (7.7). Hence

$$\varepsilon_{max} = \frac{\sigma_{max}}{E}$$

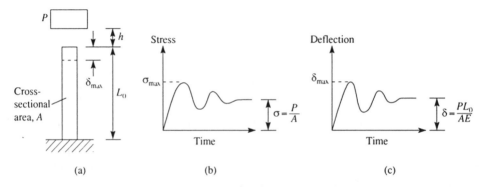

(a)   (b)   (c)

Fig. 7.17   Stress and deformation of a pile under impact loading

Using Eq. (7.4) we obtain

$$\delta_{max} = \varepsilon_{max} L_0 = \frac{\sigma_{max}}{E} L_0 \qquad (7.34)$$

Substituting for δ_{max} in Eq. (7.33) we have

$$P\left(h + \frac{\sigma_{max}}{E} L_0\right) = \frac{\sigma_{max}^2}{2E} \times AL_0$$

Rearranging we obtain a quadratic equation in σ_{max}:

$$\sigma_{max}^2 - \frac{2P}{A} \sigma_{max} - \frac{2PhE}{AL_0} = 0$$

the solution of which is

$$\sigma_{max} = \frac{P}{A}\left(1 + \sqrt{1 + \frac{2AEh}{PL_0}}\right) \qquad (7.36)$$

Note that in this particular case σ_{max} is a compressive stress. The negative root of Eq. (7.35) is discarded since σ_{max} clearly cannot be a tensile stress.

Having determined σ_{max}, the corresponding deflection, δ_{max}, may be found using Eq. (7.34). Alternatively we could have substituted for σ_{max} in Eq. (7.33) from Eq. (7.34) and obtained δ_{max} directly.

Suddenly applied loads

A special case of impact loading is derived from the previous case by equating h to zero; the load then becomes a *suddenly applied load*. The physical situation may be imagined by supposing that the weight P in Fig. 7.17(a) is in contact with the top of the column but is supported such that the pressure between the two touching surfaces is zero. If the weight is then released the whole of P is applied to the column instantaneously. Thus, when $h = 0$ in Eq. (7.36) we obtain

$$\sigma_{max} = 2\frac{P}{A} \qquad (7.37)$$

In Eq. (7.37) the quantity P/A represents the maximum stress the column would experience if the load were gradually applied and is in fact the final stress in the column when the oscillations produced by the dynamic effect of the suddenly applied load disappear. It follows, therefore, that a given load produces twice the maximum stress and hence twice the maximum strain and deformation if suddenly applied than if it were gradually applied.

Example 7.6   A hollow cylindrical steel column 3 m high has an outside diameter of 200 mm, walls 25 mm thick and has been designed assuming a failure stress of 270 N/mm². Immediately after erection, a weight of 10 kN falls through a height of

0·8 m on to the head of the column. Determine whether or not the column requires replacing and calculate the maximum deformation the column sustains. What is the maximum value of suddenly applied load that the column is able to withstand? Young's modulus $E = 200\,000$ N/mm^2.

The cross-sectional area A of the column is given by

$$A = \frac{\pi}{4}\,(200^2 - 150^2) = 13\,744 \cdot 5 \text{ mm}^2$$

From Eq. (7.36)

$$\sigma_{max} = \frac{10 \times 10^3}{13\,744 \cdot 5}\left(1 + \sqrt{1 + \frac{2 \times 13\,744 \cdot 5 \times 200\,000 \times 800}{10 \times 10^3 \times 3 \times 10^3}}\right)$$

i.e.

$$\sigma_{max} = 279 \cdot 3 \text{ N/mm}^2$$

Although this stress only just exceeds the design failure stress and it is unlikely that any obvious signs of failure would be apparent, the safest course would be to replace the column.

The maximum instantaneous shortening of the column is obtained directly from Eq. (7.34). Thus

$$\delta_{max} = \frac{279 \cdot 3 \times 3 \times 10^3}{200\,000} = 4 \cdot 2 \text{ mm}$$

Finally, Eq. (7.37) gives the maximum suddenly applied load the column could withstand, i.e.

$$P = \frac{270 \times 13\,744 \cdot 5}{2 \times 10^3} = 1855 \cdot 5 \text{ kN}$$

7.12 Deflections of axially loaded structural members

Equation (7.28) may be used to determine deflections of axially loaded structural members having a variety of geometrical and loading configurations. For example, the column shown in Fig. 7.18(a) could be part of a skeletal structure supporting floor beams at intermediate heights that produce axial loads P_1, P_2, P_3. The normal force diagram is constructed using the method of Section 3.3 and is shown in Fig. 7.18(b); the self-weight of the column has been neglected. Thus the deflection of the length, h_3, of the column is

$$\frac{P_3 h_3}{AE}$$

Similarly, the separate deflections of the lengths h_2 and h_1 are, respectively,

$$\frac{(P_2 + P_3)h_2}{AE} \quad \text{and} \quad \frac{(P_1 + P_2 + P_3)h_1}{AE}$$

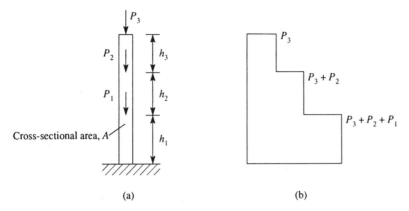

Fig. 7.18 Deflection of a column under axial loads

The total shortening of the column is then

$$\frac{1}{AE} [P_3 h_3 + (P_2 + P_3)h_2 + (P_1 + P_2 + P_3)h_1]$$

An alternative approach would be to use the principle of superposition (Section 3.7). The deflections at the top of the column due to P_1, P_2, and P_3 acting separately are, from Eq. (7.28),

$$\frac{P_1 h_1}{AE}, \qquad \frac{P_2(h_1 + h_2)}{AE} \qquad \text{and} \qquad \frac{P_3(h_1 + h_2 + h_3)}{AE}$$

The total deflection at the top of the column is then

$$\frac{1}{AE} [P_1 h_1 + P_2(h_1 + h_2) + P_3(h_1 + h_2 + h_3)]$$

which, on rearranging, becomes

$$\frac{1}{AE} [P_3 h_3 + (P_2 + P_3)h_2 + (P_1 + P_2 + P_3)h_1]$$

as before.

Changes in cross-section are also easily dealt with. Thus the deflection of the top of the column shown in Fig. 7.19(a) is

$$\frac{1}{E} \left[\frac{P_1 h_1}{A_1} + \frac{(P_1 + P_2)h_2}{A_2} \right]$$

where again the self-weight of the column is neglected.

Let us now consider the elongation, Δ, of the structural member shown in Fig. 7.20 due to its self-weight. Suppose that the density of the material of the member is ρ. The lower surface of the element, δz, supports the length, z, of the

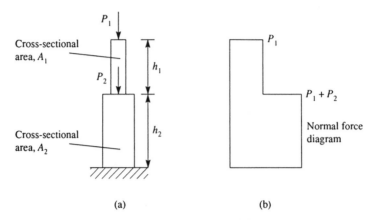

Fig. 7.19 Deflection of a column having a variable cross-section

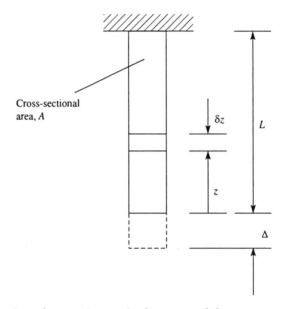

Fig. 7.20 Deflection of a member under its own weight

member. It is therefore subjected to a tensile force equal to the weight of the length z, i.e. $\rho A z$. Thus from Eq. (7.28) the elongation of δz, $\delta \Delta$, is

$$\delta \Delta = \frac{\rho A z \, \delta z}{AE}$$

It follows that the total elongation Δ is given by

$$\Delta = \int_0^L \frac{\rho A z}{AE} \, dz = \frac{\rho A}{AE} \left[\frac{z^2}{2} \right]_0^L$$

i.e.

$$\Delta = \frac{\rho A L^2}{2AE}$$

However $\rho AL = W$, the total weight of the member. Hence

$$\Delta = \frac{WL}{2AE} \qquad (7.38)$$

For a column, Eq. (7.38) would represent a shortening due to self-weight.

7.13 Deflection of a simple truss

The equality between the work done by an externally applied load and the total internal strain energy of the members of a structure may be used to determine particular deflections of simple structures.

In Fig. 7.21 a simple truss carries a gradually applied vertical load, W, at the joint A. A consideration of the equilibrium of the joint A shows that the axial forces P_{AB} and P_{AC} in the members AB and AC, respectively, are

$$P_{AB} = 1.41W \qquad \text{(tension)}$$

$$P_{AC} = W \qquad \text{(compression)}$$

The strain energy of each member is then, from Eq. (7.29)

$$U_{AB} = \frac{(1.41W)^2 \times 1.41L}{2AE} = \frac{1.41W^2L}{AE}$$

$$U_{AC} = \frac{W^2L}{2AE}$$

If the *vertical* deflection of A is Δ_v, the work done by the gradually applied load, W, is

$$\tfrac{1}{2}W\Delta_v$$

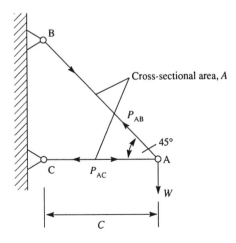

Fig. 7.21 Deflection of a simple truss

Thus equating work done to the total strain energy of the frame we have

$$\tfrac{1}{2}W\Delta_v = \frac{1.41W^2L}{AE} + \frac{W^2L}{2AE}$$

whence

$$\Delta_v = \frac{3.82WL}{AE}$$

The use of strain energy to determine deflections in this manner has limitations. In the above example Δ_v is, in fact, only the vertical component of the actual deflection of the joint A since A moves horizontally as well as vertically. Therefore we can only find the deflection of a load *in its own line of action* by this method. Furthermore, the method cannot be applied to structures subjected to more than one applied load as each load would contribute to the total work done by moving through an unknown displacement in its own line of action. There would, therefore, be as many unknown displacements as loads in the work-energy equation. We shall return to examine energy methods in much greater detail in Chapter 15.

7.14 Statically indeterminate systems

As we have seen, a statically indeterminate system is one in which support reactions or internal forces, or both, cannot be determined by applying just the force and moment equations of statical equilibrium. For example, the cantilever beam of Fig. 7.22(a) does not, theoretically, require the additional support at B to maintain its equilibrium. However, since the support is present, it will resist some of the applied load by providing a reactive force R_B. The support system now comprises three unknown reactions, R_A, R_B and M_A. It is only possible to obtain two equations of statical equilibrium, one of force and one of moment, so that the support system is statically indeterminate. Once these reactions have been determined, the internal force system in the cantilever is obtained from statics.

A different situation arises in the truss shown in Fig. 7.22(b). In this case the support reactions may be found by resolving forces and taking moments, but the forces in the members cannot be found since there are three unknown forces at each joint and only two possible equations of equilibrium. The internal forces therefore form a statically indeterminate system.

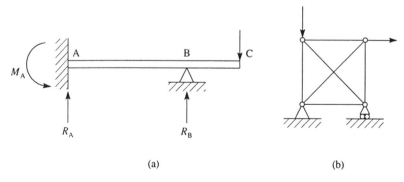

(a)   (b)

Fig. 7.22 Statically indeterminate structures

In any structural system the number of equations required for a solution is equal to the number of unknowns in the system. If the number of unknowns is greater than the possible number of equations of statical equilibrium, the structure is statically indeterminate and the excess unknowns are termed *redundancies*. Thus in Fig. 7.22(a) the support at B is a redundant support, whereas in Fig. 7.22(b) any one of the members may be regarded as the redundant member; both of these structures have a *degree of redundancy* equal to one.

Pin-jointed frame

The truss shown in Fig. 7.23 consists of three members of which only two are theoretically necessary to support a load at the joint A. Clearly, in the particular case where the load is vertical, members AB and AD could be dispensed with, but the remaining member AC would be incapable of supporting a horizontal load at A. A statically determinate structure is capable of supporting any system of loads although in the case of a two-dimensional structure, such as a plane truss, they must be applied in the plane of the structure.

The load W produces tensile forces P_{AB}, P_{AC} and P_{AD} in the members AB, AC and AD, respectively. Considering the vertical equilibrium of the joint A, we obtain

$$P_{AB} \cos \alpha + P_{AC} + P_{AD} \cos \alpha = W \tag{7.39}$$

Furthermore, from the horizontal equilibrium of joint A we have

$$P_{AB} \sin \alpha = P_{AD} \sin \alpha$$

or $$P_{AB} = P_{AD} \tag{7.40}$$

Note that Eq. (7.40) could have been obtained directly by considering the symmetry of the structure.

We now require a third equation to enable us to determine the three unknowns, P_{AB}, P_{AC} and P_{AD}. Consider the deflected shape of the truss as shown in Fig. 7.24. The joint A is displaced vertically downwards to A' causing the separate extensions, δ_{AC}, δ_{AB} and δ_{AD} in the three members. The latter two extensions are determined, to a first order of approximation, by constructing the perpendiculars AR and AQ to BA'

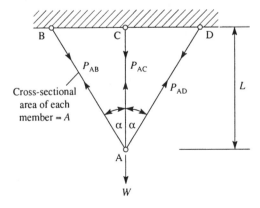

Fig. 7.23 Statically indeterminate truss

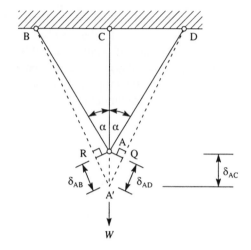

Fig. 7.24 Compatibility condition

and DA′, respectively. Clearly $\delta_{AB} = \delta_{AD}$. At all stages of the displacement the ends of the three members remain connected at A. Thus the end A of each member is displaced through the same vertical distance. We now use this *compatibility of displacement condition* to establish a relationship between these displacements and hence an additional relationship between the loads in the members.

In Fig. 7.24 the angles AA′R and AA′Q are equal to α to a first order of approximation since the displacements are small. Thus from triangle AA′R

$$\frac{\delta_{AB}}{\delta_{AC}} = \cos \alpha \tag{7.41}$$

From Eq. (7.28)

$$\delta_{AB} = \frac{P_{AB}(L/\cos \alpha)}{AE}, \qquad \delta_{AC} = \frac{P_{AC}L}{AE}$$

Substituting for δ_{AB} and δ_{AC} in Eq. (7.41) we obtain

$$P_{AB} \frac{L}{AE \cos \alpha} = P_{AC} \frac{L}{AE} \cos \alpha \tag{7.42}$$

Thus
$$P_{AB} = P_{AC} \cos^2 \alpha \tag{7.43}$$

Also, from Eq. (7.40)
$$P_{AD} = P_{AC} \cos^2 \alpha \tag{7.44}$$

We now substitute for P_{AB} and P_{AD} in Eq. (7.39) and obtain P_{AC},

i.e.
$$P_{AC} = \frac{W}{1 + 2 \cos^3 \alpha} \tag{7.45}$$

It follows from Eqs (7.43) and (7.44) that

$$P_{AB} = P_{AD} = \frac{W \cos^2 \alpha}{1 + 2 \cos^3 \alpha} \tag{7.46}$$

This method of analysis, which uses forces as the unknowns, is known as the *force* or *flexibility method*. The latter term is derived from the fact that the compatibility equation (7.42) contains the flexibilities, L/AE (see Eq. (7.28)), of the members AB and AC. An alternative method would express the unknown forces in terms of the unknown displacements (e.g. $P_{AC} = (AE/L)\delta_{AC}$) and solve for the displacements using Eqs (7.41), (7.40) and (7.39). This method is known as the *displacement* or *stiffness method*, the latter term being associated with member stiffness, AE/L.

Composite structural members

Axially loaded composite members are of direct interest in civil engineering where concrete columns are reinforced by steel bars and steel columns are frequently embedded in concrete as a fire precaution.

In Fig. 7.25 a concrete column of cross-sectional area A_C is reinforced by two steel bars having a combined cross-sectional area A_S. The modulus of elasticity of the concrete is E_C and that of the steel E_S. A load P is transmitted to the column through a plate which we shall assume is rigid so that the deflection of the concrete is equal to that of the steel. It follows that their respective strains are equal since both have the same original length. Since E_C is not equal to E_S we see from Eq. (7.7) that the compressive stresses, σ_C and σ_S, in the concrete and steel, respectively, must have different values. This also means that unless A_C and A_S have particular values, the compressive loads, P_C and P_S, in the concrete and steel are also different. The problem is therefore statically indeterminate since we can write down only one equilibrium equation, i.e.

$$P_C + P_S = P \tag{7.47}$$

However, equating displacements (the compatibility condition) we obtain, using Eq. (7.28)

$$\frac{P_C L}{A_C E_C} = \frac{P_S L}{A_S E_S} \tag{7.48}$$

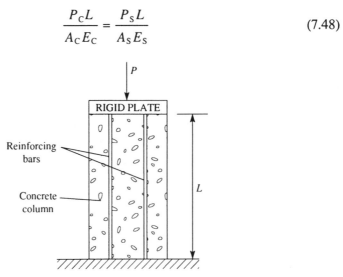

Fig. 7.25 Composite concrete column

Substituting for P_C from Eq. (7.48) in Eq. (7.47) gives

$$P_S\left(\frac{A_C E_C}{A_S E_S} + 1\right) = P$$

from which
$$P_S = \frac{A_S E_S}{A_C E_C + A_S E_S} P \qquad (7.49)$$

P_C follows directly from Eqs (7.48) and (7.49), i.e.

$$P_C = \frac{A_C E_C}{A_C E_C + A_S E_S} P \qquad (7.50)$$

The vertical displacement, δ, of the column is obtained using either side of Eq. (7.48) and the appropriate compressive load, P_C or P_S. Thus

$$\delta = \frac{PL}{A_C E_C + A_S E_S} \qquad (7.51)$$

Note that the above solution employs the flexibility method since the compatibility condition, Eq. (7.48), is written in terms of the flexibilities $L/A_C E_C$ and $L/A_S E_S$ of the concrete and steel, respectively.

The direct stresses in the steel and concrete are obtained from Eqs (7.49) and (7.50), thus

$$\sigma_S = \frac{E_S}{A_C E_C + A_S E_S} P, \qquad \sigma_C = \frac{E_C}{A_C E_C + A_S E_S} P \qquad (7.52)$$

We could, in fact, have solved directly for the stresses by writing Eqs (7.47) and (7.48) as

$$\sigma_C A_C + \sigma_S A_S = P \qquad (7.53)$$

and
$$\frac{\sigma_C L}{E_C} = \frac{\sigma_S L}{E_S} \qquad (7.54)$$

respectively.

Example 7.7 A reinforced concrete column, 5 m high, has the cross-section shown in Fig. 7.27. It is reinforced by four steel bars each 20 mm in diameter and carries a load of 1000 kN. If Young's modulus for steel is 200 000 N/mm^2 and that for concrete is 15 000 N/mm^2, calculate the stress in the steel and in the concrete and also the shortening of the column.

The total cross-sectional area, A_S, of the steel reinforcement is

$$A_S = 4 \times \frac{\pi}{4} \times 20^2 = 1257 \text{ mm}^2$$

The cross-sectional area, A_C, of the concrete is reduced due to the presence of the steel and is given by

$$A_C = 400^2 - 1257 = 158\ 743 \text{ mm}^2$$

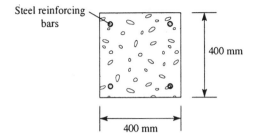

Steel reinforcing bars

400 mm

400 mm

Fig. 7.26 Reinforced concrete column of Ex. 7

Equations (7.52) then give

$$\sigma_S = \frac{200\,000 \times 1000 \times 10^3}{158\,743 \times 15\,000 + 1257 \times 200\,000} = 76{\cdot}0\ \text{N/mm}^2$$

$$\sigma_C = \frac{15\,000 \times 1000 \times 10^3}{158\,743 \times 15\,000 + 1257 \times 200\,000} = 5{\cdot}7\ \text{N/mm}^2$$

The deflection, δ, of the column is obtained using either side of Eq. (7.54). Thus

$$\delta = \frac{\sigma_C L}{E_C} = \frac{5{\cdot}7 \times 5 \times 10^3}{15\,000} = 1{\cdot}9\ \text{mm}$$

Thermal effects

It is possible for stresses to be induced by temperature changes in composite members which are additional to those produced by applied loads. These stresses arise when the components of a composite member have different rates of thermal expansion and contraction.

First, let us consider a member subjected to a uniform temperature rise, ΔT, along its length. The member expands from its original length, L_0, to a length, L_T, given by

$$L_T = L_0(1 + \alpha \Delta T)$$

where α is the coefficient of linear expansion of the material of the member. In the condition shown in Fig. 7.27 the member has been allowed to expand freely so that no stresses are induced. The increase in the length of the member is then

$$L_T - L_0 = L_0 \alpha \Delta T$$

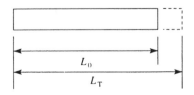

L_0

L_T

Fig. 7.27 Expansion due to temperature rise

Suppose now that expansion is completely prevented so that the final length of the member after the temperature rise is still L_0. The member has, in effect, been compressed by an amount $L_0\alpha\Delta T$, thereby producing a compressive strain, ε, which is given by (see Eq. (7.4))

$$\varepsilon = \frac{L_0\alpha\Delta T}{L_0} = \alpha\Delta T \qquad (7.55)$$

The corresponding compressive stress, σ, is from Eq. (7.7)

$$\sigma = E\alpha\Delta T \qquad (7.56)$$

In composite members the restriction on expansion or contraction is usually imposed by the attachment of one component to another. Thus in a reinforced concrete column the bond between the reinforcing steel and the concrete prevents the free expansion or contraction of either.

Consider the reinforced concrete column shown in Fig. 7.28(a) which is subjected to a temperature rise, ΔT. For simplicity we shall suppose that the reinforcement consists of a single steel bar of cross-sectional area, A_S, located along the axis of the column; the actual cross-sectional area of concrete is A_C. Young's modulus and the coefficient of linear expansion of the concrete are E_C and α_C, respectively, while the corresponding values for the steel are E_S and α_S. We shall assume that $\alpha_S > \alpha_C$.

Figure 7.28(b) shows the positions the concrete and steel would attain if they were allowed to expand freely; in this situation neither material is stressed. The displacements $L_0\alpha_C\Delta T$ and $L_0\alpha_S\Delta T$ are obtained directly from Eq. (7.55). However, since they are attached to each other, the concrete prevents the steel from expanding this full amount while the steel forces the concrete to expand further than it otherwise would; their final positions are shown in Fig. 7.28(c). It can be seen that δ_C is the effective elongation of the concrete which induces a direct tensile load, P_C. Similarly δ_S is the effective contraction of the steel which induces a compressive load, P_S. There is no externally applied load so that the resultant axial load at any section of the column is zero; thus

$$P_C(\text{tension}) = P_S(\text{compression}) \qquad (7.57)$$

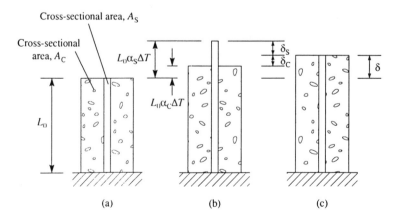

Fig. 7.28   Reinforced concrete column subjected to a temperature rise

Also, from Figs 7.28(b) and (c) we see that

$$\delta_C + \delta_S = L_0 \alpha_S \Delta T - L_0 \alpha_C \Delta T$$

or
$$\delta_C + \delta_S = L_0 \Delta T (\alpha_S - \alpha_C) \qquad (7.58)$$

From Eq. (7.28)

$$\delta_C = \frac{P_C L_0}{A_C E_C}, \qquad \delta_S = \frac{P_S L_0}{A_S E_S} \qquad (7.59)$$

Substituting for δ_C and δ_S in Eq. (7.58) we obtain

$$\frac{P_C}{A_C E_C} + \frac{P_S}{A_S E_S} = \Delta T (\alpha_S - \alpha_C) \qquad (7.60)$$

Simultaneous solution of Eqs (7.57) and (7.60) yields

$$P_C(\text{tension}) = P_S(\text{compression}) = \frac{\Delta T (\alpha_S - \alpha_C)}{\left(\dfrac{1}{A_C E_C} + \dfrac{1}{A_S E_S}\right)} \qquad (7.61)$$

or
$$P_C(\text{tension}) = P_S(\text{compression}) = \frac{\Delta T (\alpha_S - \alpha_C) A_C E_C A_S E_S}{A_C E_C + A_S E_S} \qquad (7.62)$$

The tensile stress, σ_C, in the concrete and the compressive stress, σ_S, in the steel follow directly from Eqs (7.62):

$$\sigma_C = \frac{P_C}{A_C} = \frac{\Delta T (\alpha_S - \alpha_C) E_C A_S E_S}{A_C E_C + A_S E_S},$$

$$\sigma_S = \frac{P_S}{A_S} = \frac{\Delta T (\alpha_S - \alpha_C) A_C E_C E_S}{A_C E_C + A_S E_S} \qquad (7.63)$$

From Figs 7.28(b) and (c) it can be seen that the actual elongation, δ, of the column is given by either

$$\delta = L_0 \alpha_C \Delta T + \delta_C \quad \text{or} \quad \delta = L_0 \alpha_S \Delta T - \delta_S \qquad (7.64)$$

Using the first of Eqs (7.64) and substituting for δ_C from Eqs (7.59) then P_C from Eqs (7.62) we have

$$\delta = L_0 \alpha_C \Delta T + \frac{\Delta T (\alpha_S - \alpha_C) A_C E_C A_S E_S L_0}{A_C E_C (A_C E_C + A_S E_S)}$$

which simplifies to

$$\delta = L_0 \Delta T \left(\frac{\alpha_C A_C E_C + \alpha_S A_S E_S}{A_C E_C + A_S E_S} \right) \qquad (7.65)$$

Clearly when $\alpha_C = \alpha_S = \alpha$, say, $P_C = P_S = 0$, $\sigma_C = \sigma_S = 0$ and $\delta = L_0 \alpha \Delta T$ as for unrestrained expansion.

The above analysis also applies to the case, $\alpha_C > \alpha_S$, when, as can be seen from Eqs (7.62) and (7.63) the signs of P_C, P_S, σ_C and σ_S are reversed. Thus the load and stress in the concrete become compressive, while those in the steel become tensile. A similar argument applies when ΔT specifies a temperature reduction.

Note that the flexibility method is again employed in this analysis and that Eq. (7.58) is an expression of the compatibility of displacement of the concrete and steel. Also note that the stresses could have been obtained directly by writing Eqs (7.57) and (7.58) as

$$\sigma_C A_C = \sigma_S A_S$$

and

$$\frac{\sigma_C L_0}{E_C} + \frac{\sigma_S L_0}{E_S} = L_0 \Delta T(\alpha_S - \alpha_C)$$

respectively.

Example 7.8   A rigid slab of weight 100 kN is supported on three columns each of height 4 m and cross-sectional area 300 mm^2 arranged in line. The two outer columns are fabricated from material having a Young's modulus of 80 000 N/mm^2 and a coefficient of linear expansion of $1.85 \times 10^{-5}/°C$; the corresponding values for the inner column are 200 000 N/mm^2 and $1.2 \times 10^{-5}/°C$. If the slab remains firmly attached to each column, determine the stress in each column and the displacement of the slab if the temperature is increased by 100°C.

The problem may be solved by determining separately the stresses and displacements produced by the applied load and the temperature rise; the two sets of results are then superimposed. Let subscripts o and i refer to the outer and inner columns, respectively. Using Eqs (7.52) we have

$$\sigma_i(\text{load}) = \frac{E_i}{A_o E_o + A_i E_i} P, \quad \sigma_o(\text{load}) = \frac{E_o}{A_o E_o + A_i E_i} P \qquad (i)$$

In Eqs (i)

$$A_o E_o + A_i E_i = 2 \times 300 \times 80\,000 + 300 \times 200\,000 = 108 \cdot 0 \times 10^6$$

Thus

$$\sigma_i(\text{load}) = \frac{200\,000 \times 100 \times 10^3}{108 \cdot 0 \times 10^6} = 185 \cdot 2 \text{ N/mm}^2 \text{ (compression)}$$

$$\sigma_o(\text{load}) = \frac{80\,000 \times 100 \times 10^3}{108 \cdot 0 \times 10^6} = 74 \cdot 1 \text{ N/mm}^2 \text{ (compression)}$$

Eqs (7.63) give the values of σ_i (temp.) and σ_o (temp.) produced by the temperature rise. Thus

$$\sigma_o(\text{temp.}) = \frac{\Delta T(\alpha_i - \alpha_o)E_o A_i E_i}{A_o E_o + A_i E_i},$$

$$\sigma_i(\text{temp.}) = \frac{\Delta T(\alpha_i - \alpha_o)A_o E_o E_i}{A_o E_o + A_i E_i}, \qquad (ii)$$

In Eqs (ii) $\alpha_o > \alpha_i$ so that σ_o (temp.) is a compressive stress while σ_i (temp.) is a tensile stress. Thus

$$\sigma_o(\text{temp.}) = \frac{100(1\cdot2 - 1\cdot85) \times 10^{-5} \times 80\,000 \times 300 \times 200\,000}{108\cdot0 \times 10^6}$$

$$= -28\cdot9 \text{ N/mm}^2 \text{ (i.e. compression)}$$

$$\sigma_i(\text{temp.}) = \frac{100(1\cdot2 - 1\cdot85) \times 10^{-5} \times 2 \times 300 \times 80\,000 \times 200\,000}{108\cdot0 \times 10^6}$$

$$= -57\cdot8 \text{ N/mm}^2 \text{ (i.e. tension)}$$

Superimposing the sets of stresses, we obtain the final values of stress, σ_i and σ_o, due to load and temperature change combined. Hence

$$\sigma_i = 185\cdot2 - 57\cdot8 = 127\cdot4 \text{ N/mm}^2 \qquad \text{(compression)}$$
$$\sigma_o = 74\cdot1 + 28\cdot9 = 103\cdot0 \text{ N/mm}^2 \qquad \text{(compression)}$$

The displacements due to the load and temperature change are found using Eqs (7.51) and (7.65), respectively. Hence

$$\delta(\text{load}) = \frac{100 \times 10^3 \times 4 \times 10^3}{108\cdot0 \times 10^6} = 3\cdot7 \text{ mm} \quad \text{(contraction)}$$

$$\delta(\text{temp.}) = 4 \times 10^3 \times 100$$
$$\times \left(\frac{1\cdot85 \times 10^{-5} \times 2 \times 300 \times 80\,000 + 1\cdot2 \times 10^{-5} \times 300 \times 200\,000}{108\cdot0 \times 10^6} \right)$$

$$= 6\cdot0 \text{ mm} \quad \text{(elongation)}$$

The final displacement of the slab involves an overall elongation of the columns of $6\cdot0 - 3\cdot7 = 2\cdot3$ mm.

Initial stresses and prestressing

The terms initial stress and prestressing refer to structural situations in which some or all of the components of a structure are in a state of stress *before* external loads are applied. In some cases, for example welded connections, this is an unavoidable by-product of fabrication and unless the whole connection is stress-relieved by suitable heat treatment the initial stresses are not known with any real accuracy. On the other hand, the initial stress in a component may be controlled as in a bolted connection; the subsequent applied load may or may not affect the initial stress in the bolt.

Initial stresses may be deliberately induced in a structural member so that the adverse effects of an applied load are minimized. In this category is the prestressing of beams fabricated from concrete which is particularly weak in tension. An overall state of compression is induced in the concrete so that tensile stresses due to applied loads merely reduce the level of compressive stress in the concrete rather than cause tension. Two methods of prestressing are employed, pre- and post-tensioning. In the former the prestressing tendons are positioned in the mould before the concrete is poured and loaded to the required level of tensile stress. After the concrete has set, the tendons are released and the tensile load in the tendons is transmitted, as a compressive load, to the concrete. In a post-tensioned beam, metal tubes or conduits are located in the mould at points where reinforcement is required, the concrete is poured and allowed to set. The reinforcing tendons are then passed through the conduits, tensioned and finally attached to end plates which transmit the tendon tensile load, as a compressive load, to the concrete.

Usually the reinforcement in a concrete beam supporting vertical shear loads is placed closer to either the upper or the lower surface since, as we shall see in Chapter 12, such a loading system induces tension in one part of the beam and compression in the other; clearly the reinforcement is placed in the tension zone. To demonstrate the basic principle, however, we shall investigate the case of a post-tensioned beam containing one axially loaded prestressing tendon.

Suppose that the initial load in the prestressing tendon in the concrete beam shown in Fig. 7.29 is F. In the absence of an applied load the resultant load at any section of the beam is zero so that the load in the concrete is also F but compressive. If now a tensile load, P, is applied to the beam, the tensile load in the prestressing tendon will increase by an amount ΔP_T while the compressive load in the concrete will decrease by an amount ΔP_C. From a consideration of equilibrium,

$$\Delta P_T + \Delta P_C = P \tag{7.66}$$

Furthermore, the total tensile load in the tendon is $F + \Delta P_T$ while the total compressive load in the concrete is $F - \Delta P_C$.

The tendon and concrete beam are interconnected through the end plates so that they both suffer the same elongation, δ, due to P. Thus, from Eq. (7.28)

$$\delta = \frac{\Delta P_T L}{A_T E_T} = \frac{\Delta P_C L}{A_C E_C} \tag{7.67}$$

where E_T and E_C are Young's modulus for the tendon and the concrete, respectively. From Eq. (7.67)

$$\Delta P_T = \frac{A_T E_T}{A_C E_C} \Delta P_C \tag{7.68}$$

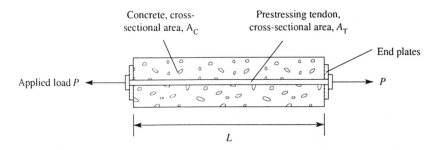

Concrete, cross-sectional area, A_C

Prestressing tendon, cross-sectional area, A_T

End plates

Applied load P

P

L

Fig. 7.29 Prestressed concrete beam

Substituting in Eq. (7.66) for ΔP_T we obtain

$$\Delta P_C \left(\frac{A_T E_T}{A_C E_C} + 1 \right) = P$$

whence
$$\Delta P_C = \frac{A_C E_C}{A_C E_C + A_T E_T} P \qquad (7.69)$$

Substituting now for ΔP_C in Eq. (7.68) from Eq. (7.69) gives

$$\Delta P_T = \frac{A_T E_T}{A_C E_C + A_T E_T} P \qquad (7.70)$$

The final loads, P_C and P_T, in the concrete and tendon, respectively, are then

$$P_C = F - \frac{A_C E_C}{A_C E_C + A_T E_T} P \quad \text{(compression)} \qquad (7.71)$$

and
$$P_T = F + \frac{A_T E_T}{A_C E_C + A_T E_T} P \quad \text{(tension)} \qquad (7.72)$$

The corresponding final stresses, σ_C and σ_T, follow directly and are given by

$$\sigma_C = \frac{P_C}{A_C} = \frac{1}{A_C} \left(F - \frac{A_C E_C}{A_C E_C + A_T E_T} P \right) \quad \text{(compression)} \qquad (7.73)$$

and
$$\sigma_T = \frac{P_T}{A_T} = \frac{1}{A_T} \left(F + \frac{A_T E_T}{A_C E_C + A_T E_T} P \right) \quad \text{(tension)} \qquad (7.74)$$

Obviously if the bracketed term in Eq. (7.73) is negative then σ_C will be a tensile stress.

Finally the elongation, δ, of the beam due to P is obtained from either of Eqs (7.67) and is

$$\delta = \frac{L}{A_C E_C + A_T E_T} P \qquad (7.75)$$

Example 7.9   A concrete beam of rectangular cross-section, 120 mm × 300 mm, is to be reinforced by six high-tensile steel prestressing tendons each having a cross-sectional area of 300 mm². If the level of prestress in the tendons is 150 N/mm², determine the corresponding compressive stress in the concrete. If the reinforced beam is subjected to an axial tensile load of 150 kN, determine the final stress in the steel and in the concrete assuming that the ratio of the elastic modulus of steel to that of concrete is 15.

The cross-sectional area, A_C, of the concrete in the beam is given by

$$A_C = 120 \times 300 - 6 \times 300 = 34\,200 \text{ mm}^2$$

The initial compressive load in the concrete is equal to the initial tensile load in the steel; thus

$$\sigma_{Ci} \times 34\,200 = 150 \times 6 \times 300 \tag{i}$$

where σ_{Ci} is the initial compressive stress in the concrete. Hence

$$\sigma_{Ci} = 7 \cdot 9 \text{ N/mm}^2$$

The final stress in the concrete and in the steel are given by Eqs (7.73) and (7.74), respectively. Hence, from Eq. (7.73)

$$\sigma_C = \frac{F}{A_C} - \frac{E_C}{A_C E_C + A_T E_T} P \tag{ii}$$

in which $F/A_C = \sigma_{Ci} = 7 \cdot 9 \text{ N/mm}^2$. Rearranging Eq. (ii) we have

$$\sigma_C = 7 \cdot 9 - \frac{1}{A_C + \left(\dfrac{E_T}{E_C}\right) A_T} P$$

or $\qquad \sigma_C = 7 \cdot 9 - \dfrac{150 \times 10^3}{34\,200 + 15 \times 6 \times 300} = 5 \cdot 4 \text{ N/mm}^2 \qquad$ (compression)

Similarly, from Eq. (7.74)

$$\sigma_T = 150 + \frac{1}{\left(\dfrac{E_C}{E_T}\right) A_C + A_T} P$$

whence $\qquad \sigma_T = 150 + \dfrac{150 \times 10^3}{\frac{1}{15} \times 34\,200 + 6 \times 300} = 186 \cdot 8 \text{ N/mm}^2 \quad$ (tension)

7.15   Thin-walled shells under internal pressure

So far we have been concerned with stress systems which involve either a single direct stress or a shear stress acting at a point in a structural member. However, as we shall see in Chapter 14, it is possible for combinations of direct and shear stresses to act simultaneously to form a *complex* stress system in two or three dimensions. As a preliminary example on combined stresses we shall investigate the direct stress system generated in the walls of a thin-walled shell which is subjected to internal pressure; the shell may be either cylindrical or spherical. Figure 7.30 shows a long, thin-walled cylindrical shell subjected to an internal pressure p. This internal pressure has a dual effect: it acts on the sealed ends of the shell, thereby inducing a *longitudinal* direct stress on cross-sections of the shell, and it also tends to separate one half of the shell from the other along a diametral plane, thus producing *circumferential* or *hoop* stresses.

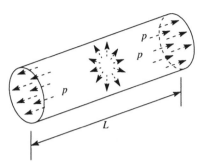

Fig. 7.30 Thin cylindrical shell under internal pressure

Suppose that d is the internal diameter of the shell and t the thickness of its walls. In Fig. 7.31 the axial load on each end of the shell due to the pressure p is

$$p \times \frac{\pi d^2}{4}$$

This load is equilibrated by an internal force corresponding to the longitudinal direct stress, σ_L, so that

$$\sigma_L \pi \, dt = p \, \frac{\pi d^2}{4}$$

which gives
$$\sigma_L = \frac{pd}{4t} \tag{7.76}$$

Now consider a unit length of the half shell formed by a diametral plane (Fig. 7.32). The force on the shell, produced by p, in the opposite direction to the circumferential stress, σ_C, is given by

$$p \times \text{projected area of the shell in the direction of } \sigma_C$$

Thus for equilibrium of the unit length of shell

$$2\sigma_C \times (1 \times t) = p \times (1 \times d)$$

whence
$$\sigma_C = \frac{pd}{2t} \tag{7.77}$$

We can now represent the state of stress at any point in the wall of the shell by considering the stress acting on the edges of a very small element of the shell wall

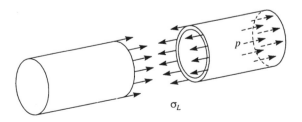

Fig. 7.31 Longitudinal stresses due to internal pressure

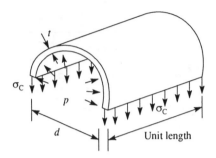

Fig. 7.32   Circumferential stress due to internal pressure

as shown in Fig. 7.33(a). The stresses comprise the longitudinal stress, σ_L, (Eq. (7.76)) and the circumferential stress, σ_C, (Eq. (7.77)). Since the element is very small, the effect of the curvature of the shell wall can be neglected so that the state of stress may be represented as a *two-dimensional* stress system acting on a plane element of thickness, t (Fig. 7.33(b)). We shall investigate this and other forms of complex stress system in Chapter 14.

   In addition to stresses, the internal pressure produces corresponding strains in the walls of the shell which lead to a change in volume. Consider the element of Fig. 7.33(b). The longitudinal strain, ε_L, is, from Eqs (7.13)

$$\varepsilon_L = \frac{\sigma_L}{E} - v\,\frac{\sigma_C}{E}$$

or, substituting for σ_L and σ_C from Eqs (7.76) and (7.77), respectively,

$$\varepsilon_L = \frac{pd}{2tE}\left(\frac{1}{2} - v\right) \tag{7.78}$$

Similarly, the circumferential strain, ε_C, is given by

$$\varepsilon_C = \frac{pd}{2tE}\left(1 - \frac{1}{2}v\right) \tag{7.79}$$

The increase in length of the shell is $\varepsilon_L L$ while the increase in circumference is $\varepsilon_C \pi d$. We see from the latter expression that the increase in circumference of the shell

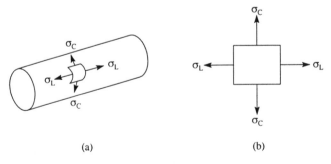

(a)   (b)

Fig. 7.33   Two-dimensional stress system

corresponds to an increase in diameter, $\varepsilon_C d$, so that the circumferential strain is equal to diametral strain (and also radial strain). The increase in volume, ΔV, of the shell is then given by

$$\Delta V = \frac{\pi}{4}(d + \varepsilon_C d)^2 (L + \varepsilon_L L) - \frac{\pi}{4} d^2 L$$

which, when second-order terms are neglected, simplifies to

$$\Delta V = \frac{\pi d^2 L}{4}(2\varepsilon_C + \varepsilon_L) \tag{7.80}$$

Substituting for ε_L and ε_C in Eq. (7.80) from Eqs (7.78) and (7.79) we obtain

$$\Delta V = \frac{\pi d^2 L}{4} \frac{pd}{tE}\left(\frac{5}{4} - \nu\right)$$

whence the volumetric strain is

$$\Delta V \Big/ \frac{\pi d^2 L}{4} = \frac{pd}{tE}\left(\frac{5}{4} - \nu\right) \tag{7.81}$$

The analysis of a spherical shell is somewhat simpler since only one direct stress is involved. It can be seen from Figs 7.34(a) and (b) that no matter which diametral plane is chosen, the tensile stress, σ, in the walls of the shell is constant. Thus for the equilibrium of the hemispherical portion shown in Fig. 7.34(b)

$$\sigma \times \pi dt = p \times \frac{\pi d^2}{4}$$

from which

$$\sigma = \frac{pd}{4t} \tag{7.82}$$

Again we have a two-dimensional state of stress acting on a small element of the shell wall (Fig. 7.34(c)) but in this case the direct stresses in the two directions are

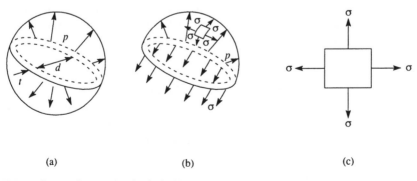

(a)   (b)   (c)

Fig. 7.34   Stress in a spherical shell

equal. Also the volumetric strain is determined in an identical manner to that for the cylindrical shell and is

$$\frac{3pd}{4tE}(1-v) \qquad\qquad (4.83)$$

Example 7.10   A thin-walled, cylindrical shell has an internal diameter of 2 m and is fabricated from plates 20 mm thick. Calculate the safe pressure in the shell if the tensile strength of the plates is 400 N/mm^2 and the factor of safety is 6. Determine also the percentage increase in the volume of the shell when it is subjected to this pressure. Take Young's modulus $E = 200\,000$ N/mm^2 and Poisson's ratio $v = 0\cdot3$.

The maximum tensile stress in the walls of the shell is the circumferential stress, σ_C, given by Eq. (7.77). Thus

$$\frac{400}{6} = \frac{p\times 2\times 10^3}{2\times 20}$$

from which $p = 1\cdot33$ N/mm^2

The volumetric strain is obtained from Eq. (7.81) and is

$$\frac{1\cdot33\times 2\times 10^3}{20\times 200\,000}\left(\frac{5}{4} - 0\cdot3\right) = 0\cdot00063$$

Hence the percentage increase in volume is 0·063%.

Problems

P.7.1   A column 3 m high has a hollow circular cross-section of external diameter 300 mm and carries an axial load of 5000 kN. If the stress in the column is limited to 150 N/mm^2 and the shortening of the column under load must not exceed 2 mm calculate the maximum allowable internal diameter. Take $E = 200\,000$ N/mm^2.

Ans.   205·6 mm.

P.7.2   A steel girder is firmly attached to a wall at each end so that changes in its length are prevented. If the girder is initially unstressed, calculate the stress induced in the girder when it is subjected to a uniform temperature rise of 30 K. The coefficient of linear expansion of the steel is 0·000 05/K and Young's modulus $E = 180\,000$ N/mm^2. (Note $L = L_0(1 + \alpha T)$.)

Ans.   270 N/mm^2 (compression).

P.7.3   A column 3 m high has a solid circular cross-section and carries an axial load of 10 000 kN. If the direct stress in the column is limited to 150 N/mm^2 determine the minimum allowable diameter. Calculate also the shortening of the column due to this load and the increase in its diameter. Take $E = 200\,000$ N/mm^2 and $v = 0\cdot3$.

Ans.   291·3 mm, 2·25 mm, 0·066 mm.

P.7.4 A structural member, 2 m long, is found to be 1·5 mm short when positioned in a framework. To enable the member to be fitted it is heated uniformly along its length. Determine the necessary temperature rise. Calculate also the residual stress in the member when it cools to its original temperature if movement of the ends of the member is prevented.

If the member has a rectangular cross-section, determine the percentage change in cross-sectional area when the member is fixed in position and at its original temperature.

Young's modulus $E = 200\ 000\ \text{N/mm}^2$, Poisson's ratio $v = 0·3$ and the coefficient of linear expansion of the material of the member is $0·000\ 012/\text{K}$.

Ans.   62·5 K, 150 N/mm² (tension), 0·045% (reduction).

P.7.5 A member of a framework is required to carry an axial tensile load of 100 kN. It is proposed that the member be comprised of two angle sections back to back in which one 18 mm diameter hole is allowed per angle for connections. If the allowable stress is 155 N/mm², suggest suitable angles.

Ans.   Required minimum area of cross-section = 645·2 mm². From steel tables, two equal angles 51 × 51 × 4·6 mm are satisfactory.

P.7.6 Two structural members, A and B, are of circular cross-section and of the same material; each has a length of 250 mm. Member A has a diameter of 25 mm for a length of 50 mm and a diameter of 20 mm for the remainder, while member B has a diameter of 25 mm for a length of 200 mm and a diameter of 20 mm for the remainder. If B receives an axial blow sufficient to produce a maximum stress of 200 N/mm², find the maximum stress produced by the same blow on A, assuming that the strain energy absorbed is the same in each case.

Ans.   175·2 N/mm².

P.7.7 A bar of circular cross-section, 2 m long, is securely held in a vertical position by its upper end. A freely sliding weight falls from a height of 30 mm on to a stop at the lower end of the bar and produces a stress of 150 N/mm². Determine the stress if the load had been applied gradually and also the maximum stress if the load had fallen from a height of 40 mm. Take $E = 200\ 000\ \text{N/mm}^2$.

Ans.   3·57 N/mm², 172·6 N/mm².

P.7.8 A column 3 m high has a hollow circular cross-section of external diameter 300 mm and carries an axial load of 5000 kN. If the stress in the column is limited to 150 N/mm² and the shortening of the column under load must not exceed 2 mm, calculate the maximum internal diameter. Calculate also the maximum shortening of the column if the load were suddenly applied. Take $E = 200\ 000\ \text{N/mm}^2$.

Ans.   206 mm, 4 mm.

P.7.9 A concrete pile 5 m long has a diameter of 200 mm and is to be driven into the ground using a weight of 2 kN. If the maximum instantaneous stress the concrete can withstand is 25 N/mm², calculate the height through which the weight should be dropped on to the head of the pile. Take $E = 15\ 000\ \text{N/mm}^2$.

Ans.   1·63 m.

P.7.10   A vertical hanger supporting the deck of a suspension bridge is formed from a steel cable 25 m long and having a diameter of 7·5 mm. If the density of the steel is 7850 kg/m³ and the load at the lower end of the hanger is 5 kN, determine the maximum stress in the cable and its elongation. Young's modulus $E = 200\ 000\ \text{N/mm}^2$.

Ans.   115·1 N/mm², 14·3 mm.

P.7.11   A concrete chimney 40 m high has a cross-sectional area (of concrete) of 0·15 m² and is stayed by three groups of four cables attached to the chimney at heights of 15 m, 25 m and 35 m. If each cable is anchored to the ground at a distance of 20 m from the base of the chimney and tensioned to a force of 15 kN, calculate the maximum stress in the chimney and the shortening of the chimney including the effect of its own weight. The density of concrete is 2500 kg/m³ and Young's modulus $E = 20\ 000\ \text{N/mm}^2$.

Ans.   1·9 N/mm², 2·2 mm.

P.7.12   A column of height h has a rectangular cross-section which tapers linearly in width from b_1 at the base of the column to b_2 at the top. The breadth of the cross-section is constant and equal to a. Determine the shortening of the column due to an axial load P.

Ans. $(Ph/[aE(b_1 - b_2)])\log_e(b_1/b_2)$.

P.7.13   Determine the vertical deflection of the 20 kN load in the truss shown in Fig. P.7.13. The cross-sectional area of the tension members is 100 mm² while that of the compression members is 200 mm². Young's modulus $E = 205\ 000\ \text{N/mm}^2$.

Ans.   4·5 mm.

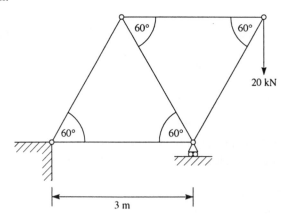

Fig. P.7.13

P.7.14   The truss shown in Fig. P.7.14 has members of cross-sectional area 1200 mm² and Young's modulus 205 000 N/mm². Determine the vertical deflection of the load.

Ans.   10·3 mm.

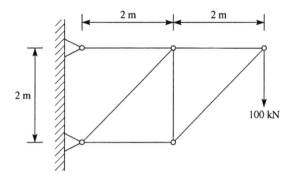

Fig. P.7.14

P.7.15 The members AD and CD of the pin-jointed framework shown in Fig. P.7.15 each has a cross-sectional area of 500 mm²; the member BD has a cross-sectional area of 250 mm². If the framework carries a vertical load of 100 kN at D, calculate the stress in each member and the vertical deflection of D. Take $E = 200\ 000$ N/mm².

Ans. $\sigma_{CD} = \sigma_{AD} = 83.4$ N/mm², $\sigma_{BD} = 111.2$ N/mm², 1.1 mm.

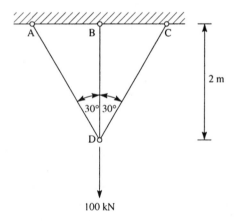

Fig. P.7.15

P.7.16 The pin-jointed framework shown in Fig. P.7.16 supports a vertical load W at the joint B. Determine the loads in the members.

Ans. $P_{BD} = 0$, $P_{BC} = W/\sqrt{2}$ (compression), $P_{BA} = W/\sqrt{2}$ (tension).

P.7.17 Three identical bars of length L are hung in a vertical position as shown in Fig. P.7.17. A rigid, weightless beam is attached to their lower ends and this in turn carries a load P. Calculate the load in each bar.

Ans. $P_1 = P/12$, $P_2 = P/3$, $P_3 = 7P/12$.

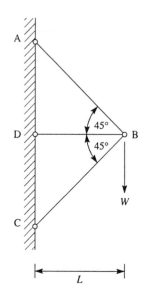

Fig. P.7.16

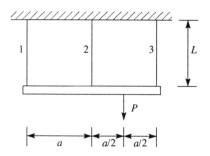

Fig. P.7.17

P.7.18   A composite column is formed by placing a steel bar, 20 mm in diameter and 200 mm long, inside an alloy cylinder of the same length whose internal and external diameters are 20 mm and 25 mm, respectively. The column is then subjected to an axial load of 50 kN. If E for steel is 200 000 N/mm² and E for the alloy is 70 000 N/mm², calculate the stress in the cylinder and in the bar, the shortening of the column and the strain energy stored in the column.

Ans.   45·8 N/mm² (cyl.), 131 N/mm² (bar), 0·13 mm, 3·3 N m.

P.7.19   A timber column, 3 m high, has a rectangular cross-section, 100 mm × 200 mm, and is reinforced over its complete length by two steel plates each 200 mm wide and 10 mm thick attached to its 200 mm wide faces. The column is designed to carry a load of 100 kN. If the failure stress of the timber is 55 N/mm² and that of the steel is 380 N/mm², check the design using a factor of safety of 3 for the timber and 2 for the steel. E (timber) = 15 000 N/mm², E (steel) = 200 000 N/mm².

Ans.   σ (timber) = 13·6 N/mm²   (allowable stress = 18·3 N/mm²),
σ (steel) = 181·8 N/mm²   (allowable stress = 190 N/mm²).

P.7.20 The composite bar shown in Fig. P.7.20 is initially unstressed. If the temperature of the bar is reduced by an amount T uniformly along its length, find an expression for the tensile stress induced. The coefficients of linear expansion of steel and aluminium are α_S and α_A per unit temperature change, respectively, while the corresponding values of Young's modulus are E_S and E_A.

Ans. $T(\alpha_S L_1 + \alpha_A L_2)/(L_1/E_S + L_2/E_A)$.

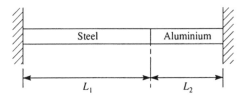

Fig. P.7.20

P.7.21 A short bar of copper, 25 mm in diameter, is enclosed centrally within a steel tube of external diameter 36 mm and thickness 3 mm. At 0°C the ends of the bar and tube are rigidly fastened together and the complete assembly heated to 80°C. Calculate the stress in the bar and in the tube if E for copper is 100 000 N/mm^2, E for steel is 200 000 N/mm^2 and the coefficients of linear expansion of copper and steel are 0·000 01/°C and 0·000 006/°C, respectively.

Ans. σ (steel) $= 28\cdot3$ N/mm^2 (tension),
σ (copper) $= 17\cdot9$ N/mm^2 (compression).

P.7.22 A structural member, 2 m long, is found to be 1·5 mm short when positioned in a framework. To enable the member to be fitted it is heated uniformly along its length. Determine the necessary temperature rise. Calculate also the residual stress in the member when it cools to its original temperature. Also, if the member has a rectangular cross-section, determine the percentage change in cross-sectional area when the member is fixed in position and at its original temperature. Take $E = 200\ 000$ N/mm^2, Poisson's ratio $\nu = 0\cdot3$ and $\alpha = 0\cdot000\ 012/°C$.

Ans. 62·5°C, 150 N/mm^2, 0·045%.

P.7.23 A bar of mild steel of diameter 75 mm is placed inside a hollow aluminium cylinder of internal diameter 75 mm and external diameter 100 mm; both bar and cylinder are the same length. The resulting composite bar is subjected to an axial compressive load of 10^6 N. If the bar and cylinder contract by the same amount, calculate the stress in each.

The temperature of the compressed composite bar is then reduced by 150°C but no change in length is permitted. Calculate the final stress in the bar and in the cylinder. Take E (steel) $= 200\ 000$ N/mm^2, E (aluminium) $= 80\ 000$ N/mm^2, α (steel) $= 0\cdot000\ 012/°C$, α (aluminium) $= 0\cdot000\ 005/°C$.

Ans.
Due to load: σ (steel) $= 172\cdot5$ N/mm^2 (compression),
σ (aluminium) $= 69\cdot0$ N/mm^2 (compression).
Final stress: σ (steel) $= 187\cdot5$ N/mm^2 (tension),
σ (aluminium) $= 9\cdot0$ N/mm^2 (compression).

P.7.24 Two structural members are connected together by a hinge which is formed as shown in Fig. P.7.24. The bolt is tightened up onto the sleeve through rigid end plates until the tensile force in the bolt is 10 kN. The distance between the head of the bolt and the nut is then 100 mm and the sleeve is 80 mm in length. If the diameter of the bolt is 15 mm and the internal and outside diameters of the sleeve are 20 mm and 30 mm, respectively, calculate the final stresses in the bolt and sleeve when an external tensile load of 5 kN is applied to the bolt.

Ans. σ (bolt) = 65·4 N/mm^2 (tension),

σ (sleeve) = 16·7 N/mm^2 (compression).

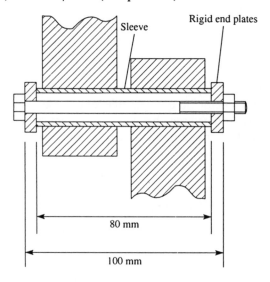

Fig. P.7.24

P.7.25 Calculate the minimum wall thickness of a cast iron water pipe having an internal diameter of 1 m under a head of 120 m. The limiting tensile strength of cast iron is 20 N/mm^2 and the density of water is 1000 kg/m^3.

Ans. 29·5 mm.

P.7.26 A thin-walled spherical shell is fabricated from steel plates and has to withstand an internal pressure of 0·75 N/mm^2. The internal diameter is 3 m and the joint efficiency 80%. Calculate the thickness of plates required using a working stress of 80 N/mm^2. (Note, effective thickness of plates = 0·8 × actual thickness).

Ans. 8·8 mm.

CHAPTER 8

Properties of Engineering Materials

It is now clear from the discussion in Chapter 7 that the structural designer requires a knowledge of the behaviour of materials under different types of load before he/she can be reasonably sure of designing a safe and, at the same time, economic structure.

One of the most important properties of a material is its strength, by which we mean the value of stress at which it fractures. Equally important in many instances, particularly in elastic design, is the stress at which yielding begins. In addition, the designer must have a knowledge of the stiffness of a material so that he/she can prevent excessive deflections occurring that could cause damage to adjacent structural members. Other factors that must be taken into consideration in design include the character of the different loads. It is common experience, for example, that a material such as cast iron fractures readily under a sharp blow whereas mild steel merely bends.

We shall, therefore, in this chapter examine some of the properties of engineering materials and the methods used to determine them. Initially, however, we shall discuss the more important materials used in civil engineering, with some reference to their different functions.

The basic and most widely used materials in civil engineering construction are steel, in its various forms, and concrete. Steel is fabricated into a variety of structural shapes for use as beams, columns, plates, connectors and to act as reinforcement in the comparatively weak tensile zones of concrete beams. Concrete itself is used in the construction of beams, columns, floor slabs and foundations and decoratively as wall cladding. Generally, as we have noted, structural concrete is reinforced by steel bars in its weak tensile zones and is sometimes used to encase steel columns as a precaution against fire damage. Instances of unreinforced structural concrete are few and are usually restricted to gravity structures such as dams and comparatively lightly loaded foundations.

In addition to steel and concrete, timber is employed extensively in civil engineering as formwork during the construction of concrete structures and in its own right as a structural material in light roof trusses and decorative beams. Frequently timber beams and arches are laminated to eliminate the less desirable characteristics of timber such as cracking, shrinkage and warping. Non-structurally, timber is found in floors, ceilings, wall panels, etc.

Of other materials in general use, masonry, ceramics and plastics are the most common. Masonry is used to support compressive loads as columns or walls and is also used to form in-fill panels in steel or concrete skeletal structures. Ceramics and plastics fulfil mainly non-structural roles and are frequently used decoratively as wall, floor or ceiling cladding.

8.1 Classification of engineering materials

Engineering materials may be grouped into two distinct categories, ductile materials and brittle materials, which exhibit very different properties under load. We shall define the properties of ductility and brittleness and also some additional properties which may depend upon the applied load or which are basic characteristics of the material.

Ductility

A material is said to be *ductile* if it is capable of withstanding large strains under load before fracture occurs. These large strains are accompanied by a visible change in cross-sectional dimensions and therefore give warning of impending failure. Materials in this category include mild steel, aluminium and some of its alloys, copper and polymers.

Brittleness

A brittle material exhibits little deformation before fracture, the strain normally being below 5%. Brittle materials therefore may fail suddenly without visible warning. Included in this group are concrete, cast iron, high-strength steels, timber and ceramics.

Elastic materials

A material is said to be *elastic* if deformations disappear completely on removal of the load. All known engineering materials are, in addition, *linearly elastic* within certain limits of stress so that strain, within these limits, is directly proportional to stress.

Plasticity

A material is perfectly *plastic* if no strain disappears after the removal of load. Ductile materials are *elastoplastic* and behave in an elastic manner until the *elastic limit* is reached after which they behave plastically. When the stress is relieved the elastic component of the strain is recovered but the plastic strain remains as a *permanent set*.

Isotropic materials

In many materials the elastic properties are the same in all directions at each point in the material although they may vary from point to point; such a material is known as *isotropic*. An isotropic material having the same properties at all points is known as *homogeneous*, for example mild steel.

Anisotropic materials

Materials having varying elastic properties in different directions are known as *anisotropic*.

Orthotropic materials

Although a structural material may possess different elastic properties in different directions, this variation may be limited, as in the case of timber which has just two values of Young's modulus, one in the direction of the grain and one perpendicular to the grain. A material whose elastic properties are limited to three different values in three mutually perpendicular directions is known as *orthotropic*.

8.2 Testing of engineering materials

The properties of engineering materials are determined mainly by the mechanical testing of specimens machined to prescribed sizes and shapes. The testing may be static or dynamic in nature depending on the particular property being investigated. Possibly the most common mechanical static tests are tensile and compressive tests which are carried out on a wide range of materials. Ferrous and non-ferrous metals are subjected to both forms of test, while compression tests are usually carried out on many non-metallic materials such as concrete, timber and brick which are normally used in compression. Other static tests include bending, shear and hardness tests, while the toughness of a material, in other words its ability to withstand shock loads, is determined by impact tests.

Tensile tests

Tensile tests are normally carried out on metallic materials and, in addition, timber. Test pieces are machined from a batch of material, their dimensions being specified by Codes of Practice. They are commonly circular in cross-section, although flat test pieces having rectangular cross-sections are used when the batch of material is in the form of a plate. A typical test piece would have the dimensions specified in Fig. 8.1. Usually the diameter of a central portion of the test piece is fractionally less than that of the remainder to ensure that the test piece fractures between the gauge points.

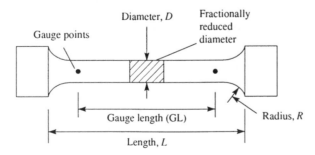

Fig. 8.1 Standard cylindrical test piece

Before the test begins, the mean diameter of the test piece is obtained by taking measurements at several sections using a micrometer screw gauge. Gauge points are punched at the required gauge length, the test piece is placed in the testing machine and a suitable strain measuring device, usually an extensometer, is attached to the test piece at the gauge points so that the extension is measured over the given gauge length. Increments of load are applied and the corresponding extensions recorded. This procedure continues until yield (see Section 8.3) occurs, when the extensometer is removed as a precaution against the damage which would be caused if the test piece fractured unexpectedly. Subsequent extensions are measured by dividers placed in the gauge points until, ultimately, the test piece fractures. The final gauge length and the diameter of the test piece in the region of the fracture are measured so that the percentage elongation and percentage reduction in area may be calculated. The two parameters give a measure of the ductility of the material.

A stress–strain curve is drawn (see Figs 8.8 and 8.12), the stress normally being calculated on the basis of the original cross-sectional area of the test piece, i.e. a *nominal stress* as opposed to an *actual stress* (which is based on the actual area of cross-section). For ductile materials there is a marked difference in the latter stages of the test as a considerable reduction in cross-sectional area occurs between yield and fracture. From the stress–strain curve the ultimate stress, the yield stress and Young's modulus, E, are obtained (see Section 7.7).

There are a number of variations on the basic tensile test described above. Some of these depend upon the amount of additional information required and some upon the choice of equipment. Thus there is a wide range of strain measuring devices to choose from, extending from different makes of mechanical extensometer, e.g. Huggenberger, Lindley, Cambridge, to the electrical resistance strain gauge. The last would normally be used on flat test pieces, one on each face to eliminate the effects of possible bending. At the same time a strain gauge could be attached in a direction perpendicular to the direction of loading so that lateral strains are measured. The ratio lateral strain/longitudinal strain is Poisson's ratio, v, (Section 7.8).

Testing machines are usually driven hydraulically. More sophisticated versions employ load cells to record load and automatically plot load against extension or stress against strain on a pen recorder as the test proceeds, an advantage when investigating the distinctive behaviour of mild steel at yield.

Compression tests

A compression test is similar in operation to a tensile test, with the obvious difference that the load transmitted to the test piece is compressive rather than tensile. This is achieved by placing the test piece between the platens of the testing machine and reversing the direction of loading. Test pieces are normally cylindrical and are limited in length to eliminate the possibility of failure being caused by instability (Chapter 18). Again contractions are measured over a given gauge length by a suitable strain measuring device.

Variations in test pieces occur when only the ultimate strength of the material in compression is required. For this purpose concrete test pieces may take the form of cubes having edges approximately 10 cm long, while mild steel test pieces are still cylindrical in section but are of the order of 1 cm long.

Bending tests

Many structural members are subjected primarily to bending moments. Bending tests are therefore carried out on simple beams constructed from the different materials to determine their behaviour under this type of load.

Two forms of loading are employed, the choice depending upon the type specified in Codes of Practice for the particular material. In the first a simply supported beam is subjected to a 'two-point' loading system as shown in Fig. 8.2(a). Two concentrated loads are applied symmetrically to the beam, producing zero shear force and constant bending moment in the central span of the beam (Figs 8.2(b) and (c)). The condition of pure bending is therefore achieved in the central span (see Section 9.1).

The second form of loading system consists of a single concentrated load at mid-span (Fig. 8.3(a)) which produces the shear force and bending moment diagrams shown in Figs 8.3(b) and (c).

The loads may be applied manually by hanging weights on the beam or by a testing machine. Deflections are measured by a dial gauge placed underneath the beam. From the recorded results a load–deflection diagram is plotted.

For most ductile materials the test beams continue to deform without failure and fracture does not occur. Thus plastic properties, for example the ultimate strength in bending, cannot be determined for such materials. In the case of brittle materials, including cast iron, timber and various plastics, failure does occur, so that plastic properties can be evaluated. For such materials the ultimate strength in bending is defined by the *modulus of rupture*. This is taken to be the maximum direct stress in

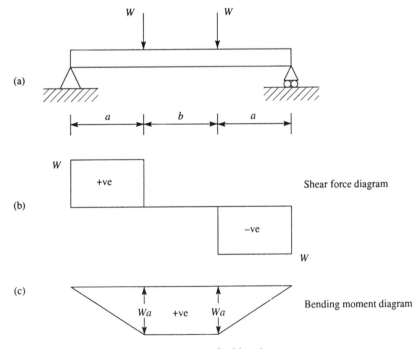

Fig. 8.2 Bending test on a beam, 'two-point' load

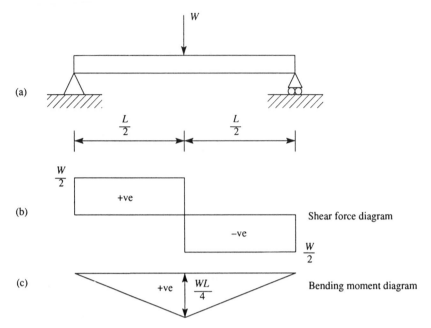

Fig. 8.3   Bending test on a beam, single load

bending, $\sigma_{z,u}$, corresponding to the ultimate moment M_u, and is assumed to be related to M_u by the elastic relationship

$$\sigma_{z,u} = \frac{M_u}{I}\, y_{max} \quad \text{(see Eq. 9.9)}$$

Other bending tests are designed to measure the ductility of a material and involve the bending of a bar round a pin. The angle of bending at which the bar starts to crack is then taken as an indication of its ductility.

Shear tests

Two main types of shear test are used to determine the shear properties of materials. One type investigates the direct or transverse shear strength of a material and is used in connection with the shear strength of bolts, rivets and beams. A typical arrangement is shown diagrammatically in Fig. 8.4 where the test piece is clamped to a block and the load applied through the shear tool until failure occurs. In the arrangement shown the test piece is subjected to double shear, whereas if it extended only partially across the gap in the block it would be subjected to single shear. In either case the average shear strength is taken as the maximum load divided by the shear resisting area.

The other type of shear test is used to evaluate the basic shear properties of a material such as the shear modulus, G (Eq. (7.9)), the shear stress at yield and the ultimate shear stress. In the usual form of test a solid circular-section test piece is placed in a torsion machine and twisted by controlled increments of torque. The

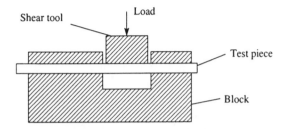

Fig. 8.4 Shear test

corresponding angles of twist are recorded and torque–twist diagrams plotted from which the shear properties of the material are obtained. The method is similar to that used to determine the tensile properties of a material from a tensile test and uses relationships derived in Chapter 11.

Hardness tests

The machinability of a material and its resistance to scratching or penetration are determined by its 'hardness'. There also appears to be a connection between the hardness of some materials and their tensile strength so that hardness tests may be used to determine the properties of a finished structural member where tensile and other tests would be impracticable. Hardness tests are also used to investigate the effects of heat treatment, hardening and tempering and of cold forming. Two types of hardness test are in common use: *indentation tests* and *scratch and abrasion tests*.

Indentation tests may be subdivided into two classes: static and dynamic. Of the static tests the *Brinell* is the most common. In this a hardened steel ball is pressed into the material under test by a static load acting for a fixed period of time. The load in kg divided by the spherical area of the indentation in mm² is called the *Brinell Hardness Number* (BHN). Thus in Fig. 8.5, if D is the diameter of the ball, F the load in kg, h the depth of the indentation, and d the diameter of the indentation, then

$$\text{BHN} = \frac{F}{\pi Dh} = \frac{2F}{\pi D[D - \sqrt{D^2 - d^2}]}$$

In practice the hardness number of a given material is found to vary with F and D so that for uniformity the test is standardized. For steel and hard materials $F = 3000$ kg

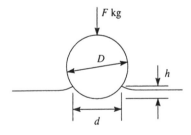

Fig. 8.5 Brinell hardness test

and $D = 10$ mm while for soft materials $F = 500$ kg and $D = 10$ mm; in addition the load is usually applied for 15 s.

In the Brinell test the dimensions of the indentation are measured by means of a microscope. To avoid this rather tedious procedure, direct reading machines have been devised of which the *Rockwell* is typical. The indenting tool, again a hardened sphere, is first applied under a definite light load. This indenting tool is then replaced by a diamond cone with a rounded point which is then applied under a specified indentation load. The difference between the depth of the indentation under the two loads is taken as a measure of the hardness of the material and is read directly from the scale.

A typical dynamic hardness test is performed by the *Shore Scleroscope* which consists of a small hammer approximately 20 mm long and 6 mm in diameter fitted with a blunt, rounded, diamond point. The hammer is guided by a vertical glass tube and allowed to fall freely from a height of 25 cm onto the specimen, which it indents before rebounding. A certain proportion of the energy of the hammer is expended in forming the indentation so that the height of the rebound, which depends upon the energy still possessed by the hammer, is taken as a measure of the hardness of the material.

A number of tests have been devised to measure the 'scratch hardness' of materials. In one test, the smallest load in grams which, when applied to a diamond point, produces a scratch visible to the naked eye on a polished specimen of material is called its hardness number. In other tests the magnitude of the load required to produce a definite width of scratch is taken as the measure of hardness. Abrasion tests, involving the shaking over a period of time of several specimens placed in a container, measure the resistance to wear of some materials. In some cases there appears to be a connection between wear and hardness number although the results show no level of consistency.

Impact tests

It has been found that certain materials, particularly heat-treated steels, are susceptible to failure under shock loading whereas an ordinary tensile test on the same material would show no abnormality. Impact tests measure the ability of

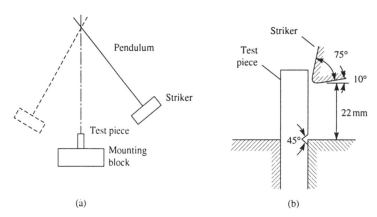

(a)   (b)

Fig. 8.6   Izod impact test

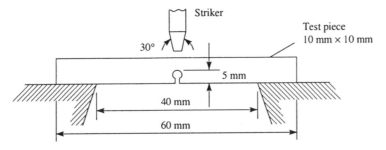

Fig. 8.7 Charpy impact test

materials to withstand shock loads and provide an indication of their *toughness*. Two main tests are in use, the *Izod* and the *Charpy*.

Both tests rely on a striker or weight attached to a pendulum. The pendulum is released from a fixed height, the weight strikes a notched test piece and the angle through which the pendulum then swings is a measure of the toughness of the material. The arrangement for the Izod test is shown diagrammatically in Fig. 8.6(a). The specimen and the method of mounting are shown in detail in Fig. 8.6(b). The Charpy test is similar in operation except that the test piece is supported in a different manner as shown in the plan view in Fig. 8.7.

8.3 Stress–strain curves

We shall now examine in detail the properties of the different materials used in civil engineering construction from the viewpoint of the results obtained from tensile and compression tests.

Low carbon steel (mild steel)

A nominal stress–strain curve for mild steel, a ductile material, is shown in Fig. 8.8. From 0 to 'a' the stress–strain curve is linear, the material in this range obeying

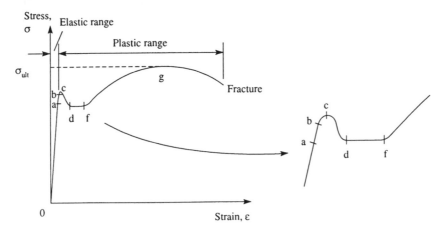

Fig. 8.8 Stress–strain curve for mild steel

Hooke's law. Beyond 'a', the *limit of proportionality*, stress is no longer proportional to strain and the stress–strain curve continues to 'b', the *elastic limit*, which is defined as the maximum stress that can be applied to a material without producing a permanent plastic deformation or *permanent set* when the load is removed. In other words, if the material is stressed beyond 'b' and the load then removed, a residual strain exists at zero load. For many materials it is impossible to detect a difference between the limit of proportionality and the elastic limit. From 0 to 'b' the material is said to be in the *elastic range* while from 'b' to fracture the material is in the *plastic range*. The transition from the elastic to the plastic range may be explained by considering the arrangement of crystals in the material. As the load is applied, slipping occurs between the crystals which are aligned most closely to the direction of load. As the load is increased, more and more crystals slip with each equal load increment until appreciable strain increments are produced and the plastic range is reached.

A further increase in stress from 'b' results in the mild steel reaching its *upper yield point* at 'c' followed by a rapid fall in stress to its *lower yield point* at 'd'. The existence of a lower yield point for mild steel is a peculiarity of the tensile test wherein the movement of the ends of the test piece produced by the testing machine does not proceed as rapidly as its plastic deformation; the load therefore decreases, as does the stress. From 'd' to 'f' the strain increases at a roughly constant value of stress until *strain hardening* (see Section 8.4) again causes an increase in stress. This increase in stress continues, accompanied by a large increase in strain to 'g', the *ultimate stress*, σ_{ult}, of the material. At this point the test piece begins, visibly, to 'neck' as shown in Fig. 8.9. The material in the test piece in the region of the 'neck' is almost perfectly plastic at this stage and from thence, onwards to fracture, there is a reduction in nominal stress.

For mild steel, yielding occurs at a stress of the order of $300 \ \text{N/mm}^2$. At fracture the strain (i.e. the elongation) is of the order of 30%. The gradient of the linear portion of the stress–strain curve gives a value for Young's modulus in the region of $200 \ 000 \ \text{N/mm}^2$.

The characteristics of the fracture are worthy of examination. In a cylindrical test piece the two halves of the fractured test piece have ends which form a 'cup and cone' (Fig. 8.10). The actual failure planes in this case are inclined at approximately 45° to the axis of loading and coincide with planes of maximum shear stress

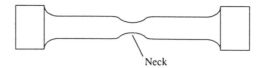

Neck

Fig. 8.9 'Necking' of a test piece in the plastic range

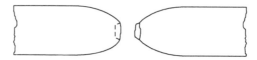

Fig. 8.10 'Cup-and-cone' failure of a mild steel test piece

(Section 14.2). Similarly, if a flat tensile specimen of mild steel is polished and then stressed, a pattern of fine lines appears on the polished surface at yield. These lines, which were first discovered by Lüder in 1854, intersect approximately at right angles and are inclined at 45° to the axis of the specimen, thereby coinciding with planes of maximum shear stress. These forms of yielding and fracture suggest that the crystalline structure of the steel is relatively weak in shear with yielding taking the form of the sliding of one crystal plane over another rather than the tearing apart of two crystal planes.

The behaviour of mild steel in compression is very similar to its behaviour in tension, particularly in the elastic range. In the plastic range it is not possible to obtain ultimate and fracture loads since, due to compression, the area of cross-section increases as the load increases producing a 'barrelling' effect as shown in Fig. 8.11. This increase in cross-sectional area tends to decrease the true stress, thereby increasing the load resistance. Ultimately a flat disc is produced. For design purposes the ultimate stresses of mild steel in tension and compression are assumed to be the same.

Aluminium

Aluminium and some of its alloys are also ductile materials, although their stress–strain curves do not have the distinct yield stress of mild steel. A typical stress–strain curve is shown in Fig. 8.12. The points 'a' and 'b' again mark the limit

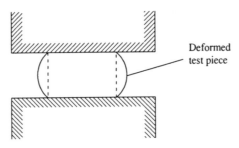

Fig. 8.11   'Barrelling' of a mild steel test piece in compression

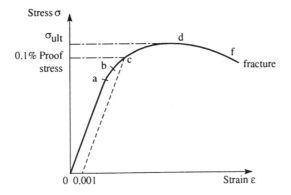

Fig. 8.12   Stress–strain curve for aluminium

of proportionality and elastic limit, respectively, but are difficult to determine experimentally. Instead a *proof stress* is defined which is the stress required to produce a given permanent strain on removal of the load. Thus, in Fig. 8.12, a line drawn parallel to the linear portion of the stress–strain curve from a strain of $0 \cdot 001$ (i.e. a strain of $0 \cdot 1\%$) intersects the stress–strain curve at the $0 \cdot 1\%$ proof stress. For elastic design this, or the $0 \cdot 2\%$ proof stress, is taken as the working stress.

Beyond the limit of proportionality the material extends plastically, reaching its ultimate stress, σ_{ult}, at 'd' before finally fracturing under a reduced nominal stress at 'f'.

A feature of the fracture of aluminium alloy test pieces is the formation of a 'double cup' as shown in Fig. 8.13, implying that failure was initiated in the central portion of the test piece while the outer surfaces remained intact. Again considerable 'necking' occurs.

In compression tests on aluminium and its ductile alloys similar difficulties are encountered to those experienced with mild steel. The stress–strain curve is very similar in the elastic range to that obtained in a tensile test but the ultimate strength in compression cannot be determined; in design its value is assumed to coincide with that in tension.

Brittle materials

These include cast iron, high-strength steel, concrete, timber, ceramics, glass, etc. The plastic range for brittle materials extends to only small values of strain. A typical stress–strain curve for a brittle material under tension is shown in Fig. 8.14. Little or no yielding occurs and fracture takes place very shortly after the elastic limit is reached.

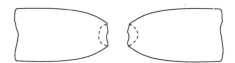

Fig. 8.13 'Double-cup' failure of an aluminium alloy test piece

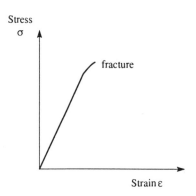

Fig. 8.14 Stress–strain curve for a brittle material

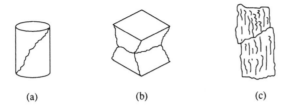

(a) (b) (c)

Fig. 8.15 Failure of brittle materials

The fracture of a cylindrical test piece takes the form of a single failure plane approximately perpendicular to the direction of loading with no visible 'necking' and an elongation of the order of 2–3%.

In compression the stress–strain curve for a brittle material is very similar to that in tension except that failure occurs at a much higher value of stress; for concrete the ratio is of the order of 10:1. This is thought to be due to the presence of microscopic cracks in the material, giving rise to high stress concentrations which are more likely to have a greater effect in reducing tensile strength than compressive strength.

The form of the fracture of brittle materials under compression is clear and visible. A cast-iron cylinder, for example, cracks on a diagonal plane as shown in Fig. 8.15(a) while failure of a concrete cube is shown in Fig. 8.15(b) where failure planes intersect at approximately 45° along each vertical face. Fig. 8.15(c) shows a typical failure of a rectangular block of timber in compression. Failure in all these cases is due primarily to a breakdown in shear on planes inclined to the direction of compression.

All the stress–strain curves described in the preceding discussion are those produced in tensile or compression tests in which the strain is applied at a negligible rate. A rapid strain application would result in significant changes in the apparent properties of the materials giving possible variations in yield stress of up to 100%.

8.4 Strain hardening

The stress–strain curve for a material is influenced by the *strain history*, or the loading and unloading of the material, within the plastic range. Thus in Fig. 8.16 a

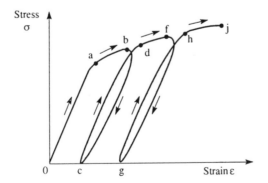

Fig. 8.16 Strain hardening of a material

test piece is initially stressed in tension beyond the yield stress at, 'a', to a value at 'b'. The material is then unloaded to 'c' and reloaded to 'f' producing an increase in yield stress from the value at 'a' to the value at 'd'. Subsequent unloading to 'g' and loading to 'j' increases the yield stress still further to the value at 'h'. This increase in strength resulting from the loading and unloading is known as *strain hardening*. It can be seen from Fig. 8.16 that the stress–strain curve during the unloading and loading cycles forms loops, the shaded areas in Fig. 8.16. These indicate that strain energy is lost during the cycle, the energy being dissipated in the form of heat produced by internal friction. This energy loss is known as *mechanical hysteresis* and the loops as *hysteresis loops*. Although the ultimate stress is increased by strain hardening it is not influenced to the same extent as yield stress. The increase in strength produced by strain hardening is accompanied by decreases in toughness and ductility.

8.5 Creep and relaxation

We have seen in Chapter 7 that a given load produces a calculable value of stress in a structural member and hence a corresponding value of strain once the full value of the load is transferred to the member. However, after this initial or 'instantaneous' stress and its corresponding value of strain have been attained, a great number of structural materials continue to deform slowly and progressively under load over a period of time. This behaviour is known as *creep*. A typical creep curve is shown in Fig. 8.17.

Some materials such as plastics and rubber exhibit creep at room temperatures but most structural materials require high temperatures or long-duration loading at moderate temperatures. In some 'soft' metals, such as zinc and lead, creep occurs over a relatively short period of time, whereas materials such as concrete may be subject to creep over a period of years. Creep occurs in steel to a slight extent at normal temperatures but becomes very important at temperatures above 316°C.

Closely related to creep is *relaxation*. Whereas creep involves an increase in strain under constant stress, relaxation is the decrease in stress experienced over a period of time by a material subjected to a constant strain.

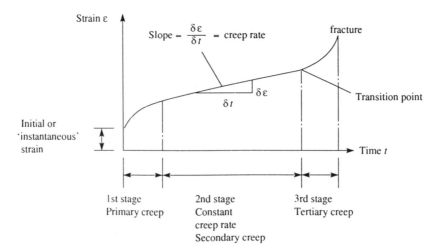

Fig. 8.17 Typical creep curve

8.6 Fatigue

Structural members are frequently subjected to repetitive loading over a long period of time. Thus, for example, the members of a bridge structure suffer variations in loading possibly thousands of times a day as traffic moves over the bridge. In these circumstances a structural member may fracture at a level of stress substantially below the ultimate stress for non-repetitive static loads; this phenomenon is known as *fatigue*.

Fatigue cracks are most frequently initiated at sections in a structural member where changes in geometry, for example holes, notches or sudden changes in section, cause *stress concentrations*. Designers seek to eliminate such areas by ensuring that rapid changes in section are as smooth as possible. Thus at re-entrant corners, fillets are provided as shown in Fig. 8.18.

Other factors which affect the failure of a material under repetitive loading are the type of loading (fatigue is primarily a problem with repeated tensile stresses due, probably, to the fact that microscopic cracks can propagate more easily under tension), temperature, the material, surface finish (machine marks are potential crack propagators), corrosion and residual stresses produced by welding.

Frequently in structural members an alternating stress, σ_{alt}, is superimposed on a static or mean stress, σ_{mean}, as illustrated in Fig. 8.19. The value of σ_{alt} is the most important factor in determining the number of cycles of load that produce failure. The stress, σ_{alt}, that can be withstood for a specified number of cycles is called the

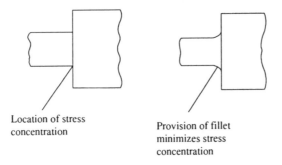

Location of stress concentration

Provision of fillet minimizes stress concentration

Fig. 8.18 Stress concentration location

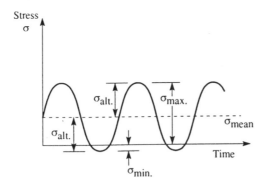

Fig. 8.19 Alternating stress in fatigue loading

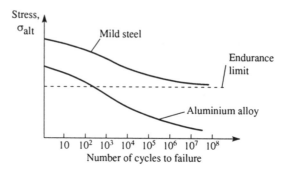

Fig. 8.20   Stress–endurance curves

fatigue strength of the material. Some materials, such as mild steel, possess a stress level that can be withstood for an indefinite number of cycles. This stress is known as the *endurance limit* of the material; no such limit has been found for aluminium and its alloys. Fatigue data are frequently presented in the form of an *S–n* curve or stress–endurance curve as shown in Fig. 8.20.

In many practical situations the amplitude of the alternating stress varies and is frequently random in nature. The *S–n* curve does not, therefore, apply directly and an alternative means of predicting failure is required. *Miner's cumulative damage theory* suggests that failure will occur when

$$\frac{n_1}{N_1} + \frac{n_2}{N_2} + \cdots + \frac{n_r}{N_r} = 1 \tag{8.1}$$

where n_1, n_2, ..., n_r are the number of applications of stresses σ_{alt}, σ_{mean} and N_1, N_2, ..., N_r are the number of cycles to failure of stresses σ_{alt}, σ_{mean}.

8.7   Design methods

In Section 8.3 we examined stress–strain curves for different materials and saw that, generally, there are two significant values of stress: the yield stress, σ_Y, and the ultimate stress, $\sigma_{ult.}$. Either of these two stresses may be used as the basis of design which must ensure, of course, that a structure will adequately perform the role for which it is constructed. In any case the maximum stress in a structure should be kept below the elastic limit of the material otherwise a permanent set will result when the loads are applied and then removed.

Two design approaches are possible. The first, known as *elastic design*, uses either the yield stress (for ductile materials), or the ultimate stress (for brittle materials) and establishes a *working* or *allowable stress* within the elastic range of the material by applying a suitable factor of safety whose value depends upon a number of considerations. These include the type of material, the type of loading (fatigue loading would require a larger factor of safety than static loading which is obvious from Section 8.6) and the degree of complexity of the structure. Therefore for materials such as steel, the working stress, σ_W, is given by

$$\sigma_W = \frac{\sigma_Y}{n} \tag{8.2}$$

where n is the factor of safety, a typical value being 1·65. For a brittle material such as concrete, the working stress would be given by

$$\sigma_W = \frac{\sigma_{ult}}{n} \qquad (8.3)$$

in which n is of the order of 2·5.

Elastic design has been superseded for concrete by *limit state* or *ultimate load* design and for steel by *plastic design* (or limit, or ultimate load design). In this approach the structure is designed with a given factor of safety against complete collapse which is assumed to occur in a concrete structure when the stress reaches σ_{ult} and occurs in a steel structure when the stress at one or more points reaches σ_Y (see Section 9.10). In the design process working or actual loads are determined and then factored to give the required ultimate or collapse load of the structure. Knowing σ_{ult} (for concrete) or σ_Y (for steel) the appropriate section may then be chosen for the structural member.

The factors of safety used in ultimate load design depend upon several parameters. These may be grouped into those related to the material of the member and those related to loads. Thus in the ultimate load design of a reinforced concrete beam the values of σ_{ult} for concrete and σ_Y for the reinforcing steel are factored by *partial safety factors* to give *design strengths* that allow for variations of workmanship or quality of control in manufacture. Typical values for these partial safety factors are 1·5 for concrete and 1·15 for the reinforcement. Note that the design strength in both cases is less than the actual strength. In addition, as stated above, design loads are obtained in which the actual loads are increased by multiplying the latter by a partial safety factor which depends upon the type of load being considered.

As well as strength, structural members must possess sufficient stiffness, under normal working loads, to prevent deflections being excessive and thereby damaging adjacent parts of the structure. Another consideration related to deflection is the appearance of a structure which can be adversely affected if large deflections cause cracking of protective and/or decorative coverings. This is particularly critical in reinforced concrete beams where the concrete in the tension zone of the beam cracks; this does not affect the strength of the beam since the tensile stresses are withstood by the reinforcement. However, if deflections are large the crack widths will be proportionately large and the surface finish and protection afforded by the concrete to the reinforcement would be impaired.

Codes of Practice limit deflections of beams either by specifying maximum span/depth ratios or by fixing the maximum deflection in terms of the span. A typical limitation for a reinforced concrete beam is that the total deflection of the beam should not exceed span/250. An additional proviso is that the deflection that takes place after the construction of partitions and finishes should not exceed span/350 or 20 mm, whichever is the lesser. A typical value for a steel beam is span/360.

It is clear that the deflections of beams under normal working loads occur within the elastic range of the material of the beam no matter whether elastic or ultimate load theory has been used in their design. Deflections of beams, therefore, are checked using elastic analysis.

Table 8.1

Material	Density (kN/m³)	Modulus of elasticity, E (N/mm²)	Shear modulus, G (N/mm²)	Yield stress, σ_Y (N/mm²)	Ultimate stress, σ_{ult} (N/mm²)	Poisson's ratio v
Aluminium alloy	27·0	70 000	40 000	290	440	0·33
Brass	82·5	103 000	41 000	103	276	
Bronze	87·0	103 000	45 000	138	345	
Cast iron	72·3	103 000	41 000		552 (comp.)	0·25
					138 (tens.)	
Concrete (med. strength)	22·8	21 400			20·7 (comp.)	0·13
Copper	80·6	117 000	41 000	245	345	
Steel (mild)	77·0	200 000	79 000	250	410–550	0·27
Steel (high carbon)	77·0	200 000	79 000	414	690	0·27
Timber	6·0	12 000			58 (comp.)	

8.8 Material properties

Table 8.1 lists some typical properties of the more common engineering materials.

Problems

P.8.1 Describe a simple tensile test and show, with the aid of sketches, how measures of the ductility of the material of the specimen may be obtained. Sketch typical stress–strain curves for mild steel and an aluminium alloy showing their important features.

P.8.2 A bar of metal 25 mm in diameter is tested on a length of 250 mm. In tension the following results were recorded:

Load (kN)	10·4	31·2	52·0	72·8
Extension (mm)	0·036	0·089	0·140	0·191

A torsion test gave the following results:

Torque (kN m)	0·051	0·152	0·253	0·354
Angle of twist (deg)	0·24	0·71	1·175	1·642

Represent these results in graphical form and hence determine Young's modulus, E, the modulus of rigidity, G, Poisson's ratio, v, and the bulk modulus, K, for the metal.
(Note: see Chapter 11 for torque–angle of twist relationship).

Ans. $E \approx 205\ 000\ \text{N/mm}^2$, $G \approx 80\ 700\ \text{N/mm}^2$, $v \approx 0\cdot27$, $K \approx 148\ 500\ \text{N/mm}^2$.

P.8.3 The actual stress–strain curve for a particular material is given by $\sigma = C\varepsilon^n$ where C is a constant. Assuming that the material suffers no change in volume during plastic deformation, derive an expression for the nominal stress–strain curve and show that this has a maximum when $\varepsilon = n/(1 - n)$.

Ans. σ (nominal) $= C\varepsilon^n/(1 + \varepsilon)$.

P.8.4 A structural member is to be subjected to a series of cyclic loads which produce different levels of alternating stress as shown below. Determine whether or not a fatigue failure is probable.

Table P.8.4

Loading	No. of cycles	No. of cycles to failure
1	10^4	5×10^4
2	10^5	10^6
3	10^6	24×10^7
4	10^7	12×10^7

Ans. Not probable $(n_1/N_1 + n_2/N_2 + \cdots = 0 \cdot 39)$.

CHAPTER 9

Bending of Beams

We have seen in Chapter 3 that bending moments in beams are produced by the action of either pure bending moments or shear loads. Reference to problem P.3.4 also shows that two symmetrically placed concentrated shear loads on a simply supported beam induce a state of pure bending, i.e. bending without shear, in the central portion of the beam. It is also possible, as we shall see in Section 9.2, to produce bending moments by applying loads parallel to but offset from the centroidal axis of a beam. Initially, however, we shall concentrate on beams subjected to pure bending moments and consider the corresponding internal stress distributions.

9.1 Symmetrical bending

Although symmetrical bending is a special case of the bending of beams of arbitrary cross-section, it is advantageous to investigate the former first, so that the more complex general case may be more easily understood.

Symmetrical bending arises in beams which have either singly or doubly symmetrical cross-sections; examples of both types are shown in Fig. 9.1.

Suppose that a length of beam, of rectangular cross-section, say, is subjected to a pure, sagging bending moment, M (see Section 3.2), applied in a vertical plane. The length of beam will bend into the shape shown in Fig. 9.2(a) in which the upper surface is concave and the lower convex. It can be seen that the upper longitudinal fibres of the beam are compressed while the lower fibres are stretched. It follows that between these two extremes there is a fibre that remains unchanged in length.

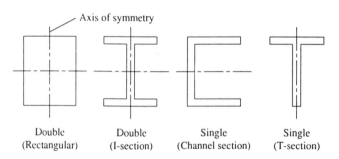

| Double | Double | Single | Single |
| (Rectangular) | (I-section) | (Channel section) | (T-section) |

Fig. 9.1 Symmetrical section beams

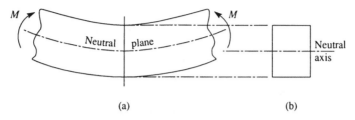

Fig. 9.2 Beam subjected to a pure sagging bending moment

Thus the direct stress varies through the depth of the beam from compression in the upper fibres to tension in the lower. Clearly the direct stress is zero for the fibre that does not change in length. The surface that contains this fibre and runs through the length of the beam is known as the *neutral surface* or *neutral plane*; the line of intersection of the neutral surface and any cross-section of the beam is termed the *neutral axis* (Fig. 9.2(b)).

The problem, therefore, is to determine the variation of direct stress through the depth of the beam, the values of the stresses and subsequently to find the corresponding beam deflection.

Assumptions

The primary assumption made in determining the direct stress distribution produced by pure bending is that plane cross-sections of the beam remain plane and normal to the longitudinal fibres of the beam after bending. We shall also assume that the material of the beam is linearly elastic, i.e. it obeys Hooke's law, and that the material of the beam is homogeneous. Cases of composite beams are considered in Chapter 12.

Direct stress distribution

Consider a length of beam (Fig. 9.3(a)) that is subjected to a pure, sagging bending moment, M, applied in a vertical plane; the beam cross-section has a vertical axis of symmetry as shown in Fig. 9.3(b). The bending moment will cause the length of

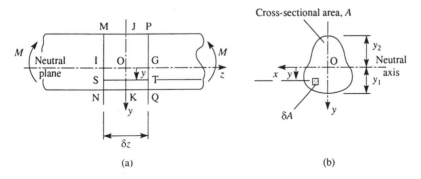

Fig. 9.3 Bending of a symmetrical section beam

beam to bend in a similar manner to that shown in Fig. 9.2(a) so that a neutral plane will exist which is, as yet, unknown distances y_1 and y_2 from the bottom and top of the beam, respectively. Coordinates of all points in the beam are referred to axes $Oxyz$ (see Section 3.2) in which the origin O lies in the neutral plane of the beam. We shall now investigate the behaviour of an elemental length, δz, of the beam formed by parallel sections MIN and PGQ (Fig. 9.3(a)) and also the fibre ST of cross-sectional area δA a distance y from the neutral plane. Clearly, before bending takes place $MP = IG = ST = NQ = \delta z$.

The application of the bending moment M causes the length of beam to bend about a centre of curvature, C, with a radius of curvature, R, measured to the neutral plane (Fig. 9.4(a)). The previously parallel plane sections MIN and PGQ remain plane according to our assumption but are now inclined at an angle $\delta\theta$ to each other. The length MP is now shorter than δz, while NQ and ST are longer. The length IG, being in the neutral plane, remains equal to δz in length, although curved. Since the fibre ST is stretched, it suffers a direct tensile strain, ε_z (parallel to the z axis of the beam), and a corresponding stress, σ_z. From Fig. 9.4(a) it can be seen that the increase in length of ST is $(R + y)\delta\theta - \delta z$ or $(R + y)\delta\theta - R\,\delta\theta$, since $\delta z = IG = R\,\delta\theta$. Thus

$$\varepsilon_z = \frac{(R + y)\delta\theta - R\,\delta\theta}{R\,\delta\theta} = \frac{y}{R} \tag{9.1}$$

We have assumed that the material of the beam obeys Hooke's law so that the direct stress, σ_z, in the fibre ST is related to ε_z by Eq. (7.7); thus

$$\sigma_z = E\,\frac{y}{R} \tag{9.2}$$

The normal force on the fibre ST, i.e. on its cross-section, is $\sigma_z\,\delta A$. However, since

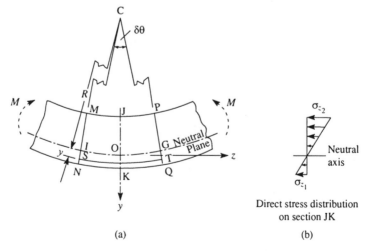

(a)

(b)

Fig. 9.4 Deflected shape of a symmetrical section beam subjected to a pure bending moment

the direct stress is caused by a pure bending moment, the resultant normal force on the complete cross-section of the beam must be zero, i.e.

$$\int_A \sigma_z \, dA = 0 \tag{9.3}$$

Substituting for σ_z in Eq. (9.3) from Eq. (9.2) gives

$$\frac{E}{R} \int_A y \, dA = 0 \tag{9.4}$$

in which both E and R are constants for a beam of a given material subjected to a given bending moment. Thus

$$\int_A y \, dA = 0 \tag{9.5}$$

Equation (9.5) states that the first moment of the area of the cross-section of the beam with respect to the neutral axis, i.e. the x axis, is equal to zero. Thus we see that *the neutral axis passes through the centroid of area of the cross-section*. Since the y axis in this case is also an axis of symmetry, it must also pass through the centroid of the cross-section. Hence the origin, O, of the coordinate axes, coincides with the centroid of area of the cross-section.

The moment about the neutral axis of the normal force $\sigma_z \delta A$, acting on the cross-section of the fibre ST is $\sigma_z y \delta A$. The integral of all such moments over the complete cross-section of the beam must equal the applied moment, M. Thus

$$M = \int_A \sigma_z y \, dA$$

which becomes, on substituting for σ_z from Eq. (9.2)

$$M = \frac{E}{R} \int_A y^2 \, dA \tag{9.6}$$

The term $\int_A y^2 \, dA$ is the *second moment of area* of the cross-section of the beam about the neutral axis and is given the symbol I. Rewriting Eq. (9.6) we have

$$M = \frac{EI}{R} \tag{9.7}$$

or, combining this expression with Eq. (9.2),

$$\frac{M}{I} = \frac{E}{R} = \frac{\sigma_z}{y} \tag{9.8}$$

From Eq. (9.8) we see that

$$\sigma_z = \frac{My}{I} \tag{9.9}$$

The direct stress, σ_z, at any point in the cross-section of a beam is therefore directly proportional to the distance of the point from the neutral axis and so varies linearly through the depth of the beam as shown, for the section JK, in Fig. 9.4(b). Clearly, for a positive, or sagging, bending moment σ_z is positive, i.e. tensile, when y is positive and compressive (i.e. negative) when y is negative. Thus in Fig. 9.4(b)

$$\sigma_{z,1} = \frac{My_1}{I} \text{ (tension)}, \qquad \sigma_{z,2} = \frac{My_2}{I} \text{ (compression)} \tag{9.10}$$

Furthermore, we see from Eq. (9.7) that the curvature, $1/R$, of the beam is given by

$$\frac{1}{R} = \frac{M}{EI} \tag{9.11}$$

and is therefore directly proportional to the applied bending moment and inversely proportional to the product EI which is known as the *flexural rigidity* of the beam.

Elastic section modulus

Equations (9.10) may be written in the form

$$\sigma_{z,1} = \frac{M}{Z_{e,1}}, \qquad \sigma_{z,2} = \frac{M}{Z_{e,2}} \tag{9.12}$$

in which the terms $Z_{e,1}(=I/y_1)$ and $Z_{e,2}(=I/y_2)$ are known as the *elastic section moduli* of the cross-section. For a beam section having the x axis as an axis of symmetry $y_1 = y_2$ and $Z_{e,1} = Z_{e,2} = Z_e$, say,

$$\sigma_{z,1} = \sigma_{z,2} = \frac{M}{Z_e} \tag{9.13}$$

Expressing the extremes of direct stress in a beam section in this form is extremely useful in elastic design where, generally, a beam of a given material is required to support a given bending moment. The maximum allowable stress in the material of the beam is known and a minimum required value for the section modulus, Z_e, can be calculated. A suitable beam section may then be chosen from handbooks listing properties and dimensions, including section moduli, of standard structural shapes.

The selection of a beam cross-section depends upon many factors; these include the type of loading and construction, the material of the beam and several others. However, for a beam subjected to bending and fabricated from material that has the same failure stress in compression as in tension, it is logical to choose a doubly symmetrical beam section having its centroid (and therefore its neutral axis) at mid-depth. Also it can be seen from Fig. 9.4(b) that the greatest values of direct stress occur at points furthest from the neutral axis so that the most efficient section is one in which most of the material is located as far as possible from the neutral axis. Such a section is the I-section shown in Fig. 9.1.

Example 9.1   A simply supported beam, 6 m long, is required to carry a uniformly distributed load of 10 kN/m. If the allowable direct stress in tension and compression is 155 N/mm², select a suitable cross-section for the beam.

From Fig. 3.15(d) we see that the maximum bending moment in a simply supported beam of length L carrying a uniformly distributed load of intensity w is given by

$$M_{max} = \frac{wL^2}{8} \qquad \text{(i)}$$

Therefore in this case

$$M_{max} = \frac{10 \times 6^2}{8} = 45 \text{ kN m}$$

The required section modulus of the beam is now obtained using Eq. (9.13), thus

$$Z_{e,min} = \frac{M_{max}}{\sigma_{z,max}} = \frac{45 \times 10^6}{155} = 290\,323 \text{ mm}^3$$

From tables of structural steel sections it can be seen that a Universal Beam, 254 mm × 102 mm × 28 kg/m, has a section modulus (about a centroidal axis parallel to its flanges) of 307 600 mm³. This is the smallest beam section having a section modulus greater than that required and allows a margin for the increased load due to the self-weight of the beam. However, we must now check that the allowable stress is not exceeded due to self-weight. The total load intensity produced by the applied load and self-weight is

$$10 + \frac{28 \times 9\cdot81}{10^3} = 10\cdot3 \text{ kN/m}$$

Hence, from Eq. (i)

$$M_{max} = \frac{10\cdot3 \times 6^2}{8} = 46\cdot4 \text{ kN m}$$

Therefore from Eq. (9.13)

$$\sigma_{z,max} = \frac{46\cdot4 \times 10^3 \times 10^3}{307\,600} = 150\cdot8 \text{ N/mm}^2$$

The allowable stress is 155 N/mm² so that the Universal Beam, 254 mm × 102 mm × 28 kg/m, is satisfactory.

Example 9.2 The cross-section of a beam has the dimensions shown in Fig. 9.5(a). If the beam is subjected to a sagging bending moment of 100 kN m applied in a vertical plane, determine the distribution of direct stress through the depth of the section.

The cross-section of the beam is doubly symmetrical so that the centroid, G, of the section, and therefore the origin of axes, coincides with the mid-point of the web. Furthermore, the bending moment is applied to the beam section in a vertical

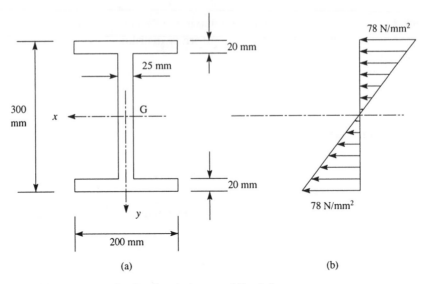

Fig. 9.5   Direct stress distribution in beam of Ex. 9.2

plane so that the x axis becomes the neutral axis of the beam section; we therefore need to calculate the second moment of area, I_x, about this axis. Thus

$$I_x = \frac{200 \times 300^3}{12} - \frac{175 \times 260^3}{12} = 193 \cdot 7 \times 10^6 \, \text{mm}^4 \text{ (see Section 9.6)}$$

From Eq. (9.9) the distribution of direct stress, σ_z, is given by

$$\sigma_z = \frac{100 \times 10^6}{193 \cdot 7 \times 10^6} y = 0 \cdot 52 y \qquad \text{(i)}$$

The direct stress therefore varies linearly over the depth of the section from a value

$$0 \cdot 52 \times (-150) = -78 \, \text{N/mm}^2 \quad \text{(compression)}$$

at the top of the beam to

$$0 \cdot 52 \times (+150) = +78 \, \text{N/mm}^2 \quad \text{(tension)}$$

at the bottom as shown in Fig. 9.5(b).

Example 9.3   Now determine the distribution of direct stress in the beam of Ex. 9.2 if the bending moment is applied in a horizontal plane and in a clockwise sense about Gy when viewed in the direction yG.

In this case the beam will bend about the vertical y axis which therefore becomes the neutral axis of the section. Thus Eq. (9.9) becomes

$$\sigma_z = \frac{M}{I_y} x \qquad \text{(i)}$$

where I_y is the second moment of area of the beam section about the y axis. Again from Section 9.6

$$I_y = 2 \times \frac{20 \times 200^3}{12} + \frac{260 \times 25^3}{12} = 27 \cdot 0 \times 10^6 \, \text{mm}^4$$

Hence, substituting for M and I_y in Eq. (i),

$$\sigma_z = \frac{100 \times 10^6}{27 \cdot 0 \times 10^6} \, x = 3 \cdot 7x$$

We have not specified a sign convention for bending moments applied in a horizontal plane; clearly in this situation the sagging/hogging convention loses its meaning. However, a physical appreciation of the problem shows that the left-hand edges of the beam are in tension while the right-hand edges are in compression. Again the distribution is linear and varies from $3 \cdot 7 \times (+100) = 370 \, \text{N/mm}^2$ (tension) at the left-hand edges of each flange to $3 \cdot 7 \times (-100) = -370 \, \text{N/mm}^2$ (compression) at the right-hand edges.

We note that the maximum stresses in this example are very much greater than those in Ex. 9.2. This is due to the fact that the bulk of the material in the beam section is concentrated in the region of the neutral axis where the stresses are low. The use of an I-section in this manner would therefore be structurally inefficient.

Example 9.4 The beam section of Ex. 9.2 is subjected to a bending moment of 100 kN m applied in a plane parallel to the longitudinal axis of the beam but inclined at 30° to the left of vertical. The sense of the bending moment is clockwise when viewed from the left-hand edge of the beam section. Determine the distribution of direct stress.

The bending moment is first resolved into two components, M_x in a vertical plane and M_y in a horizontal plane. Equation (9.9) may then be written in two forms:

$$\sigma_z = \frac{M_x}{I_x} y \quad \text{and} \quad \sigma_z = \frac{M_y}{I_y} x \tag{i}$$

The separate distributions can then be determined and superimposed. A more direct method is to combine the two equations (i) to give the total direct stress at any point (x, y) in the section. Thus

$$\sigma_z = \frac{M_x}{I_x} y + \frac{M_y}{I_y} x \tag{ii}$$

Now

$$\begin{rcases} M_x = 100 \cos 30° = 86 \cdot 6 \, \text{kN m} \\ M_y = 100 \sin 30° = 50 \cdot 0 \, \text{kN m} \end{rcases} \tag{iii}$$

M_x is, in this case, a negative bending moment producing tension in the upper half of the beam where y is negative. Also M_y produces tension in the left-hand half of the beam where x is positive; we shall therefore call M_y a positive bending moment. Substituting the values of M_x and M_y from Eqs (iii) but with the appropriate sign in

Eq. (ii) together with the values of I_x and I_y from Exs 9.2 and 9.3 we obtain

$$\sigma_z = -\frac{86\cdot6 \times 10^6}{193\cdot7 \times 10^6} y + \frac{50\cdot0 \times 10^6}{27\cdot0 \times 10^6} x \qquad \text{(iii)}$$

or

$$\sigma_z = -0\cdot45y + 1\cdot85x \qquad \text{(iv)}$$

Equation (iv) gives the value of direct stress at any point in the cross-section of the beam and may also be used to determine the distribution over any desired portion. Thus on the upper edge of the top flange $y = -150$ mm, $100 \text{ mm} \geqslant x \geqslant -100$ mm, so that the direct stress varies linearly with x. At the top left-hand corner of the top flange

$$\sigma_z = -0\cdot45 \times (-150) + 1\cdot85 \times (+100) = +252\cdot5 \text{ N/mm}^2 \quad \text{(tension)}$$

At the top right-hand corner

$$\sigma_z = -0\cdot45 \times (-150) + 1\cdot85 \times (-100) = -117\cdot5 \text{ N/mm}^2 \quad \text{(compression)}$$

The distributions of direct stress over the outer edge of each flange and along the vertical axis of symmetry are shown in Fig. 9.6. Note that the neutral axis of the beam section does not in this case coincide with either the x or y axis, although it still passes through the centroid of the section. Its inclination, α, to the x axis, say, can be found by setting $\sigma_z = 0$ in Eq. (iv). Thus

$$0 = -0\cdot45y + 1\cdot85x$$

or

$$\frac{y}{x} = \frac{1\cdot85}{0\cdot45} = 4\cdot11 = \tan\alpha$$

which gives

$$\alpha = 76\cdot3°$$

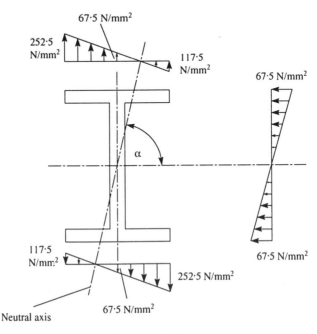

Fig. 9.6 Direct stress distribution in beam of Ex. 9.4

Note that α may be found in general terms from Eq. (ii) by again setting $\sigma_z = 0$. Hence

$$\frac{y}{x} = -\frac{M_y I_x}{M_x I_y} = \tan \alpha \tag{9.14}$$

9.2 Combined bending and axial load

In many practical situations beams and columns are subjected to combinations of axial loads and bending moments. For example, the column shown in Fig. 9.7 supports a beam seated on a bracket attached to the column. The loads on the beam produce a vertical load, P, on the bracket, the load being offset a distance e from the neutral plane of the column. The action of P on the column is therefore equivalent to an axial load, P, plus a bending moment, Pe. The direct stress at any point in the cross-section of the column is therefore the algebraic sum of the direct stress due to the axial load and the direct stress due to bending.

Consider now a length of beam having a vertical plane of symmetry and subjected to a tensile load, P, which is offset by positive distances e_y and e_x from the x and y axes, respectively (Fig. 9.8). It can be seen that P is equivalent to an axial load P plus bending moments Pe_y and Pe_x about the x and y axes, respectively. The moment Pe_y is a positive or sagging bending moment while the moment Pe_x induces tension in the region where x is positive; Pe_x is therefore also regarded as a positive moment. Thus at any point (x, y) the direct stress, σ_z, due to the combined force system is, using Eqs (7.1) and (9.9),

$$\sigma_z = \frac{P}{A} + \frac{Pe_y}{I_y} y + \frac{Pe_x}{I_y} x \tag{9.15}$$

Equation (9.15) gives the value of σ_z at any point (x, y) in the beam section for any combination of signs of P, e_x, e_y.

Example 9.5

A beam has the cross-section shown in Fig. 9.9(a). It is subjected to a normal tensile force, P, whose line of action passes through the centroid of the horizontal flange.

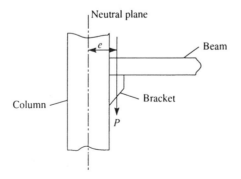

Fig. 9.7 Combined bending and axial load on a column

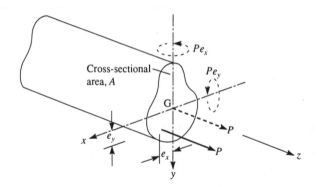

Fig. 9.8   Combined bending and axial load on a beam section

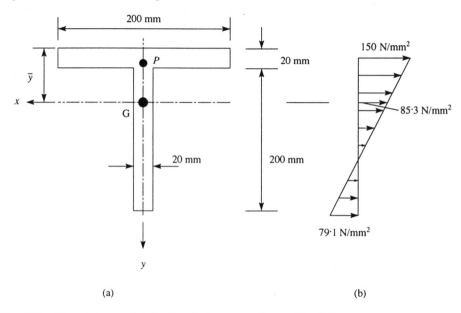

Fig. 9.9   Direct stress distribution in beam section of Ex. 9.5

Calculate the maximum allowable value of P if the maximum direct stress is limited to ± 150 N/mm².

The first step in the solution of the problem is to determine the position of the centroid, G, of the section. Thus, taking moments of areas about the top edge of the flange we have

$$(200 \times 20 + 200 \times 20)\bar{y} = 200 \times 20 \times 10 + 200 \times 20 \times 120$$

from which
$$\bar{y} = 65 \text{ mm}$$

The second moment of area of the section about the x axis is then obtained using the methods of Section 9.6 and is

$$I_x = \frac{200 \times 65^3}{3} - \frac{180 \times 45^3}{3} + \frac{20 \times 155^3}{3} = 37{\cdot}7 \times 10^6 \text{ mm}^4$$

Since the line of action of the load intersects the y axis, e_x in Eq. (9.15) is zero so that

$$\sigma_z = \frac{P}{A} + \frac{Pe_y}{I_x} y \qquad\qquad (i)$$

Also $e_y = -55$ mm so that $Pe_y = -55 P$ and Eq. (i) becomes

$$\sigma_z = P\left(\frac{1}{8000} - \frac{55}{37\cdot7 \times 10^6} y\right)$$

or

$$\sigma_z = P(1\cdot25 \times 10^{-4} - 1\cdot46 \times 10^{-6} y) \qquad\qquad (ii)$$

It can be seen from Eq. (ii) that σ_z varies linearly through the depth of the beam from a tensile value at the top of the flange where y is negative to either a tensile or compressive value at the bottom of the leg depending on whether the bracketed term is positive or negative. Therefore at the top of the flange

$$+ 150 = P[1\cdot25 \times 10^{-4} - 1\cdot46 \times 10^{-6} \times (-65)]$$

which gives the limiting value of P as 682 kN.

At the bottom of the leg of the section $y = +155$ mm, so that the right-hand side of Eq. (ii) becomes

$$P[1\cdot25 \times 10^{-4} - 1\cdot46 \times 10^{-6} \times (+155)] \equiv -1\cdot01 \times 10^{-4}P$$

which is negative for a tensile value of P. Hence the resultant direct stress at the bottom of the leg is compressive so that, for a limiting value of P,

$$-150 = -1\cdot01 \times 10^{-4}P$$

from which

$$P = 1485 \text{ kN}$$

We see therefore that the maximum allowable value of P is 682 kN, giving the direct stress distribution shown in Fig. 9.9(b).

Core of a rectangular section

In some structures, such as brick-built chimneys and gravity dams which are fabricated from brittle materials, it is inadvisable for tension to be developed in any cross-section. Clearly, from our previous discussion, it is possible for a compressive load that is offset from the neutral axis of a beam section to induce a resultant tensile stress in some regions of the cross-section if the tensile stress due to bending in those regions is greater than the compressive stress produced by the axial load. We therefore require to impose limits on the eccentricity of such a load so that no tensile stresses are induced.

Consider the rectangular section shown in Fig. 9.10 subjected to an eccentric compressive load, P, applied parallel to the longitudinal axis in the positive xy quadrant. Note that if P were inclined at some angle to the longitudinal axis, then we need only consider the component of P normal to the section since the in-plane

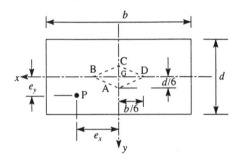

Fig. 9.10 Core of a rectangular section

component would induce only shear stresses. Since P is a compressive load and therefore negative, Eq. (9.15) becomes

$$\sigma_z = -\frac{P}{A} - \frac{Pe_y}{I_x} y - \frac{Pe_x}{I_y} x \qquad (9.16)$$

In the region of the cross-section where x and y are negative, tension will develop if

$$\left| \frac{Pe_y}{I_x} y + \frac{Pe_x}{I_y} x \right| > \left| \frac{P}{A} \right|$$

The limiting case arises when the direct stress is zero at the corner of the section, i.e. when $x = -b/2$ and $y = -d/2$. Therefore, substituting these values in Eq. (9.16) we have

$$0 = -\frac{P}{A} - \frac{Pe_y}{I_x} \left(-\frac{d}{2} \right) - \frac{Pe_x}{I_y} \left(-\frac{b}{2} \right)$$

or, since $A = bd$, $I_x = bd^3/12$, $I_y = db^3/12$ (see Section 9.6)

$$0 = -bd + 6be_y + 6de_x$$

which gives

$$be_y + de_x = \frac{bd}{6}$$

Rearranging we obtain

$$e_y = -\frac{d}{b} e_x + \frac{d}{6} \qquad (9.17)$$

Equation (9.17) defines the line AB in Fig. 9.10 which sets the limit for the eccentricity of P from both the x and y axes. It follows that P can be applied at any point in the region GAB for there to be no tension developed anywhere in the section.

Since the section is doubly symmetrical, a similar argument applies to the regions GBC, GCD and GDA; the rhombus ABCD is known as the *core of the section* and has diagonals BD $= b/3$ and AC $= d/3$.

Core of a circular section

Bending, produced by an eccentric load P, in a circular cross-section always takes place about a diameter that is perpendicular to the radius on which P acts. It is therefore logical to take this diameter and the radius on which P acts as the coordinate axes of the section (Fig. 9.11).

Suppose that P in Fig. 9.11 is a compressive load. The direct stress, σ_z, at any point (x, y) is given by Eq. (9.15) in which $e_y = 0$. Hence

$$\sigma_z = -\frac{P}{A} - \frac{Pe_x}{I_y} x \qquad (9.18)$$

Tension will occur in the region where x is negative if

$$\left| \frac{Pe_x}{I_y} x \right| > \left| \frac{P}{A} \right|$$

The limiting case occurs when $\sigma_z = 0$ and $x = -R$; hence

$$0 = -\frac{P}{A} - \frac{Pe_x}{I_y} (-R)$$

Now $A = \pi R^2$ and $I_y = \pi R^4/4$ (see Section 9.6)

so that

$$0 = -\frac{1}{\pi R^2} + \frac{4e_x}{\pi R^3}$$

from which

$$e_x = \frac{R}{4}$$

Thus the core of a circular section is a circle of radius $R/4$.

Example 9.6   A free-standing masonry wall is 7 m high, 0·6 m thick and has a density of 2000 kg/m³. Calculate the maximum, uniform, horizontal wind pressure that can occur without tension developing at any point in the wall.

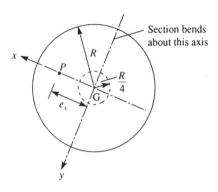

Fig. 9.11   Core of a circular section beam

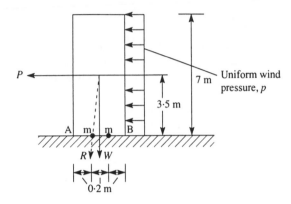

Fig. 9.12   Masonry wall of Ex. 9.6

Consider a 1 m length of wall. The forces acting are the horizontal resultant, P, of the uniform wind pressure, p, and the weight, W, of the 1 m length of wall (Fig. 9.12).

Clearly the base section is the one that experiences the greatest compressive normal load due to self-weight and also the greatest bending moment due to wind pressure. It is also the most critical section since the bending moment that causes tension is a function of the square of the height of the wall, whereas the weight causing compression is a linear function of wall height. From Fig. 9.10 it is clear that the resultant, R, of P and W must lie within the central 0·2 m of the base section, i.e. within the middle third of the section, for there to be no tension developed anywhere in the base cross-section. The limiting case arises when R passes through m, one of the middle third points, in which case the direct stress at B is zero and the moment of R (and therefore the sum of the moments of P and W) about m is zero. Hence

$$3\cdot5\,P = 0\cdot1\,W \tag{i}$$

where $$P = p \times 7 \times 1 \text{ N} \quad \text{if } p \text{ is in N/m}^2$$

and $$W = 2000 \times 9\cdot81 \times 0\cdot6 \times 7 \text{ N}$$

Substituting for P and W in Eq. (i) and solving for p gives

$$p = 336\cdot3 \text{ N/m}^2$$

9.3   Anticlastic bending

Consider the rectangular beam section in Fig. 9.13(a); the direct stress distribution in the section due to a positive bending moment applied in a vertical plane varies from compression in the upper half of the beam to tension in the lower half (Fig. 9.13(b)). However, due to the Poisson effect (see Section 7.8) the compressive stress produces a lateral elongation of the upper fibres of the beam section while the tensile stress produces a lateral contraction of the lower. The section does not therefore remain rectangular but distorts as shown in Fig. 9.13(c); the effect is known as *anticlastic bending*.

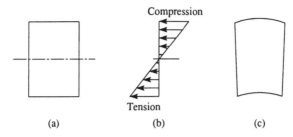

Fig. 9.13 Anticlastic bending of a beam section

Anticlastic bending is of interest in the analysis of thin-walled box beams in which the cross-sections are maintained by stiffening ribs. The prevention of anticlastic distortion induces local variations in stress distributions in the webs and covers of the box beam and also in the stiffening ribs.

9.4 Strain energy in bending

A positive bending moment applied to a length of beam causes the upper longitudinal fibres to be compressed and the lower ones to stretch as shown in Fig. 9.4(a). The bending moment therefore does work on the length of beam and this work is absorbed by the beam as strain energy.

Suppose that the bending moment, M, in Fig. 9.4(a) is gradually applied so that when it reaches its final value the angle subtended at the centre of curvature by the element δz is $\delta\theta$. From Fig. 9.4(a) we see that

$$R\,\delta\theta = \delta z$$

Substituting in Eq. (9.7) for R we obtain

$$M = \frac{EI_x}{\delta z}\,\delta\theta \tag{9.19}$$

so that $\delta\theta$ is a linear function of M. It follows that the work done by the gradually applied moment M is $M\,\delta\theta/2$ subject to the condition that the limit of proportionality is not exceeded. The strain energy, δU, of the elemental length of beam is therefore given by

$$\delta U = \tfrac{1}{2}M\,\delta\theta \tag{9.20}$$

or, substituting for $\delta\theta$ from Eq. (9.19) in Eq. (9.20),

$$\delta U = \frac{1}{2}\frac{M^2}{EI_x}\,\delta z$$

The total strain energy, U, due to bending in a beam of length L is therefore

$$U = \int_L \frac{M^2}{2EI_x}\,dz \tag{9.21}$$

9.5 Unsymmetrical bending

Frequently in civil engineering construction beam sections do not possess any axes of symmetry. Typical examples are shown in Fig. 9.14 where the angle section has legs of unequal length and the Z-section possesses anti- or skew symmetry about a horizontal axis through its centroid, but not symmetry. We shall now develop the theory of bending for beams of arbitrary cross-section.

Assumptions

We shall again assume, as in the case of symmetrical bending, that plane sections of the beam remain plane after bending and that the material of the beam is homogeneous and linearly elastic.

Sign conventions and notation

Since we are now concerned with the general case of bending we may apply loading systems to a beam in any plane. However, no matter how complex these loading systems are, they can always be resolved into components in planes containing the three coordinate axes of the beam. We shall use an identical system of axes to that shown in Fig. 3.6, but our notation for loads must be extended and modified to allow for the general case.

Figure 9.15 shows the symbols adopted, positive directions and senses for loads and moments applied externally to a beam and also the positive directions of the components u, v and w of the displacement of any point in the beam cross-section

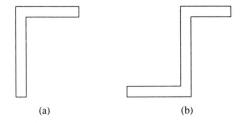

(a)   (b)

Fig. 9.14 Unsymmetrical beam sections

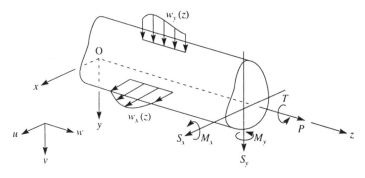

Fig. 9.15 Sign conventions and notation

parallel to the x, y and z axes, respectively. The convention for axial load, P, and torque, T, is identical to that in Fig. 3.6 but externally applied shear loads are now given the symbol S with an appropriate suffix, x or y, to indicate direction; similarly for the distributed loads $w_x(z)$ and $w_y(z)$. The suffixes used to designate the components M_x and M_y of an applied bending moment indicate the axes about which they act. Thus M_x is a bending moment in a vertical plane acting about the x axis of the beam section. Although M_x in Fig. 9.15 is a sagging bending moment and therefore in agreement with our previous convention, we need to extend the definition of a positive bending moment to include M_y which is applied in a horizontal plane. Thus we shall define M_x and M_y as positive when they each induce tensile stresses in the positive xy quadrant of the beam section.

Although positive directions and senses for externally applied forces and moments have been fixed, it can be seen from Fig. 9.16 that positive internal forces and moments form one of two different systems depending on which face of an internal section is considered. Thus if we refer internal forces and moments to that face of a section that is seen when a view is taken in the direction zO, then positive internal forces and moments are in the same direction and sense as the externally applied loads, whereas on the opposite face they form an opposite system. The former system has the advantage that axial and shear forces are always positive in the positive directions of the appropriate axes whether they are internal or external. It must be realized, however, that internal stress resultants then become *equivalent* to externally applied forces and moments and are not in equilibrium with them as would be the case if the opposite face were considered.

Direct stress distribution

Consider a beam having the arbitrary cross-section shown in Fig. 9.17. The face of the section shown is that which is seen when viewed in the direction zO so that the components M_x and M_y of the internal bending moment are positive. Suppose that the origin O of our system of axes is positioned at some point on the neutral axis of the beam section; the location and inclination α of the neutral axis to the x axis are both, as yet, unknown.

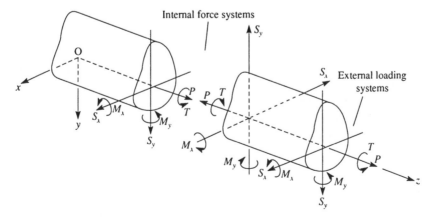

Fig. 9.16 Positive internal force system

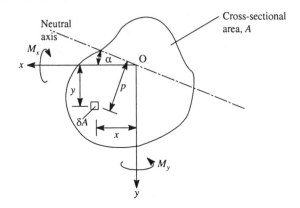

Fig. 9.17　Bending of an unsymmetrical section beam

We have seen in Section 9.1 that a beam bends about the neutral axis of its cross-section so that the radius of curvature, R, of the beam is perpendicular to the neutral axis. Therefore by direct comparison with Eq. (9.2) it can be seen that the direct stress, σ_z, on the element, δA, a perpendicular distance p from the neutral axis, is given by

$$\sigma_z = E\,\frac{p}{R} \qquad (9.22)$$

The beam section is subjected to a pure bending moment so that the resultant direct load on the section is zero. Hence

$$\int_A \sigma_z\, \mathrm{d}A = 0$$

Replacing σ_z in this equation from Eq. (9.22) we have

$$\int_A E\,\frac{p}{R}\, \mathrm{d}A = 0$$

or, for a beam of a given material subjected to a given bending moment,

$$\int_A P\, \mathrm{d}A = 0 \qquad (9.23)$$

Qualitatively Eq. (9.23) states that the first moment of area of the beam section about the neutral axis is zero. It follows that in problems involving the pure bending of beams the neutral axis always passes through the centroid of the beam section. We shall therefore choose the centroid, G, of a section as the origin of axes.

From Fig. 9.17 we see that

$$p = x \sin \alpha + y \cos \alpha \qquad (9.24)$$

so that from Eq. (9.22)

$$\sigma_z = \frac{E}{R}(x \sin \alpha + y \cos \alpha) \qquad (9.25)$$

The moment resultants of the direct stress distribution are equivalent to M_x and M_y so that

$$M_x = \int_A \sigma_z y \, dA, \qquad M_y = \int_A \sigma_z x \, dA \qquad (9.26)$$

Substituting for σ_z from Eq. (9.25) in Eqs (9.26), we obtain

$$\left.\begin{aligned}
M_x &= \frac{E \sin \alpha}{R} \int_A xy \, dA + \frac{E \cos \alpha}{R} \int_A y^2 \, dA \\
M_y &= \frac{E \sin \alpha}{R} \int_A x^2 \, dA + \frac{E \cos \alpha}{R} \int_A xy \, dA
\end{aligned}\right\} \qquad (9.27)$$

In Eqs (9.27)

$$\int_A xy \, dA = I_{xy}, \qquad \int_A y^2 \, dA = I_x, \qquad \int_A x^2 \, dA = I_y$$

where I_{xy} is the product second moment of area of the beam section about the x and y axes, I_x is the second moment of area about the x axis and I_y is the second moment of area about the y axis. Equations (9.27) may therefore be rewritten

$$\left.\begin{aligned}
M_x &= \frac{E \sin \alpha}{R} I_{xy} + \frac{E \cos \alpha}{R} I_x \\
M_y &= \frac{E \sin \alpha}{R} I_y + \frac{E \cos \alpha}{R} I_{xy}
\end{aligned}\right\} \qquad (9.28)$$

Solving Eqs (9.28) for $(E \sin \alpha)/R$ and $(E \cos \alpha)/R$ and then substituting in Eq. (9.25), we have

$$\sigma_z = \left(\frac{M_y I_x - M_x I_{xy}}{I_x I_y - (I_{xy})^2} \right) x + \left(\frac{M_x I_y - M_y I_{xy}}{I_x I_y - (I_{xy})^2} \right) y \qquad (9.29)$$

Equation (9.29) may be written in the more convenient form

$$\sigma_z = \frac{\overline{M}_x}{I_x} y + \frac{\overline{M}_y}{I_y} x \qquad (9.30)$$

where

$$\left.\begin{aligned}
\overline{M}_x &= \frac{M_x - M_y I_{xy}/I_y}{1 - (I_{xy})^2/I_x I_y} \\
\overline{M}_y &= \frac{M_y - M_x I_{xy}/I_x}{1 - (I_{xy})^2/I_x I_y}
\end{aligned}\right\} \qquad (9.31)$$

In the case where the beam section has either Ox or Oy (or both) as an axis of symmetry, then I_{xy} is zero (see Section 9.6) and Ox, Oy are *principal axes*. Equations (9.31) then reduce to

$$\overline{M}_x = M_x, \qquad \overline{M}_y = M_y$$

and Eq. (9.30) becomes

$$\sigma_z = \frac{M_x}{I_x} y + \frac{M_y}{I_y} x \quad \text{(compare with Eq. (ii) of Ex. 9.4)} \tag{9.32}$$

which is the result for symmetrical bending.

Position of the neutral axis

The direct stress at all points on the neutral axis of the beam section is zero. Thus, from Eq. (9.30)

$$0 = \frac{\overline{M}_x}{I_x} y_{\text{N.A.}} + \frac{\overline{M}_y}{I_y} x_{\text{N.A.}}$$

where $x_{\text{N.A.}}$ and $y_{\text{N.A.}}$ are the coordinates of any point on the neutral axis. Thus

$$\frac{y_{\text{N.A.}}}{x_{\text{N.A.}}} = -\frac{\overline{M}_y}{\overline{M}_x} \frac{I_x}{I_y}$$

or, referring to Fig. 9.17

$$\tan \alpha = \frac{\overline{M}_y}{\overline{M}_x} \frac{I_x}{I_y} \tag{9.33}$$

since α is positive when $y_{\text{N.A.}}$ is negative and $x_{\text{N.A.}}$ is positive.

9.6 Calculation of section properties

It will be helpful at this stage to discuss the calculation of the various section properties required in the analysis of beams subjected to bending. Initially, however, two useful theorems are quoted.

Parallel axes theorem

Consider the beam section shown in Fig. 9.18 and suppose that the second moment of area, I_G, about an axis through its centroid G is known. The second moment of area, I_N, about a parallel axis, NN, a distance b from the centroidal axis is then given by

$$I_N = I_G + Ab^2 \tag{9.34}$$

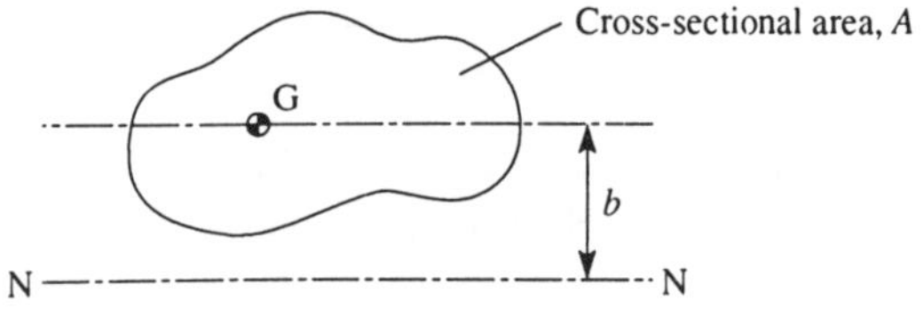

Fig. 9.18 Parallel axes theorem

Theorem of perpendicular axes

In Fig. 9.19 the second moments of area, I_x and I_y, of the section about Ox and Oy are known. The second moment of area about an axis through O perpendicular to the plane of the section (i.e. a *polar second moment of area*) is then

$$I_0 = I_x + I_y \qquad (9.35)$$

Second moments of area of standard sections

Many sections in use in civil engineering such as those illustrated in Fig. 9.1 may be regarded as comprising of a number of rectangular shapes. The problem of determining the properties of such sections is simplified if the second moments of area of the rectangular components are known and use is made of the parallel axes theorem. Thus, for the rectangular section of Fig. 9.20,

$$I_x = \int_A y^2 \, dA = \int_{-d/2}^{d/2} by^2 \, dy = b \left[\frac{y^3}{3} \right]_{-d/2}^{d/2}$$

which gives

$$I_x = \frac{bd^3}{12} \qquad (9.36)$$

Similarly

$$I_y = \frac{db^3}{12} \qquad (9.37)$$

Frequently it is useful to know the second moment of area of a rectangular section about an axis which coincides with one of its edges. Thus in Fig. 9.20, and using the parallel axes theorem,

$$I_N = \frac{bd^3}{12} + bd\left(-\frac{d}{2}\right)^2 = \frac{bd^3}{3} \qquad (9.38)$$

Example 9.7 Determine the second moments of area, I_x and I_y, of the I-section shown in Fig. 9.21.

Using Eq. (9.36)

$$I_x = \frac{bd^3}{12} - \frac{(b-t_w)d_w^3}{12}$$

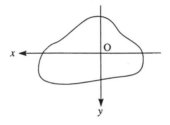

Fig. 9.19 Theorem of perpendicular axes

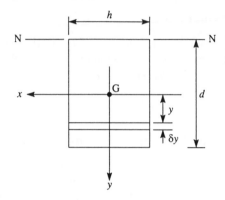

Fig. 9.20   Second moments of area of a rectangular section

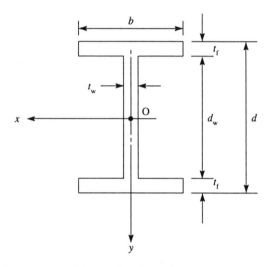

Fig. 9.21   Second moments of area of an I-section

Alternatively, using the parallel axes theorem in conjunction with Eq. (9.36)

$$I_x = 2\left[\frac{bt_f^3}{12} + bt_f\left(\frac{d_w + t_f}{2}\right)^2\right] + \frac{t_w d_w^3}{12}$$

Also, from Eq. (9.37),

$$I_y = 2\frac{t_f b^3}{12} + \frac{d_w t_w^3}{12}$$

It is also useful to determine the second moment of area, about a diameter, of a circular section. In Fig. 9.22 where the x and y axes pass through the centroid of the section,

$$I_x = \int_A y^2 \, dA = \int_{-d/2}^{d/2} 2\left(\frac{d}{2}\cos\theta\right)y^2 \, dy \qquad (9.39)$$

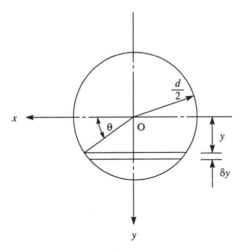

Fig. 9.22 Second moments of area of a circular section

Integration of Eq. (9.39) is simplified if an angular variable, θ, is used. Thus

$$I_x = \int_{-\pi/2}^{\pi/2} d \cos \theta \left(\frac{d}{2} \sin \theta\right)^2 \frac{d}{2} \cos \theta \, d\theta$$

i.e.

$$I_x = \frac{d^4}{8} \int_{-\pi/2}^{\pi/2} \cos^2 \theta \sin^2 \theta \, d\theta$$

which gives

$$I_x = \frac{\pi d^4}{64} \tag{9.40}$$

Clearly from symmetry

$$I_y = \frac{\pi d^4}{64} \tag{9.41}$$

Using the theorem of perpendicular axes, the polar second moment of area, I_o, is given by

$$I_o = I_x + I_y = \frac{\pi d^4}{32} \tag{9.42}$$

Product second moment of area

The product second moment of area, I_{xy}, of a beam section with respect to x and y axes is defined by

$$I_{xy} = \int_A xy \, dA \tag{9.43}$$

Thus each element of area in the cross-section is multiplied by the product of its coordinates and the integration is taken over the complete area. Although second moments of area are always positive since elements of area are multiplied by the square of one of their coordinates, it is possible for I_{xy} to be negative if the section lies predominantly in the second and fourth quadrants of the axes system. Such a situation would arise in the case of the Z-section of Fig. 9.23(a) where the product second moment of area of each flange is clearly negative.

A special case arises when one (or both) of the coordinate axes is an axis of symmetry so that for any element of area, δA, having the product of its coordinates positive, there is an identical element for which the product of its coordinates is negative (Fig. 9.23(b)). Summation (i.e. integration) over the entire section of the product second moment of area of all such pairs of elements results in a zero value for I_{xy}.

We have shown previously that the parallel axes theorem may be used to calculate second moments of area of beam sections comprising geometrically simple components. The theorem can be extended to the calculation of product second moments of area. Let us suppose that we wish to calculate the product second moment of area, I_{xy}, of the section shown in Fig. 9.23(c) about axes xy when I_{XY} about its own, say centroidal, axes system GXY is known. From Eq. (9.43)

$$I_{xy} = \int_A xy \, dA$$

or

$$I_{xy} = \int_A (X - a)(Y - b) \, dA$$

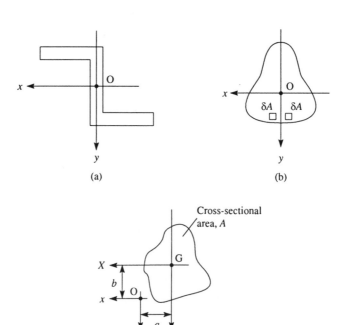

Fig. 9.23 Product second moment of area

which, on expanding, gives

$$I_{xy} = \int_A XY \, \mathrm{d}A - b \int_A X \, \mathrm{d}A - a \int_A Y \, \mathrm{d}A + ab \int_A \mathrm{d}A$$

If X and Y are centroidal axes then $\int_A X \, \mathrm{d}A = \int_A Y \, \mathrm{d}A = 0$. Hence

$$I_{xy} = I_{XY} + abA \tag{9.44}$$

It can be seen from Eq. (9.44) that if either GX or GY is an axis of symmetry then $I_{XY} = 0$ and

$$I_{xy} = abA \tag{9.45}$$

Thus for a section component having an axis of symmetry that is parallel to either of the section reference axes the product second moment of area is the product of the coordinates of its centroid multiplied by its area.

Example 9.8 A beam having the cross-section shown in Fig. 9.24 is subjected to a hogging bending moment of 1500 N m in a vertical plane. Calculate the maximum direct stress due to bending stating the point at which it acts.

The position of the centroid, G, of the section may be found by taking moments of areas about some convenient point. Thus

$$(120 \times 8 + 80 \times 8)\bar{y} = 120 \times 8 \times 4 + 80 \times 8 \times 48$$

which gives
$$\bar{y} = 21 \cdot 6 \text{ mm}$$

and
$$(120 \times 8 + 80 \times 8)\bar{x} = 80 \times 8 \times 4 + 120 \times 8 \times 24$$

giving
$$\bar{x} = 16 \text{ mm}$$

The second moments of area referred to axes Gxy are now calculated.

$$I_x = \frac{120 \times (8)^3}{12} + 120 \times 8 \times (17 \cdot 6)^2 + \frac{8 \times (80)^3}{12} + 80 \times 8 \times (26 \cdot 4)^2$$

$$= 1 \cdot 09 \times 10^6 \text{ mm}^4$$

$$I_y = \frac{8 \times (120)^3}{12} + 120 \times 8 \times (8)^2 + \frac{80 \times (8)^3}{12} + 80 \times 8 \times (12)^2$$

$$= 1 \cdot 31 \times 10^6 \text{ mm}^4$$

$$I_{xy} = 120 \times 8 \times (-8) \times (-17 \cdot 6) + 80 \times 8 \times (+12) \times (+26 \cdot 4)$$

$$= 0 \cdot 34 \times 10^6 \text{ mm}^4$$

Since $M_x = -1500$ N m and $M_y = 0$ we have, from Eqs. (9.31)

$$\bar{M}_x = -1630 \text{ N m} \quad \text{and} \quad \bar{M}_y = +505 \text{ N m}$$

Substituting these values and the appropriate second moments of area in Eq. (9.30), we obtain

$$\sigma_z = -1 \cdot 5y + 0 \cdot 39x \tag{i}$$

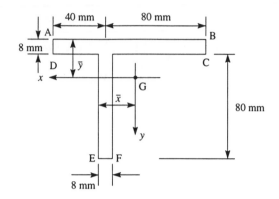

Fig. 9.24   Beam section of Ex. 9.8

Inspection of Eq. (i) shows that σ_z is a maximum at F where $x = 8$ mm, $y = 66.4$ mm. Hence

$$\sigma_{z,\max} = -96 \text{ N/mm}^2 \quad \text{(compressive)}$$

Approximations for thin-walled sections

Modern civil engineering structures frequently take the form of thin-walled cellular box beams which combine the advantages of comparatively low weight and high strength, particularly in torsion. Other forms of thin-walled structure consist of 'open' section beams such as a plate girder which is constructed from thin plates stiffened against instability.

There is no clearly defined line separating 'thick' and 'thin-walled' sections; the approximations allowed in the analysis of thin-walled sections become increasingly inaccurate the 'thicker' a section becomes. However, as a guide, it is generally accepted that the approximations are reasonably accurate for sections for which the ratio

$$\frac{t_{\max}}{b} \leq 0.1$$

where $t_{\max}$ is the maximum thickness in the section and b is a typical cross-sectional dimension.

In the calculation of the properties of thin-walled sections we shall assume that the thickness, t, of the section is small compared with its cross-sectional dimensions so that squares and higher powers of t are neglected. The section profile may then be represented by the mid-line of its wall. Thus stresses are calculated at points on the mid-line and assumed to be constant across the thickness.

Example 9.9   Calculate the second moment of area, I_x, of the channel section shown in Fig. 9.25(a).

The centroid of the section is located midway between the flanges; its horizontal position is not needed since only I_x is required. Thus

$$I_x = 2\left(\frac{bt^3}{12} + bth^2\right) + t\,\frac{[2(h - t/2)]^3}{12}$$

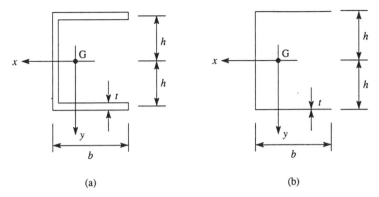

Fig. 9.25   Calculation of the second moment of area of a thin-walled channel section

which, on expanding, becomes

$$I_x = 2\left(\frac{bt^3}{12} + bth^2\right) + \frac{t}{12}\left[(2)^3\left(h^3 - \frac{3h^2t}{2} + \frac{3ht^2}{4} - \frac{t^3}{8}\right)\right]$$

Neglecting powers of t^2 and upwards we obtain

$$I_x = 2bth^2 + t\,\frac{(2h)^3}{12}$$

It is unnecessary for such calculations to be carried out in full since the final result may be obtained almost directly by regarding the section as being represented by a single line as shown in Fig. 9.25(b).

Example 9.10   A thin-walled beam has the cross-section shown in Fig. 9.26. Determine the direct stress distribution produced by a hogging bending moment M_x.

The section is antisymmetrical with its centroid at the mid-point of the vertical web. The direct stress distribution is therefore given by Eq. (9.30), viz.

$$\sigma_z = \frac{\overline{M}_x}{I_x}\,y + \frac{\overline{M}_y}{I_y}\,x \tag{i}$$

where, in this case (see Eqs (9.31)),

$$\overline{M}_x = \frac{-M_x}{1 - (I_{xy})^2/I_x I_y}, \qquad \overline{M}_y = \frac{M_x I_{xy}/I_x}{1 - (I_{xy})^2/I_x I_y} \tag{ii}$$

The section properties are calculated using the previously specified approximations for thin-walled sections; thus

$$I_x = 2\,\frac{ht}{2}\left(\frac{h}{2}\right)^2 + \frac{th^3}{12} = \frac{h^3 t}{3}$$

$$I_y = 2\,\frac{t}{3}\left(\frac{h}{2}\right)^3 = \frac{h^3 t}{12}$$

$$I_{xy} = \frac{ht}{2}\left(\frac{h}{4}\right)\left(-\frac{h}{2}\right) + \frac{ht}{2}\left(-\frac{h}{4}\right)\left(\frac{h}{2}\right) = -\frac{h^3 t}{8}$$

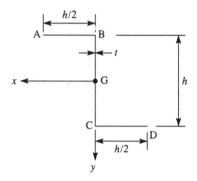

Fig. 9.26   Beam section of Ex. 9.10

Substituting these values in Eqs (ii) we obtain

$$\bar{M}_x = -2 \cdot 29\, M_x, \quad \bar{M}_y = -0 \cdot 86\, M_x$$

These expressions, in turn, when substituted in Eq. (i) give

$$\sigma_z = -\frac{M_x}{h^3 t}(6 \cdot 86 y + 10 \cdot 3 x) \tag{iii}$$

On the top flange $y = -h/2$, $h/2 \geqslant x \geqslant 0$ and the distribution of direct stress is given by

$$\sigma_z = \frac{M_x}{h^3 t}(3 \cdot 43 h - 10 \cdot 3 x)$$

which is linear. Hence

$$\sigma_{z,A} = -\frac{1 \cdot 72 M_x}{h^2 t} \quad \text{(compressive)}$$

$$\sigma_{z,B} = +\frac{3 \cdot 43 M_x}{h^2 t} \quad \text{(tensile)}$$

In the web $-h/2 \leqslant y \leqslant h/2$ and $x = 0$ so that Eq. (iii) reduces to

$$\sigma_z = -\frac{6 \cdot 86 M_x}{h^3 t} y$$

Again the distribution is linear and varies from

$$\sigma_{z,B} = +\frac{3 \cdot 43 M_x}{h^2 t} \quad \text{(tensile)}$$

to

$$\sigma_{z,C} = -\frac{3 \cdot 43 M_x}{h^2 t} \quad \text{(compressive)}$$

The distribution in the lower flange may be deduced from antisymmetry. The complete distribution is as shown in Fig. 9.27.

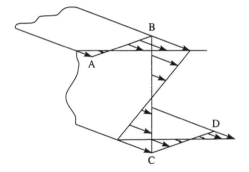

Fig. 9.27 Distribution of direct stress in beam section of Ex. 9.10

Second moments of area of inclined and curved thin-walled sections

Thin-walled sections frequently have inclined or curved walls which complicate the calculation of section properties. Consider the inclined thin section of Fig. 9.28. The second moment of area of an element δs about a horizontal axis through its centroid G is equal to $t\delta s y^2$. Therefore the total second moment of area of the section about Gx, I_x, is given by

$$I_x = \int_{-a/2}^{a/2} ty^2 \, ds = \int_{-a/2}^{a/2} t(s \sin \beta)^2 \, ds$$

i.e.
$$I_x = \frac{a^3 t \sin^2 \beta}{12}$$

Similarly
$$I_y = \frac{a^3 t \cos^2 \beta}{12}$$

The product second moment of area of the section about Gxy is

$$I_{xy} = \int_{-a/2}^{a/2} txy \, ds = \int_{-a/2}^{a/2} t(s \cos \beta)(s \sin \beta) \, ds$$

i.e.
$$I_{xy} = \frac{a^3 t \sin 2\beta}{24}$$

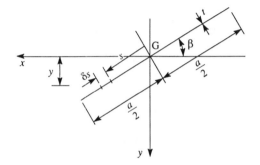

Fig. 9.28 Second moments of area of an inclined thin-walled section

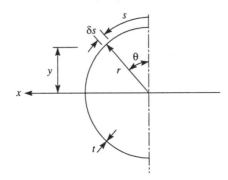

Fig. 9.29 Second moment of area of a semicircular thin-walled section

Properties of thin-walled curved sections are found in a similar manner. Thus I_x for the semicircular section of Fig. 9.29 is

$$I_x = \int_0^{\pi r} t y^2 \, \mathrm{d}s$$

Expressing y and s in terms of a single variable θ simplifies the integration; hence

$$I_x = \int_0^{\pi} t(-r \cos \theta)^2 r \, \mathrm{d}\theta$$

from which

$$I_x = \pi r^3 t / 2$$

9.7   Principal axes and principal second moments of area

In any beam section there is a set of axes, neither of which need necessarily be an axis of symmetry, for which the product second moment of area is zero. Such axes are known as *principal axes* and the second moments of area about these axes are termed principal second moments of area.

Consider the arbitrary beam section shown in Fig. 9.30. Suppose that the second moments of area I_x, I_y and the product second moment of area, I_{xy}, about arbitrary axes Oxy are known. By definition

$$I_x = \int_A y^2 \, \mathrm{d}A, \qquad I_y = \int_A x^2 \, \mathrm{d}A, \qquad I_{xy} = \int_A xy \, \mathrm{d}A \qquad (9.46)$$

The corresponding second moments of area about axes $Ox_1 y_1$, are

$$I_{x(1)} = \int_A y_1^2 \, \mathrm{d}A, \qquad I_{y(1)} = \int_A x_1^2 \, \mathrm{d}A, \qquad I_{x(1),y(1)} = \int_A x_1 y_1 \, \mathrm{d}A \qquad (9.47)$$

From Fig. 9.30

$$x_1 = x \cos \phi + y \sin \phi, \; y_1 = y \cos \phi - x \sin \phi$$

Substituting for y_1 in the first of Eqs (9.47)

$$I_{x(1)} = \int_A (y \cos \phi - x \sin \phi)^2 \, \mathrm{d}A$$

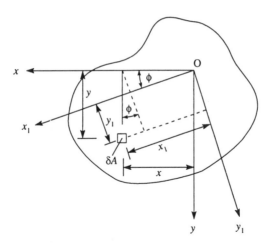

Fig. 9.30 Principal axes in a beam of arbitrary section

Expanding, we obtain

$$I_{x(1)} = \cos^2 \phi \int_A y^2 \, dA + \sin^2 \phi \int_A x^2 \, dA - 2 \cos \phi \sin \phi \int_A xy \, dA$$

which gives, using Eqs (9.46)

$$I_{x(1)} = I_x \cos^2 \phi + I_y \sin^2 \phi - I_{xy} \sin 2\phi \tag{9.48}$$

Similarly

$$I_{y(1)} = I_y \cos^2 \phi + I_x \sin^2 \phi + I_{xy} \sin 2\phi \tag{9.49}$$

and

$$I_{x(1),y(1)} = \left(\frac{I_x - I_y}{2}\right) \sin 2\phi + I_{xy} \cos 2\phi \tag{9.50}$$

Equations (9.48)–(9.50) give the second moments of area and product second moment of area about axes inclined at an angle ϕ to the x axis. In the special case where $Ox_1 y_1$ are principal axes, Ox_p, y_p, $I_{x(p),y(p)} = 0$, $\phi = \phi_p$ and Eqs (9.48) and (9.49) become

$$I_{x(p)} = I_x \cos^2 \phi_p + I_y \sin^2 \phi_p - I_{xy} \sin 2\phi_p \tag{9.51}$$

and

$$I_{y(p)} = I_y \cos^2 \phi_p + I_x \sin^2 \phi_p + I_{xy} \sin 2\phi_p \tag{9.52}$$

respectively. Furthermore, since $I_{x(1),y(1)} = I_{x(p),y(p)} = 0$, Eq. (9.50) gives

$$\tan 2\phi_p = \frac{2I_{xy}}{I_y - I_x} \tag{9.53}$$

The angle ϕ_p may be eliminated from Eqs (9.51) and (9.52) by first determining $\cos 2\phi_p$ and $\sin 2\phi_p$ using Eq. (9.53). Thus

$$\cos 2\phi_p = \frac{(I_y - I_x)/2}{\sqrt{[(I_y - I_x)/2]^2 + I_{xy}^2}}$$

$$\sin 2\phi_p = \frac{I_{xy}}{\sqrt{[(I_y - I_x)/2]^2 + I_{xy}^2}}$$

Rewriting Eq. (9.51) in terms of $\cos 2\phi_p$ and $\sin 2\phi_p$ we have

$$I_{x(p)} = \frac{I_x}{2}(1 + \cos 2\phi_p) + \frac{I_y}{2}(1 - \cos 2\phi_p) - I_{xy}\sin 2\phi_p$$

Substituting for $\cos 2\phi_p$ and $\sin 2\phi_p$ from the above we obtain

$$I_{x(p)} = \frac{I_x + I_y}{2} - \frac{1}{2}\sqrt{(I_x - I_y)^2 + 4I_{xy}^2} \tag{9.54}$$

Similarly

$$I_{y(p)} = \frac{I_x + I_y}{2} + \frac{1}{2}\sqrt{(I_x - I_y)^2 + 4I_{xy}^2} \tag{9.55}$$

Note that the solution of Eq. (9.53) gives two values for the inclination of the principal axes, ϕ_p and $\phi_p + \pi/2$, corresponding to the axes Ox_p and Oy_p.

The results of Eqs (9.48)–(9.55) may be represented graphically by Mohr's circle, a powerful method of solution for this type of problem. We shall discuss Mohr's circle in detail in Chapter 14 in connection with the analysis of complex stress and strain.

Principal axes may be used to provide an apparently simpler solution to the problem of unsymmetrical bending. Referring components of bending moment and section properties to principal axes having their origin at the centroid of a beam section, we see that Eqs (9.31) reduce to

$$\overline{M}_{x(p)} = M_{x(p)}, \quad \overline{M}_{y(p)} = M_{y(p)}$$

since $I_{x(p),y(p)} = 0$. Equation (9.30) then takes the form

$$\sigma_z = \frac{M_{x(p)}}{I_{x(p)}}y_p + \frac{M_{y(p)}}{I_{y(p)}}x_p \tag{9.56}$$

However, it must be appreciated that before $I_{x(p)}$ and $I_{y(p)}$ can be determined, I_x, I_y and I_{xy} must be known together with ϕ_p. Furthermore, the coordinates x, y of a point in the beam section must be transferred to the principal axes as must the components, M_x and M_y, of bending moment. Thus unless the position of the principal axes is obvious by inspection, the amount of computation required by the above method is far greater than direct use of Eq. (9.30) and an arbitrary, but convenient, set of centroidal axes.

9.8 Effect of shear forces on the theory of bending

So far our analysis has been based on the assumption that plane sections remain plane after bending. This assumption is only strictly true if the bending moments are produced by pure bending action rather than by shear loads, as is very often the case in practice. The presence of shear loads induces shear stresses in the cross-section of a beam which, as shown by elasticity theory, cause the cross-section to deform into the shape of a shallow inverted 's'. However, shear stresses in beams, the cross-sectional dimensions of which are small in relation to their length, are comparatively low in value so that the assumption of plane sections remaining plane after bending may be used with reasonable accuracy.

9.9 Load, shear force and bending moment relationships, general case

In Section 3.5 we derived load, shear force and bending moment relationships for loads applied in the vertical plane of a beam whose cross-section was at least singly symmetrical. These relationships are summarized in Eqs (3.8) and may be extended to the more general case in which loads are applied in both the horizontal (xz) and vertical (yz) planes of a beam of arbitrary cross-section. Thus for loads applied in a horizontal plane Eqs (3.8) become

$$\frac{\partial^2 M_y}{\partial z^2} = \frac{\partial S_x}{\partial z} = -w_x(z) \tag{9.57}$$

and for loads applied in a vertical plane Eqs (3.8) become

$$\frac{\partial^2 M_x}{\partial z^2} = \frac{\partial S_y}{\partial z} = -w_y(z) \tag{9.58}$$

We defined in Eqs (9.31) the parameters $\bar{M}_x$ and $\bar{M}_y$. We shall find it useful to establish similar parameters $\bar{S}_x$, $\bar{S}_y$, $\bar{w}_x$ and $\bar{w}_y$ in terms of the applied shear loads and distributed load intensities. Let us suppose that $\bar{S}$ bears the same relationship to $\bar{M}_y$ as S_x does to M_y (see Eqs (9.57)). Then

$$\bar{S}_x = \frac{\partial \bar{M}_y}{\partial z} = \frac{\partial M_y/\partial z - (\partial M_x/\partial z)I_{xy}/I_x}{1 - I_{xy}^2/I_x I_y}$$

or

$$\bar{S}_x = \frac{S_x - S_y I_{xy}/I_x}{1 - I_{xy}^2/I_x I_y} \tag{9.59}$$

Similarly

$$\bar{S}_y = \frac{S_y - S_x I_{xy}/I_y}{1 - I_{xy}^2/I_x I_y} \tag{9.60}$$

from Eqs (9.58)

The parameters $\bar{w}_x$ and $\bar{w}_y$ are related to load intensities by similarly derived expressions. Thus

$$\bar{w}_x = \frac{w_x - w_y I_{xy}/I_x}{1 - I_{xy}^2/I_x I_y} \tag{9.61}$$

$$\bar{w}_y = \frac{w_y - w_x I_{xy}/I_y}{1 - I_{xy}^2/I_x I_y} \tag{9.62}$$

The parameters $\bar{M}_x$, $\bar{M}_y$, $\bar{S}_x$, $\bar{S}_y$, $\bar{w}_x$ and $\bar{w}_y$ are often termed *effective* bending moments, shear forces and load intensities since they behave, in beam analysis, in an identical manner to actual bending moments, shear forces and load intensities.

9.10 Plastic bending

One of the primary assumptions of the preceding analysis of beams subjected to bending is that the stresses produced by the applied loads lie within the limit of proportionality of the material of the beam. Design, based on this elastic analysis, uses a working or allowable stress derived from the yield stress of the material of the beam with an appropriate factor of safety (see Section 8.7).

An alternative and increasingly favoured method of design of steel structures is to determine the working loads on a structure and then multiply these loads by a load factor (see Section 8.7) to obtain the ultimate loads which measure the *required maximum* strength of the structure. The components are then designed to have this required maximum strength. The problem, therefore, in the analysis of such structures is not to determine stresses due to applied loads as in elastic analysis but to determine loads that produce collapse. Clearly, when this occurs, the stress at one or more points in a structure will have exceeded the elastic limit so that the material at these points will be in a plastic state (see Section 8.3). We shall now, therefore, investigate the *plastic analysis* of beams subjected to bending.

Generally the problem is complex and is governed by the form of the stress–strain curve in tension and compression of the material of the beam. Fortunately mild steel beams, which are used extensively in civil engineering construction, possess structural properties that lend themselves to a relatively simple analysis of plastic bending.

We have seen in Section 8.3, Fig. 8.8, that mild steel obeys Hooke's law up to a sharply defined yield stress and then undergoes large strains during yielding until strain hardening causes an increase in stress. For the purpose of plastic analysis we shall neglect the upper and lower yield points and idealize the stress–strain curve as shown in Fig. 9.31. We shall also neglect the effects of strain hardening, but since this provides an increase in strength of the steel it is on the safe side to do so. Finally we shall assume that both Young's modulus, E, and the yield stress, σ_Y, have the same values in tension and compression and that plane sections remain plane after bending. The last assumption may be shown experimentally to be very nearly true.

Plastic bending of beams having a singly symmetrical cross-section

This is the most general case we shall discuss since the plastic bending of beams of arbitrary section is complex and is still being researched.

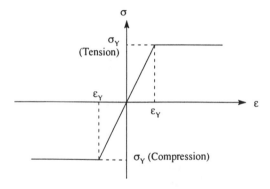

Fig. 9.31 Idealized stress–strain curve for mild steel

Consider the length of beam shown in Fig. 9.32(a) subjected to a positive bending moment, M, and possessing the singly symmetrical cross-section shown in Fig. 9.32(b). If M is sufficiently small the length of beam will bend elastically, producing at any section mm, the linear direct stress distribution of Fig. 9.32(c) where the stress, σ, at a distance y from the neutral axis of the beam is given by Eq. (9.9). In this situation the *elastic neutral axis* of the beam section passes through the centroid of area of the section (Eq. (9.5)).

Suppose now that M is increased. A stage will be reached where the maximum direct stress in the section, i.e. at the point furthest from the elastic neutral axis, is equal to the yield stress, σ_Y (Fig. 9.33(b)). The corresponding value of M is called the *yield moment*, M_Y, and is given by Eq. (9.9); thus

$$M_Y = \frac{\sigma_Y I}{y_2} \tag{9.63}$$

If the bending moment is further increased, the strain at the extremity y_2 of the section increases and exceeds the yield strain, ε_Y. However, due to plastic yielding the stress remains constant and equal to σ_Y as shown in the idealized stress–strain curve of Fig. 9.31. At some further value of M the stress at the lower extremity of the section also reaches the yield stress, σ_Y (Fig. 9.33(c)). Subsequent increases in bending moment cause the regions of plasticity at the extremities of the beam section to extend inwards, producing a situation similar to that shown in Fig. 9.33(d); at this stage the central portion or 'core' of the beam section remains elastic while the outer portions are plastic. Finally, with further increases in bending

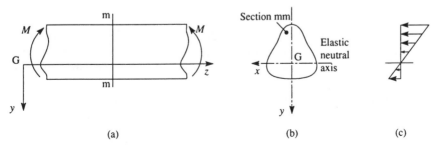

(a) (b) (c)

Fig. 9.32 Direct stress due to bending in a singly symmetrical section beam

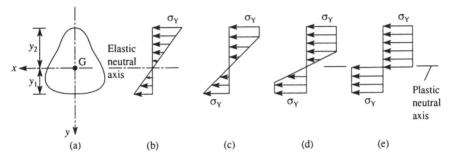

(a) (b) (c) (d) (e)

Fig. 9.33 Yielding of a beam section due to bending

moment the elastic core is reduced to a negligible size and the beam section is more or less completely plastic. Thus for all practical purposes the beam has reached its ultimate moment resisting capacity; the value of bending moment at this stage is known as the *plastic moment*, M_P, of the beam. The stress distribution corresponding to this moment may be idealized into two rectangular portions as shown in Fig. 9.33(e).

The problem now, therefore, is to determine the plastic moment, M_P. First, however, we must investigate the position of the neutral axis of the beam section when the latter is in its fully plastic state. One of the conditions used in establishing that the elastic neutral axis coincides with the centroid of a beam section was that stress is directly proportional to strain (Eq. (9.2)). It is clear that this is no longer the case for the stress distributions of Figs 9.33(c), (d) and (e). In Fig. 9.33(e) the beam section above the *plastic neutral axis* is subjected to a uniform compressive stress, σ_Y, while below the neutral axis the stress is tensile and also equal to σ_Y. Suppose that the area of the beam section below the plastic neutral axis is A_1, and that above, A_2 (Fig. 9.34(a)). Since M_P is a pure bending moment the total direct load on the beam section must be zero. Thus from Fig. 9.34

$$\sigma_Y A_1 = \sigma_Y A_2$$

so that

$$A_1 = A_2 \tag{9.64}$$

Therefore if the total cross-sectional area of the beam section is A,

$$A_1 = A_2 = \frac{A}{2} \tag{9.65}$$

and we see that the plastic neutral axis divides the beam section into two equal areas. Clearly for doubly symmetrical sections or for singly symmetrical sections in which the plane of the bending moment is perpendicular to the axis of symmetry, the elastic and plastic neutral axes coincide.

The plastic moment, M_P, can now be found by taking moments of the resultants of the tensile and compressive stresses about the neutral axis. These stress resultants

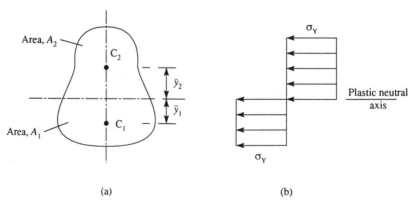

(a)   (b)

Fig. 9.34   Position of the plastic neutral axis in a beam section

act at the centroids C_1 and C_2 of the areas A_1 and A_2, respectively. Thus from Fig. 9.34

$$M_P = \sigma_Y A_1 \bar{y}_1 + \sigma_Y A_2 \bar{y}_2$$

or, using Eq. (9.65)

$$M_P = \sigma_Y \frac{A}{2} (\bar{y}_1 + \bar{y}_2) \tag{9.66}$$

Equation (9.66) may be written in a similar form to Eq. (9.13); thus

$$M_P = \sigma_Y Z_P \tag{9.67}$$

where

$$Z_P = \frac{A(\bar{y}_1 + \bar{y}_2)}{2} \tag{9.68}$$

Z_P is known as the *plastic modulus* of the cross-section. Note that the elastic modulus, Z_e, has two values for a beam of singly symmetrical cross-section (Eqs (9.12)) whereas the plastic modulus is single-valued.

Shape factor

The ratio of the plastic moment of a beam to its yield moment is known as the *shape factor*, f. Thus

$$f = \frac{M_P}{M_Y} = \frac{\sigma_Y Z_P}{\sigma_Y Z_e} = \frac{Z_P}{Z_e} \tag{9.69}$$

where Z_P is given by Eq. (9.68) and Z_e is the minimum elastic section modulus, I/y_2. It can be seen from Eq. (9.69) that f is solely a function of the geometry of the beam cross-section.

Example 9.11 Determine the yield moment, the plastic moment and the shape factor for a rectangular section beam of breadth b and depth d.

The elastic and plastic neutral axes of a rectangular cross-section coincide (Eq. (9.65)) and pass through the centroid of area of the section. Thus, from Eq. (9.63)

$$M_Y = \frac{\sigma_Y bd^3/12}{d/2} = \sigma_Y \frac{bd^2}{6} \tag{i}$$

and from Eq. (9.66)

$$M_P = \sigma_Y \frac{bd}{2} \left(\frac{d}{4} + \frac{d}{4} \right) = \sigma_Y \frac{bd^2}{4} \tag{ii}$$

Substituting for M_P and M_Y in Eqs (9.69) we obtain

$$f = \frac{M_P}{M_Y} = \frac{3}{2} \tag{iii}$$

Note that the plastic collapse of a rectangular section beam occurs at a bending moment that is 50% greater than the moment at initial yielding of the beam.

Example 9.12   Determine the shape factor for the I-section beam shown in Fig. 9.35(a).

Again, as in Ex. 9.11, the elastic and plastic neutral axes coincide with the centroid, G, of the section.

In the fully plastic condition the stress distribution in the beam is that shown in Fig. 9.35(b). The total direct force in the upper flange is

$$\sigma_Y bt_f \quad \text{(compression)}$$

and its moment about Gx is

$$\sigma_Y bt_f \left(\frac{d}{2} - \frac{t_f}{2}\right) \equiv \frac{\sigma_Y bt_f}{2}(d - t_f) \tag{i}$$

Similarly the total direct force in the web above Gx is

$$\sigma_Y t_w \left(\frac{d}{2} - t_f\right) \quad \text{(compression)}$$

and its moment about Gx is

$$\sigma_Y t_w \left(\frac{d}{2} - t_f\right) \frac{1}{2} \left(\frac{d}{2} - t_f\right) \equiv \frac{\sigma_Y t_w}{8}(d - 2t_f)^2 \tag{ii}$$

The lower half of the section is in tension and contributes the same moment about Gx so that the total plastic moment, M_P, of the complete section is given by

$$M_P = \sigma_Y [bt_f(d - t_f) + \tfrac{1}{4}t_w(d - 2t_f)^2] \tag{iii}$$

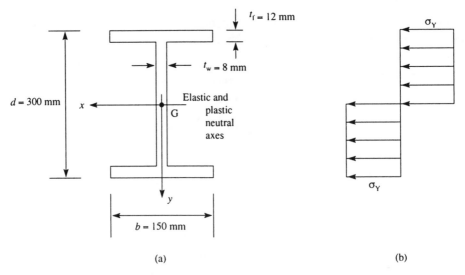

(a)   (b)

Fig. 9.35   Beam section of Ex. 9.12

Comparing Eqs (9.67) and (iii) we see that Z_P is given by

$$Z_P = bt_f(d - t_f) + \tfrac{1}{4}t_w(d - 2t_f)^2 \qquad \text{(iv)}$$

Alternatively we could have obtained Z_P from Eq. (9.68).

The second moment of area, I, of the section about the common neutral axis is

$$I = \frac{bd^3}{12} - \frac{(b - t_w)(d - 2t_f)^3}{12}$$

so that the elastic modulus Z_e is given by

$$Z_e = \frac{I}{d/2} = \frac{2}{d}\left[\frac{bd^3}{12} - \frac{(b - t_w)(d - 2t_f)^3}{12}\right] \qquad \text{(v)}$$

Substituting the actual values of the dimensions of the section in Eqs (iv) and (v) we obtain

$$Z_P = 150 \times 12(300 - 12) + \tfrac{1}{4} \times 8(300 - 2 \times 12)^2 = 6{\cdot}7 \times 10^5 \text{ mm}^3$$

and $\quad Z_e = \dfrac{2}{300}\left[\dfrac{150 \times 300^3}{12} - \dfrac{(150 - 8)(300 - 24)^3}{12}\right] = 5{\cdot}9 \times 10^5 \text{ mm}^3$

Therefore from Eqs (9.69)

$$f = \frac{M_P}{M_Y} = \frac{Z_P}{Z_e} = \frac{6{\cdot}7 \times 10^5}{5{\cdot}9 \times 10^5} = 1{\cdot}14$$

and we see that the fully plastic moment is only 14% greater than the moment at initial yielding.

Example 9.13 Determine the shape factor of the T-section shown in Fig. 9.36.

In this case the elastic and plastic neutral axes are not coincident. Suppose that the former is a depth y_e from the upper surface of the flange and the latter a depth y_P. The elastic neutral axis passes through the centroid of the section, the location of

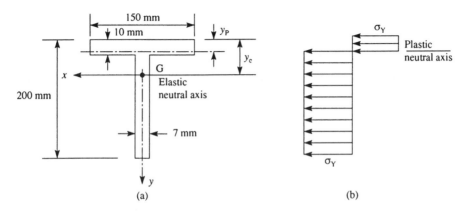

Fig. 9.36 Beam section of Ex. 9.13

which is found in the usual way. Hence, taking moments of areas about the upper surface of the flange

$$(150 \times 10 + 190 \times 7)y_e = 150 \times 10 \times 5 + 190 \times 7 \times 105$$

which gives

$$y_e = 52 \cdot 0 \text{ mm}$$

The second moment of area of the section about the elastic neutral axis is then, using Eq. (9.38)

$$I = \frac{150 \times 52^3}{3} - \frac{143 \times 42^3}{3} + \frac{7 \times 148^3}{3} = 11 \cdot 1 \times 10^6 \text{ mm}^4$$

Therefore

$$Z_e = \frac{11 \cdot 1 \times 10^6}{148} = 75\,000 \text{ mm}^3$$

Note that we choose the least value for Z_e since the stress will be a maximum at a point furthest from the elastic neutral axis.

The plastic neutral axis divides the section into equal areas (see Eq. (9.65)). Inspection of Fig. 9.36 shows that the flange area is greater than the web area so that the plastic neutral axis must lie within the flange. Hence

$$150y_P = 150(10 - y_P) + 190 \times 7$$

from which

$$y_P = 9 \cdot 4 \text{ mm}$$

Equation (9.68) may be interpreted as the first moment, about the plastic neutral axis, of the area above the plastic neutral axis plus the first moment of the area below the plastic neutral axis. Hence

$$Z_P = 150 \times 9 \cdot 4 \times 4 \cdot 7 + 150 \times 0 \cdot 6 \times 0 \cdot 3 + 190 \times 7 \times 95 \cdot 6 = 133\,800 \text{ mm}^3$$

The shape factor f is, from Eqs (9.69)

$$f = \frac{M_P}{M_Y} = \frac{Z_P}{Z_e} = \frac{133\,800}{75\,000} = 1 \cdot 78$$

Moment–curvature relationships

From Eqs (9.8) we see that the curvature k of a beam subjected to elastic bending is given by

$$k = \frac{1}{R} = \frac{M}{EI} \tag{9.70}$$

At yield, when M is equal to the yield moment, M_Y

$$k_Y = \frac{M_Y}{EI} \tag{9.71}$$

Thus the moment–curvature relationship for a beam in the linear elastic range may be expressed in non-dimensional form by combining Eqs (9.70) and (9.71), i.e.

$$\frac{M}{M_Y} = \frac{k}{k_Y} \tag{9.72}$$

This relationship is represented by the linear portion of the moment–curvature diagram shown in Fig. 9.37. When the bending moment is greater than M_Y part of the beam becomes fully plastic and the moment–curvature relationship is non-linear. As the plastic region in the beam section extends inwards towards the neutral axis the curve becomes flatter as rapid increases in curvature are produced by small increases in moment. Finally, the moment–curvature curve approaches the horizontal line $M = M_P$ as an asymptote when, theoretically, the curvature is infinite at the collapse load. From Eqs (9.69) we see that when $M = M_P$, the ratio $M/M_Y = f$, the shape factor. Clearly the equation of the non-linear portion of the moment–curvature diagram depends upon the particular cross-section being considered.

Suppose a beam of rectangular cross-section is subjected to a bending moment which produces fully plastic zones in the outer portions of the section (Fig. 9.38(a)); the depth of the elastic core is d_e. The total bending moment, M, corresponding to the stress distribution of Fig. 9.38(b) is given by

$$M = 2\sigma_Y b \frac{1}{2} (d - d_e) \frac{1}{2} \left(\frac{d}{2} + \frac{d_e}{2} \right) + 2 \frac{\sigma_Y}{2} b \frac{d_e}{2} \frac{2}{3} \frac{d_e}{2}$$

which simplifies to

$$M = \frac{\sigma_Y b d^2}{12} \left(3 - \frac{d_e^2}{d^2} \right) = \frac{M_Y}{2} \left(3 - \frac{d_e^2}{d^2} \right) \tag{9.73}$$

Note that when $d_e = d$, $M = M_Y$ and when $d_e = 0$, $M = 3M_Y/2 = M_P$ as derived in Ex. 9.11.

The curvature of the beam at the section shown may be found using Eq. (9.2) and applying this equation to a point on the outer edge of the elastic core. Thus

$$\sigma_Y = E \frac{d_e}{2R}$$

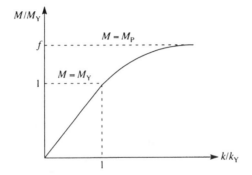

Fig. 9.37 Moment–curvature diagram for a beam

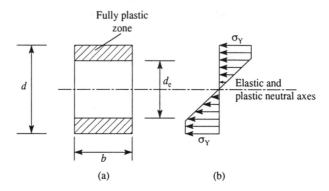

Fig. 9.38 Plastic bending of a rectangular-section beam

or
$$k = \frac{1}{R} = \frac{2\sigma_Y}{Ed_e}$$
(9.74)

The curvature of the beam at yield is obtained from Eq. (9.71), i.e.

$$k_Y = \frac{M_Y}{EI} = \frac{2\sigma_Y}{Ed}$$
(9.75)

Combining Eqs (9.74) and (9.75) we obtain

$$\frac{k}{k_Y} = \frac{d}{d_e}$$
(9.76)

Substituting for d_e/d in Eq. (9.73) from Eq. (9.76) we have

$$M = \frac{M_Y}{2}\left(3 - \frac{k_Y^2}{k^2}\right)$$

whence
$$\frac{k}{k_Y} = \frac{1}{\sqrt{3 - 2M/M_Y)}}$$
(9.77)

Equation (9.77) gives the moment–curvature relationship for a rectangular section beam for $M_Y \leqslant M \leqslant M_P$, i.e. for the non-linear portion of the moment–curvature diagram of Fig. 9.37 for the particular case of a rectangular section beam. Corresponding relationships for beams of different section are found in a similar manner.

We have seen that for bending moments in the range $M_Y \leqslant M \leqslant M_P$ a beam section comprises fully plastic regions and a central elastic core. Thus yielding occurs in the plastic regions with no increase in stress whereas in the elastic core increases in deformation are accompanied by increases in stress. The deformation of the beam is therefore controlled by the elastic core, a state sometimes termed *contained plastic flow*. As M approaches M_P the moment–curvature diagram is asymptotic to the line $M = M_P$ so that large increases in deformation occur without any increase in moment, a condition known as *unrestricted plastic flow*.

Plastic hinges

The presence of unrestricted plastic flow at a section of a beam leads us to the concept of the formation of *plastic hinges* in beams and other structures.

Consider the simply supported beam shown in Fig. 9.39(a); the beam carries a concentrated load, W, at mid-span. The bending moment diagram (Fig. 9.39(b)) is triangular in shape with a maximum moment equal to $WL/4$. If W is increased in value until $WL/4 = M_P$, the mid-span section of the beam will be fully plastic with regions of plasticity extending towards the supports as the bending moment decreases; no plasticity occurs in beam sections for which the bending moment is less than M_Y. Clearly, unrestricted plastic flow now occurs at the mid-span section where large increases in deformation take place with no increase in load. The beam therefore behaves as two rigid beams connected by a *plastic hinge* which allows them to rotate relative to each other. The value of W given by $W = 4M_P/L$ is the *collapse load* for the beam.

The length, L_P, of the plastic region of the beam may be found using the fact that at each section bounding the region the bending moment is equal to M_Y. Thus

$$M_Y = \frac{W}{2}\left(\frac{L - L_P}{2}\right)$$

Substituting for $W(=4M_P/L)$ we obtain

$$M_Y = \frac{M_P}{L}(L - L_P)$$

from which

$$L_P = L\left(1 - \frac{M_Y}{M_P}\right)$$

or, from Eqs (9.69),

$$L_P = L\left(1 - \frac{1}{f}\right) \tag{9.78}$$

For a rectangular section beam $f = 1\cdot5$ (see Ex. 9.11), giving $L_P = L/3$. For the I-section beam of Ex. 9.12, $f = 1\cdot14$ and $L_P = 0\cdot12L$ so that the plastic region in this case is much smaller than that of a rectangular section beam; this is generally true for I-section beams.

It is clear from the above that plastic hinges form at sections of maximum bending moment.

Plastic analysis of beams

We can now use the concept of plastic hinges to determine the collapse or ultimate load of beams in terms of their individual yield moment, M_P, which may be found for a particular beam section using Eq. (9.67).

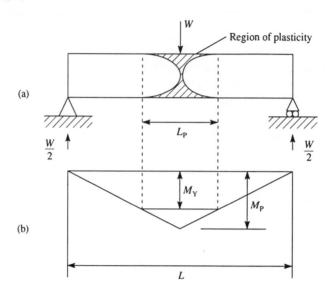

Fig. 9.39 Formation of a plastic hinge in a simply supported beam

For the case of the simply supported beam of Fig. 9.39 we have seen that the formation of a single plastic hinge is sufficient to produce failure; this is true for all statically determinate systems. Having located the position of the plastic hinge, at which the moment is equal to M_P, the collapse load is found from simple statics. Thus for the beam of Fig. 9.39, taking moments about the mid-span section, we have

$$\frac{W_U}{2}\frac{L}{2} = M_P$$

or
$$W_U = \frac{4M_P}{L} \quad \text{(as deduced before)}$$

where W_U is the ultimate value of the load W.

Example 9.14   Determine the ultimate load for a simply supported, rectangular section beam, breadth b, depth d, having a span L and subjected to a uniformly distributed load of intensity w.

The maximum bending moment occurs at mid-span and is equal to $wL^2/8$ (see Section 3.4). Thus the plastic hinge forms at mid-span when this bending moment is equal to M_P, the corresponding ultimate load intensity being w_U. Thus

$$\frac{w_U L^2}{8} = M_P \tag{i}$$

From Ex. 9.11, Eq. (ii)

$$M_P = \sigma_Y \frac{bd^2}{4}$$

so that
$$w_U = \frac{8M_P}{L^2} = \frac{2\sigma_Y bd^2}{L^2}$$

where σ_Y is the yield stress of the material of the beam.

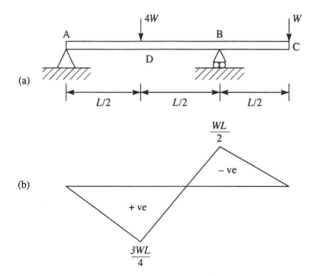

Fig. 9.40 Beam of Ex. 9.15

Example 9.15 The simply supported beam ABC shown in Fig. 9.40(a) has a cantilever overhang and supports loads of $4W$ and W. Determine the value of W at collapse in terms of the plastic moment, M_P, of the beam.

The bending moment diagram for the beam is constructed using the method of Section 3.4 and is shown in Fig. 9.40(b). Clearly as W is increased a plastic hinge will form first at D, the point of application of the $4W$ load. Thus, at collapse

$$\tfrac{3}{4} W_U L = M_P$$

so that
$$W_U = \frac{4M_P}{3L}$$

where W_U is the value of W that causes collapse.

The formation of a plastic hinge in a statically determinate beam produces large, increasing deformations which ultimately result in failure with no increase in load. In this condition the beam behaves as a mechanism with different lengths of beam rotating relative to each other about the plastic hinge. The terms *failure mechanism* or *collapse mechanism* are often used to describe this state.

In a statically indeterminate system the formation of a single plastic hinge does not necessarily mean collapse. Consider the propped cantilever shown in Fig. 9.41(a). The bending moment diagram may be drawn after the reaction at C has been determined by any suitable method of analysis of statically indeterminate beams (see Chapter 16) and is shown in Fig. 9.41(b).

As the value of W is increased a plastic hinge will form first at A where the bending moment is greatest. However, this does not mean that the beam will collapse. Instead it behaves as a statically determinate beam with a point load at B and a moment M_P at A. Further increases in W eventually result in the formation of a

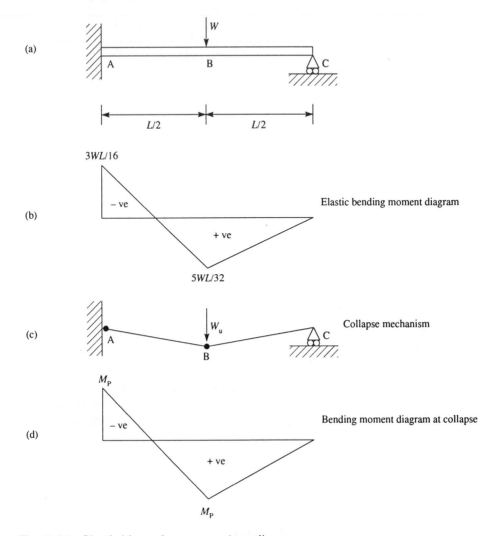

Fig. 9.41   Plastic hinges in a propped cantilever

second plastic hinge at B (Fig. 9.41(c)) when the bending moment at B reaches the value M_P. The beam now behaves as a mechanism and failure occurs with no further increase in load. The bending moment diagram for the beam is now as shown in Fig. 9.41(d) with values of bending moment of $-M_P$ at A and M_P at B. Comparing the bending moment diagram at collapse with that corresponding to the elastic deformation of the beam (Fig. 9.41(b)) we see that a redistribution of bending moment has occurred. This is generally the case in statically indeterminate systems whereas in statically determinate systems the bending moment diagrams in the elastic range and at collapse have identical shapes (see Figs 9.39(b) and 9.40(b)). In the beam of Fig. 9.41 the elastic bending moment diagram has a maximum at A. After the formation of the plastic hinge at A the bending moment remains constant while the bending moment at B increases until the second plastic hinge forms. Thus this redistribution of moments tends to increase the ultimate strength of statically indeterminate structures since failure at one section leads to other portions of the structure supporting additional load.

Having located the positions of the plastic hinges and using the fact that the moment at these hinges is M_P, we may determine the ultimate load, W_U, by statics. Therefore taking moments about A we have

$$M_P = W_U \frac{L}{2} - R_C L \qquad (9.79)$$

where R_C is the vertical reaction at the support C. Now considering the equilibrium of the length BC we obtain

$$R_C \frac{L}{2} = M_P \qquad (9.80)$$

Eliminating R_C from Eqs (9.79) and (9.80) gives

$$W_U = \frac{6M_P}{L} \qquad (9.81)$$

Note that in this particular problem it is unnecessary to determine the elastic bending moment diagram to solve for the ultimate load which is obtained using statics alone. This is a convenient feature of plastic analysis and leads to a much simpler solution of statically indeterminate structures than an elastic analysis. Furthermore, the magnitude of the ultimate load is not affected by structural imperfections such as a sinking support, whereas the same kind of imperfection would have an appreciable effect on the elastic behaviour of a structure. Note also that the principle of superposition (Section 3.7), which is based on the linearly elastic behaviour of a structure, does not hold for plastic analysis. In fact the plastic behaviour of a structure depends upon the order in which the loads are applied as well as their final values. We therefore assume in plastic analysis that all loads are applied simultaneously and that the ratio of the loads remains constant during loading.

An alternative and powerful method of analysis uses the principle of virtual work (see Section 15.2), which states that for a structure that is in equilibrium and that is given a small virtual displacement, the sum of the work done by the internal forces is equal to the work done by the external forces.

Consider the propped cantilever of Fig. 9.41(a); its collapse mechanism is shown in Fig. 9.41(c). At the instant of collapse the cantilever is in equilibrium with plastic

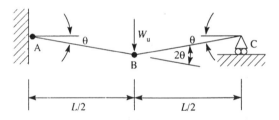

Fig. 9.42 Virtual displacements in propped cantilever of Fig. 9.41

hinges at A and B where the moments are each M_P as shown in Fig. 9.41(d). Suppose that AB is given a small rotation, θ. From geometry, BC also rotates through an angle θ as shown in Fig. 9.42; the vertical displacement of B is then $\theta L/2$. The external forces on the cantilever which do work during the virtual displacement are comprised solely of W_U since the vertical reactions at A and C are not displaced. The internal forces which do work consist of the plastic moments, M_P, at A and B and which resist rotation. Hence

$$W_U \theta \frac{L}{2} = (M_P)_A \theta + (M_P)_B 2\theta \quad \text{(see Section 15.1)}$$

from which $W_U = \dfrac{6M_P}{L}$ as before.

We have seen that the plastic hinges form at beam sections where the bending moment diagram attains a peak value. It follows that for beams carrying a series of point loads, plastic hinges are located at the load positions. However, in some instances several collapse mechanisms are possible, each giving different values of ultimate load. For example, if the propped cantilever of Fig. 9.41(a) supports two point loads as shown in Fig. 9.43(a), three possible collapse mechanisms are possible (Figs 9.43(b), (c) and (d)). Each possible collapse mechanism should be analysed and the lowest ultimate load selected.

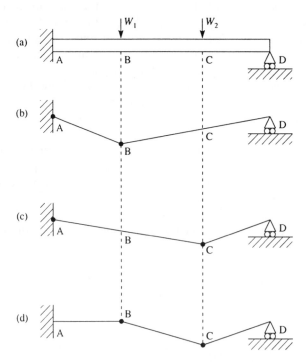

Fig. 9.43 Possible collapse mechanisms in a propped cantilever supporting two concentrated loads

Plastic design of beams

It is now clear that the essential difference between the plastic and elastic methods of design is that the former produces a structure having a more or less uniform factor of safety against collapse of all its components, whereas the latter produces a uniform factor of safety against yielding. The former method in fact gives an indication of the true factor of safety against collapse of the structure which may occur at loads only marginally greater than the yield load, depending on the cross-sections used. For example, a rectangular section mild steel beam has an ultimate strength 50% per cent greater than its yield strength (see Ex. 9.11), whereas for an I-section beam the margin is in the range 10–20% (see Ex. 9.12). It is also clear that each method of design will produce a different section for a given structural component. This distinction may be more readily understood by referring to the redistribution of bending moment produced by the plastic collapse of a statically indeterminate beam.

Two approaches to the plastic design of beams are indicated by the previous analysis. The most direct method would calculate the working loads, determine the required strength of the beam by the application of a suitable load factor, obtain by a suitable analysis the required plastic moment in terms of the ultimate load and finally, knowing the yield stress of the material of the beam, determine the required plastic section modulus. An appropriate beam section is then selected from a handbook of structural sections. The alternative method would assume a beam section, calculate the plastic moment of the section and hence the ultimate load for the beam. This value of ultimate load is then compared with the working loads to determine the actual load factor, which would then be checked against the prescribed value.

Example 9.16 The propped cantilever of Fig. 9.41(a) is 10 m long and is required to carry a load of 100 kN at mid-span. If the yield stress of mild steel is 300 N/mm², suggest a suitable section using a load factor against failure of 1·5.

The required ultimate load of the beam is $1\cdot5 \times 100 = 150$ kN. Thus from Eq. (9.81) the required plastic moment M_P is given by

$$M_P = \frac{150 \times 10}{6} = 250 \text{ kN m}$$

From Eq. (9.67) the minimum plastic modulus of the beam section is

$$Z_P = \frac{250 \times 10^6}{300} = 833\,300 \text{ mm}^3$$

Referring to an appropriate handbook we see that a Universal Beam, 406 mm × 140 mm × 46 kg/m, has a plastic modulus of 886·3 cm³. This section therefore possesses the required ultimate strength and includes a margin to allow for its self-weight. Note that unless some allowance has been made for self-weight in the estimate of the working loads the design should be rechecked to include this effect.

Effect of axial load on plastic moment

We shall investigate the effect of axial load on plastic moment with particular reference to an I-section beam, one of the most common structural shapes, which is subjected to a positive bending moment and a compressive axial load, P, (Fig. 9.44(a)).

If the beam section were subjected to its plastic moment only, the stress distribution shown in Fig. 9.44(b) would result. However, the presence of the axial load causes additional stresses which cannot, obviously, be greater than σ_Y. Thus the region of the beam section supporting compressive stresses is increased in area while the region subjected to tensile stresses is decreased in area. Clearly some of the compressive stresses are due to bending and some due to axial load so that the modified stress distribution is as shown in Fig. 9.44(c).

Since the beam section is doubly symmetrical it is reasonable to assume that the area supporting the compressive stress due to bending is equal to the area supporting the tensile stress due to bending, both areas being symmetrically arranged about the original plastic neutral axis. Thus from Fig. 9.44(d) the reduced plastic moment, $M_{\text{P.R}}$, is given by

$$M_{\text{P.R}} = \sigma_Y (Z_\text{P} - Z_\text{a}) \tag{9.82}$$

where Z_a is the plastic section modulus for the area on which the axial load is assumed to act. From Eq. (9.68)

$$Z_\text{a} = \frac{2at_\text{w}}{2}\left(\frac{a}{2} + \frac{a}{2}\right) = a^2 t_\text{w}$$

Also

$$P = 2at_\text{w}\sigma_Y$$

so that

$$a = \frac{P}{2t_\text{w}\sigma_Y}$$

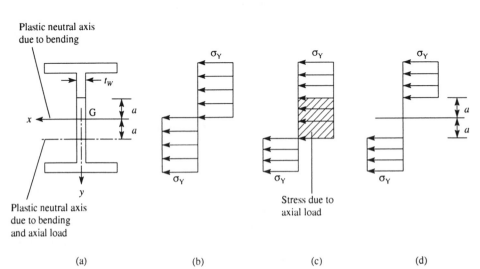

Stress due to axial load

(a)   (b)   (c)   (d)

Fig. 9.44 Combined bending and axial compression

Substituting for Z_a, in Eq. (9.82) and then for a, we obtain

$$M_{P,R} = \sigma_Y\left(Z_P - \frac{P^2}{4t_w\sigma_Y^2}\right) \tag{9.83}$$

Let σ_a be the mean axial stress due to P taken over the complete area, A, of the beam section. Then

$$P = \sigma_a A$$

Substituting for P in Eq. (9.83)

$$M_{P,R} = \sigma_Y\left(Z_P - \frac{A^2}{4t_w}\frac{\sigma_a^2}{\sigma_Y^2}\right) \tag{9.84}$$

Thus the reduced plastic section modulus may be expressed in the form

$$Z_{P,R} = Z_P - Kn^2 \tag{9.85}$$

where K is a constant that depends upon the geometry of the beam section and n is the ratio of the mean axial stress to the yield stress of the material of the beam.

Equations (9.84) and (9.85) are applicable as long as the neutral axis lies in the web of the beam section. In the rare case when this is not so, reference should be made to advanced texts on structural steel design. In addition the design of beams carrying compressive loads is influenced by considerations of local and overall instability, as we shall see in Chapter 18.

Problems

P.9.1 A girder 10 m long has the cross-section shown in Fig. P.9.1(a) and is simply supported over a span of 6 m (see Fig. P.9.1(b)). If the maximum direct stress in the girder is limited to 150 N/mm², determine the maximum permissible uniformly distributed load that may be applied to the girder.

Ans. 84·3 N/m.

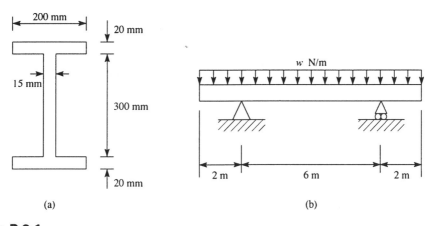

(a)   (b)

Fig. P.9.1

P.9.2   A 230 mm × 300 mm timber cantilever of rectangular cross-section projects 2·5 m from a wall and carries a load of 13 300 N at its free end. Calculate the maximum direct stress in the beam due to bending.

   Ans.   9·6 N/mm^2.

P.9.3   A floor carries a uniformly distributed load of 16 kN/m^2 and is supported by joists 300 mm deep and 110 mm wide; the joists in turn are simply supported over a span of 4 m. If the maximum stress in the joists is not to exceed 7 N/mm^2, determine the distance apart, centre to centre, at which the joists must be spaced.

   Ans.   0·36 m.

P.9.4   A wooden mast 15 m high tapers linearly from 250 mm diameter at the base to 100 mm at the top. At what point will the mast break under a horizontal load applied at the top? If the maximum permissible stress in the wood is 35 N/mm^2, calculate the magnitude of the load that will cause failure.

   Ans.   5 m from the top, 2320 N.

P.9.5   A main beam in a steel framed structure is 5 m long and simply supported at each end. The beam carries two cross-beams at distances of 1·5 m and 3 m from one end, each of which transmits a load of 20 kN to the main beam. Design the main beam using an allowable stress of 155 N/mm^2; make adequate allowance for the effect of self-weight.

   Ans.   Universal Beam, 254 mm × 102 mm × 25 kg/m.

P.9.6   A short column, whose cross-section is shown in Fig. P.9.6 is subjected to a compressive load, *P*, at the centroid of one of its flanges. Find the value of *P* such that the maximum compressive stress does not exceed 150 N/mm^2.

   Ans.   845 kN.

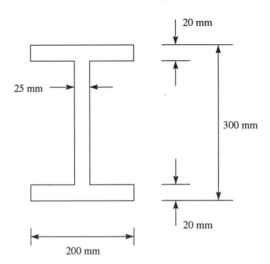

Fig. P.9.6

P.9.7   A vertical chimney built in brickwork has a uniform rectangular cross-section as shown in Fig. P.9.7(a) and is built to a height of 15 m. The brickwork has a density of 2000 kg/m³ and the wind pressure is equivalent to a uniform horizontal pressure of 750 N/m² acting over one face. Calculate the stress at each of the points A and B at the base of the chimney.

Ans.   (A) 0·11 N/mm² (compression), (B) 0·48 N/mm² (compression).

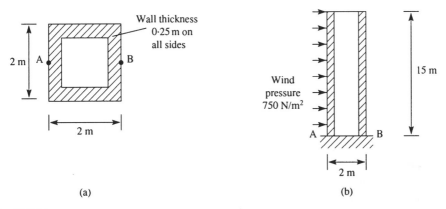

(a)   (b)

Fig. P.9.7

P.9.8   A cantilever beam of length 2 m has the cross-section shown in Fig. P.9.8. If the beam carries a uniformly distributed load of 5 kN/m together with a compressive axial load of 100 kN applied at its free end, calculate the maximum direct stress in the cross-section of the beam.

Ans.   121·5 N/mm² (compression) at the built-in end and at the bottom of the leg.

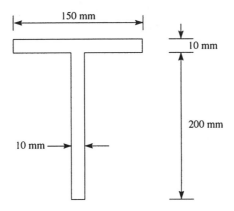

Fig. P.9.8

P.9.9   The section of a thick beam has the dimensions shown in Fig. P.9.9. Calculate the section properties I_x, I_y and I_{xy} referred to horizontal and vertical axes through the centroid of the section. Determine also the direct stress at the point A due to a bending moment $M_y = 55$ N m.

Ans.   114 N/mm² (tension).

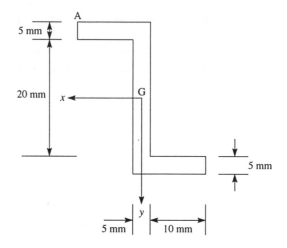

Fig. P.9.9

P.9.10   A beam possessing the thick section shown in Fig. P.9.10 is subjected to a bending moment of 12 kN m applied in a plane inclined at 30° to the right of vertical and in a sense such that its components M_x and M_y are both negative. Calculate the magnitude and position of the maximum direct stress in the beam cross-section.

Ans.   89·6 N/mm² (tension) at A.

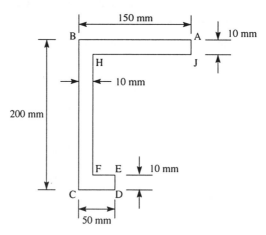

Fig. P.9.10

P.9.11   The cross-section of a beam/floor slab arrangement is shown in Fig. P.9.11. The complete section is simply supported over a span of 10 m and, in addition to its self-weight, carries a concentrated load of 25 kN acting vertically downwards at mid-span. If the density of concrete is 2000 kg/m³, calculate the maximum direct stress at the point A in its cross-section.

Ans.   5·4 N/mm² (tension).

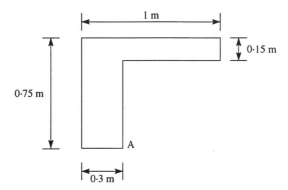

Fig. P.9.11

P.9.12 A precast concrete beam has the cross-section shown in Fig. P.9.12 and carries a vertically downward uniformly distributed load of 100 kN/m over a simply supported span of 4 m. Calculate the maximum direct stress in the cross-section of the beam, indicating clearly the point at which it acts.

Ans. -27.2 N/mm² (compression) at B.

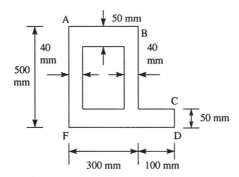

Fig. P.9.12

P.9.13 A thin-walled, cantilever beam of unsymmetrical cross-section supports shear loads at its free end as shown in Fig. P.9.13. Calculate the value of direct

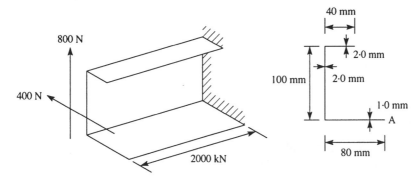

Fig. P.9.13

stress at the extremity of the lower flange (point A) at a section half-way along the beam if the position of the shear loads is such that no twisting of the beam occurs.

Ans.   194 N/mm² (tension).

P.9.14   A thin-walled cantilever with walls of constant thickness *t* has the cross-section shown in Fig. P.9.14. The cantilever is loaded by a vertical force *P* at the tip and a horizontal force 2*P* at the mid-section. Determine the direct stress at the points A and B in the cross-section at the built-in end.

Ans.   (A) − 1·84 *PL/td²*, (B) 0·1 *PL/td²*.

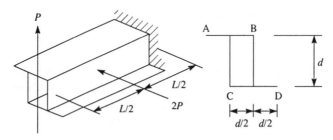

Fig. P.9.14

P.9.15   A cold-formed, thin-walled beam section of constant thickness has the profile shown in Fig. P.9.15. Calculate the position of the neutral axis and the maximum direct stress for a bending moment of 350 kN m applied about the horizontal axis G*x*.

Ans.   α = 51°40′, ± 10 N/mm².

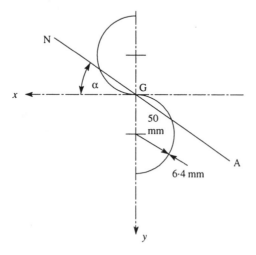

Fig. P.9.15

P.9.16   Determine the plastic moment and shape factor of a beam of solid circular cross-section having a radius *r* and yield stress σ_Y.

Ans. $M_P = 1 \cdot 33 \, \sigma_Y r^3$, $f = 1 \cdot 69$.

P.9.17 Determine the plastic moment and shape factor for a thin-walled box girder whose cross-section has a breadth b, depth d and a constant wall thickness t. Calculate f for $b = 200$ mm, $d = 300$ mm.

Ans. $M_P = \sigma_Y td(2b + d)/2$, $f = 1·17$.

P.9.18 A beam having the cross-section shown in Fig. P.9.18 is fabricated from mild steel which has a yield stress of 300 N/mm^2. Determine the plastic moment of the section and its shape factor.

Ans. $256·5$ kN m, $1·52$.

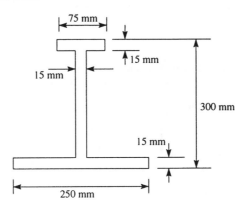

Fig. P.9.18

P.9.19 A cantilever beam of length 6 m has an additional support at a distance of 2 m from its free end as shown in Fig. P.9.19. Determine the minimum value of W at which collapse occurs if the section of the beam is identical to that of Fig. P.9.18. State clearly the form of the collapse mechanism corresponding to this ultimate load.

Ans. $128·3$ kN, plastic hinge at C.

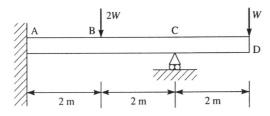

Fig. P.9.19

P.9.20 A beam of length L is rigidly built-in at each end and carries a uniformly distributed load of intensity w along its complete span. Determine the ultimate strength of the beam in terms of the plastic moment, M_P, of its cross-section.

Ans. $16\,M_P/L^2$.

P.9.21   A simply supported beam has a cantilever overhang and supports loads as shown in Fig. P.9.21. Determine the collapse load of the beam, stating the position of the corresponding plastic hinge.

Ans. $2M_p/L$, plastic hinge at D.

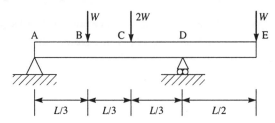

Fig. P.9.21

P.9.22   Determine the ultimate strength of the propped cantilever shown in Fig. P.9.22 and specify the corresponding collapse mechanism.

Ans. $W = 4M_p/L$, plastic hinges at A and C.

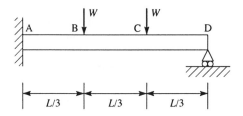

Fig. P.9.22

P.9.23   The working loads, W, on the propped cantilever of Fig. P.9.22 are each 150 kN and its span is 6 m. If the yield stress of mild steel is 300 N/mm², suggest a suitable section for the beam using a load factor of 1·75 against collapse.

Ans.   Universal Beam, 406 mm × 152 mm × 67 kg/m.

CHAPTER 10

Shear of Beams

Beams, as we saw in Chapter 3, are subjected to loads which induce internal shear forces in the planes of their cross-sections. These shear forces are distributed in a manner that depends to a large extent upon the geometry of the beam section. We shall now investigate this distribution of shear stress, beginning with the general case of an unsymmetrical section.

10.1 Shear stress distribution in a beam of unsymmetrical section

Consider an elemental length, δz, of a beam of arbitrary section subjected to internal shear forces S_x and S_y as shown in Fig. 10.1(a). The origin of the axes xyz coincides with the centroid G of the beam section. Let us suppose that the lines of action of S_x and S_y are such that no twisting of the beam occurs (see Section 10.4). The shear stresses induced are therefore due solely to shearing action and are not contributed to by torsion.

Imagine now that a 'slice' of width b_0 is taken through the length of the element. Let τ be the average shear stress along the edge, b_0, of the slice in a direction perpendicular to b_0 and in the plane of the cross-section (Fig. 10.1(b)); note that τ is not necessarily the absolute value of shear stress at this position. We saw in Chapter 7 that shear stresses on given planes induce equal, complementary shear stresses on planes perpendicular to the given planes. Thus, τ on the cross-sectional face of the slice induces shear stresses τ on the flat longitudinal face of the slice. In addition shear loads, as we saw in Chapter 3, produce internal bending moments which, in turn, give rise to direct stresses in beam cross-sections. Therefore on any filament, $\delta A'$, of the slice there is a direct stress σ_z at the section z and a direct stress $\sigma_z + (\partial \sigma_z / \partial z)\delta z$ at the section $z + \delta z$ (Fig. 10.1(b)). The slice is therefore in equilibrium in the z direction under the combined action of the direct stress due to bending and the complementary shear stress, τ. Hence

$$\tau b_0 \, \delta z + \int_{A'} \sigma_z \, dA' - \int_{A'} \left(\sigma_z + \frac{\partial \sigma_z}{\partial z} \delta z \right) dA' = 0$$

which, when simplified, becomes

$$\tau b_0 = \int_{A'} \frac{\partial \sigma_z}{\partial z} \, dA' \tag{10.1}$$

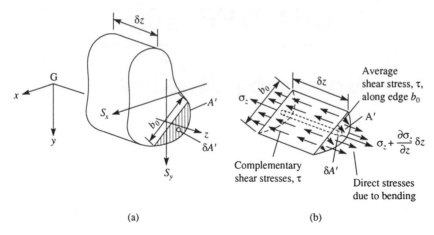

Fig. 10.1   Determination of shear stress distribution in a beam of arbitrary cross-section

We shall assume (see Section 9.8) that the direct stresses produced by the bending action of shear loads are given by the theory developed for the pure bending of beams. Therefore, for a beam of unsymmetrical section and for coordinates referred to axes through the centroid of the section

$$\sigma_z = \frac{\overline{M}_x}{I_x} y + \frac{\overline{M}_y}{I_y} x \qquad \text{(i.e. Eq. (9.30))}$$

Hence

$$\frac{\partial \sigma_z}{\partial z} = \frac{\partial \overline{M}_x}{\partial z} \frac{y}{I_x} + \frac{\partial \overline{M}_y}{\partial z} \frac{x}{I_y}$$

From Section 9.9

$$\frac{\partial \overline{M}_x}{\partial z} = \overline{S}_y, \quad \frac{\partial \overline{M}_y}{\partial z} = \overline{S}_x$$

so that

$$\frac{\partial \sigma_z}{\partial z} = \frac{\overline{S}_y}{I_x} y + \frac{\overline{S}_x}{I_y} x$$

Substituting for $\partial \sigma_z / \partial z$ in Eq. (10.1) we obtain

$$\tau b_0 = \int_{A'} \frac{\overline{S}_y}{I_x} y \, dA' + \int_{A'} \frac{\overline{S}_x}{I_y} x \, dA'$$

whence

$$\tau = \frac{\overline{S}_y}{b_0 I_x} \int_{A'} y \, dA' + \frac{\overline{S}_x}{b_0 I_y} \int_{A'} x \, dA' \qquad (10.2)$$

The slice may be taken so that the average shear stress in any chosen direction can be determined.

10.2   Shear stress distribution in symmetrical sections

Generally in civil engineering we are not concerned with shear stresses in unsymmetrical sections except where they are of the thin-walled type (see Sections 10.4 and 10.5). 'Thick' beam sections usually possess at least one axis of symmetry and are subjected to shear loads in that direction.

Suppose that the beam section shown in Fig. 10.2 is subjected to a single shear load S_y. Since the y axis is an axis of symmetry, it follows that $I_{xy} = 0$ (Section 9.6). Therefore Eqs (9.59) and (9.60) reduce to

$$\bar{S}_x = S_x = 0, \quad \bar{S}_y = S_y$$

and Eq. (10.2) becomes

$$\tau = \frac{S_y}{b_0 I_x} \int_{A'} y \, dA' \tag{10.3}$$

Clearly the important shear stresses in the beam section of Fig. 10.2 are in the direction of the load. To find the distribution of this shear stress throughout the depth of the beam we therefore take the slice, b_0, in a direction parallel to and at any distance y from the x axis. The integral term in Eq. (10.3) represents, mathematically, the first moment of the shaded area A' about the x axis. We may therefore rewrite Eq. (10.3) as

$$\tau = \frac{S_y A' \bar{y}}{b_0 I_x} \tag{10.4}$$

where $\bar{y}$ is the distance of the centroid of the area A' from the x axis. Alternatively, if the value of $\bar{y}$ is not easily determined, say by inspection, then $\int_{A'} y \, dA'$ may be found by calculating the first moment of area about the x axis of an elemental strip of length b, width δy_1, (Fig. 10.2), and integrating over the area A'. Equation (10.3) then becomes

$$\tau = \frac{S_y}{b_0 I_x} \int_y^{y_{max}} b y_1 \, dy_1 \tag{10.5}$$

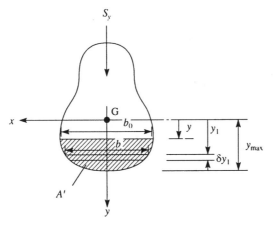

Fig. 10.2   Shear stress distribution in a symmetrical section beam

Either of Eqs (10.4) or (10.5) may be used to determine the distribution of vertical shear stress in a beam section possessing at least a horizontal or vertical axis of symmetry and subjected to a vertical shear load. The corresponding expressions for the horizontal shear stress due to a horizontal load are, by direct comparison with Eqs (10.4) and (10.5),

$$\tau = \frac{S_x A' \bar{x}}{b_0 I_y}, \qquad \tau = \frac{S_x}{b_0 I_y} \int_x^{x_{max}} b x_1 \, dx_1 \qquad (10.6)$$

in which b_0 is the length of the edge of a vertical slice.

Example 10.1   Determine the distribution of vertical shear stress in the beam section shown in Fig. 10.3(a) due to a vertical shear load S_y.

In this example the value of $\bar{y}$ for the slice A' is found easily by inspection so that we may use Eq. (10.4). From Fig. 10.3(a) we see that

$$b_0 = b, \qquad I_x = \frac{bd^3}{12}, \qquad A' = b\left(\frac{d}{2} - y\right), \qquad \bar{y} = \frac{1}{2}\left(\frac{d}{2} + y\right)$$

Hence

$$\tau = \frac{12 S_y}{b^2 d^3} b\left(\frac{d}{2} - y\right) \frac{1}{2}\left(\frac{d}{2} + y\right)$$

which simplifies to

$$\tau = \frac{6 S_y}{bd^3}\left(\frac{d^2}{4} - y^2\right) \qquad (10.7)$$

The distribution of vertical shear stress is therefore parabolic as shown in Fig. 10.3(b) and varies from $\tau = 0$ at $y = \pm d/2$ to $\tau = \tau_{max} = 3S_y/2bd$ at the neutral axis ($y = 0$) of the beam section. Note that $\tau_{max} = 1 \cdot 5 \tau_{av}$, where τ_{av}, the average vertical shear stress over the section, is given by $\tau_{av} = S_y/bd$.

Example 10.2   Determine the distribution of vertical shear stress in the I-section beam of Fig. 10.4(a) produced by a vertical shear load, S_y.

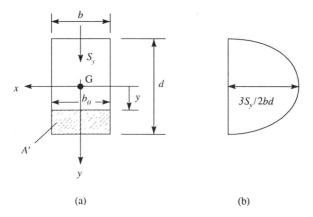

(a)   (b)

Fig. 10.3   Shear stress distribution in a rectangular section beam

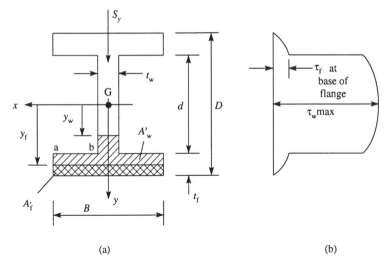

(a)   (b)

Fig. 10.4 Shear stress distribution in an I-section beam

It is clear from Fig. 10.4(a) that the geometry of each of the areas A_f' and A_w' formed by taking a slice of the beam in the flange (at $y = y_f$) and in the web (at $y = y_w$), respectively, are different and will therefore lead to different distributions of shear stress. First we shall consider the flange. The area A_f' is rectangular so that the distribution of vertical shear stress, τ_f, in the flange is, by direct comparison with Ex. 10.1,

$$\tau_f = \frac{S_y}{BI_x} \frac{B}{2} \left(\frac{D}{2} - y_f \right) \left(\frac{D}{2} + y_f \right)$$

or
$$\tau_f = \frac{S_y}{2I_x} \left(\frac{D^2}{4} - y_f^2 \right) \qquad (10.8)$$

where I_x is the second moment of area of the complete section about the centroidal axis Gx and is obtained by the methods of Section 9.6.

A difficulty arises in the interpretation of Eq. (10.8) which indicates a parabolic distribution of vertical shear stress in the flanges increasing from $\tau_f = 0$ at $y_f = \pm D/2$ to a value

$$\tau_f = \frac{S_y}{8I_x} (D^2 - d^2) \qquad (10.9)$$

at $y_f = \pm d/2$. However, the shear stress must also be zero at the inner surfaces ab, etc., of the flanges. Equation (10.8) therefore may only be taken to give an indication of the vertical shear stress distribution in the flanges *in the vicinity of the web*. Clearly if the flanges are thin so that d is close in value to D then τ_f *in the flanges* at the extremities of the web is small, as indicated in Fig. 10.4(b).

The area A_w' formed by taking a slice in the web at $y = y_w$ comprises two rectangles which may therefore be treated separately in determining $A'\bar{y}$ for the web.

Thus

$$
\tau_w = \frac{S_y}{t_w I_x} \left[B \left(\frac{D}{2} - \frac{d}{2} \right) \frac{1}{2} \left(\frac{D}{2} + \frac{d}{2} \right) + t_w \left(\frac{d}{2} - y_w \right) \frac{1}{2} \left(\frac{d}{2} + y_w \right) \right]
$$

which simplifies to

$$
\tau_w = \frac{S_y}{t_w I_x} \left[\frac{B}{8} (D^2 - d^2) + \frac{t_w}{2} \left(\frac{d^2}{4} - y_w^2 \right) \right] \tag{10.10}
$$

or

$$
\tau_w = \frac{S_y}{I_x} \left[\frac{B}{8 t_w} (D^2 - d^2) + \frac{1}{2} \left(\frac{d^2}{4} - y_w^2 \right) \right] \tag{10.11}
$$

Again the distribution is parabolic and increases from

$$
\tau_w = \frac{S_y}{I_x} \frac{B}{8 t_w} (D^2 - d^2) \tag{10.12}
$$

at $y_w = \pm d/2$ to a maximum value, $\tau_{w,max}$, given by

$$
\tau_{w,max} = \frac{S_y}{I_x} \left[\frac{B}{8 t_w} (D^2 - d^2) + \frac{d^2}{8} \right] \tag{10.13}
$$

at $y = 0$. Note that the value of τ_w at the extremities of the web (Eq. (10.12)) is greater than the corresponding values of τ_f by a factor B/t_w. The complete distribution is shown in Fig. 10.4(b).

The value of $\tau_{w,max}$ (Eq. (10.13)) is not very much greater than that of τ_w at the extremities of the web. In design checks on shear stress values in I-section beams it is usual to assume that the maximum shear stress in the web is equal to the shear load divided by the web area. In most cases the result is only slightly different from the value given by Eq. (10.13). A typical value given in Codes of Practice for the maximum allowable value of shear stress in the web of an I-section, mild steel beam is 100 N/mm^2; this is applicable to sections having web thicknesses not exceeding 40 mm.

We have been concerned so far in this example with the distribution of vertical shear stress. We now consider the situation that arises if we take the slice across one of the flanges at $x = x_f$ as shown in Fig. 10.5(a). Equations (10.4) and (10.5) still apply, but in this case $b_0 = t_f$. Thus, using Eq. (10.4),

$$
\tau_{f(h)} = \frac{S_y}{t_f I_x} t_f \left(\frac{B}{2} - x_f \right) \frac{1}{2} \left(\frac{D}{2} + \frac{d}{2} \right)
$$

where $\tau_{f(h)}$ is the distribution of horizontal shear stress in the flange. Simplifying the above equation we obtain

$$
\tau_{f(h)} = \frac{S_y (D + d)}{4 I_x} \left(\frac{B}{2} - x_f \right) \tag{10.14}
$$

Equation (10.14) shows that the horizontal shear stress varies linearly in the flanges from zero at $x_f = B/2$ to $S_y (D + d) B / 8 I_x$ at $x_f = 0$.

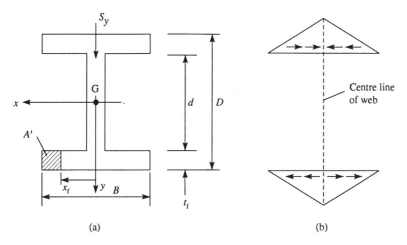

Fig. 10.5 Distribution of horizontal shear stress in the flanges of an I-section beam

We have defined a positive shear stress as being directed away from the edge b_0 of the slice towards the interior of the slice (Fig. 10.1(b)). Since Eq. (10.14) is always positive, then $\tau_{f(h)}$ in the lower flange is directed towards the outer edges of the flange. By a similar argument $\tau_{f(h)}$ in the upper flange is negative since $\bar{y}$ is negative for any slice and $\tau_{f(h)}$ is therefore directed towards the web. The distribution is shown in Fig. 10.5(b).

From Eq. (10.12) we see that the shear stress at the extremities of the web multiplied by the web thickness is

$$\tau_w t_w = \frac{S_y}{I_x} \frac{B}{8} (D+d)(D-d) = \frac{S_y}{I_x} \frac{B}{8} (D+d)2t_f \tag{10.15}$$

The product of horizontal flange stress and flange thickness at the extremities of the web is, from Eq. (10.14)

$$\tau_{f(h)} t_f = \frac{S_y}{I_x} \frac{B}{8} (D+d)t_f \tag{10.16}$$

Comparing Eqs (10.15) and (10.16) we see that

$$\tau_w t_w = 2\tau_{f(h)} t_f \tag{10.17}$$

The product *stress × thickness* gives the *shear force per unit length* in the walls of the section and is known as the *shear flow*, a particularly useful parameter when considering thin-walled sections. In the above example we note that $\tau_{f(h)} t_f$ is the shear flow at the extremities of the web produced by considering one half of the complete flange. From symmetry there is an equal shear flow at the extremities of the web from the other half of the flange. Equation (10.17) therefore expresses the equilibrium of the shear flows at the web/flange junctions. We shall return to a more detailed consideration of shear flow when investigating the shear of thin-walled sections.

In 'thick' I-section beams the horizontal flange shear stress is not of great importance since, as can be seen from Equation (10.17), it is of the order of half the magnitude of the vertical shear stress at the extremities of the web if $t_w \approx t_f$. In thin-walled I-sections (and other sections too) this horizontal shear stress can produce shear distortions of sufficient magnitude to redistribute the direct stresses due to bending, thereby seriously affecting the accuracy of the basic bending theory described in Chapter 9. This phenomenon is known as *shear lag*.

Example 10.3   Determine the distribution of vertical shear stress in a beam of circular cross-section when it is subjected to a shear force S_y (Fig. 10.6).

The area A' of the slice in this problem is a segment of a circle and therefore does not lend itself to the simple treatment of the previous two examples. We shall therefore use Eq. (10.5) to determine the distribution of vertical shear stress. Thus

$$\tau = \frac{S_y}{b_0 I_x} \int_y^{D/2} by_1 \, dy_1 \tag{10.18}$$

where

$$I_x = \frac{\pi D^4}{64} \tag{Eq. (9.40)}$$

Integration of Eq. (10.18) is simplified if angular variables are used; thus, from Fig. 10.6,

$$b_0 = 2 \times \frac{D}{2} \cos \theta, \qquad b = 2 \times \frac{D}{2} \cos \phi, \qquad y_1 = \frac{D}{2} \sin \phi, \qquad dy_1 = \frac{D}{2} \cos \phi \, d\phi$$

Hence Eq. (10.18) becomes

$$\tau = \frac{16 S_y}{\pi D^2 \cos \theta} \int_\theta^{\pi/2} \cos^2 \phi \sin \phi \, d\phi$$

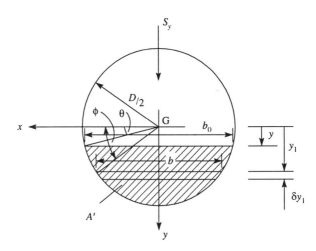

Fig. 10.6   Distribution of shear stress in a beam of circular cross-section

Integrating we obtain

$$\tau = \frac{16\,S_y}{\pi D^2 \cos\theta} \left[-\frac{\cos^3\phi}{3} \right]_\theta^{\pi/2}$$

which gives

$$\tau = \frac{16\,S_y}{3\pi D^2} \cos^2\theta$$

But

$$\cos^2\theta = 1 - \sin^2\theta = 1 - \left(\frac{y}{D/2}\right)^2$$

Therefore

$$\tau = \frac{16\,S_y}{3\pi D^2}\left(1 - \frac{4y^2}{D^2}\right) \qquad\qquad (10.19)$$

The distribution of shear stress is parabolic with values of $\tau = 0$ at $y = \pm D/2$ and $\tau = \tau_{\mathrm{max}} = 16S_y/3\pi D^2$ at $y = 0$, the neutral axis of the section.

10.3 Strain energy due to shear

Consider a small rectangular element of material of side δz, δy and thickness t subjected to a shear stress and complementary shear stress system, τ (Fig. 10.7(a)); τ produces a shear strain γ in the element so that distortion occurs as shown in Fig. 10.7(b), where displacements are relative to the side CD. The horizontal displacement of the side AB is $\gamma\delta y$ so that the shear force on the face AB moves through this distance and therefore does work. If the shear loads producing the shear stress are gradually applied, then the work done by the shear force on the element and hence the strain energy stored, δU, is given by

$$\delta U = \tfrac{1}{2}\tau t\,\delta z\gamma\,\delta y$$

or

$$\delta U = \tfrac{1}{2}\tau\gamma t\,\delta z\,\delta y$$

Now $\gamma = \tau/G$, where G is the shear modulus and $t\,\delta z\,\delta y$ is the volume of the element. Hence

$$\delta U = \frac{1}{2}\frac{\tau^2}{G} \times \text{volume of element}$$

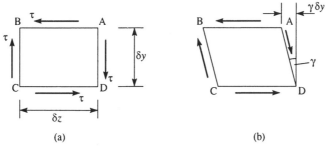

(a) (b)

Fig. 10.7 Determination of strain energy due to shear

The total strain energy, U, due to shear in a structural member in which the shear stress, τ, is uniform is then given by

$$U = \frac{\tau^2}{2G} \times \text{volume of member} \qquad (10.20)$$

10.4   Shear stress distribution in thin-walled open section beams

In considering the shear stress distribution in thin-walled open section beams we shall make identical assumptions regarding the calculation of section properties as were made in Section 9.6. In addition we shall assume that shear stresses in the plane of the cross-section and parallel to the tangent at any point on the beam wall are constant across the thickness (Fig. 10.8(a)), whereas shear stresses normal to the tangent are negligible (Fig. 10.8(b)). The validity of the latter assumption is evident when it is realized that these normal shear stresses must be zero on the inner and outer surfaces of the section and that the walls are thin. We shall further assume that the wall thickness can vary round the section but is constant along the length of the member.

Figure 10.9 shows a length of a thin-walled beam of arbitrary section subjected to shear loads S_x and S_y which are applied such that no twisting of the beam occurs. In addition to shear stresses, direct stresses due to the bending action of the shear loads are present so that an element $\delta s \times \delta z$ of the beam wall is in equilibrium under the stress system shown in Fig. 10.10(a). The shear stress τ is assumed to be positive in the positive direction of s, the distance round the profile of the section measured from an open edge. Although we have specified that the thickness t may vary with s, this variation is small for most thin-walled sections so that we may reasonably make

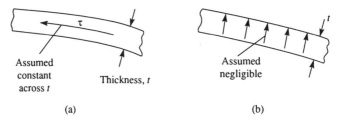

Assumed
constant
across t Thickness, t

Assumed
negligible

(a)   (b)

Fig. 10.8   Assumptions in thin-walled open section beams

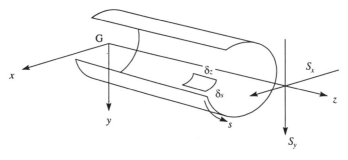

Fig. 10.9   Shear of a thin-walled open section beam

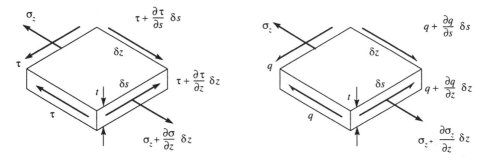

Fig. 10.10 Equilibrium of beam element

the approximation that t is constant over the length δs. As stated in Ex. 10.2 it is convenient, when considering thin-walled sections, to work in terms of shear flow to which we assign the symbol $q(=\tau t)$. Figure 10.10(b) shows the shear stress system of Fig. 10.10(a) represented in terms of q. Thus for equilibrium of the element in the z direction

$$\left(\sigma_z + \frac{\partial \sigma_z}{\partial z}\,\delta z\right) t\,\delta s - \sigma_z t\,\delta s + \left(q + \frac{\partial q}{\partial s}\,\delta s\right)\delta z - q\,\delta z = 0$$

which gives

$$\frac{\partial q}{\partial s} + t\,\frac{\partial \sigma_z}{\partial z} = 0 \tag{10.21}$$

Again we assume that the direct stresses are given by Eq. (9.30), so that

$$\frac{\partial \sigma_z}{\partial z} = \frac{\partial \overline{M}_x}{\partial z}\,\frac{y}{I_x} + \frac{\partial \overline{M}_y}{\partial z}\,\frac{x}{I_y}$$

which becomes

$$\frac{\partial \sigma_z}{\partial z} = \frac{\overline{S}_y}{I_x}\,y + \frac{\overline{S}_x}{I_y}\,x \qquad \text{(Section 9.9)}$$

Substituting in Eq. (10.21) we obtain

$$\frac{\partial q}{\partial s} = -\frac{\overline{S}_y}{I_x}\,ty - \frac{\overline{S}_x}{I_y}\,tx$$

Integrating this expression from $s = 0$ (where $q = 0$ on the open edge of the section) to any point s we have

$$q_s = -\frac{\overline{S}_y}{I_x}\int_0^s ty\,\mathrm{d}s - \frac{\overline{S}_x}{I_y}\int_0^s tx\,\mathrm{d}s \tag{10.22}$$

The shear stress at any point in the beam section wall is obtained by dividing the shear flow q_s by the appropriate wall thickness. Thus

$$\tau_s = -\frac{\overline{S}_y}{t_s I_x}\int_0^s ty\,\mathrm{d}s - \frac{\overline{S}_x}{t_s I_y}\int_0^s tx\,\mathrm{d}s \tag{10.23}$$

Note the similarity to Eq. (10.2) for the shear stress distribution in a 'thick' beam.

Example 10.4   Determine the shear flow distribution in the thin-walled Z-section beam shown in Fig. 10.11 produced by a shear load S_y applied in the plane of the web.

The origin for our system of reference axes coincides with the centroid of the section at the mid-point of the web. The centroid is also the centre of antisymmetry of the section so that the shear load, applied through this point, causes no twisting of the section and the shear flow distribution is given by Eq. (10.22) in which

$$\bar{S}_y = \frac{S_y}{1 - I_{xy}^2/I_x I_y}, \qquad \bar{S}_x = \frac{-S_y I_{xy}/I_x}{1 - I_{xy}^2/I_x I_y} \qquad \text{(i)}$$

The second moments of area of the section about the x and y axes have previously been calculated in Ex. 9.10 and are

$$I_x = \frac{h^3 t}{3}, \qquad I_y = \frac{h^3 t}{12}, \qquad I_{xy} = -\frac{h^3 t}{8}$$

Substituting these values in Eq. (i) we obtain

$$\bar{S}_y = 2{\cdot}28\, S_y, \qquad \bar{S}_x = 0{\cdot}86\, S_y$$

whence, from Eq. (10.22),

$$q_s = -\frac{S_y}{h^3} \int_0^s (6{\cdot}84\, y + 10{\cdot}32\, x)\, \mathrm{d}s$$

On the upper flange AB, $y = -h/2$ and $x = h/2 - s_A$ where $0 \leqslant s_A \leqslant h/2$. Therefore

$$q_{AB} = \frac{S_y}{h^3} \int_0^{s_A} (10{\cdot}32\, s_A - 1{\cdot}74 h)\, \mathrm{d}s_A$$

which gives
$$q_{AB} = \frac{S_y}{h^3} (5{\cdot}16 s_A^2 - 1{\cdot}74 h s_A) \qquad \text{(ii)}$$

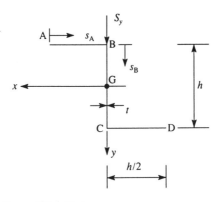

Fig. 10.11   Beam section of Ex. 10.4

Thus at $A(s_A = 0)$, $q_A = 0$ and at $B(s_A = h/2)$, $q_B = 0.42S_y/h$. Note that the order of the suffixes of q in Eq. (ii) denotes the positive direction of q(and s_A). An examination of Eq. (ii) shows that the shear flow distribution on the upper flange is parabolic with a change of sign (i.e. direction) at $s_A = 0.34h$. For values of $s_A < 0.34h$, q_{AB} is negative and is therefore in the opposite direction to s_A. Furthermore, q_{AB} has a turning value between $s_A = 0$ and $s_A = 0.34h$ at a value of s_A given by

$$\frac{dq_{AB}}{ds_A} = 10.32s_A - 1.74h = 0$$

i.e. at $s_A = 0.17h$. The corresponding value of q_{AB} is then, from Eq. (ii), $q_{AB} = -0.15S_y/h$.

In the web BC, $y = -h/2 + s_B$ where $0 \leqslant s_B \leqslant h$ and $x = 0$. Thus

$$q_{BC} = -\frac{S_y}{h^3} \int_0^{s_B} (6.84s_B - 3.42h)\, ds_B + q_B \tag{iii}$$

Note that in Eq. (iii), q_{BC} is not zero when $s_B = 0$ but equal to the value obtained by inserting $s_A = h/2$ in Eq. (ii), i.e. $q_B = 0.42S_y/h$. Integrating the first two terms on the right-hand side of Eq. (iii) we obtain

$$q_{BC} = -\frac{S_y}{h^3} (3.42s_B^2 - 3.42hs_B - 0.42h^2) \tag{iv}$$

Equation (iv) gives a parabolic shear flow distribution in the web, symmetrical about Gx and with a maximum value at $s_B = h/2$ equal to $1.28S_y/h$; q_{AB} is positive at all points in the web.

The shear flow distribution in the lower flange may be deduced from antisymmetry; the complete distribution is shown in Fig. 10.12.

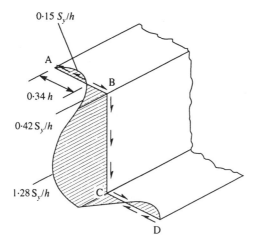

Fig. 10.12 Shear flow distribution in beam section of Ex. 10.4

Shear centre

We have specified in the previous analysis that the lines of action of the shear loads S_x and S_y must not cause twisting of the section. For this to be the case, S_x and S_y must pass through the *shear centre* of the section. Clearly in many practical situations this is not so and torsion as well as shear is induced. These problems may be simplified by replacing the shear loads by shear loads acting through the shear centre, plus a pure torque, as illustrated in Fig. 10.13 for the simple case of a channel section subjected to a vertical shear load S_y applied in the line of the web. The shear stresses corresponding to the separate loading cases are then added by superposition.

Where a section possesses an axis of symmetry, the shear centre must lie on this axis. For cruciform, T and angle sections of the type shown in Fig. 10.14 the shear centre is located at the intersection of the walls since the resultant internal shear loads all pass through this point. In fact in any beam section in which the walls are straight and intersect at just one point, that point is the shear centre of the section.

Example 10.5

Determine the position of the shear centre of the thin-walled channel section shown in Fig. 10.15.

The shear centre S lies on the horizontal axis of symmetry at some distance x_S say, from the web. If an arbitrary shear load, S_y, is applied through the shear centre, then the shear flow distribution is given by Eq. (10.22) and the moment about any point in the cross-section produced by these shear flows is *equivalent* to the moment of the applied shear load about the same point; S_y appears on both sides of the resulting equation and may therefore be eliminated to leave x_S as the unknown.

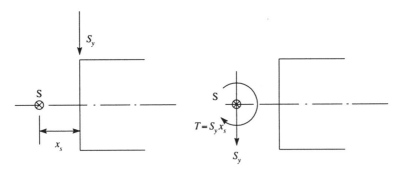

Fig. 10.13 Replacement of a shear load by a shear load acting through the shear centre plus a torque

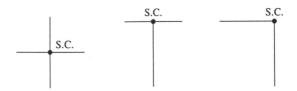

Fig. 10.14 Special cases of shear centre position

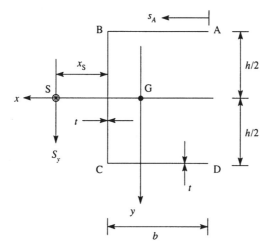

Fig. 10.15 Channel section beam of Ex. 10.5

For the channel section, Gx is an axis of symmetry so that $I_{xy} = 0$, giving $\bar{S}_y = S_y$ and $\bar{S}_x = S_x = 0$. Equation (10.22) therefore simplifies to

$$q_s = -\frac{S_y}{I_x}\int_0^s ty\,ds$$

where

$$I_x = \frac{th^3}{12} + 2bt\left(\frac{h}{2}\right)^2 = \frac{th^3}{12}\left(1 + 6\frac{b}{h}\right)$$

Substituting for I_x and noting that t is constant round the section, we have

$$q_s = -\frac{12S_y}{h^3(1 + 6b/h)}\int_0^s y\,ds \tag{i}$$

The solution of this type of problem may be reduced in length by giving some thought to what is required. We are asked, in this case, to obtain the position of the shear centre and not a complete shear flow distribution. From symmetry it can be seen that the moments of the resultant shear forces on the upper and lower flanges about the mid-point of the web are numerically equal and act in the same sense. Furthermore, the moment of the web shear about the same point is zero. Therefore it is only necessary to obtain the shear flow distribution on either the upper or lower flange for a solution. Alternatively, the choice of either flange/web junction as the moment centre leads to the same conclusion.

On the upper flange, $y = -h/2$ so that from Eq. (i) we obtain

$$q_{AB} = \frac{6S_y}{h^2(1 + 6b/h)}s_A \tag{ii}$$

Equating the anticlockwise moments of the internal shear forces about the mid-point of the web to the anticlockwise moment of the applied shear load about the same point gives

$$S_y x_S = 2 \int_0^b q_{AB} \frac{h}{2} \, ds_A$$

Substituting for q_{AB} from Eq. (ii) we have

$$S_y x_S = 2 \int_0^b \frac{6S_y}{h^2(1 + 6b/h)} \frac{h}{2} s_A \, ds_A$$

from which

$$x_S = \frac{3b^2}{h(1 + 6b/h)}$$

In the case of an unsymmetrical section, the coordinates (x_S, y_S) of the shear centre referred to some convenient point in the cross-section are obtained by first determining x_S in a similar manner to that described above and then calculating y_S by applying a shear load S_x through the shear centre. It should be noted that in each of the separate applications of S_y and S_x both $\bar{S}_y$ and $\bar{S}_x$ have values.

10.5 Shear stress distribution in thin-walled closed section beams

The shear flow and shear stress distributions in a closed section, thin-walled beam are determined in a manner similar to that described in Section 10.4 for an open section beam but with two important differences. First, the shear loads may be applied at points in the cross-section other than the shear centre so that shear and torsion occur simultaneously. We shall see that a solution may be obtained for this case without separating the shear and torsional effects, although such an approach is an acceptable alternative, particularly if the position of the shear centre is required. Secondly, it is not generally possible to choose an origin for s that coincides with a known value of shear flow. A closed section beam under shear is therefore singly redundant as far as the internal force system is concerned and requires an equation

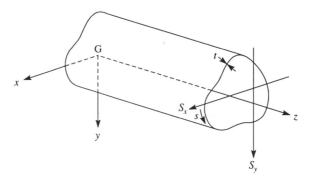

Fig. 10.16 Shear of a thin-walled closed section beam

additional to the equilibrium equation (10.21). Identical assumptions are made regarding section properties, wall thickness and shear stress distribution as were made for the open section beam.

The thin-walled beam of arbitrary closed section shown in Fig. 10.16 is subjected to shear loads S_x and S_y applied through any point in the cross-section. These shear loads produce direct and shear stresses on any element in the beam wall identical to those shown in Figs 10.10(a) and (b). The equilibrium equation (10.21) is therefore applicable and is

$$\frac{\partial q}{\partial s} + t \frac{\partial \sigma_z}{\partial z} = 0$$

By an identical procedure of substitution for σ_z as for an open section beam we obtain

$$\int_0^s \frac{\partial q}{\partial s} \, ds = - \frac{\bar{S}_y}{I_x} \int_0^s ty \, ds - \frac{\bar{S}_x}{I_y} \int_0^s tx \, ds$$

If, at the origin for s, the shear flow q has the unknown value $q_{s,0}$ then integration of the above equation gives

$$q_s - q_{s,0} = - \frac{\bar{S}_y}{I_x} \int_0^s ty \, ds - \frac{\bar{S}_x}{I_y} \int_0^s tx \, ds$$

or

$$q_s = - \frac{\bar{S}_y}{I_x} \int_0^s ty \, ds - \frac{\bar{S}_x}{I_y} \int_0^s tx \, ds + q_{s,0} \tag{10.24}$$

It is clear from a comparison of Eqs (10.24) and (10.22) that the first two terms of the right-hand side of Eq. (10.24) represent the shear flow distribution in an open section beam with the shear loads applied through its shear centre. We shall denote this 'open section' or 'basic' shear flow distribution by q_b and rewrite Eq. (10.24) as

$$q_s = q_b + q_{s,0}$$

We obtain q_b by supposing that the closed section beam is 'cut' at some convenient point, thereby producing an 'open section' beam as shown in Fig. 10.17(b); we take the 'cut' as the origin for s. The shear flow distribution round this 'open section' beam is given by Eq. (10.22), i.e.

$$q_b = - \frac{\bar{S}_y}{I_x} \int_0^s ty \, ds - \frac{\bar{S}_x}{I_y} \int_0^s tx \, ds$$

Eq. (10.22) is valid only if the shear loads produce no twist; in other words, S_x and S_y must be applied through the shear centre of the 'open section' beam. Thus by 'cutting' the closed section beam to determine q_b we are, in effect, transferring the line of action of S_x and S_y to the shear centre, $S_{s,0}$, of the resulting 'open section' beam. The implication is, therefore, that when we 'cut' the section we must simultaneously introduce a pure torque to compensate for the transference of S_x and S_y. We shall show in Chapter 11 that the application of a pure torque to a closed section beam results in a constant shear flow round the walls of the beam. In this case $q_{s,0}$, which is effectively a constant shear flow round the section, corresponds to

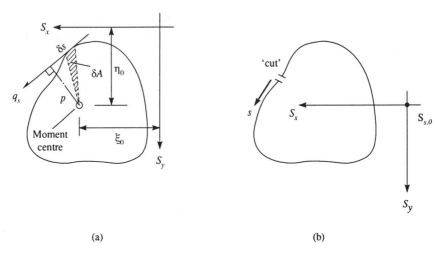

(a)   (b)

Fig. 10.17 Determination of shear flow value at the origin for *s* in a closed section beam

the pure torque produced by the shear load transference. Clearly different positions of the 'cut' will result in different values for $q_{s,0}$ since the corresponding 'open section' beams have different shear centre positions.

It is immaterial whether S_x and S_y in Fig. 10.17 are externally applied loads or internal shear forces since we have stipulated in Section 9.5 that when internal force systems are those acting on that face of the section that is seen when viewed in the direction zO they act in the same sense as externally applied loads. S_x and S_y are therefore the stress resultants of the internal shear flows q_s. Thus, equating internal and external anticlockwise moments in Fig. 10.17(a), we have

$$S_x \eta_0 - S_y \xi_0 = \oint p q_s \, ds = \oint p q_b \, ds + q_{s,0} \oint p \, ds$$

where $\oint$ denotes integration taken completely round the section. In Fig. 10.17(a) the elemental area δA is given by

$$\delta A = \tfrac{1}{2} p \, \delta s$$

Thus
$$\oint p \, ds = 2 \oint dA$$

or
$$\oint p \, ds = 2A$$

where A is the area enclosed by the mid-line of the section wall. Hence

$$S_x \eta_0 - S_y \xi_0 = \oint p q_b \, ds + 2A q_{s,0} \tag{10.25}$$

If the moment centre coincides with the lines of action of S_x and S_y then Eq. (10.25) reduces to

$$0 = \oint p q_b \, ds + 2A q_{s,0} \tag{10.26}$$

The unknown shear flow $q_{s,0}$ follows from either of Eqs (10.25) or (10.26). Note that the signs of the moment contributions of S_x and S_y on the left-hand side of Eq. (10.25) depend upon the position of their lines of action relative to the moment centre. The values given in Eq. (10.25) apply only to Fig. 10.17(a) and could change for different moment centres and/or differently positioned shear loads.

Shear centre

A complication arises in the determination of the position of the shear centre of a closed section beam since the line of action of the arbitrary shear load (applied through the shear centre as in Ex. 10.5) must be known before $q_{s,0}$ can be determined from either of Eqs (10.25) or (10.26). However, before the position of the shear centre can be found, $q_{s,0}$ must be obtained. Thus an alternative method of determining $q_{s,0}$ is required. We therefore consider the rate of twist of the beam which, when the shear loads act through the shear centre, is zero.

Consider an element, $\delta s \times \delta z$, of the wall of the beam subjected to a system of shear and complementary shear stresses as shown in Fig. 10.18(a). These shear stresses induce a shear strain, γ, in the element which is given by

$$\gamma = \phi_1 + \phi_2$$

irrespective of whether direct stresses (due to bending action) are present or not. If the linear displacements of the sides of the element in the s and z directions are δv_t (i.e. a tangential displacement) and δw, respectively, then as both δs and δz become infinitely small

$$\gamma = \frac{\partial w}{\partial s} + \frac{\partial v_t}{\partial z} \tag{10.27}$$

Suppose now that the beam section is given a small angle of twist, θ, about its centre of twist, R. If we assume that the shape of the cross-section of the beam is unchanged by this rotation (i.e. it moves as a rigid body), then from Fig. 10.18(b) it can be seen that the tangential displacement, v_t, of a point in the wall of the beam section is given by

$$v_t = p_R \theta$$

Hence

$$\frac{\partial v_t}{\partial z} = p_R \frac{\partial \theta}{\partial z}$$

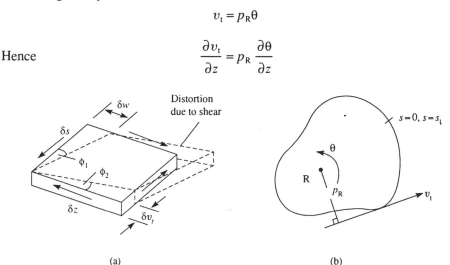

(a)   (b)

Fig. 10.18 Rate of twist in a thin-walled closed section beam

Since we are assuming that the section rotates as a rigid body, it follows that θ is a function of z only so that the above equation may be written

$$\frac{\partial v_t}{\partial z} = p_R \frac{d\theta}{dz}$$

Substituting for $\partial v_t/\partial z$ in Eq. (10.27) we have

$$\gamma = \frac{\partial w}{\partial s} + p_R \frac{d\theta}{dz}$$

Now

$$\gamma = \frac{\tau}{G} = \frac{q_s}{Gt}$$

Thus

$$\frac{q_s}{Gt} = \frac{\partial w}{\partial s} + p_R \frac{d\theta}{dz}$$

Integrating both sides of this equation completely round the cross-section of the beam, i.e. from $s = 0$ to $s = s_1$ (see Fig. 10.18(b)),

$$\oint \frac{q_s}{Gt} \, ds = \oint \frac{\partial w}{\partial s} \, ds + \frac{d\theta}{dz} \oint p_R \, ds$$

which gives

$$\oint \frac{q_s}{Gt} \, ds = [w]_{s=0}^{s=s_1} + \frac{d\theta}{dz} 2A$$

The axial displacement, w, must have the same value at $s = 0$ and $s = s_1$. Therefore the above expression reduces to

$$\frac{d\theta}{dz} = \frac{1}{2A} \oint \frac{q_s}{Gt} \, ds \qquad (10.28)$$

For shear loads applied through the shear centre, $d\theta/dz = 0$ so that

$$0 = \oint \frac{q_s}{Gt} \, ds$$

which may be written

$$0 = \oint \frac{1}{Gt} (q_b + q_{s,0}) \, ds$$

Hence

$$q_{s,0} = -\frac{\oint (q_b/Gt) \, ds}{\oint ds/Gt} \qquad (10.29)$$

If G is constant then Eq. (10.29) simplifies to

$$q_{s,0} = -\frac{\oint (q_b/t) \, ds}{\oint ds/t} \qquad (10.30)$$

Example 10.6 A thin-walled, closed section beam has the singly symmetrical, trapezoidal cross-section shown in Fig. 10.19. Calculate the distance of the shear centre from the wall AD. The shear modulus G is constant throughout the section.

The shear centre lies on the horizontal axis of symmetry so that it is only necessary to apply a shear load S_y through S to determine x_S. Furthermore the axis of symmetry coincides with the centroidal reference axis Gx so that $I_{xy} = 0$, $\bar{S}_y = S_y$ and $\bar{S}_x = S_x = 0$. Equation (10.24) therefore simplifies to

$$q_s = -\frac{S_y}{I_x} \int_0^s ty\, ds + q_{s,0} \tag{i}$$

Note that in Eq. (i) only the second moment of area about the x axis and coordinates of points referred to the x axis are required so that it is unnecessary to calculate the position of the centroid on the x axis. It will not, in general, and in this case in particular, coincide with S.

The second moment of area of the section about the x axis is given by

$$I_x = \frac{12 \times 600^3}{12} + \frac{8 \times 300^3}{12} + 2\left[\int_0^{800} 10\left(150 + \frac{150}{800}s\right)^2 ds\right]$$

from which $I_x = 1074 \times 10^6$ mm^4. Alternatively, the second moment of area of each inclined wall about an axis through its own centroid may be found using the method described in Section 9.6 and then transferred to the x axis by the parallel axes theorem.

We now obtain the q_b shear flow distribution by 'cutting' the beam section at the mid-point O of the wall CB. Thus, since $y = s_A$ we have

$$q_{b,OB} = -\frac{S_y}{I_x} \int_0^{S_A} 8s_A\, ds_A$$

which gives

$$q_{b,OB} = -\frac{S_y}{I_x} 4s_A^2 \tag{ii}$$

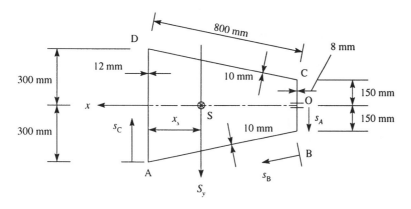

Fig. 10.19 Closed section beam of Ex. 10.6

Thus
$$q_{b,B} = -\frac{S_y}{I_x} \times 9 \times 10^4$$

For the wall BA where $y = 150 + 150s_B/800$

$$q_{b,BA} = -\frac{S_y}{I_x} \left[\int_0^{s_B} 10\left(150 + \frac{150}{800} s_B\right) ds_B + 9 \times 10^4 \right]$$

from which

$$q_{b,BA} = -\frac{S_y}{I_x} \left(1500s_B + \frac{15}{16} s_B^2 + 9 \times 10^4 \right) \qquad (iii)$$

whence
$$q_{b,A} = -\frac{S_y}{I_x} \times 189 \times 10^4$$

In the wall AD, $y = 300 - s_C$ so that

$$q_{b,AD} = -\frac{S_y}{I_x} \left[\int_0^{s_C} 12(300 - s_C) ds_C + 189 \times 10^4 \right]$$

which gives

$$q_{b,AD} = -\frac{S_y}{I_x} (3600s_C - 6s_C^2 + 189 \times 10^4) \qquad (iv)$$

The remainder of the q_b distribution follows from symmetry.

The shear load S_y is applied through the shear centre of the section so that we must use Eq. (10.30) to determine $q_{s,0}$. Now

$$\oint \frac{ds}{t} = \frac{600}{12} + \frac{2 \times 800}{10} + \frac{300}{8} = 247 \cdot 5$$

Hence

$$q_{s,0} = -\frac{2}{247 \cdot 5} \left(\int_0^{150} \frac{q_{b,OB}}{8} ds_A + \int_0^{800} \frac{q_{b,BA}}{10} ds_B + \int_0^{300} \frac{q_{b,AD}}{12} ds_C \right) \qquad (v)$$

Substituting for $q_{b,OB}$, $q_{b,AD}$ and $q_{b,AD}$ in Eq. (v) from Eqs (ii), (iii) and (iv), respectively, we obtain

$$q_{s,0} = \frac{2S_y}{247 \cdot 5 I_x} \left[\int_0^{150} \frac{s_A^2}{2} ds_A + \int_0^{800} \left(150s_B + \frac{15}{160} s_B^2 + 9 \times 10^3\right) ds_B \right.$$
$$\left. + \int_0^{300} \left(300s_C - \frac{1}{2} s_C^2 + \frac{189 \times 10^4}{12}\right) ds_C \right]$$

from which

$$q_{s,0} = \frac{S_y}{I_x} \times 1 \cdot 04 \times 10^6$$

Taking moments about the mid-point of the wall AD we have

$$S_y x_s = 2\left(\int_0^{150} 786 q_{OB} \, ds_A + \int_0^{800} 294 q_{BA} \, ds_B \right) \qquad \text{(vi)}$$

Noting that $q_{OB} = q_{b,OB} + q_{s,0}$ and $q_{BA} = q_{b,BA} + q_{s,0}$ we rewrite Eq. (vi) as

$$S_y x_s = \frac{2S_y}{I_x} \left[\int_0^{150} 786(-4s_A{}^2 + 1\cdot04 \times 10^6) \, ds_A \right.$$

$$\left. + \int_0^{800} 294(-1500 s_B - \tfrac{15}{16} s_B{}^2 + 0\cdot95 \times 10^6) \, ds_B \right] \qquad \text{(vii)}$$

Integrating Eq. (vii) and eliminating S_y gives

$$x_S = 282 \text{ mm.}$$

Problems

P.10.1 A cantilever has the inverted T-section shown in Fig. P.10.1. It carries a vertical shear load of 4 kN in a downward direction. Determine the distribution of vertical shear stress in its cross-section.

Ans. In web: $\tau = 0\cdot004 \, (44^2 - y^2) \, \text{N/mm}^2.$

In flange: $\tau = 0\cdot004 \, (26^2 - y^2) \, \text{N/mm}^2.$

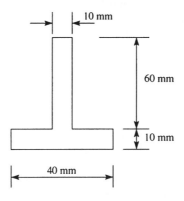

Fig. P.10.1

P.10.2 An I-section beam having the cross-sectional dimensions shown in Fig. P.10.2 carries a vertical shear load of 80 kN. Calculate and sketch the distribution of vertical shear stress across the beam section and determine the percentage of the total shear load carried by the web.

Ans. τ (base of flanges) $= 1\cdot1 \, \text{N/mm}^2$, τ (ends of web) $= 11\cdot0 \, \text{N/mm}^2$,

τ (neutral axis) $= 15\cdot7 \, \text{N/mm}^2$, 95.5%.

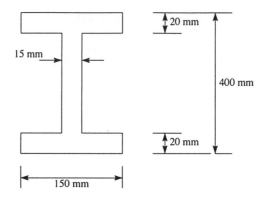

Fig. P.10.2

P.10.3 A doubly symmetrical I-section beam is reinforced by a flat plate attached to the upper flange as shown in Fig. P.10.3. If the resulting compound beam is subjected to a vertical shear load of 200 kN, determine the distribution of shear stress in the portion of the cross-section that extends from the top of the plate to the neutral axis. Calculate also the shear force per unit length of beam resisted by the shear connection between the plate and the flange of the I-section beam.

Ans. τ (top of plate) = 0
τ (bottom of plate) = 0·69 N/mm^2
τ (top of flange) = 1·36 N/mm^2
τ (bottom of flange) = 1·79 N/mm^2
τ (top of web) = 14·3 N/mm^2
τ (neutral axis) = 15·25 N/mm^2
Shear force per unit length = 272 kN/m.

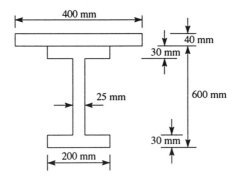

Fig. P.10.3

P.10.4 A timber beam has a rectangular cross-section, 150 mm wide by 300 mm deep, and is simply supported over a span of 4 m. The beam is subjected to a two point loading at the quarter span points. If the beam fails in shear when the total of the two concentrated loads is 180 kN, determine the maximum shear stress at failure.

Ans. 3 N/mm^2.

P.10.5 A beam has the singly symmetrical thin-walled cross-section shown in Fig. P.10.5. Each wall of the section is flat and has the same length, a, and thickness, t. Determine the shear flow distribution round the section due to a vertical shear load, S_y, applied through the shear centre and find the distance of the shear centre from the point C.

Ans. $q_{AB} = 3S_y(2as_A - s_A^2/2)/16a^3 \sin \alpha$

$q_{BC} = 3S_y(3/2 + s_B/a - s_B^2/2a^2)/16a \sin \alpha$

S.C. is $5a \cos \alpha/8$ from C.

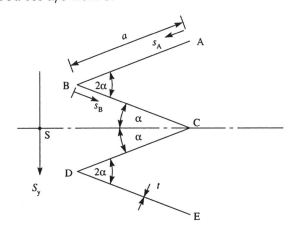

Fig. P.10.5

P.10.6 Define the term 'shear centre' of a thin-walled open section and determine the position of the shear centre of the thin-walled open section shown in Fig. P.10.6.

Ans. $2 \cdot 66r$ from centre of semicircular wall.

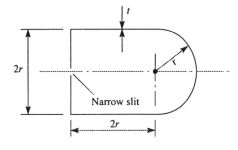

Fig. P.10.6

P.10.7 Determine the position of the shear centre of the cold-formed, thin-walled section shown in Fig. P.10.7. The thickness of the section is constant throughout.

Ans. $87 \cdot 5$ mm above centre of semicircular wall.

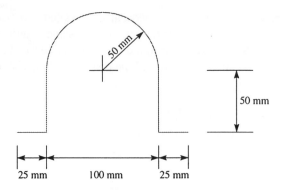

Fig. P.10.7

P.10.8 Determine the position of the shear centre of the cold-formed, thin-walled channel section shown in Fig. P.10.8.

Ans. 1·24*r* from mid-point of web.

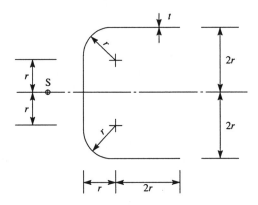

Fig. P.10.8

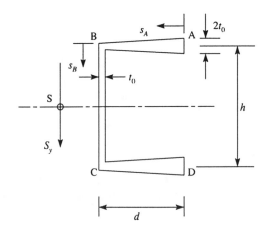

Fig. P.10.9

P.10.9 The thin-walled channel section shown in Fig. P.10.9 has flanges that decrease linearly in thickness from $2t_0$ at the tip to t_0 at their junction with the web. The web has a constant thickness t_0. Determine the distribution of shear flow round the section due to a shear load S_y applied through the shear centre S. Determine also the position of the shear centre.

Ans. $q_{AB} = S_y t_0 h(s_A - s_A^2/4d)/I_x$
$q_{BC} = S_y t_0 (hs_B - s_B^2 + 3hd/2)/2I_x$

where $I_x = t_0 h^2 (h + 9d)/12$, $h/2$ from mid-point of web.

P.10.10 Calculate the position of the shear centre of the thin-walled unsymmetrical channel section shown in Fig. P.10.10.

Ans.   23·3 mm from web BC.
76·5 mm from flange CD.

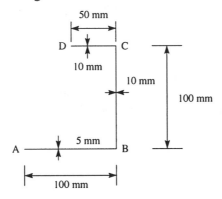

Fig. P.10.10

P.10.11 The closed, thin-walled, hexagonal section shown in Fig. P.10.11 supports a shear load of 30 kN applied along one side. Determine the shear flow distribution round the section if the walls are of constant thickness throughout.

Ans. $q_{OB} = 155 - 0·006s_A^2$, $q_{BC} = 140 - 0·6s_B - 0·003s_B^2$,
$q_{CD} = 50 - 1·2s_C + 0·003s_C^2$, $q_{DE} = 0·006s_D^2 - 0·6s_D - 40$.

Remainder of distribution follows by symmetry. All shear flows in N/mm.

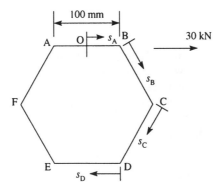

Fig. P.10.11

P.10.12 A closed section, thin-walled beam has the shape of a quadrant of a circle and is subjected to a shear load S applied tangentially to its curved side as shown in Fig. P.10.12. If the walls are of constant thickness throughout determine the shear flow distribution round the section.

Ans. $q_{OA} = S(\cos\theta - 0\cdot45)/0\cdot62r$

$q_{AB} = S(0\cdot35s^2 - 0\cdot707rs + 0.257r^2)/0\cdot62r^3$.

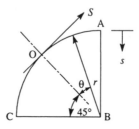

Fig. P.10.12

P.10.13 An overhead crane runs on tracks supported by a thin-walled beam whose closed cross-section has the shape of an isosceles triangle (Fig. P.10.13). If the walls of the section are of constant thickness throughout determine the position of its shear centre.

Ans. 0·71 m from horizontal wall.

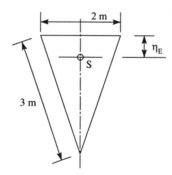

Fig. P.10.13

P.10.14 A box girder has the singly symmetrical trapezoidal cross-section shown in Fig. P.10.14. It supports a vertical shear load of 500 kN applied through its shear centre and in a direction perpendicular to its parallel sides. Calculate the shear flow distribution and the maximum shear stress in the section.

Ans. $q_{OA} = 3\sqrt{3}S_y s_A \times 10^3/I_x$

$q_{AB} = \sqrt{3}S_y(3 - 5s_B^2/2 + 5s_B/2) \times 10^3/I_x$

$q_{BC} = \sqrt{3}S_y(3 - 2s_C) \times 10^3/I_x$, I_x in m⁴; s_A, etc., in m

$\tau_{max} = 32\ \text{N/mm}^2$.

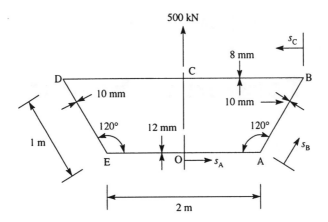

Fig. P.10.14

CHAPTER 11

Torsion of Beams

Torsion in beams arises generally from the action of shear loads whose points of application do not coincide with the shear centre of the beam section. Examples of practical situations where this occurs are shown in Fig. 11.1 where, in Fig. 11.1(a), a concrete encased I-section steel beam supports an offset masonry wall and in Fig. 11.1(b) a floor slab, cast integrally with its supporting reinforced concrete beams, causes torsion of the beams as it deflects under load. Relevant Codes of Practice either imply or demand that torsional stresses and deflections be checked and provided for in design.

The solution of torsion problems is complex particularly in the case of beams of solid section and arbitrary shape for which exact solutions do not exist. Use is then made of empirical formulae which are conveniently expressed in terms of correction factors based on the geometry of a particular shape of cross-section. The simplest case involving the torsion of solid section beams (as opposed to hollow cellular sections) is that of a circular section shaft or bar. This case therefore forms an instructive introduction to the more complex cases of the torsion of solid section, thin-walled open section and thin-walled closed section beams.

11.1 Torsion of solid and hollow circular-section bars

Figure 11.2(a) shows a circular-section bar of length L subjected to equal and opposite torques, T, at each end. The torque at any section of the bar is therefore equal to T and is constant along its length. We shall assume that cross-sections

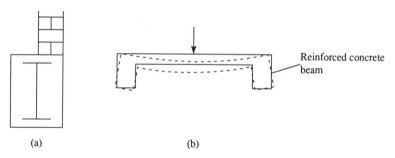

(a)   (b)

Fig. 11.1 Causes of torsion in beams

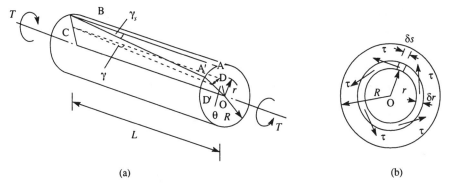

Fig. 11.2   Torsion of a solid circular-section bar

remain plane during twisting, that radii remain straight during twisting and that all normal cross-sections equal distances apart suffer the same relative rotation.

Consider the generator AB on the surface of the bar and parallel to its longitudinal axis. Due to twisting, the end A is displaced to A' so that the radius OA rotates through a small angle, θ, to OA'. The shear strain, γ_s, on the surface of the bar is then equal to the angle ABA' in radians so that

$$\gamma_s = \frac{AA'}{L} = \frac{R\theta}{L}$$

Similarly the shear strain, γ, at any radius r is given by the angle DCD' so that

$$\gamma = \frac{DD'}{L} = \frac{r\theta}{L}$$

The shear stress, τ, at the radius r is related to the shear strain γ by Eq. (7.9). Thus

$$\gamma = \frac{\tau}{G} = \frac{r\theta}{L}$$

or, rearranging

$$\frac{\tau}{r} = G\,\frac{\theta}{L} \tag{11.1}$$

Consider now any cross-section of the bar as shown in Fig. 11.2(b). The shear stress, τ, on an annulus of radius r and width δr is tangential to the annulus, is in the plane of the cross-section and is constant round the annulus since the cross-section of the bar is perfectly symmetrical. The shear force on the element δs of the annulus is then $\tau\,\delta s\,\delta r$ and its moment about the centre, O, of the section is $\tau\,\delta s\,\delta r\,r$. Summing the moments on all such elements of the annulus we obtain the torque, δT, on the annulus, i.e.

$$\delta T = \int_0^{2\pi r} \tau\,\delta r\,r\,ds$$

which gives

$$\delta T = 2\pi r^2 \tau\,\delta r$$

The total torque on the bar is now obtained by summing the torques from each annulus in the cross-section. Thus

$$T = \int_0^R 2\pi r^2 \tau \, dr \qquad (11.2)$$

Substituting for τ in Eq. (11.2) from Eq. (11.1) we have

$$T = \int_0^R 2\pi r^3 G \frac{\theta}{L} \, dr$$

which gives

$$T = \frac{\pi R^4}{2} G \frac{\theta}{L}$$

or

$$T = JG \frac{\theta}{L} \qquad (11.3)$$

where $J = \pi R^4/2 (= \pi D^4/32)$ is defined as the polar second moment of area of the cross-section (see Eq. (9.42)). Combining Eqs (11.1) and (11.3) we have

$$\frac{T}{J} = \frac{\tau}{r} = G \frac{\theta}{L} \qquad (11.4)$$

Note that for a given torque acting on a given bar the shear stress is a maximum at the outer surface of the bar. Note also that these shear stresses induce complementary shear stresses on planes parallel to the axis of the bar but not on the actual surface (Fig. 11.3).

Torsion of a circular section hollow bar

The preceding analysis may be applied directly to a hollow bar of circular section having outer and inner radii R_o and R_i, respectively. Equation (11.2) then becomes

$$T = \int_{R_i}^{R_o} 2\pi r^2 \tau \, dr$$

Substituting for τ from Eq. (11.1) we have

$$T = \int_{R_i}^{R_o} 2\pi r^3 G \frac{\theta}{L} \, dr$$

whence

$$T = \frac{\pi}{2} (R_o^4 - R_i^4) G \frac{\theta}{L}$$

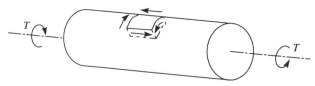

Fig. 11.3 Shear and complementary shear stresses on the surface of a circular-section bar subjected to torsion

Thus the polar second moment of area, J, is given by

$$J = \frac{\pi}{2} (R_o{}^4 - R_i{}^4) \tag{11.5}$$

Statically indeterminate circular-section bars under torsion

In many instances bars subjected to torsion are supported in such a way that the support reactions are statically indeterminate. These reactions must be determined, however, before values of maximum stress and angle of twist can be obtained.

Figure 11.4(a) shows a bar of uniform circular cross-section firmly supported at each end and subjected to a concentrated torque at a point B along its length. From equilibrium we have

$$T = T_A + T_C \tag{11.6}$$

A second equation is obtained by considering the compatibility of displacement at B of the two lengths AB and BC. Thus the angle of twist at B in AB must equal the angle of twist at B in BC, i.e.

$$\theta_{B(AB)} = \theta_{B(BC)}$$

or using Eq. (11.3)

$$\frac{T_A L_{AB}}{GJ} = \frac{T_C L_{BC}}{GJ}$$

whence

$$T_A = T_C \frac{L_{BC}}{L_{AB}}$$

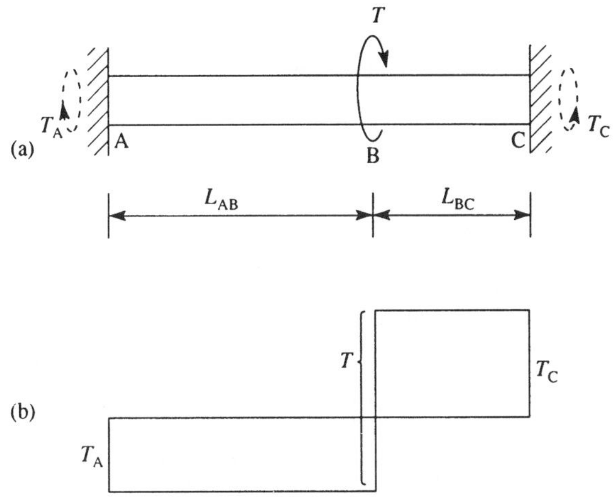

Fig. 11.4 Torsion of a circular-section bar with built-in ends

Substituting in Eq. (11.6) for T_A we obtain

$$T_A = T_C \left(\frac{L_{BC}}{L_{AB}} + 1 \right)$$

which gives

$$T_C = \frac{L_{AB}}{L_{AB} + L_{BC}} T \qquad (11.7)$$

Hence

$$T_A = \frac{L_{BC}}{L_{AB} + L_{BC}} T \qquad (11.8)$$

The distribution of torque along the length of the bar is shown in Fig. 11.4(b). Note that if $L_{AB} > L_{BC}$, T_C is the maximum torque in the bar.

Example 11.1   A bar of circular cross-section is 2·5 m long (Fig. 11.5). For 2 m of its length its diameter is 200 mm while for the remaining 0·5 m its diameter is 100 mm. If the bar is firmly supported at its ends and subjected to a torque of 50 kN m applied at its change of section, calculate the maximum stress in the bar and the angle of twist at the point of application of the torque. Take $G = 80\,000$ N/mm^2.

In this problem Eqs (11.7) and (11.8) cannot be used directly since the bar changes section at B. Thus from equilibrium

$$T = T_A + T_C \qquad (i)$$

and from the compatibility of displacement at B in the lengths AB and BC

$$\theta_{B(AB)} = \theta_{B(BC)}$$

or using Eq. (11.3)

$$\frac{T_A L_{AB}}{GJ_{AB}} = \frac{T_C L_{BC}}{GJ_{BC}}$$

whence

$$T_A = \frac{L_{BC}}{L_{AB}} \frac{J_{AB}}{J_{BC}} T_C \qquad (ii)$$

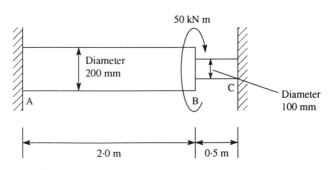

Fig. 11.5   Bar of Ex. 11.1

Substituting in Eq. (i) we obtain

$$T = T_C \left(\frac{L_{BC}}{L_{AB}} \frac{J_{AB}}{J_{BC}} + 1 \right)$$

or

$$50 = T_C \left[\frac{0 \cdot 5}{2 \cdot 0} \times \left(\frac{200 \times 10^{-3}}{100 \times 10^{-3}} \right)^4 + 1 \right]$$

from which

$$T_C = 10 \text{ kN m}$$

Hence, from Eq. (i)

$$T_A = 40 \text{ kN m}$$

Although the maximum torque occurs in the length AB, the length BC has the smaller diameter. It can be seen from Eqs (11.4) that shear stress is directly proportional to torque and inversely proportional to diameter (or radius) cubed. We therefore conclude that in this case the maximum shear stress occurs in the length BC of the bar and is given by

$$\tau_{max} = \frac{10 \times 10^6 \times 100 \times 32}{2 \times \pi \times 100^4} = 50 \cdot 9 \text{ N/mm}^2$$

Also the rotation at B is given by either

$$\theta_B = \frac{T_A L_{AB}}{GJ_{AB}} \quad \text{or} \quad \theta_B = \frac{T_C L_{BC}}{GJ_{BC}}$$

Using the first of these expressions we have

$$\theta_B = \frac{40 \times 10^6 \times 2 \times 10^3 \times 32}{80\,000 \times \pi \times 200^4} = 0 \cdot 0064 \text{ radians}$$

or

$$\theta_B = 0 \cdot 37°$$

11.2 Strain energy due to torsion

It can be seen from Eq. (11.3) that for a bar of a given material, a given length, L, and radius, R, the angle of twist is directly proportional to the applied torque.

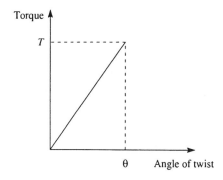

Fig. 11.6 Torque–angle of twist relationship for a gradually applied torque

Therefore a torque–angle of twist graph is linear and for a gradually applied torque takes the form shown in Fig. 11.6. The work done by a gradually applied torque, T, is equal to the area under the torque–angle of twist curve and is given by

$$\text{Work done} = \tfrac{1}{2} T\theta$$

The corresponding strain energy stored, U, is therefore also given by

$$U = \tfrac{1}{2} T\theta$$

Substituting for T and θ from Eqs (11.4) in terms of the maximum shear stress, τ_{max}, on the surface of the bar we have

$$U = \frac{1}{2} \frac{\tau_{max} J}{R} \times \frac{\tau_{max} L}{GR}$$

or

$$U = \frac{1}{4} \frac{\tau_{max}}{G} \pi R^2 L \quad \text{since} \quad J = \frac{\pi R^4}{2}$$

Hence

$$U = \frac{\tau_{max}^2}{4G} \times \text{volume of bar} \tag{11.9}$$

Alternatively, in terms of the applied torque T we have

$$U = \tfrac{1}{2} T\theta = \frac{T^2 L}{2GJ} \tag{11.10}$$

11.3   Plastic torsion of circular-section bars

Equations (11.4) apply only if the shear stress–shear strain curve for the material of the bar in torsion is linear. Stresses greater than the yield shear stress, τ_Y, induce plasticity in the outer region of the bar and this extends radially inwards as the torque is increased. It is assumed, in the plastic analysis of a circular-section bar subjected to torsion, that cross-sections of the bar remain plane and that radii remain straight.

For a material such as mild steel which has a definite yield point the shear stress–shear strain curve may be idealized in a similar manner to that for direct stress (see Fig. 9.31) as shown in Fig. 11.7. Thus, after yield, the shear strain increases at a

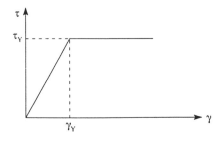

Fig. 11.7   Idealized shear stress — shear strain curve for a mild steel bar

more or less constant value of shear stress. It follows that the shear stress in the plastic region of a mild steel bar is constant and equal to τ_Y. Figure 11.8 illustrates the various stages in the development of full plasticity in a mild steel bar of circular section. In Fig. 11.8(a) the maximum stress at the outer surface of the bar has reached the yield stress, τ_Y. Equations (11.4) still apply, therefore, so that at the outer surface of the bar

$$\frac{T_Y}{J} = \frac{\tau_Y}{R}$$

or
$$T_Y = \frac{\pi R^3}{2} \tau_Y \tag{11.11}$$

where T_Y is the torque producing yield. In Fig. 11.8(b) the torque has increased above the value T_Y so that the plastic region extends inwards to a radius r_e. Within r_e the material remains elastic and forms an *elastic core*. At this stage the total torque is the sum of the contributions from the elastic core and the plastic zone, i.e.

$$T = \frac{\tau_Y J_e}{r_e} + \int_{r_e}^{R} 2\pi r^2 \tau_Y \, dr$$

where J_e is the polar second moment of area of the elastic core and the contribution from the plastic zone is derived in an identical manner to Eq. (11.2) but in which $\tau = \tau_Y = $ constant. Hence

$$T = \frac{\tau_Y \pi r_e^3}{2} + \frac{2}{3} \pi \tau_Y (R^3 - r_e^3)$$

which simplifies to

$$T = \frac{2\pi R^3}{3} \tau_Y \left(1 - \frac{r_e^3}{4R^3}\right) \tag{11.12}$$

Note that for a given value of torque, Eq. (11.12) fixes the radius of the elastic core of the section. In stage three (Fig. 11.8(c)) the cross-section of the bar is completely

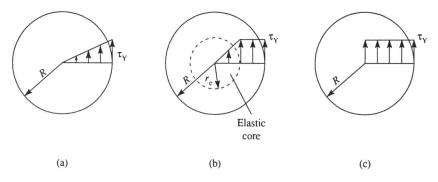

(a)   (b)   (c)

Fig. 11.8   Plastic torsion of a circular-section bar

plastic so that r_e in Eq. (11.12) is zero and the ultimate torque or fully plastic torque, T_P, is given by

$$T_P = \frac{2\pi R^3}{3} \tau_Y \tag{11.13}$$

Comparing Eqs (11.11) and (11.13) we see that

$$\frac{T_P}{T_Y} = \frac{4}{3} \tag{11.14}$$

so that only a one-third increase in torque is required after yielding to bring the bar to its ultimate load-carrying capacity.

Since we have assumed that radii remain straight during plastic torsion, the angle of twist of the bar must be equal to the angle of twist of the elastic core which may be obtained directly from Eq. (11.3). Thus for a bar of length L and shear modulus G,

$$\theta = \frac{TL}{GJ_e} = \frac{2TL}{\pi G r_e^4} \tag{11.15}$$

or, in terms of the shear stress, τ_Y, at the outer surface of the elastic core

$$\theta = \frac{\tau_Y L}{G r_e} \tag{11.16}$$

Either of Eqs (11.15) or (11.16) shows that θ is inversely proportional to the radius, r_e, of the elastic core. Clearly, when the bar becomes fully plastic, $r_e \to 0$ and θ becomes, theoretically, infinite. In practical terms this means that twisting continues with no increase in torque in the fully plastic state.

11.4   Torsion of a thin-walled closed section beam

Although the analysis of torsion problems is generally complex and in some instances relies on empirical methods for a solution, the torsion of a thin-walled beam of arbitrary closed section is relatively straightforward.

Figure 11.9(a) shows a thin-walled closed section beam subjected to a torque, T. The thickness, t, is constant along the length of the beam but may vary round the

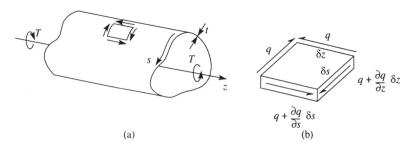

(a)   (b)

Fig. 11.9 Torsion of a thin-walled closed section beam

cross-section. The torque T induces a stress system in the walls of the beam which consists solely of shear stresses if the applied loading comprises only a pure torque. In some cases structural or loading discontinuities or the method of support produce a system of direct stresses in the walls of the beam even though the loading consists of torsion only. These effects, known as axial constraint effects, are considered in more advanced texts.

The shear stress system on an element of the beam wall may be represented in terms of the shear flow, q, (see Section 10.4) as shown in Fig. 11.9(b). Again we are assuming that the variation of t over the side δs of the element may be neglected. For equilibrium of the element in the z direction we have

$$\left(q + \frac{\partial q}{\partial s}\,\delta s\right)\delta z - q\,\delta z = 0$$

which gives

$$\frac{\partial q}{\partial s} = 0 \tag{11.17}$$

Considering equilibrium in the s direction,

$$\left(q + \frac{\partial q}{\partial z}\,\delta z\right)\delta s - q\,\delta s = 0$$

from which

$$\frac{\partial q}{\partial z} = 0 \tag{11.18}$$

Equations (11.17) and (11.18) may only be satisfied simultaneously by a constant value of q. We deduce, therefore, that the application of a pure torque to a thin-walled closed section beam results in the development of a constant shear flow in the beam wall. However, the shear stress, τ, may vary round the cross-section since we allow the wall thickness, t, to be a function of s.

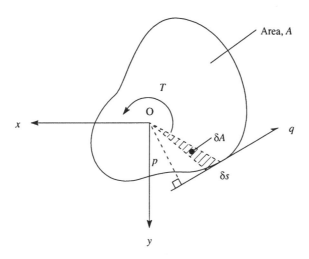

Fig. 11.10 Torque–shear flow relationship in a thin-walled closed section beam

The relationship between the applied torque and this constant shear flow may be derived by considering the torsional equilibrium of the section shown in Fig. 11.10. The torque produced by the shear flow acting on the element, δs, of the beam wall is $q \, \delta s \, p$. Hence

$$T = \oint pq \, ds$$

or, since $q = $ constant

$$T = q \oint p \, ds \tag{11.19}$$

We have seen in Section 10.5 that $\oint p \, ds = 2A$ where A is the area enclosed by the mid-line of the beam wall. Hence

$$T = 2Aq \tag{11.20}$$

The theory of the torsion of thin-walled closed section beams is known as the *Bredt-Batho theory* and Eq. (11.20) is often referred to as the *Bredt-Batho formula*.
It follows from Eq. (11.20) that

$$\tau = \frac{q}{t} = \frac{T}{2At} \tag{11.21}$$

and that the maximum shear stress in a beam subjected to torsion will occur at the section where the torque is a maximum and at the point in that section where the thickness is a minimum. Thus

$$\tau_{max} = \frac{T_{max}}{2At_{min}} \tag{11.22}$$

In Section 10.5 we derived an expression (Eq. (10.28)) for the rate of twist, $d\theta/dz$, in a shear-loaded thin-walled closed section beam. Equation (10.28) also applies to the case of a closed section beam under torsion in which the shear flow is constant if it is assumed that, as in the case of the shear-loaded beam, cross-sections remain undistorted after loading. Thus, rewriting Eq. (10.28) for the case $q_s = q = $ constant, we have

$$\frac{d\theta}{dz} = \frac{q}{2A} \oint \frac{ds}{Gt} \tag{11.23}$$

Substituting for q from Eq. (11.20) we obtain

$$\frac{d\theta}{dz} = \frac{T}{4A^2} \oint \frac{ds}{Gt} \tag{11.24}$$

or, if G, the shear modulus, is constant round the section

$$\frac{d\theta}{dz} = \frac{T}{4A^2 G} \oint \frac{ds}{t} \tag{11.25}$$

Example 11.2   A thin-walled circular-section beam has a diameter of 200 mm and is 2 m long; it is firmly restrained against rotation at each end. A concentrated torque of 30 kN m is applied to the beam at its mid-span point. If the maximum shear stress in the beam is limited to 200 N/mm^2 and the maximum angle of twist to $2°$, calculate the minimum thickness of the beam walls. Take $G = 25\,000 \text{ N/mm}^2$.

The minimum thickness of the beam corresponding to the maximum allowable shear stress of 200 N/mm^2 is obtained directly using Eq. (11.22) in which $T_{max} = 15$ kN m. Thus

$$t_{min} = \frac{15 \times 10^6 \times 4}{2 \times \pi \times 200^2 \times 200} = 1{\cdot}2 \text{ mm}$$

The rate of twist along the beam is given by Eq. (11.25) in which

$$\oint \frac{ds}{t} = \frac{\pi \times 200}{t_{min}}$$

Hence

$$\frac{d\theta}{dz} = \frac{T}{4A^2 G} \times \frac{\pi \times 200}{t_{min}} \tag{i}$$

Taking the origin for z at one of the fixed ends and integrating Eq. (i) for half the length of the beam we obtain

$$\theta = \frac{T}{4A^2 G} \times \frac{200\,\pi}{t_{min}} z + C_1$$

where C_1 is a constant of integration. At the fixed end where $z = 0$, $\theta = 0$ so that $C_1 = 0$. Hence

$$\theta = \frac{T}{4A^2 G} \times \frac{200\,\pi}{t_{min}} z$$

The maximum angle of twist occurs at the mid-span of the beam where $z = 1$ m. Hence

$$t_{min} = \frac{15 \times 10^6 \times 200 \times \pi \times 1 \times 10^3 \times 180}{4 \times (\pi \times 200^2/4)^2 \times 25\,000 \times 2 \times \pi} = 2{\cdot}7 \text{ mm}$$

The minimum allowable thickness that satisfies both conditions is therefore 2·7 mm.

11.5 Torsion of solid section beams

Generally, by solid section beams, we mean beam sections in which the walls do not form a closed loop system. Examples of such sections are shown in Fig. 11.11. An obvious exception is the hollow circular section bar which is, however, a special case of the solid circular section bar. The prediction of stress distributions and angles of twist produced by the torsion of such sections is complex and relies on the St. Venant warping function or Prandtl stress function methods of solution. Both of these methods are based on the theory of elasticity which may be found in advanced texts devoted solely to this topic. Even so, exact solutions exist for only a few practical cases, one of which is the circular-section bar.

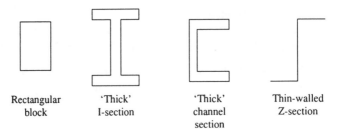

| Rectangular block | 'Thick' I-section | 'Thick' channel section | Thin-walled Z-section |

Fig. 11.11 Examples of solid beam sections

In all torsion problems, however, it is found that the torque, T, and the rate of twist, $d\theta/dz$, are related by the equation

$$T = GJ \frac{d\theta}{dz} \tag{11.26}$$

where G is the shear modulus and J is the *torsion constant*. For a circular-section bar J is the polar second moment of area of the section (see Eq. (11.3)) while for a thin-walled closed section beam J, from Eq. (11.25), is seen to be equal to $4A^2/\oint(ds/t)$. It is J in fact that distinguishes one torsion problem from another.

For 'thick' sections of the type shown in Fig. 11.11 J is obtained empirically in terms of the dimensions of the particular section. For example, the torsion constant of the 'thick' I-section shown in Fig. 11.12 is given by

$$J = 2J_1 + J_2 + 2\alpha D^4$$

where

$$J_1 = \frac{bt_f^{\,3}}{3}\left[1 - 0.63\,\frac{t_f}{b}\left(1 - \frac{t_f^{\,4}}{12b^4}\right)\right]$$

$$J_2 = \tfrac{1}{3}dt_w^{\,3}$$

$$\alpha = \frac{t_1}{t_2}\left(0.15 + 0.1\,\frac{r}{t_f}\right)$$

in which $t_1 = t_f$ and $t_2 = t_w$ if $t_f < t_w$, or $t_1 = t_w$ and $t_2 = t_f$ if $t_f > t_w$.

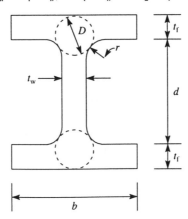

Fig. 11.12 Torsion constant for a 'thick' I-section beam

It can be seen from the above that J_1 and J_2, which are the torsion constants of the flanges and web, respectively, are each equal to one-third of the product of their length and their thickness cubed multiplied, in the case of the flanges, by an empirical constant. The torsion constant for the complete section is then the sum of the torsion constants of the components plus a contribution from the material at the web/flange junction. If the section were thin-walled, $t_f \ll b$ and D^4 would be negligibly small, in which case

$$J \simeq 2\frac{bt_f^3}{3} + \frac{dt_w^3}{3}$$

Generally, for thin-walled sections the torsion constant J may be written as

$$J = \tfrac{1}{3}\sum st^3 \tag{11.27}$$

in which s is the length and t the thickness of each component in the cross-section or, if t varies with s,

$$J = \frac{1}{3}\int_{\text{section}} t^3 \, \mathrm{d}s \tag{11.28}$$

The shear stress distribution in a thin-walled open section beam may be shown to be related to the rate of twist by the expression

$$\tau = 2Gn\frac{\mathrm{d}\theta}{\mathrm{d}z} \tag{11.29}$$

where n is the distance to any point in the section wall measured normally from its mid-line. The distribution is therefore linear across the thickness as shown in Fig. 11.13 and is zero at the mid-line of the wall. An alternative expression for shear stress distribution is obtained, in terms of the applied torque, by substituting for $\mathrm{d}\theta/\mathrm{d}z$ in Eq. (11.29) from Eq. (11.26). Thus

$$\tau = 2n\frac{T}{J} \tag{11.30}$$

It is clear from either of Eqs (11.29) or (11.30) that the maximum value of shear stress occurs at the outer surfaces of the wall when $n = \pm t/2$. Hence

$$\tau_{\max} = \pm Gt\frac{\mathrm{d}\theta}{\mathrm{d}z} = \pm\frac{Tt}{J} \tag{11.31}$$

The positive and negative signs in Eqs (11.31) indicate the direction of the shear stress in relation to the assumed direction for s.

The behaviour of closed and open section beams under torsional loads is similar in that they twist and develop internal shear stress systems. However, the manner in which each resists torsion is different. It is clear from the preceding discussion that a pure torque applied to a beam section produces a closed, continuous shear stress system since the resultant of any other shear stress system would generally be a shear force unless, of course, the system were self-equilibrating. In a closed section

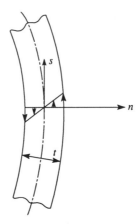

Fig. 11.13 Shear stress distribution due to torsion in a thin-walled open section beam

beam this closed loop system of shear stresses is allowed to develop in a continuous path round the cross-section, whereas in an open section beam it can only develop within the thickness of the walls; examples of both systems are shown in Fig. 11.14. Here, then, lies the basic difference in the manner in which torsion is resisted by closed and open section beams and the reason for the comparatively low torsional stiffness of thin-walled open sections. Clearly the development of a closed loop system of shear stresses in an open section is restricted by the thinness of the walls.

Example 11.3   The thin-walled section shown in Fig. 11.15 is symmetrical about a horizontal axis through O. The thickness t_0 of the centre web CD is constant, while the thickness of the other walls varies linearly from t_0 at points C and D to zero at the open ends A, F, G and H. Determine the torsion constant J for the section and also the maximum shear stress produced by a torque T.

Since the thickness of the section varies round its profile except for the central web, we use both Eqs (11.27) and (11.28) to determine the torsion constant. Thus,

$$J = \frac{2at_0^3}{3} + 2 \times \frac{1}{3} \int_0^a \left(\frac{s_A t_0}{a} \right)^3 ds_A + 2 \times \frac{1}{3} \int_0^{3a} \left(\frac{s_B t_0}{3a} \right)^3 ds_B$$

which gives

$$J = \frac{4at_0^3}{3}$$

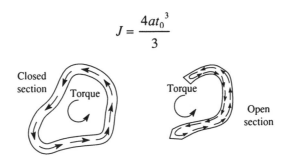

Fig. 11.14 Shear stress development in closed and open section beams subjected to torsion

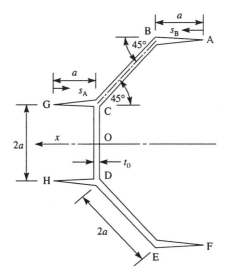

Fig. 11.15 Beam section of Ex. 11.3

The maximum shear stress is now obtained using Eq. (11.31), i.e.

$$\tau_{max} = \pm \frac{Tt_0}{J} = \pm \frac{3Tt_0}{4at_0{}^3} = \pm \frac{3T}{4at_0{}^2}$$

11.6 Warping of cross-sections under torsion

Although we have assumed that the shapes of closed and open beam sections remain undistorted during torsion, they do not remain plane. Thus, for example, the cross-section of a rectangular section box beam, although remaining rectangular when twisted, warps out of its plane as shown in Fig. 11.16(a), as does the channel section of Fig. 11.16(b). The calculation of warping displacements is covered in more advanced texts and is clearly of importance if a beam is, say, built into a rigid foundation at one end. In such a situation the warping is supressed and direct tensile and compressive stresses are induced which must be investigated in design particularly if a beam is of concrete where even low tensile stresses can cause severe cracking.

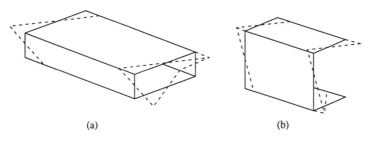

(a)   (b)

Fig. 11.16 Warping of beam sections due to torsion

Some beam sections do not warp under torsion; these include solid (and hollow) circular-section bars and square box sections of constant thickness.

Problems

P.11.1   The solid bar of circular cross-section shown in Fig. P.11.1 is subjected to a torque of 1 kN m at its free end and a torque of 3 kN m at its change of section. Calculate the maximum shear stress in the bar and the angle of twist at its free end. $G = 70\,000$ N/mm².

Ans.   40·6 N/mm², 0·6°.

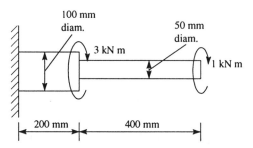

Fig. P.11.1

P.11.2   A hollow circular-section shaft 2 m long is firmly supported at each end and has an outside diameter of 80 mm. The shaft is subjected to a torque of 12 kN m applied at a point 1·5 m from one end. If the shear stress in the shaft is limited to 150 N/mm² and the angle of twist to 1·5°, calculate the maximum allowable internal diameter. The shear modulus $G = 80\,000$ N/mm².

Ans.   63·8 mm.

P.11.3   A bar ABCD of circular cross-section having a diameter of 50 mm is firmly supported at each end and carries two concentrated torques at B and C as shown in Fig. P.11.3. Calculate the maximum shear stress in the bar and the maximum angle of twist. Take $G = 70\,000$ N/mm².

Ans.   66·2 N/mm² in CD, 2·3° at B.

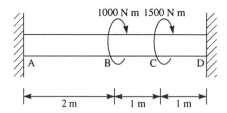

Fig. P.11.3

P.11.4   A bar ABCD has a circular cross-section of 75 mm diameter over half its length and 50 mm diameter over the remaining half of its length. A torque of 1 kN m is applied at C mid-way between B and D as shown in Fig. P.11.4. Sketch the distribution of torque along the length of the bar and calculate the maximum shear stress and the maximum angle of twist in the bar.

Ans. $\tau_{max} = 23 \cdot 7$ N/mm^2 in CD, $0 \cdot 4°$ at C.

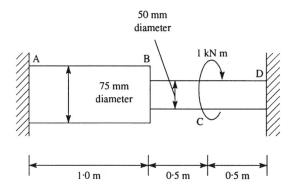

Fig. P.11.4

P.11.5 A thin-walled rectangular section box girder carries a uniformly distributed torque loading of 1 kN m/mm over the outer half of its length as shown in Fig. P.11.5. Calculate the maximum shear stress in the walls of the box girder and also the distribution of angle of twist along its length; illustrate your answer with a sketch. Take $G = 70\ 000$ N/mm^2.

Ans. 133 N/mm^2. In AB, $\theta = 218 \times 10^{-6}z$ degrees.
In BC, $\theta = 0 \cdot 109 \times 10^{-6}(4000z - z^2/2) - 0 \cdot 218$ degrees.

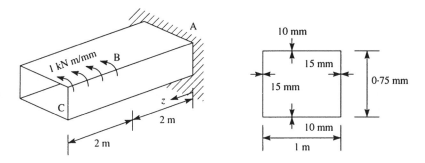

Fig. P.11.5

P.11.6 The thin-walled box section beam ABCD shown in Fig. P.11.6 is attached at each end to supports which allow rotation of the ends of the beam in the longitudinal vertical plane of symmetry but prevent rotation of the ends in vertical planes perpendicular to the longitudinal axis of the beam. The beam is subjected to a uniform torque loading of 20 N m/mm over the portion BC of its span. Calculate the maximum shear stress in the cross-section of the beam and the distribution of angle of twist along its length.

Ans. 71·4 N/mm^2, $\theta_B = \theta_C = 0 \cdot 36°$, θ at mid-span $= 0 \cdot 73°$.

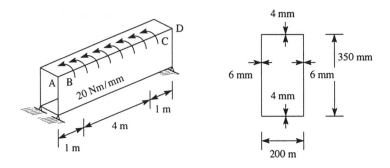

Fig. P.11.6

P.11.7 Figure P.11.7 shows a thin-walled cantilever box-beam having a constant width of 50 mm and a depth which decreases linearly from 200 mm at the built-in end to 150 mm at the free end. If the beam is subjected to a torque of 1 kN m at its free end, plot the angle of twist of the beam at 500 mm intervals along its length and determine the maximum shear stress in the beam section. Take $G = 25\,000$ N/mm^2.

Ans. $\tau_{max} = 33 \cdot 3$ N/mm^2.

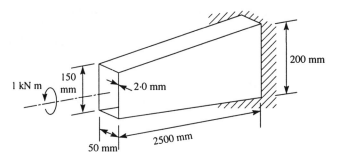

Fig. P.11.7

P.11.8 The cold-formed section shown in Fig. P.11.8 is subjected to a torque of 50 N m. Calculate the maximum shear stress in the section and its rate of twist. $G = 25\,000$ N/mm^2.

Ans. $\tau_{max} = 220 \cdot 6$ N/mm^2, $d\theta/dz = 0.0044$ rad/mm.

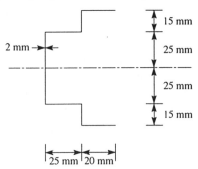

Fig. P.11.8

P.11.9 The thin-walled angle section shown in Fig. P.11.9 supports shear loads that produce both shear and torsional effects. Determine the maximum shear stress in the cross-section of the angle, stating clearly the point at which it acts.

Ans. 17.7 N/mm² on the inside of flange BC at 16·2 mm from point B.

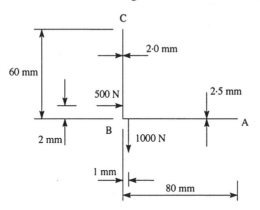

Fig. P.11.9

P.11.10 Figure P.11.10 shows the cross-section of a thin-walled inwardly lipped channel. The lips are of constant thickness while the flanges increase linearly in thickness from 1·27 mm, where they meet the lips, to 2·54 mm at their junctions with the web. The web has a constant thickness of 2·54 mm and the shear modulus *G* is 26 700 N/mm². Calculate the maximum shear stress in the section and also its rate of twist if it is subjected to a torque of 100 N m.

Ans. $\tau_{max} = 297 \cdot 2$ N/mm², $d\theta/dz = 0 \cdot 0044$ rad/mm.

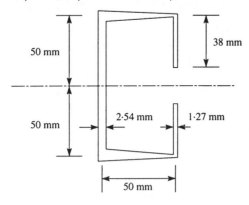

Fig. P.11.10

CHAPTER 12

Composite Beams

Frequently in civil engineering construction beams are fabricated from comparatively inexpensive materials of low strength which are reinforced by small amounts of high-strength material such as steel. In this way a timber beam of rectangular section may have steel plates bolted to its sides or to its top and bottom surfaces. Again, concrete beams are reinforced in their weak tension zones and also, if necessary, in their compression zones, by steel reinforcing bars. Other instances arise where steel beams support concrete floor slabs in which the strength of the concrete may be allowed for in the design of the beams. The design of reinforced concrete beams and concrete-and-steel beams is covered by Codes of Practice and relies, as in the case of steel beams, on ultimate load analysis. The design of steel reinforced timber beams is not covered by a code, and we shall therefore limit the analysis of this type of beam to an elastic approach.

12.1 Steel reinforced timber beams

The timber joist of breadth b and depth d shown in Fig. 12.1 is reinforced by two steel plates bolted to its sides, each plate being of thickness t and depth d. Let us suppose that the beam is bent to a radius R at this section by a positive bending moment, M. Clearly, since the steel plates are firmly attached to the sides of the timber joist, both are bent to the same radius, R. Thus, from Eq. (9.7), the bending moment, M_t, carried by the timber joist is

$$M_t = \frac{E_t I_t}{R} \tag{12.1}$$

where E_t is Young's modulus for the timber and I_t is the second moment of area of the timber section about the centroidal axis, Gx. Similarly for the steel plates

$$M_s = \frac{E_s I_s}{R} \tag{12.2}$$

in which I_s is the combined second moment of area about Gx of the two plates. The total bending moment is then

$$M = M_t + M_s = \frac{1}{R}(E_t I_t + E_s I_s)$$

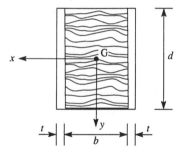

Fig. 12.1 Steel-reinforced timber beam

from which

$$\frac{1}{R} = \frac{M}{E_t I_t + E_s I_s} \tag{12.3}$$

From a comparison of Eqs (12.3) and (9.7) we see that the composite beam behaves as a homogeneous beam of bending stiffness EI where

$$EI = E_t I_t + E_s I_s$$

or

$$EI = E_t \left(I_t + \frac{E_s}{E_t} I_s \right) \tag{12.4}$$

The composite beam may therefore be treated wholly as a timber beam having a total second moment of area

$$I_t + \frac{E_s}{E_t} I_s$$

This is equivalent to replacing the steel reinforcing plates by timber 'plates' each having a thickness $(E_s/E_t)t$ as shown in Fig. 12.2(a). Alternatively, the beam may be

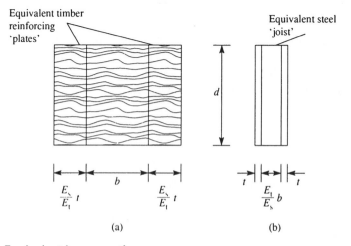

Fig. 12.2 Equivalent beam sections

transformed into a wholly steel beam by writing Eq. (12.4) as

$$EI = E_s \left(\frac{E_t}{E_s} I_t + I_s \right)$$

so that the second moment of area of the equivalent steel beam is

$$\frac{E_t}{E_s} I_t + I_s$$

which is equivalent to replacing the timber joist by a steel 'joist' of breadth $(E_t/E_s)b$ (Fig. 12.2(b)). Note that the transformed sections of Fig. 12.2 apply only to the case of bending about the horizontal axis, Gx. Note also that the depth, d, of the beam is unchanged by either transformation.

The direct stress due to bending in the timber joist is obtained using Eq. (9.9), i.e.

$$\sigma_t = \frac{M_t y}{I_t} \tag{12.5}$$

From Eqs (12.1) and (12.3)

$$M_t = \frac{E_t I_t}{E_t I_t + E_s I_s} M$$

or

$$M_t = \frac{M}{1 + \dfrac{E_s I_s}{E_t I_t}} \tag{12.6}$$

Substituting in Eq. (12.5) from Eqn. (12.6) we have

$$\sigma_t = \frac{My}{I_t + \dfrac{E_s}{E_t} I_s} \tag{12.7}$$

Equation (12.7) could in fact have been deduced directly from Eq. (9.9) since $I_t + (E_s/E_t)I_s$ is the second moment of area of the equivalent timber beam of Fig. 12.2(a). Similarly, by considering the equivalent steel beam of Fig. 12.2(b), we obtain the direct stress distribution in the steel, i.e.

$$\sigma_s = \frac{My}{I_s + \dfrac{E_t}{E_s} I_t} \tag{12.8}$$

Example 12.1   A beam is formed by connecting two timber joists each 100 mm × 400 mm with a steel plate 12 mm × 300 mm placed symmetrically between them (Fig. 12.3). If the beam is subjected to a bending moment of 50 kN m, determine the maximum stresses in the steel and in the timber. The ratio of Young's modulus for steel to that of timber is 12:1.

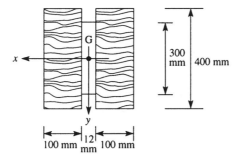

Fig. 12.3 Steel-reinforced timber beam of Ex. 12.1

The second moments of area of the timber and steel about the centroidal axis, Gx, are

$$I_t = 2 \times 100 \times \frac{400^3}{12} = 1067 \times 10^6 \text{ mm}^4$$

and

$$I_s = 12 \times \frac{300^3}{12} = 27 \times 10^6 \text{ mm}^4$$

respectively. Therefore, from Eq. (12.7) we have

$$\sigma_t = \pm \frac{50 \times 10^6 \times 200}{1067 \times 10^6 + 12 \times 27 \times 10^6} = \pm 7 \cdot 2 \text{ N/mm}^2$$

and from Eq. (12.8)

$$\sigma_s = \pm \frac{50 \times 10^6 \times 150}{27 \times 10^6 + 1067 \times 10^6/12} = \pm 64 \cdot 7 \text{ N/mm}^2$$

Consider now the steel-reinforced timber beam of Fig. 12.4(a) in which the steel plates are attached to the top and bottom surfaces of the timber. The section may be transformed into an equivalent timber beam (Fig. 12.4(b)) or steel beam (Fig. 12.4(c)) by the methods used for the beam of Fig. 12.1. The direct stress distributions are then obtained from Eqs (12.7) and (12.8). There is, however, one important difference between the beam of Fig. 12.1 and that of Fig. 12.4(a). In the latter case, when the beam is subjected to shear loads, the connection between the timber and steel must resist horizontal complementary shear stresses as shown in Fig. 12.5. Generally, it is sufficiently accurate to assume that the timber joist resists all the vertical shear and then calculate an average value of shear stress, τ_{av}. Thus

$$\tau_{av} = \frac{S_y}{bd}$$

so that, based on this approximation, the horizontal complementary shear stress is S_y/bd and the shear force per unit length resisted by the timber/steel connection is S_y/d.

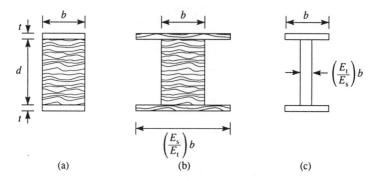

Fig. 12.4   Reinforced timber beam with steel plates attached to its top and bottom surfaces

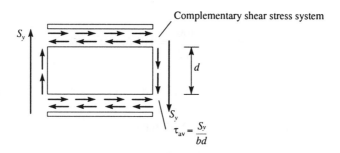

Fig. 12.5   Shear stresses between steel plates and timber beam

Example 12.2   A timber joist 100 mm × 200 mm is reinforced on its top and bottom surfaces by steel plates 15 mm thick × 100 mm wide. The composite beam is simply supported over a span of 4 m and carries a uniformly distributed load of 10 kN/m. Determine the maximum direct stress in the timber and in the steel and also the shear force per unit length transmitted by the timber/steel connection. Take $E_s/E_t = 15$.

The second moments of area of the timber and steel about a horizontal axis through the centroid of the beam are

$$I_t = \frac{100 \times 200^3}{12} = 66.7 \times 10^6 \text{ mm}^4$$

and $I_s = 2 \times 15 \times 100 \times 107.5^2 = 34.7 \times 10^6 \text{ mm}^4$

respectively. The maximum bending moment in the beam occurs at mid-span and is

$$M_{max} = \frac{10 \times 4^2}{8} = 20 \text{ kN m}$$

From Eq. (12.7)

$$\sigma_{t, max} = \pm \frac{20 \times 10^6 \times 100}{66.7 \times 10^6 + 15 \times 34.7 \times 10^6} = \pm 3.4 \text{ N/mm}^2$$

and from Eq. (12.8)

$$\sigma_{s,\max} = \pm \frac{20 \times 10^6 \times 115}{34 \cdot 7 \times 10^6 + 66 \cdot 7 \times 10^6/15} = \pm 58 \cdot 8 \ N/mm^2$$

The maximum shear force in the beam occurs at the supports and is equal to $10 \times 4/2 = 20$ kN. The average shear stress in the timber joist is then

$$\tau_{av} = \frac{20 \times 10^3}{100 \times 200} = 1 \ N/mm^2$$

It follows that the shear force per unit length in the timber/steel connection is $1 \times 100 = 100$ N/mm or 100 kN/m. Note that this value is an approximation for design purposes since, as we saw in Chapter 10, the distribution of shear stress through the depth of a beam of rectangular section is not uniform.

12.2 Reinforced concrete beams

As we have noted in Chapter 8, concrete is a brittle material which is weak in tension. It follows that a beam comprised solely of concrete would have very little bending strength since the concrete in the tension zone of the beam would crack at very low values of load. Concrete beams are therefore reinforced in their tension zones (and sometimes in their compression zones) by steel bars embedded in the concrete. Generally, whether the beam is precast or forms part of a slab/beam structure, the bars are positioned in a mould (usually fabricated from timber and called formwork) into which the concrete is poured. On setting, the concrete shrinks and grips the steel bars; the adhesion or *bond* between the bars and the concrete transmits bending and shear loads from the concrete to the steel.

In the design of reinforced concrete beams the elastic method has been superseded by the ultimate load method. We shall, however, for completeness, consider both methods.

Elastic theory

Consider the concrete beam section shown in Fig. 12.6(a). The beam is subjected to a bending moment, M, and is reinforced in its tension zone by a number of steel bars of total cross-sectional area A_s. The centroid of the reinforcement is at a depth d_1 from the upper surface of the beam; d_1 is known as the *effective depth* of the beam. The bending moment, M, produces compression in the concrete above the neutral axis whose position is at some, as yet unknown, depth, n, below the upper surface of the beam. Below the neutral axis the concrete is in tension and is assumed to crack so that its contribution to the bending strength of the beam is negligible. Thus all tensile forces are resisted by the reinforcing steel.

The reinforced concrete beam section may be conveniently analysed by the method employed in Section 12.1 for steel reinforced beams. The steel reinforcement is, therefore, transformed into an equivalent area, mA_s, of concrete in which m, the *modular ratio*, is given by

$$m = \frac{E_s}{E_c}$$

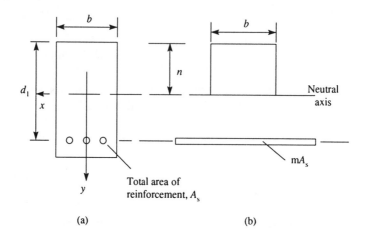

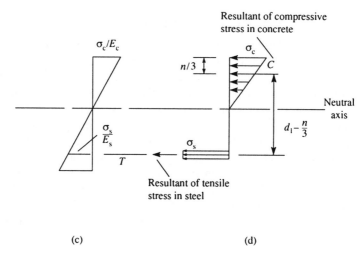

Fig. 12.6   Reinforced concrete beam

where E_s and E_c are Young's moduli for steel and concrete, respectively. The transformed section is shown in Fig. 12.6(b). Taking moments of areas about the neutral axis we have

$$bn\frac{n}{2} = mA_s(d_1 - n)$$

which, when rearranged, gives a quadratic equation in n, i.e.

$$\frac{bn^2}{2} + mA_s n - mA_s d_1 = 0 \tag{12.9}$$

whence

$$n = \frac{mA_s}{b}\left(\sqrt{1 + \frac{2bd_1}{mA_s}} - 1\right) \tag{12.10}$$

Note that the negative solution of Eq. (12.9) has no practical significance and is therefore ignored.

The second moment of area, I_c, of the transformed section is

$$I_c = \frac{bn^3}{3} + mA_s(d_1 - n)^2 \tag{12.11}$$

so that the maximum stress, σ_c, induced in the concrete is

$$\sigma_c = \frac{Mn}{I_c} \tag{12.12}$$

The stress, σ_s, in the steel may be deduced from the strain diagram (Fig. 12.6(c)) which is linear throughout the depth of the beam since the beam section is assumed to remain plane during bending. Thus

$$\frac{\sigma_s/E_s}{d_1 - n} = \frac{\sigma_c/E_c}{n}$$

from which

$$\sigma_s = \sigma_c \frac{E_s}{E_c}\left(\frac{d_1 - n}{n}\right) = \sigma_c m\left(\frac{d_1}{n} - 1\right) \tag{12.13}$$

Substituting for σ_c from Eq. (12.12) we obtain

$$\sigma_s = \frac{mM}{I_c}(d_1 - n) \tag{12.14}$$

Frequently, instead of determining stresses in a given beam section subjected to a given applied bending moment, we wish to calculate the moment of resistance of a beam when either the stress in the concrete or the steel reaches a maximum allowable value. Equations (12.12) and (12.14) may be used to solve this type of problem but an alternative and more direct method considers moments due to the resultant loads in the concrete and steel. Thus, from the stress diagram of Fig. 12.6(d)

$$M = C\left(d_1 - \frac{n}{3}\right)$$

so that

$$M = \frac{\sigma_c}{2} bn\left(d_1 - \frac{n}{3}\right) \tag{12.15}$$

Alternatively, taking moments about the centroid of the concrete stress diagram

$$M = T\left(d_1 - \frac{n}{3}\right)$$

or

$$M = \sigma_s A_s\left(d_1 - \frac{n}{3}\right) \tag{12.16}$$

Equation (12.16) may also be used in conjunction with Eq. (12.13) to 'design' the area of reinforcing steel in a beam section subjected to a given bending moment so that the stresses in the concrete and steel attain their maximum allowable values simultaneously. Such a section is known as a *critical* or *economic* section. The position of the neutral axis is obtained directly from Eq. (12.13) in which σ_s, σ_c, m and d_1 are known. The required area of steel is then determined from Eq. (12.16).

Example 12.3   A rectangular section reinforced concrete beam has a breadth of 200 mm and is 350 mm deep to the centroid of the steel reinforcement which consists of two steel bars each having a diameter of 20 mm. If the beam is subjected to a bending moment of 30 kN m, calculate the stress in the concrete and in the steel. The modular ratio m is 15.

The area A_s of the steel reinforcement is given by

$$A_s = 2 \times \frac{\pi}{4} \times 20^2 = 628 \cdot 3 \text{ mm}^2$$

The position of the neutral axis is obtained from Eq. (12.10) and is

$$n = \frac{15 \times 628 \cdot 3}{200} \left(\sqrt{1 + \frac{2 \times 200 \times 350}{15 \times 628 \cdot 3}} - 1 \right) = 140 \cdot 5 \text{ mm}$$

Now using Eq. (12.11)

$$I_c = \frac{200 \times 140 \cdot 5^3}{3} + 15 \times 628 \cdot 3(350 - 140 \cdot 5)^2 = 598 \cdot 5 \times 10^6 \text{ mm}^4$$

The maximum stress in the concrete follows from Eq. (12.12), i.e.

$$\sigma_c = \frac{30 \times 10^6 \times 140 \cdot 5}{598 \cdot 5 \times 10^6} = 7 \cdot 0 \text{ N/mm}^2$$

and from Eq. (12.14)

$$\sigma_s = \frac{15 \times 30 \times 10^6}{598 \cdot 5 \times 10^6} (350 - 140 \cdot 5) = 157 \cdot 5 \text{ N/mm}^2$$

Example 12.4   A reinforced concrete beam has a rectangular section of breadth 250 mm and a depth of 400 mm to the steel reinforcement, which consists of three 20 mm diameter bars. If the maximum allowable stresses in the concrete and steel are 7·0 N/mm² and 140 N/mm², respectively, determine the moment of resistance of the beam. The modular ratio $m = 15$.

The area, A_s, of steel reinforcement is

$$A_s = 3 \times \frac{\pi}{4} \times 20^2 = 942 \cdot 5 \text{ mm}^2$$

From Eq. (12.10)

$$n = \frac{15 \times 942 \cdot 5}{250} \left(\sqrt{1 + \frac{2 \times 250 \times 400}{15 \times 942 \cdot 5}} - 1 \right) = 163 \cdot 5 \text{ mm}$$

The maximum bending moment that can be applied such that the permissible stress in the concrete is not exceeded is given by Eq. (12.15). Thus

$$M = \frac{7}{2} \times 250 \times 163 \cdot 5 \left(400 - \frac{163 \cdot 5}{3}\right) \times 10^{-6} = 49 \cdot 4 \text{ kN m}$$

Similarly, from Eq. (12.16) the stress in the steel limits the applied moment to

$$M = 140 \times 942 \cdot 5 \left(400 - \frac{163 \cdot 5}{3}\right) \times 10^{-6} = 45 \cdot 6 \text{ kN m}$$

The steel is therefore the limiting material and the moment of resistance of the beam is 45·6 kN m.

Example 12.5 A rectangular section reinforced concrete beam is required to support a bending moment of 40 kN m and is to have dimensions of breadth 250 mm and effective depth 400 mm. The maximum allowable stresses in the steel and concrete are 120 N/mm² and 6·5 N/mm², respectively; the modular ratio is 15. Determine the required area of reinforcement such that the limiting stresses in the steel and concrete are attained simultaneously.

Using Eq. (12.13) we have

$$120 = 6 \cdot 5 \times 15 \left(\frac{400}{n} - 1\right)$$

from which $n = 179 \cdot 3$ mm.

The required area of steel is now obtained from Eq (12.16); hence

$$A_s = \frac{M}{\sigma_s(d_1 - n/3)}$$

i.e. $$A_s = \frac{40 \times 10^6}{120(400 - 179 \cdot 3/3)} = 979 \cdot 7 \text{ mm}^2$$

It may be seen from Ex. 12.4 that for a beam of given cross-sectional dimensions, increases in the area of steel reinforcement do not result in increases in the moment of resistance after a certain value has been attained. When this stage is reached the concrete becomes the limiting material, so that additional steel reinforcement only serves to reduce the stress in the steel. However, the moment of resistance of a beam of a given cross-section may be increased above the value corresponding to the limiting concrete stress by the addition of steel in the compression zone of the beam.

Figure 12.7(a) shows a concrete beam reinforced in both its tension and compression zones. The centroid of the compression steel of area A_{sc} is at a depth d_2 below the upper surface of the beam, while the tension steel of area A_{st} is at a depth d_1. The section may again be transformed into an equivalent concrete section as shown in Fig. 12.7(b). However, when determining the second moment of area of the transformed section it must be remembered that the area of concrete in the

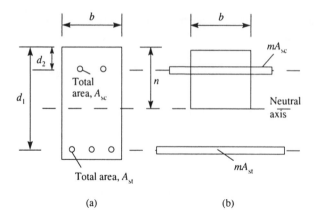

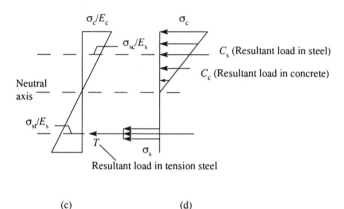

Fig. 12.7   Reinforced concrete beam with steel in tension and compression zones

compression zone is reduced due to the presence of the steel. Thus taking moments of areas about the neutral axis we have

$$\frac{bn^2}{2} - A_{sc}(n - d_2) + mA_{sc}(n - d_2) = mA_{st}(d_1 - n)$$

or, rearranging,

$$\frac{bn^2}{2} + (m - 1)A_{sc}(n - d_2) = mA_{st}(d_1 - n) \tag{12.17}$$

It can be seen from Eq. (12.17) that multiplying A_{sc} by $(m - 1)$ in the transformation process rather than m automatically allows for the reduction in the area of concrete caused by the presence of the compression steel. Thus the second moment of area of the transformed section is

$$I_c = \frac{bn^3}{3} + (m - 1)A_{sc}(n - d_2)^2 + mA_{st}(d_1 - n)^2 \tag{12.18}$$

The maximum stress in the concrete is then

$$\sigma_c = \frac{Mn}{I_c} \quad \text{(see Eq. (12.12))}$$

The stress in the tension steel and in the compression steel are obtained from the strain diagram of Fig. 12.7(c). Hence

$$\frac{\sigma_{sc}/E_s}{n - d_2} = \frac{\sigma_c/E_c}{n} \tag{12.19}$$

whence

$$\sigma_{sc} = \frac{m(n - d_2)}{n} \sigma_c = \frac{mM(n - d_2)}{I_c} \tag{12.20}$$

and

$$\sigma_{st} = \frac{mM}{I_c}(d_1 - n) \text{ as before} \tag{12.21}$$

An alternative expression for the moment of resistance of the beam is derived by taking moments of the resultant steel and concrete loads about the compressive reinforcement. Therefore from the stress diagram of Fig. 12.7(d)

$$M = T(d_1 - d_2) - C_c\left(\frac{n}{3} - d_2\right)$$

whence

$$M = \sigma_{st} A_{st}(d_1 - d_2) - \frac{\sigma_c}{2} bn\left(\frac{n}{3} - d_2\right) \tag{12.22}$$

Example 12.6 A rectangular section concrete beam is 180 mm wide and has a depth of 360 mm to its tensile reinforcement. It is subjected to a bending moment of 45 kN m and carries additional steel reinforcement in its compression zone at a depth of 40 mm from the upper surface of the beam. Determine the necessary areas of reinforcement if the stress in the concrete is limited to 8·5 N/mm² and that in the steel to 140 N/mm². The modular ratio $E_s/E_c = 15$.

Assuming that the stress in the tensile reinforcement and that in the concrete attain their limiting values we can determine the position of the neutral axis using Eq. (12.13). Thus

$$140 = 8{\cdot}5 \times 15\left(\frac{360}{n} - 1\right)$$

from which

$$n = 171{\cdot}6 \text{ mm}$$

Substituting this value of n in Eq. (12.22) we have

$$45 \times 10^6 = 140 A_{st}(360 - 40) + \frac{8{\cdot}5}{2} \times 180 \times 171{\cdot}6\left(\frac{171{\cdot}6}{3} - 40\right)$$

which gives

$$A_{st} = 954 \text{ mm}^2$$

We can now use Eq. (12.17) to determine A_{sc} or, alternatively, we could equate the load in the tensile steel to the combined compressive load in the concrete and compression steel. Substituting for n and A_{st} in Eq. (12.17) we have

$$\frac{180 \times 171 \cdot 6^2}{2} + (15 - 1)A_{sc}(171 \cdot 6 - 40) = 15 \times 954(360 - 171 \cdot 6)$$

from which
$$A_{sc} = 24 \cdot 9 \text{ mm}^2$$

The stress in the compression steel may be obtained from Eq. (12.20), i.e.

$$\sigma_{sc} = 15 \frac{(171 \cdot 6 - 40)}{171 \cdot 6} \times 8 \cdot 5 = 97 \cdot 8 \text{ N/mm}^2$$

In many practical situations reinforced concrete beams are cast integrally with floor slabs, as shown in Fig. 12.8. Clearly, the floor slab contributes to the overall strength of the structure so that the part of the slab adjacent to a beam may be regarded as forming part of the beam. The result is a T-beam whose flange, or the major portion of it, is in compression. The assumed width, B, of the flange cannot be greater than L, the distance between the beam centres; in most instances B is specified in Codes of Practice.

It is usual to assume in the analysis of T-beams that the neutral axis lies within the flange or coincides with its under surface. In either case the beam behaves as a rectangular section concrete beam of width B and effective depth d_1 (Fig. 12.9). Therefore, the previous analysis of rectangular section beams still applies.

Ultimate load theory

We have previously noted in this chapter and also in Chapter 8 that the modern design of reinforced concrete structures relies on ultimate load theory. The

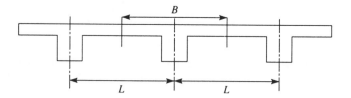

Fig. 12.8 Slab-reinforced concrete beam arrangement

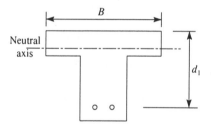

Fig. 12.9 Analysis of a reinforced concrete T-beam

calculated moment of resistance of a beam section is therefore based on the failure strength of concrete in compression and the yield strength of the steel reinforcement in tension modified by suitable factors of safety. Typical values are 1·5 for concrete (based on its 28-day cube strength) and 1·15 for steel. However, failure of the concrete in compression could occur suddenly in a reinforced concrete beam, whereas failure of the steel by yielding would be gradual. It is therefore preferable that failure occurs in the reinforcement rather than in the concrete. Thus, in design, the capacity of the concrete is underestimated to ensure that the preferred form of failure occurs. A further factor affecting the design stress for concrete stems from tests in which it has been found that concrete subjected to compressive tests due to bending always fails before attaining a compressive stress equal to the 28-day cube strength. The characteristic strength of concrete in compression is therefore taken as two-thirds of the 28-day cube strength. A typical design strength for concrete in compression is then

$$\frac{\sigma_{cu}}{1 \cdot 5} \times 0 \cdot 67 = 0 \cdot 45 \sigma_{cu}$$

where σ_{cu} is the 28-day cube strength. The corresponding figure for steel is

$$\frac{\sigma_Y}{1 \cdot 5} = 0 \cdot 87 \sigma_Y$$

In the ultimate load analysis of reinforced concrete beams it is assumed that plane sections remain plane during bending and that there is no contribution to the bending strength of the beam from the concrete in tension. From the first of these assumptions we deduce that the strain varies linearly through the depth of the beam as shown in Fig. 12.10(b). However, the stress diagram in the concrete is not linear but has the rectangular–parabolic shape shown in Fig. 12.10(c). Design charts in Codes of Practice are based on this stress distribution, but for direct calculation purposes a reasonably accurate approximation can be made in which the rectangular–parabolic stress distribution of Fig. 12.10(c) is replaced by an equivalent rectangular distribution as shown in Fig. 12.11(b) in which the compressive stress in the concrete is assumed to extend down to the mid-effective

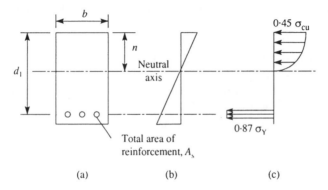

(a)   (b)   (c)

Fig. 12.10 Stress and strain distributions in a reinforced concrete beam

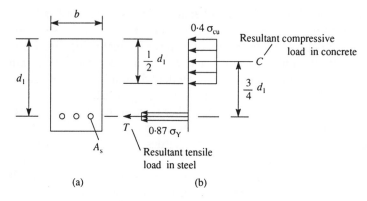

Fig. 12.11   Approximation of stress distribution in concrete

depth of the section at the maximum condition, i.e. at the ultimate moment of resistance, M_u, of the section.

M_u is then given by

$$M_u = C \tfrac{3}{4} d_1 = 0 \cdot 40 \sigma_{cu} b \tfrac{1}{2} d_1 \tfrac{3}{4} d_1$$

which gives

$$M_u = 0 \cdot 15 \sigma_{cu} b (d_1)^2 \tag{12.23}$$

or

$$M_u = T \tfrac{3}{4} d_1 = 0 \cdot 87 \sigma_Y A_s \tfrac{3}{4} d_1$$

from which

$$M_u = 0 \cdot 65 \sigma_Y A_s d_1 \tag{12.24}$$

whichever is the lesser. For applied bending moments less than M_u a rectangular stress block may be assumed for the concrete in which the stress is $0 \cdot 4 \sigma_{cu}$ but in which the depth of the neutral axis must be calculated. For beam sections in which the applied bending moment is greater than M_u, compressive reinforcement is required.

Example 12.7 A reinforced concrete beam having an effective depth of 600 mm and a breadth of 250 mm is subjected to a bending moment of 350 kN m. If the 28-day cube strength of the concrete is 30 N/mm² and the yield stress in tension of steel is 400 N/mm², determine the required area of reinforcement.

First it is necessary to check whether or not the applied moment exceeds the ultimate moment of resistance provided by the concrete. Hence, using Eq. (12.23)

$$M_u = 0 \cdot 15 \times 30 \times 250 \times 600^2 \times 10^{-6} = 405 \text{ kN m}$$

Since this is greater than the applied moment, the beam section does not require compression reinforcement.

We now assume the stress distribution shown in Fig. 12.12 in which the neutral axis of the section is at a depth n below the upper surface of the section. Thus, taking moments about the tensile reinforcement we have

$$350 \times 10^6 = 0 \cdot 4 \times 30 \times 250 n \left(600 - \frac{n}{2} \right)$$

from which

$$n = 243 \cdot 3 \text{ mm}.$$

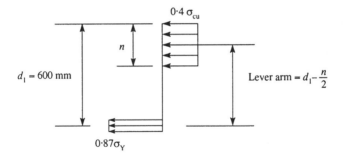

Fig. 12.12 Stress distribution in beam of Ex. 12.7

The lever arm is therefore equal to $600 - 243 \cdot 3/2 = 478 \cdot 4$ mm. Now taking moments about the centroid of the concrete we have

$$0 \cdot 87 \times 400 \times A_s \times 478 \cdot 4 = 350 \times 10^6$$

which gives $\qquad A_s = 2102 \cdot 3$ mm^2.

Example 12.8   A reinforced concrete beam of breadth 250 mm is required to have an effective depth as small as possible. Design the beam and reinforcement to support a bending moment of 350 kN m assuming that $\sigma_{cu} = 30$ N/mm^2 and $\sigma_Y = 400$ N/mm^2.

In this example the effective depth of the beam will be as small as possible when the applied moment is equal to the ultimate moment of resistance of the beam. Thus, using Eq. (12.23)

$$350 \times 10^6 = 0 \cdot 15 \times 30 \times 250 \times d_1^2$$

which gives $\qquad d_1 = 557 \cdot 8$ mm.

This is not a practical dimension since it would be extremely difficult to position the reinforcement to such accuracy. We therefore assume $d_1 = 558$ mm. Since the section is stressed to the limit, we see from Fig. 12.11(b) that the lever arm is

$$\tfrac{3}{4} d_1 = \tfrac{3}{4} \times 558 = 418 \cdot 5 \text{ mm}.$$

Hence, from Eq. (12.24)

$$350 \times 10^6 = 0 \cdot 87 \times 400 A_s \times 418 \cdot 5$$

from which $\qquad A_s = 2403 \cdot 2$ mm^2

A comparison of Exs 12.7 and 12.8 shows that the reduction in effective depth is only made possible by an increase in the area of steel reinforcement.

We have noted that the ultimate moment of resistance of a beam section of given dimensions can only be increased by the addition of compression reinforcement. However, although the design stress for tension reinforcement is $0 \cdot 87 \, \sigma_Y$, compression reinforcement is designed to a stress of $0 \cdot 72 \, \sigma_Y$ to avoid the possibility of the reinforcement buckling between the binders or stirrups. The method of designing a beam section to include compression reinforcement is simply treated as an extension of the singly reinforced case and is best illustrated by an example.

Example 12.9   A reinforced concrete beam has a breadth of 300 mm and an effective depth to the tension reinforcement of 618 mm. Compression reinforcement, if required, will be placed at a depth of 60 mm. If $\sigma_{cu} = 30$ N/mm^2 and $\sigma_Y = 410$ N/mm^2, design the steel reinforcement if the beam is to support a bending moment of 650 kN m.

The ultimate moment of resistance provided by the concrete is obtained using Eq. (12.23) and is

$$M_u = 0 \cdot 15 \times 30 \times 300 \times 618^2 \times 10^{-6} = 515 \cdot 6 \text{ kN m}.$$

This is less than the applied moment so that compression reinforcement is required to resist the excess moment of $650 - 515 \cdot 6 = 134 \cdot 4$ kN m. Thus, if A_{sc} is the area of compression reinforcement,

$$134 \cdot 4 \times 10^6 = \text{lever arm} \times 0 \cdot 72 \times 410 A_{sc}$$

i.e.

$$134 \cdot 4 \times 10^6 = (618 - 60) \times 0 \cdot 72 \times 410 A_{sc}$$

which gives

$$A_{sc} = 815 \cdot 9 \text{ mm}^2.$$

The tension reinforcement, A_{st}, is required to resist the moment of 515·6 kN m (as though the beam were singly reinforced) plus the excess moment of 134·4 kN m. Hence

$$A_{st} = \frac{515 \cdot 6 \times 10^6}{0 \cdot 75 \times 618 \times 0 \cdot 87 \times 410} + \frac{134 \cdot 4 \times 10^6}{(618 - 60) \times 0 \cdot 87 \times 410}$$

from which

$$A_{st} = 3793 \cdot 8 \text{ mm}^2.$$

The ultimate load analysis of reinforced concrete T-beams is simplified in a similar manner to the elastic analysis by assuming that the neutral axis does not lie below the lower surface of the flange. The ultimate moment of a T-beam therefore corresponds to a neutral axis position coincident with the lower surface of the flange as shown in Fig. 12.13(a). Thus M_u is the lesser of the two values given by

$$M_u = 0 \cdot 4 \sigma_{cu} B h_f \left(d_1 - \frac{h_f}{2} \right) \tag{12.25}$$

or

$$M_u = 0 \cdot 87 \sigma_Y A_s \left(d_1 - \frac{h_f}{2} \right) \tag{12.26}$$

For T-beams subjected to bending moments less than M_u, the neutral axis lies within the flange and must be found before, say, the amount of tension reinforcement can be determined. Compression reinforcement is rarely required in T-beams due to the comparatively large areas of concrete in compression.

Example 12.10   A reinforced concrete T-beam has a flange width of 1200 mm and an effective depth of 618 mm; the thickness of the flange is 150 mm. Determine the required area of reinforcement if the beam is required to resist a bending moment of 500 kN m. Take $\sigma_{cu} = 30$ N/mm^2 and $\sigma_Y = 410$ N/mm^2.

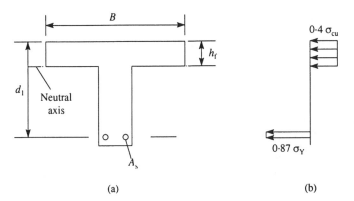

Fig. 12.13 Ultimate load analysis of a reinforced concrete T-beam

M_u for this beam section may be determined using Eq. (12.25), i.e.

$$M_u = 0.4 \times 30 \times 1200 \times 150 \left(618 - \frac{150}{2}\right) \times 10^{-6} = 1173 \text{ kN m}$$

Since this is greater than the applied moment, we deduce that the neutral axis lies within the flange. Thus from Fig. 12.14

$$500 \times 10^6 = 0.4 \times 30 \times 1200n \left(618 - \frac{n}{2}\right)$$

the solution of which gives

$$n = 59 \text{ mm}$$

Now taking moments about the centroid of the compression concrete we have

$$500 \times 10^6 = 0.87 \times 410 \times A_s \left(618 - \frac{59}{2}\right)$$

which gives $A_s = 2381.9 \text{ mm}^2$.

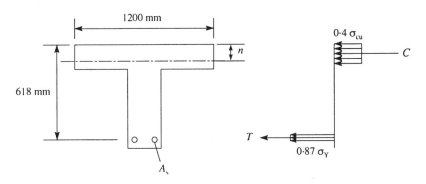

Fig. 12.14 Reinforced concrete T-beam of Ex. 12.10

12.3   Steel and concrete beams

In many instances concrete slabs are supported on steel beams, the two being joined together by shear connectors to form a composite structure. We therefore have a similar situation to that of the reinforced concrete T-beam in which the flange of the beam is concrete but the leg is a standard steel section.

Ultimate load theory is used to analyse steel and concrete beams with stress limits identical to those applying in the ultimate load analysis of reinforced concrete beams; plane sections are also assumed to remain plane.

Consider the steel and concrete beam shown in Fig. 12.15(a) and let us suppose that the neutral axis lies within the concrete flange. We ignore the contribution of the concrete in the tension zone of the beam to its bending strength, so that the assumed stress distribution takes the form shown in Fig. 12.15(b). A convenient method of designing the cross-section to resist a bending moment, M, is to assume the lever arm to be $(h_c + h_s)/2$ and then to determine the area of steel from the moment equation

$$M = 0.87\sigma_Y A_s \frac{(h_c + h_s)}{2} \tag{12.27}$$

The available compressive force in the concrete slab, $0.4\sigma_{cu}bh_c$, is then checked to ensure that it exceeds the tensile force, $0.87\sigma_Y A_s$, in the steel . If it does not, the neutral axis of the section lies within the steel and A_s given by Eq. (12.27) will be too small. If the neutral axis lies within the concrete slab the moment of resistance of the beam is determined by first calculating the position of the neutral axis. Thus, since the compressive force in the concrete is equal to the tensile force in the steel

$$0.4\sigma_{cu}bn_1 = 0.87\sigma_Y A_s \tag{12.28}$$

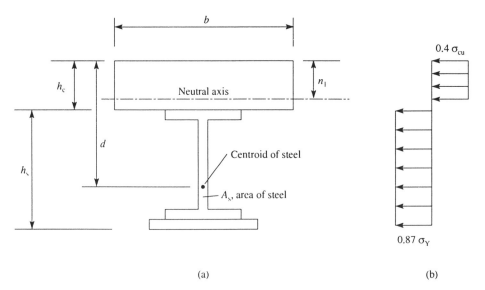

(a)   (b)

Fig. 12.15   Ultimate load analysis of a steel and concrete beam, neutral axis within the concrete

Then, from Fig. 12.15

$$M_u = 0.87\sigma_Y A_s\left(d - \frac{n_1}{2}\right)$$ (12.29)

If the neutral axis lies within the steel, the stress distribution shown in Fig. 12.16(b) is assumed in which the compressive stress in the steel above the neutral axis is the resultant of the tensile stress and twice the compressive stress. Thus, if the area of steel in compression is A_{sc}, we have, equating compressive and tensile forces

$$0.4\sigma_{cu}bh_c + 2 \times (0.87\sigma_Y)A_{sc} = 0.87\sigma_Y A_s$$ (12.30)

which gives A_{sc} and hence h_{sc}. Now taking moments

$$M_u = 0.87\sigma_Y A_s\left(d - \frac{h_c}{2}\right) - 2 \times (0.87\sigma_Y)A_{sc}\left(h_{sc} - \frac{h_c}{2}\right)$$ (12.31)

Example 12.11 A concrete slab 150 mm thick is 1·8 m wide and is to be supported by a steel beam. The total depth of the steel/concrete composite beam is limited to 562 mm. Find a suitable beam section if the composite beam is required to resist a bending moment of 709 kN m. Take $\sigma_{cu} = 30$ N/mm^2 and $\sigma_Y = 350$ N/mm^2.

Using Eq. (12.27)

$$A_s = \frac{2 \times 709 \times 10^6}{0.87 \times 350 \times 562} = 8286 \text{ mm}^2$$

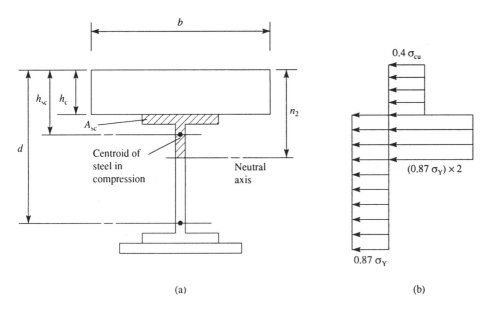

(a)   (b)

Fig. 12.16 Ultimate load analysis of a steel and concrete beam, neutral axis within the steel

The tensile force in the steel is then

$$0.87 \times 350 \times 8286 \times 10^{-3} = 2523 \text{ kN}$$

and the compressive force in the concrete is

$$0.4 \times 1.8 \times 10^3 \times 150 \times 30 \times 10^{-3} = 3240 \text{ kN}.$$

The neutral axis therefore lies within the concrete slab so that the area of steel in tension is, in fact, equal to A_s. From Steel Tables we see that a Universal Beam of nominal size 406 mm × 152 mm × 67 kg/m has an actual overall depth of 412 mm and a cross-sectional area of 8530 mm². The position of the neutral axis of the composite beam incorporating this beam section is obtained from Eq. (12.28); hence,

$$0.4 \times 30 \times 1800 n_1 = 0.87 \times 350 \times 8530$$

which gives $n_1 = 120$ mm.

Substituting for n_1 in Eq. (12.29) we obtain the moment of resistance of the composite beam,

$$M_u = 0.87 \times 350 \times 8530(356 - 60) \times 10^{-6} = 769 \text{ kN m}$$

Since this is greater than the applied moment we deduce that the beam section is satisfactory.

Problems

P.12.1   A timber beam 200 mm wide by 300 mm deep is reinforced on its top and bottom surfaces by steel plates each 12 mm thick by 200 mm wide. If the allowable stress in the timber is 8 N/mm² and that in the steel is 110 N/mm², find the allowable bending moment. The ratio of the modulus of elasticity of steel to that of timber is 20.

Ans.   94·7 kN m.

P.12.2   A simply supported beam of span 3·5 m carries a uniformly distributed load of 46·5 kN/m. The beam has the box section shown in Fig. P.12.2. Determine the required thickness of the steel plates if the allowable stresses are 124 N/mm² for the steel and 8 N/mm² for the timber. The modular ratio of steel to timber is 20.

Ans.   17 mm.

P.12.3   A timber beam 150 mm wide by 300 mm deep is reinforced by a steel plate 150 mm and 12 mm thick which is securely attached to its lower surface. Determine the percentage increase in the moment of resistance of the beam produced by the steel reinforcing plate. The allowable stress in the timber is 12 N/mm² and in the steel, 150 N/mm². The modular ratio is 20.

Ans.   176%.

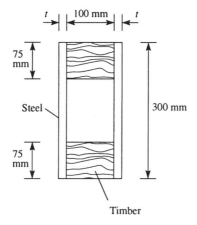

Fig. P.12.2

P.12.4   A singly reinforced rectangular concrete beam of effective span 4·5 m is required to carry a uniformly distributed load of 16·8 kN/m. The overall depth, D, is to be twice the breadth and the centre of the steel is to be at $0·1D$ from the underside of the beam. Using elastic theory find the dimensions of the beam and the area of steel reinforcement required if the stresses are limited to 8 N/mm² in the concrete and 140 N/mm² in the steel. Take $m = 15$.

Ans. $D = 404$ mm, $A_s = 992·3$ mm².

P.12.5   A reinforced concrete beam is of rectangular section 300 mm wide by 775 mm deep. It has five 25 mm diameter bars as tensile reinforcement in one layer with 25 mm cover and three 25 mm diameter bars as compression reinforcement, also in one layer with 25 mm cover. Find the moment of resistance of the section using elastic theory if the allowable stresses are 7·5 N/mm² and 125 N/mm² in the concrete and steel, respectively. The modular ratio is 16.

Ans.   214·5 kN m.

P.12.6   A reinforced concrete T-beam is required to carry a uniformly distributed load of 42 kN/m on a simply supported span of 6 m. The slab is 125 mm thick, the rib is 250 mm wide and the effective depth to the tensile reinforcement is 550 mm. The working stresses are 8·5 N/mm² in the concrete and 140 N/mm² in the steel; the modular ratio is 15. Making a reasonable assumption as to the position of the neutral axis find the area of steel reinforcement required and the breadth of the compression flange.

Ans.   2655·7 mm², 700 mm (N.A. coincides with base of slab).

P.12.7   Repeat P.12.4 using ultimate load theory assuming $\sigma_{cu} = 24$ N/mm² and $\sigma_Y = 280$ N/mm².

Ans. $D = 307·8$ mm, $A_s = 843$ mm².

P.12.8   Repeat P.12.5 using ultimate load theory and take $\sigma_{cu} = 22·5$ N/mm², $\sigma_Y = 250$ N/mm².

Ans.   222·5 kN m.

P.12.9   Repeat P.12.6 using ultimate load theory. Assume $\sigma_{cu} = 25\cdot5$ N/mm^2 and $\sigma_Y = 280$ N/mm^2.

Ans.   1592 mm^2, 304 mm (N.A. coincides with base of slab).

P.12.10   A concrete slab 175 mm thick and 2 m wide is supported by, and firmly connected to, a 457 mm × 152 mm × 74 kg/m Universal Beam whose actual depth is 461·3 mm and whose cross-sectional area is 9490 mm^2. If $\sigma_{cu} = 30$ N/mm^2 and $\sigma_Y = 350$ N/mm^2, find the moment of resistance of the resultant steel and concrete beam.

Ans.   919·5 kN m.

CHAPTER 13

Deflection of Beams

In Chapters 9, 10 and 11 we investigated the *strength* of beams in terms of the stresses produced by the action of bending, shear and torsion, respectively. An associated problem is the determination of the deflections of beams caused by different loads for, in addition to strength, a beam must possess sufficient *stiffness* so that excessive deflections do not have an adverse effect on adjacent structural members. In many cases, maximum allowable deflections are specified by Codes of Practice in terms of the dimensions of the beam, particularly the span; typical values are quoted in Section 8.7.

The design of beams from the point of view of strength has been discussed in Chapter 9, where we saw that two approaches were possible: elastic and plastic design. However, it is obvious that actual beam deflections must be limited to the elastic range of a beam, otherwise permanent distortion results. Thus in determining the deflections of beams under load, elastic theory is used.

There are several different methods of obtaining deflections in beams, the choice depending upon the type of problem being solved. For example, the double integration method gives the complete shape of a beam whereas the moment-area method can only be used to determine the deflection at a particular beam section. The latter method, however, is also useful in the analysis of statically indeterminate beams.

Generally beam deflections are caused primarily by the bending action of applied loads. In some instances, however, where a beam's cross-sectional dimensions are not small compared with its length, deflections due to shear become significant and must be calculated. We shall consider beam deflections due to shear in addition to those produced by bending. We shall also include deflections due to unsymmetrical bending.

13.1 Differential equation of symmetrical bending

In Chapter 9 we developed an expression relating the curvature, $1/R$, of a beam to the applied bending moment, M, and flexural rigidity, EI, i.e.

$$\frac{1}{R} = \frac{M}{EI} \qquad\qquad \text{(Eq. (9.11))}$$

For a beam of a given material and cross-section, EI is constant so that the curvature is directly proportional to the bending moment. We have also shown that bending moments produced by shear loads vary along the length of a beam, which implies that the curvature of the beam also varies along its length; Eq. (9.11) therefore gives the curvature at a particular section of a beam.

Consider a beam having a vertical plane of symmetry and loaded such that at a section of the beam the deflection of the neutral plane is v and the slope of the tangent to the neutral plane at this section is dv/dz (Fig. 13.1). The axes $Gxyz$ are centroidal axes so that in the unloaded condition Gz lies in the neutral plane of the beam. Also, if the applied loads produce a positive, i.e. sagging, bending moment at this section, then the upper surface of the beam is concave and the centre of curvature lies above the beam as shown. For the system of axes shown in Fig. 13.1, the sign convention usually adopted in mathematical theory gives a negative value for this curvature; thus

$$\frac{1}{R} = -\frac{\dfrac{d^2v}{dz^2}}{\left[1+\left(\dfrac{dv}{dz}\right)^2\right]^{3/2}} \tag{13.1}$$

For small deflections dv/dz is small so that $(dv/dz)^2$ is negligibly small compared with unity. Equation (13.1) then reduces to

$$\frac{1}{R} = -\frac{d^2v}{dz^2} \tag{13.2}$$

whence, from Eq. (9.11)

$$\frac{d^2v}{dz^2} = -\frac{M}{EI} \tag{13.3}$$

Double integration of Eq. (13.3) then yields the equation of the deflection curve of the neutral plane of the beam.

In the majority of problems concerned with beam deflections the bending moment varies along the length of a beam and therefore M in Eq. (13.3) must be expressed as a function of z before integration can commence. Alternatively, it may be

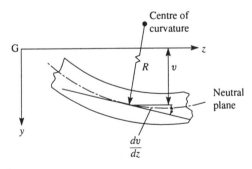

Fig. 13.1   Deflection and curvature of a beam due to bending

convenient in cases where the load is a known function of z to use the relationships of Eq. (3.8). Thus

$$\frac{d^3v}{dz^3} = -\frac{S}{EI} \qquad (13.4)$$

$$\frac{d^4v}{dz^4} = \frac{w}{EI} \qquad (13.5)$$

We shall now illustrate the use of Eqs (13.3), (13.4) and (13.5) by considering some standard cases of beam deflection.

Example 13.1 Determine the deflection curve and the deflection of the free end of the cantilever shown in Fig.13.2(a); the flexural rigidity of the cantilever is EI.

The load W causes the cantilever to deflect such that its neutral plane takes up the curved shape shown in Fig. 13.2(b); the deflection at any section Z is then v while that at its free end is v_{tip}. The axis system is chosen so that the origin coincides with the built-in end where the deflection is clearly zero.

The bending moment, M, at the section Z is, from Fig. 13.2(a)

$$M = -W(L - z) \quad \text{(i.e. hogging)} \qquad (i)$$

Substituting for M in Eq. (13.3) we obtain

$$\frac{d^2v}{dz^2} = \frac{W}{EI} \cdot (L - z)$$

or in more convenient form

$$EI \frac{d^2v}{dz^2} = W(L - z) \qquad (ii)$$

Integrating Eq. (ii) with respect to z gives

$$EI \frac{dv}{dz} = W\left(Lz - \frac{z^2}{2}\right) + C_1$$

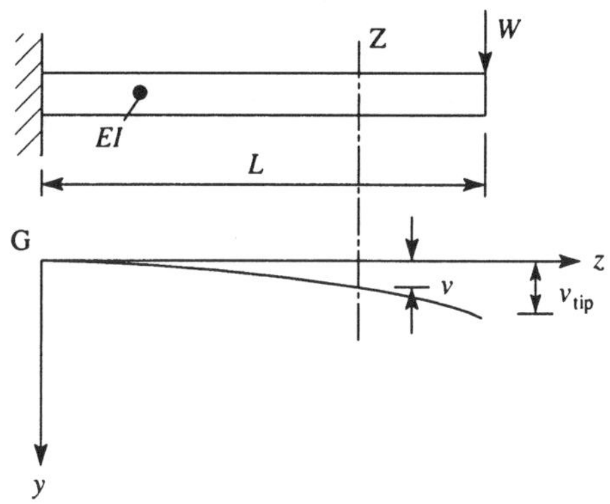

Fig. 13.2 Deflection of a cantilever beam carrying a concentrated load at its free end (Ex. 13.1)

where C_1 is a constant of integration which is obtained from the boundary condition that $dv/dz = 0$ at the built-in end where $z = 0$. Hence $C_1 = 0$ and

$$EI \frac{dv}{dz} = W\left(Lz - \frac{z^2}{2}\right) \tag{iii}$$

Integrating Eq. (iii) we obtain

$$EIv = W\left(\frac{Lz^2}{2} - \frac{z^3}{6}\right) + C_2$$

in which C_2 is again a constant of integration. At the built-in end $v = 0$ when $z = 0$ so that $C_2 = 0$. Hence the equation of the deflection curve of the cantilever is

$$v = \frac{W}{6EI}(3Lz^2 - z^3) \tag{iv}$$

The deflection, v_{tip}, at the free end is obtained by setting $z = L$ in Eq. (iv). Thus

$$v_{tip} = \frac{WL^3}{3EI} \tag{v}$$

and is clearly positive and downwards.

Example 13.2 Determine the deflection curve and the deflection of the free end of the cantilever shown in Fig. 13.3(a).

The bending moment, M, at any section Z is given by

$$M = -\frac{w}{2}(L - z)^2 \tag{i}$$

Substituting for M in Eq. (13.3) and rearranging we have

$$EI \frac{d^2v}{dz^2} = \frac{w}{2}(L - z)^2 = \frac{w}{2}(L^2 - 2Lz + z^2) \tag{ii}$$

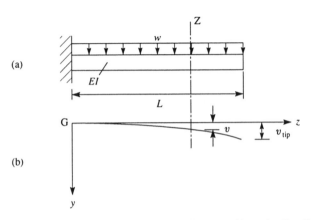

Fig. 13.3 Deflection of a cantilever beam carrying a uniformly distributed load

Integration of Eq. (ii) yields

$$EI \frac{dv}{dz} = \frac{w}{2}\left(L^2 z - Lz^2 + \frac{z^3}{3}\right) + C_1$$

When $z = 0$ at the built-in end, $v = 0$ so that $C_1 = 0$ and

$$EI \frac{dv}{dz} = \frac{w}{2}\left(L^2 z - Lz^2 + \frac{z^3}{3}\right) \qquad \text{(iii)}$$

Integrating Eq. (iii) we have

$$EIv = \frac{w}{2}\left(L^2 \frac{z^2}{2} - \frac{Lz^3}{3} + \frac{z^4}{12}\right) + C_2$$

and since $v = 0$ when $z = 0$, $C_2 = 0$. The deflection curve of the beam therefore has the equation

$$v = \frac{w}{24EI}(6L^2 z^2 - 4Lz^3 + z^4) \qquad \text{(iv)}$$

and the deflection at the free end where $z = L$ is

$$v_{tip} = \frac{wL^4}{8EI} \qquad \text{(v)}$$

which is again positive and downwards.

The applied loading in this case may be easily expressed in mathematical form so that a solution can be obtained using Eq. (13.5), i.e.

$$\frac{d^4 v}{dz^4} = \frac{w}{EI} \qquad \text{(vi)}$$

in which w = constant. Integrating Eq. (vi) we obtain

$$EI \frac{d^3 v}{dz^3} = wz + C_1$$

We note from Eq. (13.4) that

$$\frac{d^3 v}{dz^3} = -\frac{S}{EI}$$

Therefore when $z = 0$, $S = wL$ and

$$\frac{d^3 v}{dz^3} = -\frac{wL}{EI}$$

which gives

$$C_1 = -wL$$

Alternatively we could have determined C_1 from the boundary condition that when $z = L$, $S = 0$.

Hence

$$EI \frac{d^3 v}{dz^3} = w(z - L) \qquad \text{(vii)}$$

Integrating Eq. (vii) gives

$$EI \frac{d^2 v}{dz^2} = w \left(\frac{z^2}{2} - Lz \right) + C_2$$

From Eq. (13.3) we see that

$$\frac{d^2 v}{dz^2} = - \frac{M}{EI}$$

and when $z = 0$, $M = -wL^2/2$ (or when $z = L$, $M = 0$) so that

$$C_2 = \frac{wL^2}{2}$$

and

$$EI \frac{d^2 v}{dz^2} = \frac{w}{2} (z^2 - 2Lz + L^2)$$

which is identical to Eq. (ii). The solution then proceeds as before.

Example 13.3 The cantilever beam shown in Fig. 13.4(a) carries a uniformly distributed load over part of its span. Calculate the deflection of the free end.

If we assume that the cantilever is weightless then the bending moment at all sections between D and F is zero. It follows that the length DF of the beam remains straight. The deflection at D can be deduced from Eq. (v) of Ex. 13.2 and is

$$v_D = \frac{wa^4}{8EI}$$

Similarly the slope of the cantilever at D is found by substituting $z = a$ and $L = a$ in Eq. (iii) of Ex. 13.2; thus

$$\left(\frac{dv}{dz} \right)_D = \theta_D = \frac{wa^3}{6EI}$$

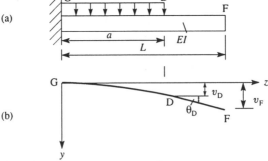

Fig. 13.4 Cantilever beam of Ex. 13.3

The deflection, v_F, at the free end of the cantilever is then given by

$$v_F = \frac{wa^4}{8EI} + (L - a)\frac{wa^3}{6EI}$$

which simplifies to

$$v_F = \frac{wa^3}{24EI}(4L - a)$$

Example 13.4 Determine the deflection curve and the mid-span deflection of the simply supported beam shown in Fig. 13.5(a).

The support reactions are each $wL/2$ and the bending moment, M, at any section Z, a distance z from the left-hand support is

$$M = \frac{wL}{2}z - \frac{wz^2}{2} \tag{i}$$

Substituting for M in Eq. (13.3) we obtain

$$EI\frac{d^2v}{dz^2} = -\frac{w}{2}(Lz - z^2) \tag{ii}$$

Integrating we have

$$EI\frac{dv}{dz} = -\frac{w}{2}\left(\frac{Lz^2}{3} - \frac{z^3}{3}\right) + C_1$$

From symmetry it is clear that at the mid-span section the gradient, $dv/dz = 0$.

Hence

$$0 = -\frac{w}{2}\left(\frac{L^3}{8} - \frac{L^3}{24}\right) + C_1$$

whence

$$C_1 = \frac{wL^3}{24}$$

Therefore

$$EI\frac{dv}{dz} = -\frac{w}{24}(6Lz^2 - 4z^3 - L^3) \tag{iii}$$

Integrating again gives

$$EIv = -\frac{w}{24}(2Lz^3 - z^4 - L^3z) + C_2$$

Since $v = 0$ when $z = 0$ (or since $v = 0$ when $z = L$) it follows that $C_2 = 0$ and the deflected shape of the beam has the equation

$$v = -\frac{w}{24EI}(2Lz^3 - z^4 - L^3z) \tag{iv}$$

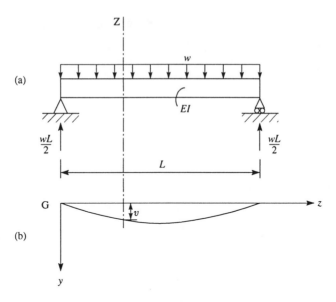

Fig. 13.5 Deflection of a simply supported beam carrying a uniformly distributed load (Ex. 13.4)

The maximum deflection occurs at mid-span where $z = L/2$ and is

$$v_{\text{mid-span}} = \frac{5wL^4}{384EI} \tag{v}$$

So far the constants of integration were determined immediately they arose. However, in some cases a relevant boundary condition, say a value of gradient, is not obtainable. The method is then to carry the unknown constant through the succeeding integration and use known values of deflection at two sections of the beam. Thus in the previous example Eq. (ii) is integrated twice to obtain

$$EIv = -\frac{w}{2}\left(\frac{Lz^3}{6} - \frac{z^4}{12}\right) + C_1 z + C_2$$

The relevant boundary conditions are $v = 0$ at $z = 0$ and $z = L$. The first of these gives $C_2 = 0$ while from the second we have $C_1 = wL^3/24$. Thus the equation of the deflected shape of the beam is

$$v = -\frac{w}{24EI}(2Lz^3 - z^4 - L^3 z)$$

as before.

Example 13.5   Figure 13.6(a) shows a simply supported beam carrying a concentrated load W at mid-span. Determine the deflection curve of the beam and the maximum deflection.

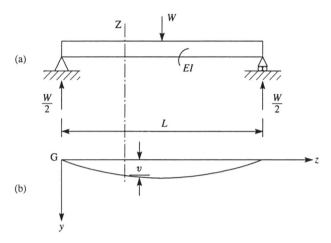

Fig. 13.6 Deflection of a simply supported beam carrying a concentrated load at mid-span (Ex. 13.5)

The support reactions are each $W/2$ and the bending moment M at a section Z a distance $z(\leqslant L/2)$ from the left-hand support is

$$M = \frac{W}{2} z \tag{i}$$

From Eq. (13.3) we have

$$EI \frac{d^2 v}{dz^2} = -\frac{W}{2} z \tag{ii}$$

Integrating we obtain

$$EI \frac{dv}{dz} = -\frac{W}{2} \frac{z^2}{2} + C_1$$

From symmetry the slope of the beam is zero at mid-span where $z = L/2$. Thus $C_1 = WL^2/16$ and

$$EI \frac{dv}{dz} = -\frac{W}{16} (4z^2 - L^2) \tag{iii}$$

Integrating Eq. (iii) we have

$$EIv = -\frac{W}{16} \left(\frac{4z^3}{3} - L^2 z \right) + C_2$$

and when $z = 0$, $v = 0$ so that $C_2 = 0$. The equation of the deflection curve is therefore

$$v = -\frac{W}{48EI} (4z^3 - 3L^2 z) \tag{iv}$$

The maximum deflection occurs at mid-span and is

$$v_{\text{mid-span}} = \frac{WL^3}{48EI} \tag{v}$$

Note that in this problem we could not use the boundary condition that $v = 0$ at $z = L$ to determine C_2 since Eq. (i) applies only for $0 \leqslant z \leqslant L/2$; it follows that Eqs (iii) and (iv) for slope and deflection apply only for $0 \leqslant z \leqslant L/2$ although the deflection curve is clearly symmetrical about mid-span.

Example 13.6 The simply supported beam shown in Fig. 13.7(a) carries a concentrated load W at a distance a from the left-hand support. Determine the deflected shape of the beam, the deflection under the load and the maximum deflection.

Considering the moment and force equilibrium of the beam we have

$$R_A = \frac{W}{L}(L - a), \qquad R_B = \frac{Wa}{L}$$

At a section Z_1, a distance z from the left-hand support where $z \leqslant a$, the bending moment is

$$M = R_A z \tag{i}$$

At the section Z_2, where $z \geqslant a$

$$M = R_A z - W(z - a) \tag{ii}$$

Substituting both expressions for M in turn in Eq. (13.3) we obtain

$$EI \frac{d^2 v}{dz^2} = -R_A z \qquad (z \leqslant a) \tag{iii}$$

and

$$EI \frac{d^2 v}{dz^2} = -R_A z + W(z - a) \qquad (z \geqslant a)$$

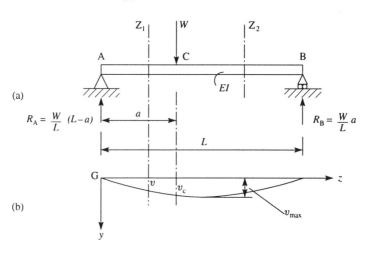

Fig. 13.7 Deflection of a simply supported beam carrying a concentrated load not at mid-span (Ex. 13.6)

Integrating Eqs (iii) and (iv) we obtain

$$EI\frac{dv}{dz} = -R_A\frac{z^2}{2} + C_1 \qquad (z \leqslant a) \tag{v}$$

$$EI\frac{dv}{dz} = -R_A\frac{z^2}{2} + W\left(\frac{z^2}{2} - az\right) + C_1' \qquad (z \geqslant a) \tag{vi}$$

and

$$EIv = -R_A\frac{z^3}{6} + C_1 z + C_2 \qquad (z \leqslant a) \tag{vii}$$

$$EIv = -R_A\frac{z^3}{6} + W\left(\frac{z^3}{6} - \frac{az^2}{2}\right) + C_1' z + C_2' \qquad (z \geqslant a) \tag{viii}$$

in which C_1, C_1', C_2, C_2' are arbitrary constants. In using the boundary conditions to determine these constants, it must be remembered that Eqs (v) and (vii) apply only for $0 \leqslant z \leqslant a$ and Eqs (vi) and (viii) apply only for $a \leqslant z \leqslant L$. At the left-hand support $v = 0$ when $z = 0$, therefore, from Eq. (vii), $C_2 = 0$. It is not possible to determine C_1, C_1' and C_2 directly since the application of further known boundary conditions does not isolate any of these constants. However, since $v = 0$ when $z = L$ we have, from Eq (viii),

$$0 = -R_A\frac{L^3}{6} + W\left(\frac{L^3}{6} - \frac{aL^2}{2}\right) + C_1'L + C_2'$$

which, after substituting $R_A = W(L-a)/L$, simplifies to

$$0 = -\frac{WaL^2}{3} + C_1'L + C_2' \tag{ix}$$

Additional equations are obtained by considering the continuity which exists at the point of application of the load; at this section Eqs (v)–(viii) apply. Thus, from Eqs (v) and (vi)

$$-R_A\frac{a^2}{2} + C_1 = -R_A\frac{a^2}{2} + W\left(\frac{a^2}{2} - a^2\right) + C_1'$$

which gives

$$C_1 = -\frac{Wa^2}{2} + C_1' \tag{x}$$

Now equating values of deflection at $z = a$ we have, from Eqs (vii) and (viii)

$$-R_A\frac{a^3}{6} + C_1 a = -R_A\frac{a^3}{6} + W\left(\frac{a^3}{6} - \frac{a^3}{2}\right) + C_1'a + C_2'$$

which yields

$$C_1 a = -\frac{Wa^3}{3} + C_1'a + C_2' \tag{xi}$$

Solution of the simultaneous equations (ix), (x) and (xi) gives

$$C_1 = \frac{Wa}{6L}(a-2L)(a-L)$$

$$C_1' = \frac{Wa}{6L}(a^2+2L^2)$$

$$C_2' = -\frac{Wa^3}{6}$$

Equations (v)–(vii) then become, respectively,

$$EI\frac{dv}{dz} = \frac{W(a-L)}{6L}[3z^2 + a(a-2L)] \qquad (z \leqslant a) \tag{xii}$$

$$EI\frac{dv}{dz} = \frac{Wa}{6L}(3z^2 - 6Lz + a^2 + 2L^2) \qquad (z \geqslant a) \tag{xiii}$$

$$EIv = \frac{W(a-L)}{6L}[z^3 + a(a-2L)z] \qquad (z \leqslant a) \tag{xiv}$$

$$EIv = \frac{Wa}{6L}[z^3 - 3Lz^2 + (a^2 + 2L^2)z - a^2 L] \qquad (z \geqslant a) \tag{xv}$$

The deflection of the beam under the load is obtained by putting $z = a$ into either of Eqs (xiv) or (xv). Thus

$$v_C = \frac{Wa^2(a-L)^2}{3EIL} \tag{xvi}$$

This is not, however, the maximum deflection of the beam. This will occur, if $a < L/2$, at some section between C and B. Its position may be found by equating dv/dz in Eq. (xiii) to zero. Hence

$$0 = 3z^2 - 6Lz + a^2 + 2L^2 \tag{xvii}$$

The solution of Eq. (xvii) is then substituted in Eq. (v) and the maximum deflection follows.

For a central concentrated load $a = L/2$ and

$$v_C = \frac{WL^3}{48EI} \qquad \text{as before}$$

Example 13.7   Determine the deflection curve of the beam AB shown in Fig. 13.8 when it carries a distributed load that varies linearly in intensity from zero at the left-hand support to w_o at the right-hand support.

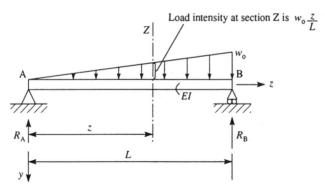

Fig. 13.8 Deflection of a simply supported beam carrying a triangularly distributed load

To find the support reactions we first take moments about B. Thus

$$R_A L = \tfrac{1}{2} w_o L \, \frac{L}{3}$$

which gives

$$R_A = \frac{w_o L}{6}$$

Resolution of vertical forces then gives

$$R_B = \frac{w_o L}{3}$$

The bending moment, M, at any section Z, a distance z from A is

$$M = R_A z - \frac{1}{2}\left(w_o \frac{z}{L}\right) z \, \frac{z}{3}$$

or

$$M = \frac{w_o}{6L}(L^2 z - z^3) \tag{i}$$

Substituting for M in Eq. (13.3) we obtain

$$EI \frac{d^2 v}{dz^2} = -\frac{w_o}{6L}(L^2 z - z^3) \tag{ii}$$

which, when integrated, becomes

$$EI \frac{dv}{dz} = -\frac{w_o}{6L}\left(L^2 \frac{z^2}{2} - \frac{z^4}{4}\right) + C_1 \tag{iii}$$

Integrating Eq. (iii) we have

$$EIv = -\frac{w_o}{6L}\left(L^2 \frac{z^3}{6} - \frac{z^5}{20}\right) + C_1 z + C_2 \tag{iv}$$

The deflection $v = 0$ at $z = 0$ and $z = L$. From the first of these conditions we obtain $C_2 = 0$, while from the second

$$0 = -\frac{w_0}{6L}\left(\frac{L^5}{6} - \frac{L^5}{20}\right) + C_1 L$$

which gives

$$C_1 = \frac{7w_0 L^4}{360}$$

The deflection curve then has the equation

$$v = \frac{w_0}{360EIL}(3z^5 - 10L^2 z^3 + 7L^4 z) \tag{v}$$

An alternative method of solution is to use Eq. (13.5) and express the applied load in mathematical form. Thus

$$EI\frac{d^4 v}{dz^4} = w = w_0\frac{z}{L} \tag{vi}$$

Integrating we obtain

$$EI\frac{d^3 v}{dz^3} = w_0\frac{z^2}{2L} + C_3$$

When $z = 0$ we see from Eq. (13.4) that

$$EI\frac{d^3 v}{dz^3} = -R_A = -\frac{w_0 L}{6}$$

Hence

$$C_3 = -\frac{w_0 L}{6}$$

and

$$EI\frac{d^3 v}{dz^3} = w_0\frac{z^2}{2L} - \frac{w_0 L}{6} \tag{vii}$$

Integrating Eq. (vii) we have

$$EI\frac{d^2 v}{dz^2} = \frac{w_0 z^3}{6L} - \frac{w_0 L}{6}z + C_4$$

Since the bending moment is zero at the supports we have

$$EI\frac{d^2 v}{dz^2} = 0 \qquad \text{when } z = 0$$

Hence $C_4 = 0$ and

$$EI\frac{d^2 v}{dz^2} = \frac{w_0}{6L}(z^3 - L^2 z)$$

as before.

13.2 Singularity functions

A comparison of Exs 13.5 and 13.6 shows that the double integration method becomes extremely lengthy when even relatively small complications such as the lack of symmetry due to an offset load are introduced. Again the addition of a second concentrated load on the beam of Ex. 13.6 would result in a total of six equations for slope and deflection producing six arbitrary constants. Clearly the computation involved in determining these constants would be tedious, even though a simply supported beam carrying two concentrated loads is a comparatively simple practical case. An alternative approach is to introduce so-called *singularity* or *half-range* functions. Such functions were first applied to beam deflection problems by Macauley in 1919 and hence the method is frequently known as *Macauley's method*.

We now introduce a quantity $[z - a]$ and define it to be zero if $(z - a) < 0$, i.e. $z < a$, and to be simply $(z - a)$ if $z > a$. The quantity $[z - a]$ is known as a singularity or half-range function and is defined to have a value only when the argument is positive in which case the square brackets behave in an identical manner to ordinary parentheses. Thus in Ex. 13.6 the bending moment at a section of the beam furthest from the origin for z may be written

$$M = R_A z - W[z - a]$$

This expression applies to both the regions AC and CB since $W[z - a]$ disappears for $z < a$. Equations (iii) and (iv) in Ex. 13.6 then become the single equation

$$EI \frac{d^2 v}{dz^2} = -R_A z + W[z - a]$$

which on integration yields

$$EI \frac{dv}{dz} = -R_A \frac{z^2}{2} + \frac{W}{2}[z - a]^2 + C_1$$

and

$$EIv = -R_A \frac{z^3}{6} + \frac{W}{6}[z - a]^3 + C_1 z + C_2$$

Note that the square brackets *must be retained* during the integration. The arbitrary constants C_1 and C_2 are found using the boundary conditions that $v = 0$ when $z = 0$

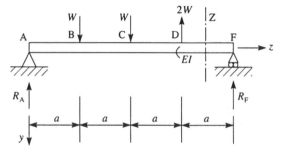

Fig. 13.9 Macauley's method for the deflection of a simply supported beam (Ex. 13.8)

and $z = L$. From the first of these and remembering that $[z - a]^3$ is zero for $z < a$, we have $C_2 = 0$. From the second we have

$$0 = -R_A \frac{L^3}{6} + \frac{W}{6} [L - a]^3 + C_1 L$$

in which $R_A = W(L - a)/L$.

Substituting for R_A gives

$$C_1 = \frac{Wa(L - a)}{6L} (2L - a)$$

whence

$$Elv = \frac{W}{6L} \{(a - L)z^3 + L[z - a]^3 + a(L - a)(2L - a)z\}$$

The deflection of the beam under the load is then

$$v_C = \frac{Wa^2(a - L)^2}{3EIL}$$

as before.

Example 13.8   Determine the position and magnitude of the maximum upward and downward deflections of the beam shown in Fig. 13.9.

A consideration of the overall equilibrium of the beam gives the support reactions; thus

$$R_A = \tfrac{3}{4} W \text{ (upward)}, \qquad R_F = \tfrac{3}{4} W \text{ (downward)}$$

Using the method of singularity functions and taking the origin of axes at the left-hand support, we write down an expression for the bending moment, M, at any section Z between D and F, *the region of the beam furthest from the origin*. Thus

$$M = R_A z - W[z - a] - W[z - 2a] + 2W[z - 3a] \tag{i}$$

Substituting for M in Eq. (13.3) we have

$$EI \frac{d^2 v}{dz^2} = -\frac{3}{4} Wz + W[z - a] + W[z - 2a] - 2W[z - 3a] \tag{ii}$$

Integrating Eq. (ii) and retaining the square brackets we obtain

$$EI \frac{dv}{dz} = -\frac{3}{8} Wz^2 + \frac{W}{2}[z - a]^2 + \frac{W}{2}[z - 2a]^2 - W[z - 3a]^2 + C_1 \tag{iii}$$

and $\quad Elv = -\dfrac{1}{8} Wz^3 + \dfrac{W}{6}[z - a]^3 + \dfrac{W}{6}[z - 2a]^3 - \dfrac{W}{3}[z - 3a]^3 + C_1 z + C_2 \quad$ (iv)

in which C_1 and C_2 are arbitrary constants. When $z = 0$ (at A), $v = 0$ and hence $C_2 = 0$. Note that the second, third and fourth terms on the right-hand side of

Eq. (iv) disappear for $z < a$. Also $v = 0$ at $z = 4a$ (F) so that, from Eq. (iv), we have

$$0 = -\frac{W}{8}64a^3 + \frac{W}{6}27a^3 + \frac{W}{6}8a^3 - \frac{W}{3}a^3 + 4aC_1$$

which gives $C_1 = \frac{5}{8}Wa^2$.

Equations (iii) and (iv) now become

$$EI\frac{dv}{dz} = -\frac{3}{8}Wz^2 + \frac{W}{2}[z-a]^2 + \frac{W}{2}[z-2a]^2 - W[z-3a]^2 + \frac{5}{8}Wa^2 \quad \text{(v)}$$

and

$$EIv = -\frac{1}{8}Wz^3 + \frac{W}{6}[z-a]^3 + \frac{W}{6}[z-2a]^3 - \frac{W}{3}[z-3a]^3 + \frac{5}{8}Wa^2z \quad \text{(vi)}$$

respectively.

To determine the maximum upward and downward deflections we need to know in which bays $dv/dz = 0$ and thereby which terms in Eq. (v) disappear when the exact positions are being located. One method is to select a bay and determine the sign of the slope of the beam at the extremities of the bay. A change of sign will indicate that the slope is zero within the bay.

By inspection of Fig. 13.9 it seems likely that the maximum downward deflection will occur in BC. At B, using Eq. (v)

$$EI\frac{dv}{dz} = -\frac{3}{8}Wa^2 + \frac{5}{8}Wa^2$$

which is clearly positive. At C

$$EI\frac{dv}{dz} = -\frac{3}{8}W4a^2 + \frac{W}{2}a^2 + \frac{5}{8}Wa^2$$

which is negative. Therefore, the maximum downward deflection does occur in BC and its exact position is located by equating dv/dz to zero for any section in BC. Thus, from Eq. (v)

$$0 = -\frac{3}{8}Wz^2 + \frac{W}{2}[z-a]^2 + \frac{5}{8}Wa^2$$

or, simplifying,

$$0 = z^2 - 8az + 9a^2 \quad \text{(vii)}$$

Solution of Eq. (vii) gives

$$z = 1{\cdot}35a$$

so that the maximum downward deflection is, from Eq. (vi)

$$EIv = -\frac{1}{8}W(1{\cdot}35a)^3 + \frac{W}{6}(0{\cdot}35a)^3 + \frac{5}{8}Wa^2(1{\cdot}35a)$$

i.e.

$$v_{max}(\text{downward}) = \frac{0{\cdot}54Wa^3}{EI}$$

In a similar manner it can be shown that the maximum upward deflection lies between D and F at $z = 3 \cdot 42a$ and that its magnitude is

$$v_{max}(\text{upward}) = \frac{0 \cdot 04 W a^3}{EI}$$

An alternative method of determining the position of maximum deflection is to select a possible bay, set $dv/dz = 0$ for that bay and solve the resulting equation in z. If the solution gives a value of z that lies within the bay, then the selection is correct, otherwise the procedure must be repeated for a second and possibly a third and a fourth bay. This method is quicker than the former if the correct bay is selected initially; if not, the equation corresponding to each selected bay must be completely solved, a procedure clearly longer than determining the sign of the slope at the extremities of the bay.

Example 13.9 Determine the position and magnitude of the maximum deflection in the beam of Fig. 13.10.

Following the method of Ex. 13.8 we determine the support reactions and find the bending moment, M, at any section Z in the bay furthest from the origin of the axes. Thus

$$M = R_A z - w \frac{L}{4} \left[z - \frac{5L}{8} \right] \tag{i}$$

Examining Eq. (i) we see that the singularity function $[z - 5L/8]$ does not become zero until $z \le 5L/8$ although Eq. (i) is only valid for $z \ge 3L/4$. To obviate this difficulty we extend the distributed load to the support D while simultaneously restoring the status quo by applying an upward distributed load of the same intensity and length as the additional load (Fig. 13.11).

At the section Z, a distance z from A, the bending moment is now given by

$$M = R_A z - \frac{w}{2} \left[z - \frac{L}{2} \right]^2 + \frac{w}{2} \left[z - \frac{3L}{4} \right]^2 \tag{ii}$$

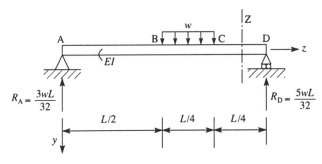

Fig. 13.10 Deflection of a beam carrying a part-span uniformly distributed load (Ex. 13.9)

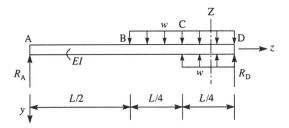

Fig. 13.11 Method of solution for a part span uniformly distributed load

Equation (ii) is now valid for all sections of the beam if the singularity functions are discarded as they become zero. Substituting Eq. (ii) into Eq. (13.3) we obtain

$$EI\frac{d^2v}{dz^2} = -\frac{3}{32}wLz + \frac{w}{2}\left[z - \frac{L}{2}\right]^2 - \frac{w}{2}\left[z - \frac{3L}{4}\right]^2 \tag{iii}$$

Integrating Eq. (iii) gives

$$EI\frac{dv}{dz} = -\frac{3}{64}wLz^2 + \frac{w}{6}\left[z - \frac{L}{2}\right]^3 - \frac{w}{6}\left[z - \frac{3L}{4}\right]^3 + C_1 \tag{iv}$$

$$EIv = -\frac{wLz^3}{64} + \frac{w}{24}\left[z - \frac{L}{2}\right]^4 - \frac{w}{24}\left[z - \frac{3L}{4}\right]^4 + C_1z + C_2 \tag{v}$$

where C_1 and C_2 are arbitrary constants. The required boundary conditions are $v = 0$ when $z = 0$ and $z = L$. From the first of these we obtain $C_2 = 0$ while the second gives

$$0 = -\frac{wL^4}{64} + \frac{w}{24}\left(\frac{L}{2}\right)^4 - \frac{w}{24}\left(\frac{L}{4}\right)^4 + C_1L$$

from which

$$C_1 = \frac{27wL^3}{2048}$$

Equations (iv) and (v) then become

$$EI\frac{dv}{dz} = -\frac{3}{64}wLz^2 + \frac{w}{6}\left[z - \frac{L}{2}\right]^3 - \frac{w}{6}\left[z - \frac{3L}{4}\right]^3 + \frac{27wL^3}{2048} \tag{vi}$$

and

$$EIv = -\frac{wLz^3}{64} + \frac{w}{24}\left[z - \frac{L}{2}\right]^4 - \frac{w}{24}\left[z - \frac{3L}{4}\right]^4 + \frac{27wL^3}{2048}z \tag{vii}$$

In this problem, the maximum deflection clearly occurs in the region BC of the beam. Thus equating the slope to zero for BC we have

$$0 = -\frac{3}{64}wLz^2 + \frac{w}{6}\left[z - \frac{L}{2}\right]^3 + \frac{27wL^3}{2048}$$

which simplifies to

$$z^3 - 1 \cdot 78Lz^2 + 0 \cdot 75zL^2 - 0 \cdot 046L^3 = 0 \tag{viii}$$

Solving Eq. (viii) by trial and error, we see that the slope is zero at $z \approx 0.6L$. Hence from Eq. (vii) the maximum deflection is

$$v_{max} = \frac{4 \cdot 53 \times 10^{-3} wL^4}{EI}$$

Example 13.10   Determine the deflected shape of the beam shown in Fig. 13.12.

In this problem an external moment M_o is applied to the beam at B. The support reactions are found in the normal way and are

$$R_A = -\frac{M_o}{L} \text{ (downwards)}, \qquad R_C = \frac{M_o}{L} \text{ (upwards)}$$

The bending moment at any section Z between B and C is then given by

$$M = R_A z + M_o \tag{i}$$

Equation (i) is valid only for the region BC and clearly does not contain a singularity function which would cause M_o to vanish for $z \leqslant b$. We overcome this difficulty by writing

$$M = R_A z + M_o [z-b]^0 \qquad \text{(Note: } [z-b]^0 = 1) \tag{ii}$$

Equation (ii) has the same value as Eq. (i) but is now applicable to all sections of the beam since $[z-b]^0$ disappears when $z \leqslant b$. Substituting for M from Eq. (ii) in Eq. (13.3) we obtain

$$EI \frac{d^2 v}{dz^2} = -R_A z - M_o [z-b]^0 \tag{iii}$$

Integration of Eq. (iii) yields

$$EI \frac{dv}{dz} = -R_A \frac{z^2}{2} - M_o [z-b] + C_1 \tag{iv}$$

and

$$EIv = -R_A \frac{z^3}{6} - \frac{M_o}{2} [z-b]^2 + C_1 z + C_2 \tag{v}$$

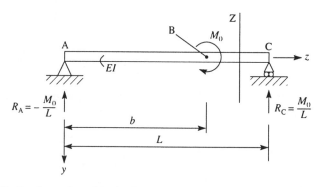

Fig. 13.12   Deflection of a simply supported beam carrying a point moment (Ex. 13.10)

where C_1 and C_2 are arbitrary constants. The boundary conditions are $v = 0$ when $z = 0$ and $z = L$. From the first of these we have $C_2 = 0$ while the second gives

$$0 = \frac{M_o}{L} \frac{L^3}{6} - \frac{M_o}{2} [L - b]^2 + C_1 L$$

from which
$$C_1 = \frac{M_o}{6L} (2L^2 - 6Lb + 3b^2)$$

The equation of the deflection curve of the beam is then

$$v = - \frac{M_o}{6EIL} \{z^3 + 3L[z - b]^2 - (2L^2 - 6Lb + 3b^2)z\} \qquad \text{(vi)}$$

13.3 Moment-area method for symmetrical bending

The double integration method and the method of singularity functions are used when the complete deflection curve of a beam is required. However, if only the deflection of a particular point is required, the moment-area method is generally more suitable.

Consider the curvature–moment equation (13.3), i.e.

$$\frac{d^2 v}{dz^2} = - \frac{M}{EI}$$

Integration of this equation between any two sections, say A and B, of a beam gives

$$\int_A^B \frac{d^2 v}{dz^2} \, dz = - \int_A^B \frac{M}{EI} \, dz \qquad (13.6)$$

or
$$\left[\frac{dv}{dz} \right]_A^B = - \int_A^B \frac{M}{EI} \, dz$$

which gives
$$\left(\frac{dv}{dz} \right)_B - \left(\frac{dv}{dz} \right)_A = - \int_A^B \frac{M}{EI} \, dz \qquad (13.7)$$

In qualitative terms Eq. (13.7) states that the change of slope between two sections A and B of a beam is numerically equal to minus the area of the M/EI diagram between those sections.

We now return to Eq. (13.3) and multiply both sides by z thereby retaining the equality. Thus

$$\frac{d^2 v}{dz^2} z = - \frac{M}{EI} z \qquad (13.8)$$

Integrating Eq. (13.8) between two sections A and B of a beam we have

$$\int_A^B \frac{d^2 v}{dz^2} z \, dz = - \int_A^B \frac{M}{EI} z \, dz \qquad (13.9)$$

The left-hand side of Eq. (13.9) may be integrated by parts and yields

$$\left[z\frac{dv}{dz}\right]_A^B - \int_A^B \frac{dv}{dz}\,dz = -\int_A^B \frac{M}{EI}\,z\,dz$$

or

$$\left[z\frac{dv}{dz}\right]_A^B - [v]_A^B = -\int_A^B \frac{M}{EI}\,z\,dz$$

Hence, inserting the limits we have

$$z_B\left(\frac{dv}{dz}\right)_B - z_A\left(\frac{dv}{dz}\right)_A - (v_B - v_A) = -\int_A^B \frac{M}{EI}\,z\,dz \qquad (13.10)$$

in which z_B and z_A represent the z coordinate of each of the sections B and A, respectively, while $(dv/dz)_B$ and $(dv/dz)_A$ are the respective slopes; v_B and v_A are the corresponding deflections. The right-hand side of Eq. (13.10) represents the moment of the area of the M/EI diagram between the sections A and B *about A*.

Equations (13.7) and (13.10) may be used to determine values of slope and deflection at any section of a beam. We note that in both equations we are concerned with the geometry of the M/EI diagram. This will be identical in shape to the bending moment diagram unless there is a change of section. Furthermore, the form of the right-hand side of both Eqs (13.7) and (13.10) allows two alternative methods of solution. In cases where the geometry of the M/EI diagram is relatively simple, we can employ a *semi-graphical* approach based on the actual geometry of the M/EI diagram. Alternatively, in complex problems, the bending moment may be expressed as a function of z and a completely analytical solution obtained. Both methods are illustrated in the following examples.

Example 13.11 Determine the slope and deflection of the free end of the cantilever beam shown in Fig. 13.13.

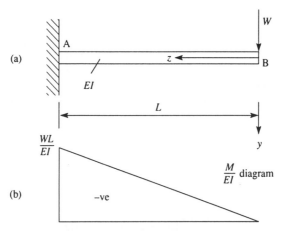

Fig. 13.13 Moment-area method for the deflection of a cantilever (Ex. 13.11)

We choose the origin of the axes at the free end B of the cantilever. Equation (13.7) then becomes

$$\left(\frac{dv}{dz}\right)_A - \left(\frac{dv}{dz}\right)_B = -\int_A^B \frac{M}{EI} \, dz$$

or, since $(dv/dz)_A = 0$

$$-\left(\frac{dv}{dz}\right)_B = -\int_0^L \frac{M}{EI} \, dz \qquad (i)$$

Generally at this stage we decide which approach is most suitable; however, both semi-graphical and analytical methods are illustrated here. Using the geometry of Fig. 13.13(b) we have

$$-\left(\frac{dv}{dz}\right)_B = -\frac{1}{2} L \left(\frac{-WL}{EI}\right)$$

which gives

$$\left(\frac{dv}{dz}\right)_B = -\frac{WL^2}{2EI}$$

(compare with the value given by Eq. (iii) of Ex. 13.1. Note the change in sign due to the different origin for z).

Alternatively, since the bending moment at any section z is $-Wz$ we have, from Eq. (i)

$$-\left(\frac{dv}{dz}\right)_B = -\int_0^L -\frac{Wz}{EI} \, dz$$

which again gives

$$\left(\frac{dv}{dz}\right)_B = -\frac{WL^2}{2EI}$$

With the origin for z at B, Eq. (13.10) becomes

$$z_A \left(\frac{dv}{dz}\right)_A - z_B \left(\frac{dv}{dz}\right)_B - (v_A - v_B) = -\int_B^A \frac{M}{EI} z \, dz \qquad (ii)$$

Since $(dv/dz)_A = 0$, $z_B = 0$ and $v_A = 0$, Eq. (ii) reduces to

$$v_B = -\int_0^L \frac{M}{EI} z \, dz \qquad (iii)$$

Again we can now decide whether to proceed semi-graphically or analytically. Using the former approach and taking the moment of the area of the M/EI diagram about B, we have

$$v_B = -\frac{1}{2} L \left(\frac{-WL}{EI}\right) \frac{2}{3} L$$

which gives

$$v_B = \frac{WL^3}{3EI} \qquad \text{(compare with Eq. (v) of Ex. 13.1)}$$

Alternatively we have

$$v_B = -\int_0^L \frac{(-Wz)}{EI} z\,dz = \int_0^L \frac{Wz^2}{EI}\,dz$$

which gives

$$v_B = \frac{WL^3}{3EI}$$

as before.

Note that if the built-in end had been selected as the origin for z, we could not have determined v_B directly since the term $z_B(dv/dz)_B$ in Eq. (ii) would not have vanished. The solution for v_B would then have consisted of two parts, first the determination of $(dv/dz)_B$ and then the calculation of v_B.

Example 13.12   Determine the maximum deflection in the simply supported beam shown in Fig. 13.14(a).

From symmetry we deduce that the beam reactions are each $wL/2$; the M/EI diagram has the geometry shown in Fig. 13.14(b).

If we take the origin of axes to be at A and consider the half-span AC, Eq. (13.10) becomes

$$z_C\left(\frac{dv}{dz}\right)_C - z_A\left(\frac{dv}{dz}\right)_A - (v_C - v_A) = -\int_A^C \frac{M}{EI} z\,dz \tag{i}$$

In this problem $(dv/dz)_C = 0$, $z_A = 0$ and $v_A = 0$; hence Eq. (i) reduces to

$$v_C = \int_0^{L/2} \frac{M}{EI} z\,dz \tag{ii}$$

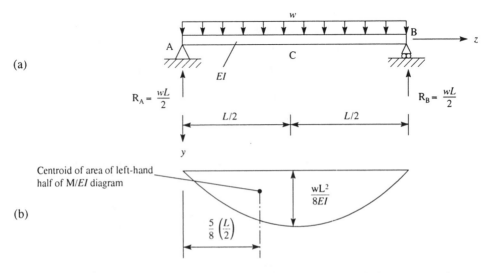

Fig. 13.14 Moment-area method for a simply supported beam carrying a uniformly distributed load

Using the geometry of the M/EI diagram, i.e. the semi-graphical approach, and taking the moment of the area of the M/EI diagram between A and C about A we have from Eq. (ii)

$$v_C = \frac{2}{3} \frac{wL^2}{8EI} \frac{L}{2} \frac{5}{8} \left(\frac{L}{2}\right)$$

which gives

$$v_C = \frac{5wL^4}{384EI}$$

(see Eq. (v) of Ex. 13.4).

For the completely analytical approach we express the bending moment M as a function of z; thus

$$M = \frac{wL}{2} z - \frac{wz^2}{2}$$

or

$$M = \frac{w}{2} (Lz - z^2)$$

Substituting for M in Eq. (ii) we have

$$v_C = \int_0^{L/2} \frac{w}{2EI} (Lz^2 - z^3) \, dz$$

which gives

$$v_C = \frac{w}{2EI} \left[\frac{Lz^3}{3} - \frac{z^4}{4} \right]_0^{L/2}$$

Hence

$$v_C = \frac{5wL^4}{384EI}$$

Example 13.13 Figure 13.15(a) shows a cantilever beam of length L carrying a concentrated load W at its free end. The section of the beam changes mid-way along its length so that the second moment of area of its cross-section is reduced by half. Determine the deflection of the free end.

In this problem the bending moment and M/EI diagrams have different geometrical shapes. Choosing the origin of axes at C, Eq. (13.10) becomes

$$z_A \left(\frac{dv}{dz}\right)_A - z_C \left(\frac{dv}{dz}\right)_C - (v_A - v_C) = - \int_C^A \frac{M}{EI} z \, dz \tag{i}$$

in which $(dv/dz)_A = 0$, $z_C = 0$, $v_A = 0$. Hence

$$v_C = - \int_0^L \frac{M}{EI} z \, dz \tag{ii}$$

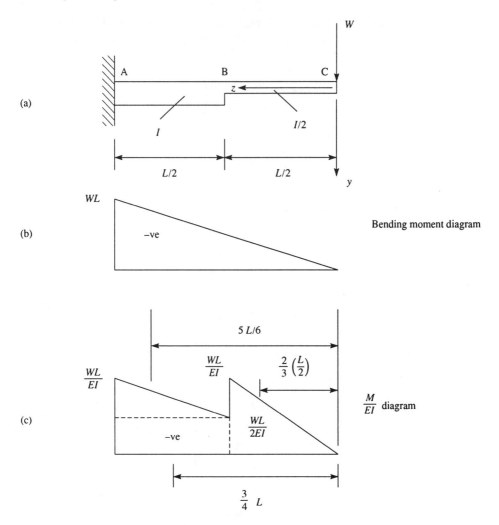

Fig. 13.15 Deflection of a cantilever of varying section

From the geometry of the M/EI diagram (Fig. 13.15(c)) and taking moments of areas about C we have

$$v_C = -\left\{\left(\frac{-WL}{2EI}\right)\frac{L}{2}\frac{3L}{4} + \frac{1}{2}\left(\frac{-WL}{2EI}\right)\frac{L}{2}\frac{5L}{6} + \frac{1}{2}\left(\frac{-WL}{EI}\right)\frac{L}{2}\frac{2}{3}\frac{L}{2}\right\}$$

which gives

$$v_C = \frac{3WL^3}{8EI}$$

Analytically we have

$$v_C = -\left[\int_0^{L/2}\frac{-Wz^2}{EI/2}\,\mathrm{d}z + \int_{L/2}^{L}\frac{-Wz^2}{EI}\,\mathrm{d}z\right]$$

or
$$v_C = \frac{W}{EI}\left\{\left[\frac{2z^3}{3}\right]_0^{L/2} + \left[\frac{z^3}{3}\right]_{L/2}^L\right\}$$

Hence
$$v_C = \frac{3WL^3}{8EI}$$

as before.

Example 13.14 The cantilever beam shown in Fig. 13.16 tapers along its length so that the second moment of area of its cross-section varies linearly from its value I_o at the free end to $2I_o$ at the built-in end. Determine the deflection at the free end when the cantilever carries a concentrated load W.

Choosing the origin of axes at the free end B we have, from Eq. (13.10),

$$z_A\left(\frac{dv}{dz}\right)_A - z_B\left(\frac{dv}{dz}\right)_B - (v_A - v_B) = -\int_B^A \frac{M}{EI_z}z\,dz \tag{i}$$

in which I_Z, the second moment of area at any section Z, is given by

$$I_Z = I_o\left(1 + \frac{z}{L}\right)$$

Also $(dv/dz)_A = 0$, $z_B = 0$, $v_A = 0$ so that Eq. (i) reduces to

$$v_B = -\int_0^L \frac{Mz}{EI_o\left(1 + \dfrac{z}{L}\right)}\,dz \tag{ii}$$

The geometry of the M/EI diagram in this case will be complicated so that the analytical approach is most suitable. Therefore since $M = -Wz$, Eq. (ii) becomes

$$v_B = \int_0^L \frac{Wz^2}{EI_o\left(1 + \dfrac{z}{L}\right)}\,dz$$

or
$$v_B = \frac{WL}{EI_o}\int_0^L \frac{z^2}{L + z}\,dz \tag{iii}$$

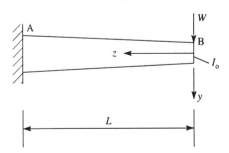

Fig. 13.16 Deflection of a cantilever of tapering section

Rearranging Eq. (iii) we have

$$v_B = \frac{WL}{EI_o}\left\{ \int_0^L (z-L)\,\mathrm{d}z + \int_0^L \frac{L^2}{L+z}\,\mathrm{d}z \right\}$$

Hence

$$v_B = \frac{WL}{EI_o}\left[\left(\frac{z^2}{2} - Lz\right) + L^2 \log_e(L-z) \right]_0^L$$

so that

$$v_B = \frac{WL^3}{EI_o}\left(-\frac{1}{2} + \log_e 2 \right)$$

whence

$$v_B = \frac{0.19WL^3}{EI_o}$$

13.4   Deflections due to unsymmetrical bending

We noted in Chapter 9 that a beam bends about its neutral axis whose inclination to arbitrary centroidal axes is determined from Eq. (9.33). Beam deflections, therefore, are always perpendicular in direction to the neutral axis.

Suppose that at some section of a beam, the deflection normal to the neutral axis (and therefore an absolute deflection) is ζ. Thus, as shown in Fig. 13.17, the centroid G is displaced to G'. If the displacement corresponds to a bending moment whose components M_x and M_y give positive values for $\bar{M}_x$ and $\bar{M}_y$, the direction of the displacement will generally be as shown in Fig. 13.17 with components

$$u = \zeta \sin \alpha, \quad v = \zeta \cos \alpha \tag{13.11}$$

The centre of curvature of the beam lies in a longitudinal plane perpendicular to the neutral axis of the beam and passing through the centroid of any section. Hence for a radius of curvature R, we see, by direct comparison with Eq. (13.2) that

$$\frac{1}{R} = -\frac{\mathrm{d}^2\zeta}{\mathrm{d}z^2} \tag{13.12}$$

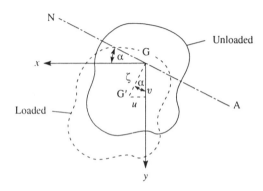

Fig. 13.17   Deflection of a beam of unsymmetrical cross-section

or, substituting from Eqs (13.11)

$$\frac{\sin \alpha}{R} = -\frac{d^2 u}{dz^2}, \qquad \frac{\cos \alpha}{R} = -\frac{d^2 v}{dz^2} \qquad (13.13)$$

We observe from the derivation of Eq. (9.30) that

$$\frac{E \sin \alpha}{R} = \frac{\overline{M}_y}{I_y}, \qquad \frac{E \cos \alpha}{R} = \frac{\overline{M}_x}{I_x} \qquad (13.14)$$

Therefore eliminating $(\sin \alpha)/R$ and $(\cos \alpha)/R$ between Eqs (13.13) and (13.14) we have

$$\frac{d^2 u}{dz^2} = -\frac{\overline{M}_y}{EI_y}, \qquad \frac{d^2 v}{dz^2} = -\frac{\overline{M}_x}{EI_x} \qquad (13.15)$$

Note the similarity between Eqs (13.15) and (13.3).

Example 13.15 Determine the horizontal and vertical components of the deflection of the free end of the cantilever shown in Fig. 13.18. The second moments of area of its unsymmetrical section are I_x, I_y and I_{xy}.

From Eqs (13.15) we have

$$\frac{d^2 u}{dz^2} = -\frac{\overline{M}_y}{EI_y}, \qquad \frac{d^2 v}{dz^2} = -\frac{\overline{M}_x}{EI_x} \qquad (i)$$

We shall concentrate initially on the vertical component of deflection, v. Since M_x varies with $z\,(M_y = 0)$, $\overline{M}_x$ must be expressed as a function of z before the second of Eqs (i) can be integrated. Now

$$\frac{\partial \overline{M}_x}{\partial z} = \overline{S}_y \qquad \text{(see Section 9.9)}$$

and $\overline{S}_y$ is constant along the length of the cantilever.

Hence $$\overline{M}_x = \overline{S}_y z + C_1$$

where C_1 is an unknown constant of integration which is determined using the boundary condition that $\overline{M}_x = 0$ when $z = L$ (note that M_x and M_y are zero at $z = L$).

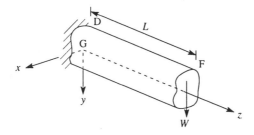

Fig. 13.18 Deflection of a cantilever of unsymmetrical cross-section carrying a concentrated load at its free end (Ex. 13.15)

Hence $C_1 = -\bar{S}_y L$ and

$$\bar{M}_x = \bar{S}_y(z - L) \qquad (ii)$$

Substituting for $\bar{M}_x$ in the second of Eqs (i) we have

$$\frac{d^2 v}{dz^2} = -\frac{\bar{S}_y}{EI_x}(z - L) \qquad (iii)$$

Integrating Eq. (iii) gives

$$\frac{dv}{dz} = -\frac{\bar{S}_y}{EI_x}\left(\frac{z^2}{2} - Lz\right) + C_2 \qquad (iv)$$

and

$$v = \frac{\bar{S}_y}{EI_x}\left(\frac{z^3}{6} - \frac{Lz^2}{2}\right) + C_2 z + C_3 \qquad (v)$$

The constants of integration, C_2 and C_3, are found from the boundary conditions that $v = 0$ when $z = 0$ and $dv/dz = 0$ when $z = 0$. From the first of these, $C_3 = 0$, while from the second $C_2 = 0$ also. Hence

$$v = \frac{\bar{S}_y}{EI_x}\left(\frac{z^3}{6} - \frac{Lz^2}{2}\right) \qquad (vi)$$

The vertical component, v_{tip}, of the deflection at the free end of the cantilever is then

$$v_{tip} = \frac{\bar{S}_y L^3}{3EI_x} \qquad (vii)$$

In the derivation of Eq. (vi) the loading has been expressed in general form using only the fact that $\bar{S}_y$ is constant along the length of the cantilever since both $S_y(= W)$ and $S_x(=0)$ are both constant along its length; by the same argument, $\bar{S}_x$ is constant. Also, we see that the boundary conditions used in evaluating C_1, C_2 and C_3 apply equally to $\bar{M}_y$, du/dz and u. The expression for the horizontal component of deflection may therefore be written down by direct comparison with Eqs (vi) and (vii). Hence

$$u = -\frac{\bar{S}_x}{EI_y}\left(\frac{z^3}{6} - \frac{Lz^2}{2}\right) \qquad (viii)$$

and

$$u_{tip} = \frac{\bar{S}_x L^3}{3EI_y} \qquad (ix)$$

Comparing Eqs (vii) and (ix) with Eq. (v) of Ex. 13.1 we see why $\bar{S}_x$, $\bar{S}_y$ are termed 'effective' shear forces.

The solution is completed by evaluating $\bar{S}_x$ and $\bar{S}_y$ in terms of the applied loading. Thus since $S_y = W$ and $S_x = 0$ we see from Eqs (9.59) and (9.60) that

$$\bar{S}_x = \frac{-WI_{xy}/I_x}{1 - I_{xy}^2/I_x I_y}, \qquad \bar{S}_y = \frac{W}{1 - I_{xy}^2/I_x I_y}$$

whence u_{tip} and v_{tip}.

Example 13.16   Determine the deflection of the free end of the cantilever beam shown in Fig. 13.19. The second moments of area of its cross-section about a horizontal and vertical system of centroidal axes are I_x, I_y and I_{xy}.

In this problem S_y, and therefore $\bar{S}_y$ and $\bar{S}_x$, are functions of z. We therefore use the relationships

$$\frac{\partial^2 \bar{M}_x}{\partial z^2} = -\bar{w}_y, \qquad \frac{\partial^2 \bar{M}_y}{\partial z^2} = -\bar{w}_x \qquad \text{(i)}$$

to determine the variation of $\bar{M}_x$ and $\bar{M}_y$ with z. Integrating the first of Eqs (i) we obtain

$$\frac{\partial \bar{M}_x}{\partial z} = -\bar{w}_y z + C_1 \qquad \text{(ii)}$$

When $z = L$, $\dfrac{\partial \bar{M}_x}{\partial z} = \bar{S}_y = 0$, hence $C_1 = \bar{w}_y L$ and

$$\frac{\partial \bar{M}_x}{\partial z} = \bar{w}_y (L - z) \qquad \text{(iii)}$$

Integrating Eq. (iii) we have

$$\bar{M}_x = \bar{w}_y \left(Lz - \frac{z^2}{2} \right) + C_2 \qquad \text{(iv)}$$

When $z = L$, $\bar{M}_x = 0$ so that $C_2 = -\bar{w}_y L^2/2$. Thus Eq. (iv) becomes

$$\bar{M}_x = \bar{w}_y \left(Lz - \frac{z^2}{2} - \frac{L^2}{2} \right) \qquad \text{(v)}$$

The remainder of the solution is identical in form to Ex. 13.15 and yields

$$v_{\text{tip}} = \frac{\bar{w}_y L^4}{8EI_x}, \qquad u_{\text{tip}} = \frac{\bar{w}_x L^4}{8EI_y} \qquad \text{(compare with Eq. (v) of Ex. 13.2)}$$

in which

$$\bar{w}_y = \frac{w}{1 - I_{xy}^2/I_x I_y}, \qquad \bar{w}_x = \frac{-w I_{xy}/I_x}{1 - I_{xy}^2/I_x I_y}$$

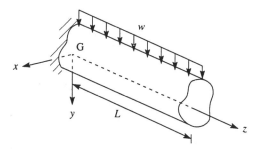

Fig. 13.19   Deflection of a cantilever of unsymmetrical cross-section carrying a uniformly distributed load (Ex. 13.16)

13.5 Moment-area method for unsymmetrical bending

We may use the concept of 'effective' bending moments to write down equations for slope and deflection of a beam subjected to unsymmetrical bending corresponding to Eqs (13.7) and (13.10) for the symmetrical case. Thus in the vertical yz plane we have, for sections A and B of a beam,

$$\left(\frac{dv}{dz}\right)_B - \left(\frac{dv}{dz}\right)_A = -\int_A^B \frac{\bar{M}_x}{EI_x}\,dz \tag{13.16}$$

and in the horizontal xz plane

$$\left(\frac{du}{dz}\right)_B - \left(\frac{du}{dz}\right)_A = -\int_A^B \frac{\bar{M}_y}{EI_y}\,dz \tag{13.17}$$

Similarly Eq. (13.10) becomes

$$z_B\left(\frac{dv}{dz}\right)_B - z_A\left(\frac{dv}{dz}\right)_A - (v_B - v_A) = -\int_A^B \frac{\bar{M}_x}{EI_x}\,z\,dz \tag{13.18}$$

and

$$z_B\left(\frac{du}{dz}\right)_B - z_A\left(\frac{du}{dz}\right)_A - (u_B - u_A) = -\int_A^B \frac{\bar{M}_y}{EI_y}\,z\,dz \tag{13.19}$$

In Eqs (13.16)–(13.19) we are concerned with the area and the moment of area of the 'effective' bending moment diagram divided by the appropriate flexural rigidity. Therefore, although the semi-graphical approach is possible, it will generally be simpler to use the relationships developed in Section 9.9 and work analytically.

Example 13.17 Determine the horizontal and vertical components of the deflection of the free end of the cantilever shown in Fig. 13.18.

Taking the origin for z at the free end, F, we rewrite Eqs (13.18) and (13.19) as

$$z_D\left(\frac{dv}{dz}\right)_D - z_F\left(\frac{dv}{dz}\right)_F - (v_D - v_F) = -\int_F^D \frac{\bar{M}_x}{EI_x}\,z\,dz \tag{i}$$

and

$$z_D\left(\frac{du}{dz}\right)_D - z_F\left(\frac{du}{dz}\right)_F - (u_D - u_F) = -\int_F^D \frac{\bar{M}_y}{EI_y}\,z\,dz \tag{ii}$$

respectively. In Eqs (i) and (ii) $(dv/dz)_D = (du/dz)_D = 0$, $z_F = 0$ and $v_D = u_D = 0$. Hence we have

$$v_F = -\int_0^L \frac{\bar{M}_x}{EI_x}\,z\,dz \tag{iii}$$

$$u_F = -\int_0^L \frac{\bar{M}_y}{EI_y}\,z\,dz \tag{iv}$$

Now

$$\frac{\partial \bar{M}_x}{\partial z} = \bar{S}_y \quad \text{(see Section 9.9)}$$

Integrating,

$$\overline{M}_x = \overline{S}_y z + C_1$$

The boundary conditions are $\overline{M}_x = 0$ (i.e. $M_x = M_y = 0$) at $z = 0$. Thus $C_1 = 0$ and

$$\overline{M}_x = \overline{S}_y z \qquad\qquad\qquad\text{(v)}$$

Substituting for $\overline{M}_x$ in Eq. (iii) we obtain

$$v_F = -\int_0^L \frac{\overline{S}_y}{EI_x} z^2 \, dz$$

which gives

$$v_F = -\frac{\overline{S}_y L^3}{3EI_x} \qquad\qquad\qquad\text{(vi)}$$

Similarly

$$u_F = -\frac{\overline{S}_x L^3}{3EI_y} \qquad\qquad\qquad\text{(vii)}$$

In Eqs (vi) and (vii)

$$\overline{S}_y = \frac{-W}{1 - I_{xy}^2/I_x I_y} \qquad \overline{S}_x = \frac{W I_{xy}/I_x}{1 - I_{xy}^2/I_x I_y} \qquad\text{(viii)}$$

Hence v_F and u_F (compare with Eqs (vii) and (ix) of Ex. 13.15).

Note that in Eqs (viii) $S_y = -W$ since the origin for z is at the free end of the beam, so that W acts on the face of the section which is seen when viewed in the direction Oz (see Fig. 9.16).

13.6 Deflection due to shear

So far in this chapter we have been concerned with deflections produced by the bending action of shear loads. These shear loads however, as we saw in Chapter 10, induce shear stress distributions throughout beam sections which in turn produce shear strains and therefore shear deflections. Generally, shear deflections are small compared with bending deflections, but in some cases of deep beams they can be comparable. In the following we shall use strain energy to derive an expression for the deflection due to shear in a beam having a cross-section which is at least singly symmetrical.

In Chapter 10 we showed that the strain energy U of a piece of material subjected to a uniform shear stress τ is given by

$$U = \frac{\tau^2}{2G} \times \text{volume} \qquad\qquad\text{(Eq. (10.20))}$$

However, we also showed in Chapter 10 that shear stress distributions are not uniform throughout beam sections. We therefore write Eq. (10.20) as

$$U = \frac{\beta}{2G} \times \left(\frac{S}{A}\right)^2 \times \text{volume} \qquad\qquad\text{(13.20)}$$

in which S is the applied shear force, A is the cross-sectional area of the beam section and β is a constant which depends upon the distribution of shear stress through the beam section; β is known as the *form factor*.

To determine β we consider an element $b_0\,\delta y$ in an elemental length δz of a beam subjected to a vertical shear load S_y (Fig. 13.20); we shall suppose that the beam section has a vertical axis of symmetry. The shear stress τ is constant across the width, b_0, of the element (see Section 10.2). The strain energy, δU, of the element $b_0\,\delta y\,\delta z$ is, from Eq. (10.20),

$$\delta U = \frac{\tau^2}{2G} \times b_0\,\delta y\,dz \tag{13.21}$$

Therefore the total strain energy U in the elemental length of beam is given by

$$U = \frac{\delta z}{2G} \int_{y_1}^{y_2} \tau^2 b_0\,dy \tag{13.22}$$

Alternatively U for the elemental length of beam is obtained using Eq. (13.20); thus

$$U = \frac{\beta}{2G} \times \left(\frac{S_y}{A}\right)^2 \times A\,\delta z \tag{13.23}$$

Equating Eqs (13.23) and (13.22) we have

$$\frac{\beta}{2G} \times \left(\frac{S_y}{A}\right)^2 \times A\,\delta z = \frac{\delta z}{2G} \int_{y_1}^{y_2} \tau^2 b_0\,dy$$

whence $$\beta = \frac{A}{S_y^2} \int_{y_1}^{y_2} \tau^2 b_0\,dy \tag{13.24}$$

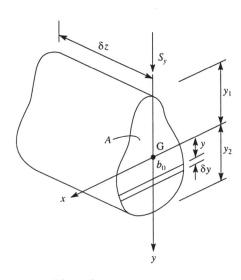

Fig. 13.20   Determination of form factor β

The shear stress distribution in a beam having a singly or doubly symmetrical cross-section and subjected to a vertical shear force, S_y, is given by Eq. (10.4), i.e.

$$\tau = \frac{S_y A' \bar{y}}{b_0 I_x}$$

Substituting this expression for τ in Eq. (13.24) we obtain

$$\beta = \frac{A}{S_y^2} \int_{y_1}^{y_2} \left(\frac{S_y A' \bar{y}}{b_0 I_x} \right)^2 b_0 \, dy$$

which gives

$$\beta = \frac{A}{I_x^2} \int_{y_1}^{y_2} \frac{(A' \bar{y})^2}{b_0} \, dy \qquad (13.25)$$

Suppose now that δv_s is the deflection due to shear in the elemental length of beam of Fig. 13.20. The work done by the shear force S_y (assuming it to be constant over the length δz and gradually applied) is then

$$\tfrac{1}{2} S_y \, \delta v_s$$

which is equal to the strain energy stored. Hence

$$\frac{1}{2} S_y \, \delta v_s = \frac{\beta}{2G} \times \left(\frac{S}{A} \right)^2 \times A \, \delta z$$

which gives

$$\delta v_s = \frac{\beta}{G} \left(\frac{S}{A} \right) \delta z$$

The total deflection due to shear in a beam of length L subjected to a vertical shear force S_y is then

$$v_s = \frac{\beta}{G} \int_L \left(\frac{S_y}{A} \right) dz \qquad (13.26)$$

Example 13.18 A cantilever beam of length L has a rectangular cross-section of breadth B and depth D and carries a vertical concentrated load, W, at its free end. Determine the deflection of the free end, including the effects of both bending and shear. The flexural rigidity of the cantilever is EI and its shear modulus G.

Using Eq. (13.25) we obtain the form factor β for the cross-section of the beam directly. Thus

$$\beta = \frac{BD}{(BD^3/12)^2} \int_{-D/2}^{D/2} \frac{1}{B} \left[B \left(\frac{D}{2} - y \right) \frac{1}{2} \left(\frac{D}{2} + y \right) \right]^2 dy \quad \text{(see Ex. 10.1)}$$

which simplifies to

$$\beta = \frac{36}{D^5} \int_{-D/2}^{D/2} \left(\frac{D^4}{16} - \frac{D^2 y^2}{2} + y^4 \right) dy$$

Integrating we obtain

$$\beta = \frac{36}{D^5} \left[\frac{D^4 y}{16} - \frac{D^2 y^3}{6} + \frac{y^5}{5} \right]_{-D/2}^{D/2}$$

which gives

$$\beta = \frac{6}{5}$$

Note that the dimensions of the cross-section do not feature in the expression for β. The form factor for any rectangular cross-section is therefore $6/5$ or $1 \cdot 2$.

Let us suppose that v_s is the vertical deflection of the free end of the cantilever due to shear. Hence, from Eq. (13.26) we have

$$v_s = \frac{6}{5G} \int_0^L \left(\frac{W}{BD} \right) dz$$

so that

$$v_s = \frac{6WL}{5GBD} \tag{i}$$

The vertical deflection due to bending of the free end of a cantilever carrying a concentrated load has previously been determined in Ex. 13.1 and is $WL^3/3EI$. The total deflection, v_T, produced by bending and shear is then

$$v_T = \frac{WL^3}{3EI} + \frac{6WL}{5GBD} \tag{ii}$$

Rewriting Eq. (ii) we obtain

$$v_T = \frac{WL^3}{3EI} \left[1 + \frac{3}{10} \frac{E}{G} \left(\frac{D}{L} \right)^2 \right] \tag{iii}$$

For many materials $(3E/10G)$ is approximately unity so that the contribution of shear to the total deflection is $(D/L)^2$ of the bending deflection. Clearly this term only becomes significant for short, deep beams.

13.7   Statically indeterminate beams

The beams we have considered so far have been supported in such a way that the support reactions could be determined using the equations of statical equilibrium; such beams are therefore *statically determinate*. However, many practical cases arise in which additional supports are provided so that there are a greater number of unknowns than the possible number of independent equations of equilibrium; the support systems of such beams are therefore *statically indeterminate*. Simple examples are shown in Fig. 13.21 where, in Fig. 13.21(a), the cantilever does not, theoretically, require the additional support at its free end and in Fig. 13.21(b) any one of the three supports is again, theoretically, *redundant*. A beam such as that shown in Fig. 13.21(b) is known as a *continuous beam* since it has more than one span and is continuous over one or more supports.

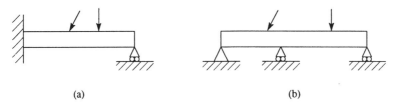

Fig. 13.21 Examples of statically indeterminate beams

We saw in Section 7.14 that additional equations are obtained in statically indeterminate systems by considering the displacements of the system. We shall therefore use the results of the previous work in this chapter to investigate methods of solving statically indeterminate beam systems. Having determined the reactions, diagrams of shear force and bending moment follow in the normal manner.

The examples given below are relatively simple cases of statically indeterminate beams. We shall investigate more complex cases in Chapter 16.

Method of superposition

In Section 3.7 we discussed the principle of superposition and saw that the combined effect of a number of forces on a structural system may be found by the addition of their separate effects. The principle may be· applied to the determination of support reactions in relatively simple statically indeterminate beams. We shall illustrate the method by examples.

Example 13.19 The cantilever AB shown in Fig. 13.22(a) carries a uniformly distributed load and is provided with an additional support at its free end. Determine the reaction at the additional support.

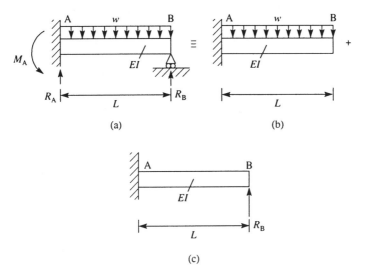

Fig. 13.22 Propped cantilever of Ex. 13.19

Suppose that the reaction at the support B is R_B. Using the principle of superposition we can represent the combined effect of the distributed load and the reaction R_B as the sum of the two loads acting separately as shown in Fig. 13.22(b) and (c). Also, since the vertical deflection of B in Fig. 13.22(a) is zero, it follows that the vertical downward deflection of B in Fig. 13.22(b) must be numerically equal to the vertically upward deflection of B in Fig. 13.22(c). Therefore using the results of Exs (13.1) and (13.2) we have

$$\left| \frac{R_B L^3}{3EI} \right| = \left| \frac{wL^4}{8EI} \right|$$

whence

$$R_B = \tfrac{3}{8} wL$$

It is now possible to determine the reactions R_A and M_A at the built-in end using the equations of simple statics. Thus taking moments about A for the beam in Fig. 13.22(a) we have

$$M_A = \frac{wL^2}{2} - R_B L = \frac{wL^2}{2} - \frac{3}{8} wL^2 = \frac{1}{8} wL^2$$

Resolving vertically

$$R_A = wL - R_B = wL - \frac{3}{8} wL = \frac{5}{8} wL$$

In the solution of Ex. 13.19 we selected R_B as the *redundancy*; in fact, any one of the three support reactions, M_A, R_A or R_B, could have been chosen. Let us suppose that M_A is taken to be the redundant reaction. We now represent the combined loading of Fig. 13.22(a) as the sum of the separate loading systems shown in Figs 13.23(a) and (b) and work in terms of the rotations of the beam at A due to the distributed load and the applied moment, M_A. Clearly, since there is no rotation at the built-in end of a cantilever, the rotations produced separately in Figs 13.23(a) and (b) must be numerically equal but opposite in direction. Using the method of Section 13.1 it may be shown that

$$\theta_A \text{ (due to } w) = \frac{wL^3}{24EI} \quad \text{(clockwise)}$$

and

$$\theta_A \text{ (due to } M_A) = \frac{M_A L}{3EI} \quad \text{(anticlockwise)}$$

Since

$$|\theta_A(M_A)| = |\theta_A(w)|$$

we have

$$M_A = \frac{wL^2}{8} \quad \text{as before.}$$

Built-in or fixed-end beams

In practice single-span beams may not be free to rotate about their supports but are connected to them in a manner that prevents rotation. Thus a reinforced concrete

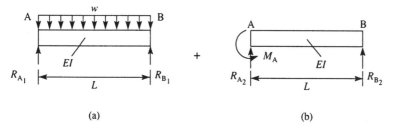

(a)   +   (b)

Fig. 13.23 Alternative solution of Ex. 13.19

beam may be cast integrally with its supports as shown in Fig. 13.24(a) or a steel beam may be bolted at its ends to steel columns (Fig. 13.24(b)). Clearly neither of the beams of Fig. 13.24(a) or (b) can be regarded as simply supported.

Consider the fixed beam of Fig. 13.25. Any system of vertical loads induces reactions of force and moment, the latter arising from the constraint against rotation provided by the supports. Thus there are four unknown reactions and only two possible equations of statical equilibrium; the beam is therefore statically indeterminate and has two redundancies. A solution is obtained by considering known values of slope and deflection at particular beam sections.

Example 13.20 Figure 13.26(a) shows a fixed beam carrying a central concentrated load, W. Determine the value of the fixed-end moments, M_A and M_B.

Since the ends A and B of the beam are prevented from rotating, moments M_A and M_B are induced in the supports; these are termed fixed-end moments. From symmetry we see that $M_A = M_B$ and $R_A = R_B = W/2$.

The beam AB in Fig. 13.26(a) may be regarded as a simply supported beam carrying a central concentrated load with moments M_A and M_B applied at the supports. The bending moment diagrams corresponding to these two loading cases

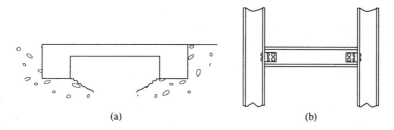

(a)   (b)

Fig. 13.24 Practical examples of fixed beams

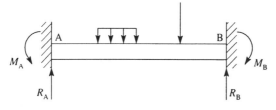

Fig. 13.25 Support reactions in a fixed beam

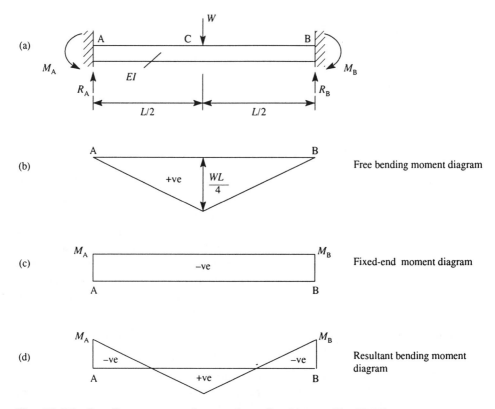

Fig. 13.26 Bending moment diagram for a fixed beam (Ex. 13.20)

are shown in Fig. 13.26(b) and (c) and are known as the *free bending moment diagram* and the *fixed-end moment diagram*, respectively. Clearly the concentrated load produces sagging (positive) bending moments, while the fixed-end moments induce hogging (negative) bending moments. The resultant or final bending moment diagram is constructed by superimposing the free and fixed-end moment diagrams as shown in Fig. 13.26(d).

The moment-area method is now used to determine the fixed-end moments, M_A and M_B. From Eq. (13.7) the change in slope between any two sections of a beam is equal to minus the area of the M/EI diagram between those sections. Therefore the net area of the bending moment diagram of Fig. 13.26(d) must be zero since the change of slope between the ends of the beam is zero. It follows that the area of the free bending moment diagram is numerically equal to the area of the fixed-end moment diagram; thus

$$M_A L = \frac{1}{2} \frac{WL}{4} L$$

Hence

$$M_A = M_B = \frac{WL}{8}$$

and the resultant bending moment diagram has principal values as shown in Fig. 13.27. Note that the maximum positive bending moment is equal in magnitude

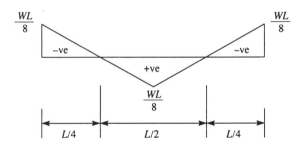

Fig. 13.27 Complete bending moment diagram for fixed beam of Ex. 13.20

to the maximum negative bending moment and that points of contraflexure (i.e. where the bending moment changes sign) occur at the quarter-span points.

Having determined the support reactions, the deflected shape of the beam may be found by any of the methods described in the previous part of this chapter.

Example 13.21 Determine the fixed-end moments and the fixed-end reactions for the beam shown in Fig. 13.28(a).

The resultant bending moment diagram is shown in Fig. 13.28(b) where the line AB represents the datum from which values of bending moment are measured. Again the net area of the resultant bending moment diagram is zero since the change in slope between the ends of the beam is zero. Hence

$$\frac{1}{2}(M_A + M_B)L = \frac{1}{2}L\frac{Wab}{L}$$

which gives

$$M_A + M_B = \frac{Wab}{L} \tag{i}$$

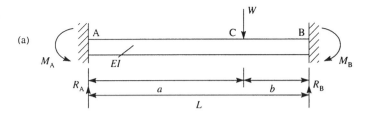

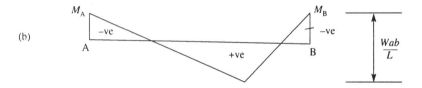

Fig. 13.28 Fixed beam of Ex. 13.28

We require a further equation to solve for M_A and M_B. This we obtain using Eq. (13.10) and taking the origin for z at A; hence we have

$$z_B \left(\frac{dv}{dz} \right)_B - z_A \left(\frac{dv}{dz} \right)_A - (v_B - v_A) = - \int_A^B \frac{M}{EI} z \, dz \qquad \text{(ii)}$$

In Eq. (ii) $(dv/dz)_B = (dv/dz)_A = 0$ and $v_B = v_A = 0$ so that

$$0 = \int_A^B \frac{M}{EI} z \, dz \qquad \text{(iii)}$$

and the moment of the area of the M/EI diagram between A and B about A is zero. Since EI is constant for the beam, we need only consider the bending moment diagram. Therefore from Fig. 13.28(b)

$$M_A L \frac{L}{2} + (M_B - M_A) \frac{L}{2} \frac{2}{3} L = \frac{1}{2} a \frac{Wab}{L} \frac{2a}{3} + \frac{1}{2} b \frac{Wab}{L} \left(a + \frac{1}{3} b \right)$$

Simplifying, we obtain

$$M_A + 2M_B = \frac{Wab}{L^2} (2a + b) \qquad \text{(iv)}$$

Solving Eqs (i) and (iv) simultaneously we obtain

$$M_A = \frac{Wab^2}{L^2}, \qquad M_B = \frac{Wa^2b}{L^2} \qquad \text{(v)}$$

We can now use statics to obtain R_A and R_B; hence, taking moments about B

$$R_A L - M_A + M_B - Wb = 0$$

Substituting for M_A and M_B from Eqs (v) we have

$$R_A L = \frac{Wab^2}{L^2} - \frac{Wa^2b}{L^2} + Wb$$

whence

$$R_A = \frac{Wb^2}{L^3} (3a + b)$$

Similarly

$$R_B = \frac{Wa^2}{L^3} (a + 3b)$$

Example 13.22   The fixed beam shown in Fig. 13.29(a) carries a uniformly distributed load of intensity w. Determine the support reactions.

From symmetry, $M_A = M_B$ and $R_A = R_B$. Again the net area of the bending moment diagram must be zero since the change of slope between the ends of the beam is zero (Eq. (13.7)). Hence

$$M_A L = \frac{2}{3} \frac{wL^2}{8} L$$

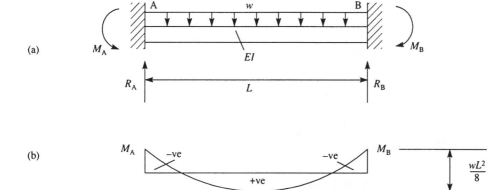

Fig. 13.29 Fixed beam carrying a uniformly distributed load (Ex. 13.22)

so that
$$M_A = M_B = \frac{wL^2}{12}$$

From statics,
$$R_A = R_B = \frac{wL}{2}$$

Example 13.23 The fixed beam of Fig. 13.30 carries a uniformly distributed load over part of its span. Determine the values of the fixed-end moments.

Consider a small element δz of the distributed load. We can use the results of Ex. 13.21 to write down the fixed-end moments produced by this elemental load since it may be regarded, in the limit as $\delta z \to 0$, as a concentrated load. Therefore from Eqs (v) of Ex. 13.21 we have

$$\delta M_A = w\,\delta z\,\frac{z(L-z)^2}{L^2}$$

The total moment at A, M_A, due to all such elemental loads is then

$$M_A = \int_a^b \frac{w}{L^2} z(L-z)^2\,\mathrm{d}z$$

which gives
$$M_A = \frac{w}{L^2}\left[\frac{L^2}{2}(b^2-a^2) - \frac{2}{3}L(b^3-a^3) + \frac{1}{4}(b^4-a^4)\right] \tag{i}$$

Similarly
$$M_B = \frac{wb^3}{L^2}\left(\frac{L}{3} - \frac{b}{4}\right) \tag{ii}$$

If the load covers the complete span, $a = 0$, $b = L$ and Eqs (i) and (ii) reduce to

$$M_A = M_B = \frac{wL^2}{12}$$

as in Ex. 13.22.

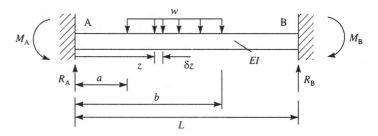

Fig. 13.30 Fixed beam with part-span uniformly distributed load (Ex. 13.23)

Fixed beam with a sinking support

In most practical situations the ends of a fixed beam would not remain perfectly aligned indefinitely. Since the ends of such a beam are prevented from rotating, a deflection of one end of the beam relative to the other induces fixed-end moments as shown in Fig. 13.31(a). These are in the same sense and for the relative displacement shown produce a total anticlockwise moment equal to $M_A + M_B$ on the beam. This moment is equilibrated by a clockwise couple formed by the force reactions at the supports. The resultant bending moment diagram is shown in Fig. 13.31(b) and, as in previous examples, its net area is zero since there is no change of slope between the ends of the beam and EI is constant (see Eq. (13.7)). This condition is satisfied by $M_A = M_B$.

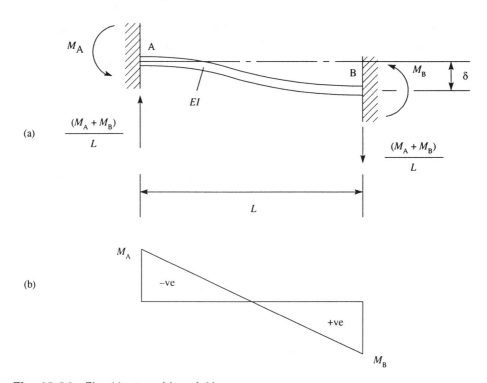

Fig. 13.31 Fixed beam with a sinking support

Let us now assume an origin for z at A; Eq. (13.10) becomes

$$z_B \left(\frac{dv}{dz}\right)_B - z_A \left(\frac{dv}{dz}\right)_A - (v_B - v_A) = -\int_A^B \frac{M}{EI} z \, dz \qquad (i)$$

in which $(dv/dz)_A = (dv/dz)_B = 0$, $v_A = 0$ and $v_B = \delta$. Hence Eq. (i) reduces to

$$\delta = \int_0^L \frac{M}{EI} z \, dz$$

Using the semi-graphical approach and taking moments of areas about A we have

$$\delta = -\frac{1}{2} \frac{L}{2} \frac{M_A}{EI} \frac{L}{6} + \frac{1}{2} \frac{L}{2} \frac{M_A}{EI} \frac{5}{6} L$$

which gives $\qquad\qquad M_A = \dfrac{6EI\,\delta}{L^2} \quad \text{(hogging)}$

It follows that $\qquad\qquad M_B = \dfrac{6EI\,\delta}{L^2} \quad \text{(sagging)}$

The effect of building in the ends of a beam is to increase both its strength and its stiffness. For example, the maximum bending moment in a simply supported beam carrying a central concentrated load W is $WL/4$ but it is $WL/8$ if the ends are built-in. A comparison of the maximum deflections shows a respective reduction from $WL^3/48EI$ to $WL^3/192EI$. It would therefore appear desirable for all beams to have their ends built-in if possible. However, in practice this is rarely done since, as we have seen, settlement of one of the supports induces additional bending moments in a beam. It is also clear that such moments can be induced during erection unless the supports are perfectly aligned. Furthermore, temperature changes can induce large stresses while live loads, which produce vibrations and fluctuating bending moments, can have adverse effects on the fixity of the supports.

One method of eliminating these difficulties is to employ a double cantilever construction. We have seen that points of contraflexure (i.e. zero bending moment) occur at sections along a fixed beam. Thus if hinges were positioned at these points the bending moment diagram and deflection curve would be unchanged but settlement of a support or temperature changes would have little or no effect on the beam.

Problems

P.13.1 The beam shown in Fig. P.13.1 is simply supported symmetrically at two points 2 m from each end and carries a uniformly distributed load of 5 kN/m together with two concentrated loads of 2 kN each at its free ends. Calculate the deflection at the mid-span point and at its free ends using the method of double integration. $EI = 43 \times 10^{12} \, \text{N}\,\text{mm}^2$.

Ans. 3·5 mm downwards, 2·1 mm upwards.

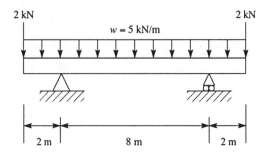

Fig. P.13.1

P.13.2   A beam AB of length L (Fig. P.13.2) is freely supported at A and at a point C which is at a distance KL from the end B. If a uniformly distributed load of intensity w per unit length acts on AC, find the value of K which will cause the upward deflection of B to equal the downward deflection mid-way between A and C.

   Ans.   0·24.

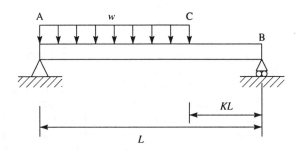

Fig. P.13.2

P.13.3   A uniform beam is simply supported over a span of 6 m. It carries a triangularly distributed load with intensity varying from 30 kN/m at the left-hand support to 90 kN/m at the right-hand support. Find the equation of the deflection curve and hence the deflection at the mid-span point. The second moment of area of the cross-section of the beam is 120×10^6 mm^4 and Young's modulus $E = 206\,000$ N/mm^2.

   Ans.   41 mm.

P.13.4   A cantilever having a flexural rigidity EI carries a distributed load that varies in intensity from w per unit length at the built-in end to zero at the free end. Find the deflection of the free end.

   Ans. $wL^4/30EI$.

P.13.5   Determine the position and magnitude of the maximum deflection of the simply supported beam shown in Fig. P.13.5 in terms of its flexural rigidity EI.

   Ans. $37·8/EI$ m at 2·9 m from left-hand support.

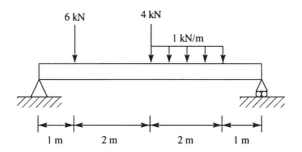

Fig. P.13.5

P.13.6 Calculate the position and magnitude (in terms of *EI*) of the maximum deflection in the beam shown in Fig. P.13.6.

Ans. 1310/*EI* m at 13·4 m from left-hand support.

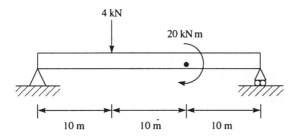

Fig. P.13.6

P.13.7 Determine the equation of the deflection curve of the beam shown in Fig. P.13.7. The flexural rigidity of the beam is *EI*.

Ans.

$$v = \frac{1}{EI}\left\{\frac{225}{6}z^{3} - \frac{100}{2}[z-1]^{2} + \frac{50}{12}[z-2]^{4} - \frac{50}{12}[z-4]^{4} - \frac{525}{6}[z-4]^{3} + 504z\right\}.$$

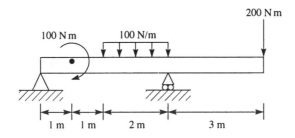

Fig. P.13.7

P.13.8 The beam shown in Fig. P.13.8 has its central portion reinforced so that its flexural rigidity is twice that of the outer portions. Use the moment-area method to determine the central deflection.

Ans. $3WL^{3}/256EI$.

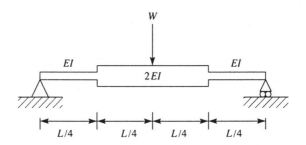

Fig. P.13.8

P.13.9   A simply supported beam of flexural rigidity *EI* carries a triangularly distributed load as shown in Fig. P.13.9. Determine the deflection of the mid-point of the beam.

Ans. $w_0 L^4 / 120EI$.

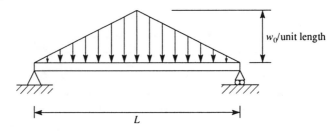

Fig. P.13.9

P.13.10   The simply supported beam shown in Fig. P.13.10 has its outer regions reinforced so that their flexural rigidity may be regarded as infinite compared with the central region. Determine the central deflection.

Ans. $7WL^3 / 384EI$.

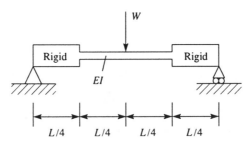

Fig. P.13.10

P.13.11   Calculate the horizontal and vertical components of the deflection at the centre of the simply supported span AB of the thick Z-section beam shown in Fig. P.13.11. Take $E = 200\ 000\ \text{N/mm}^2$.

Ans. $u = 2\cdot 43$ mm, $v = 1\cdot 75$ mm.

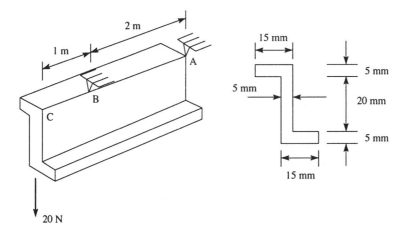

2 m

1 m

A

B

C

20 N

15 mm

5 mm

5 mm

20 mm

5 mm

15 mm

Fig. P.13.11

P.13.12   The simply supported beam shown in Fig. P.13.12 supports a uniformly distributed load of 10 N/mm in the plane of its horizontal flange. The properties of its cross-section referred to horizontal and vertical axes through its centroid are $I_x = 1{\cdot}67 \times 10^6$ mm⁴, $I_y = 0{\cdot}95 \times 10^6$ mm⁴ and $I_{xy} = 0{\cdot}74 \times 10^6$ mm⁴. Determine the magnitude and direction of the deflection at the mid-span section of the beam. Take $E = 70\ 000$ N/mm².

Ans.   52·5 mm at 23°54′ below horizontal.

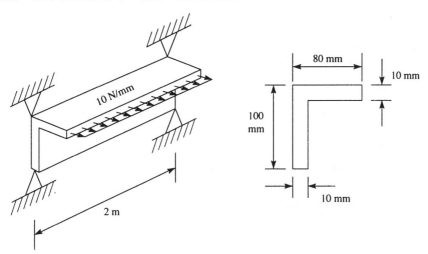

10 N/mm

2 m

80 mm

10 mm

100 mm

10 mm

Fig. P.13.12

P.13.13   A uniform cantilever of arbitrary cross-section and length L has section properties I_x, I_y, and I_{xy} with respect to the centroidal axes shown (Fig. P.13.13). It is loaded in the vertical plane by a tip load W. The tip of the beam is hinged to a horizontal link which constrains it to move in the vertical direction only (provided that the actual deflections are small). Assuming that the link is rigid and that there are no twisting effects, calculate the force in the link and the deflection of the tip of the beam.

Ans. WI_{xy}/I_x (compression if I_{xy} is positive), $WL^3/3EI_x$.

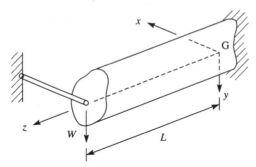

Fig. P.13.13

P.13.14   A thin-walled beam is simply supported at each end and supports a uniformly distributed load of intensity w per unit length in the plane of its lower horizontal flange (see Fig. P.13.14). Calculate the horizontal and vertical components of the deflection of the mid-span point. Take $E = 200\,000$ N/mm².

Ans. $u = -9\cdot1$ mm, $v = 5\cdot2$ mm.

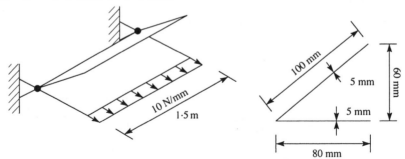

Fig. P.13.14

P.13.15   A uniform beam of arbitrary unsymmetrical cross-section and length $2L$ is built-in at one end and is simply supported in the vertical direction at a point half-way along its length. This support, however, allows the beam to deflect freely in the horizontal x direction (Fig. P.13.15). Determine the vertical reaction at the support.

Ans. $5W/2$.

P.13.16   A cantilever of length $3L$ has section second moments of area I_x, I_y, and I_{xy} referred to horizontal and vertical axes through the centroid of its cross-section. If the cantilever carries a vertically downward load W at its free end and is pinned to a support which prevents both vertical and horizontal movement at a distance $2L$ from the built-in end, calculate the magnitude of the vertical reaction at the support. Show also that the horizontal reaction is zero.

Ans. $7W/4$.

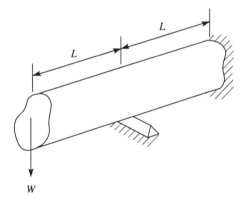

Fig. P.13.15

P.13.17 Calculate the deflection due to shear at the mid-span point of a simply supported rectangular section beam of length L which carries a vertically downward load W at mid-span. The beam has a cross-section of breadth B and depth D; the shear modulus is G.

Ans. $3WL/10GBD$.

P.13.18 Determine the deflection due to shear at the free end of a cantilever of length L and rectangular cross-section $B \times D$ which supports a uniformly distributed load of intensity w. The shear modulus is G.

Ans. $3wL^2/5GBD$.

P.13.19 A cantilever of length L has a solid circular cross-section of diameter D and carries a vertically downward load W at its free end. The modulus of rigidity of the cantilever is G. Calculate the shear stress distribution across a section of the cantilever and hence determine the deflection due to shear at its free end.

Ans. $\tau = 16W(1 - 4y^2/D^2)/3\pi D^2$, $40WL/9\pi GD^2$.

P.13.20 Show that the deflection due to shear in a rectangular section beam supporting a vertical shear load S_y is 20% greater for a shear stress distribution given by the expression

$$\tau = \frac{S_y A' \bar{y}}{b_o I_x}$$

than for a distribution assumed to be uniform.

A rectangular section cantilever beam 200 mm wide by 400 mm deep and 2 m long carries a vertically downward load of 500 kN at a distance of 1 m from its free end. Calculate the deflection at the free end taking into account both shear and bending effects. Take $E = 200\ 000$ N/mm^2 and $G = 70\ 000$ N/mm^2.

Ans. 2·06 mm.

P.13.21 The beam shown in Fig. P.13.21 is simply supported at each end and is provided with an additional support at mid-span. If the beam carries a uniformly distributed load of intensity w and has a flexural rigidity EI, use the principle of superposition to determine the reactions in the supports.

Ans. $5wL/4$ (central support), $3wL/8$ (outside supports).

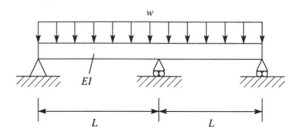

Fig. P.13.21

P.13.22 A built-in beam ACB of span L carries a concentrated load W at C a distance a from A and b from B. If the flexural rigidity of the beam is EI, use the principle of superposition to determine the support reactions.

Ans. $R_A = Wb^2(L+2a)/L^3$, $R_B = Wa^2(L+2b)/L^3$,
$M_A = Wab^2/L^2$, $M_B = Wa^2b/L^2$.

P.13.23 A beam has a second moment of area I for the central half of its span and $I/2$ for the outer quarters. If the beam carries a central concentrated load W, find the deflection at mid-span if the beam is simply supported and also the fixed-end moments when both ends of the beam are built-in.

Ans. $3WL^3/128EI, 5WL/48$.

P.13.24 A cantilever beam projects 1·5 m from its support and carries a uniformly distributed load of 16 kN/m over its whole length together with a load of 30 kN at 0·75 m from the support. The outer end rests on a prop which compresses 0·12 mm for every kN of compressive load. If the value of EI for the beam is 2000 kNm^2, determine the reaction in the prop.

Ans.   12·6 kN.

CHAPTER 14

Complex Stress and Strain

In Chapters 7, 9, 10 and 11 we have determined stress distributions produced separately by axial load, bending moment, shear force and torsion. However, in many practical situations some or all of these force systems act simultaneously so that the various stresses are combined to form complex systems which may include both direct and shear stresses. In such cases it is no longer a simple matter to predict the mode of failure of a structural member, particularly since, as we shall see, the direct and shear stresses at a point due to, say, bending and torsion are not necessarily the maximum values of direct and shear stress at that point.

Therefore as a preliminary to the investigation of the theories of elastic failure in Section 14.10 we shall examine states of stress and strain at points in structural members subjected to complex loading systems.

14.1 Representation of stress at a point

We have seen that generally stress distributions in structural members vary throughout the member. For example the direct stress in a cantilever beam carrying a point load at its free end varies along the length of the beam and throughout its depth. Suppose that we are interested in the state of stress at a point lying in the vertical plane of symmetry and on the upper surface of the beam mid-way along its span. The direct stress at this point on planes perpendicular to the axis of the beam can be calculated using Eq. (9.9). This stress may be imagined to be acting on two opposite sides of a very small thin element ABCD in the surface of the beam at the point (Fig. 14.1(a) and (b)).

Since the element is thin we can ignore any variation in direct stress across its thickness. Similarly, since the sides of the element are extremely small we can assume that σ has the same value on each opposite side BC and AD of the element and that σ is constant along these sides (in this particular case σ is constant across the width of the beam but the argument would apply if it were not). Thus we are representing the stress at a point in a structural member by a stress system acting on the sides and in the plane of a thin, very small element; such an element is known as a two-dimensional element.

Although some states of stress require representation by three-dimensional elements, we shall restrict our analysis to two-dimensional cases. Also, since two dimensional elements may be aligned in any direction in a structural member,

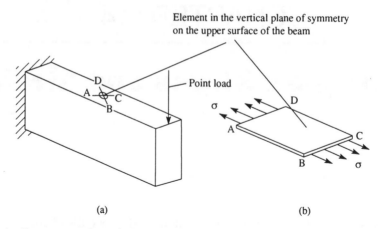

Fig. 14.1   Representation of stress at a point in a structural member

their edges will not necessarily be parallel to beam reference axes (Fig. 3.6) and
it is no longer practicable to use these axes to define directions of stress. We shall
therefore revert to a simple xy system in which the x axis of the element is
parallel to the longitudinal axis of a structural member and the y axis is
perpendicular to the longitudinal axis. In Fig. 14.1(b), therefore, the direct stress
would become σ_x.

14.2   Determination of stresses on inclined planes

Suppose that we wish to determine the direct and shear stresses at the same point in
the cantilever beam Fig. 14.1 but on a plane PQ inclined at an angle to the axis of
the beam as shown in Fig. 14.2(a). The direct stress on the sides AD and BC of the
element ABCD is σ_x in accordance with the sign convention now adopted.

Consider the triangular portion PQR of the element ABCD where QR is parallel to
the sides AD and BC. On QR there is a direct stress which must also be σ_x since
there is no variation of direct stress on planes parallel to QR between the opposite
sides of the element. On the side PQ of the triangular element let σ_n be the direct
stress and τ the shear stress. Although the stresses are uniformly distributed along
the sides of the elements it is convenient to represent them by single arrows as shown in
Fig. 14.2(b).

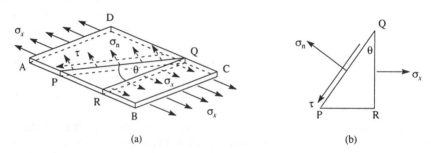

Fig. 14.2   Determination of stresses on an inclined plane

The triangular element PQR is in equilibrium under the action of forces corresponding to the stresses σ_x, σ_n and τ. Thus, resolving forces in a direction perpendicular to PQ and assuming that the element is of unit thickness we have

$$\sigma_n PQ = \sigma_x \, QR \cos \theta$$

or
$$\sigma_n = \sigma_x \, \frac{QR}{PQ} \cos \theta$$

which simplifies to
$$\sigma_n = \sigma_x \cos^2 \theta \qquad (14.1)$$

Resolving forces parallel to PQ

$$\tau PQ = \sigma_x QR \sin \theta$$

from which
$$\tau = \sigma_x \cos \theta \sin \theta$$

or
$$\tau = \frac{\sigma_x}{2} \sin 2\theta \qquad (14.2)$$

We see from Eqs (14.1) and (14.2) that although the applied load induces direct stresses only on planes perpendicular to the axis of the beam, both direct and shear stresses exist on planes inclined to the axis of the beam. Furthermore it can be seen from Eq. (14.2) that the shear stress τ is a maximum when $\theta = 45°$. This explains the mode of failure of ductile materials subjected to simple tension and other materials such as timber under compression. For example, a flat aluminium alloy test piece fails in simple tension along a line at approximately 45° to the axis of loading as illustrated in Fig. 14.3. This suggests that the crystal structure of the metal is relatively weak in shear and that failure takes the form of sliding of one crystal plane over another as opposed to the tearing apart of two crystal planes. The failure is therefore a shear failure although the test piece is in simple tension.

Biaxial stress system

A more complex stress system may be produced by a loading system such as that shown in Fig. 14.4 where a thin-walled hollow cylinder is subjected to an internal

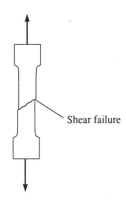
Shear failure

Fig. 14.3 Mode of failure in an aluminium alloy test piece

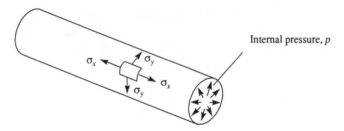

Fig. 14.4   Generation of a biaxial stress system

pressure, p. The internal pressure induces circumferential or hoop stresses σ_y, given by Eq. (7.77), on planes parallel to the axis of the cylinder and, in addition, longitudinal stresses, σ_x, on planes perpendicular to the axis of the cylinder (Eq. (7.76)). Thus any two-dimensional element of unit thickness in the wall of the cylinder and having sides perpendicular and parallel to the axis of the cylinder supports a biaxial stress system as shown in Fig. 14.4. In this particular case σ_x and σ_y each have constant values irrespective of the position of the element.

Let us consider the equilibrium of a triangular portion ABC of the element as shown in Fig. 14.5(a) and (b). Resolving forces in a direction perpendicular to AB we have

$$\sigma_n AB = \sigma_x BC \cos\theta + \sigma_y AC \sin\theta$$

or

$$\sigma_n = \sigma_x \frac{BC}{AB} \cos\theta + \sigma_y \frac{AC}{AB} \sin\theta$$

which gives

$$\sigma_n = \sigma_x \cos^2\theta + \sigma_y \sin^2\theta \tag{14.3}$$

Resolving forces parallel to AB

$$\tau AB = \sigma_x BC \sin\theta - \sigma_y AC \cos\theta$$

or

$$\tau = \sigma_x \frac{BC}{AB} \sin\theta - \sigma_y \frac{AC}{AB} \cos\theta$$

which gives

$$\tau = \left(\frac{\sigma_x - \sigma_y}{2}\right) \sin 2\theta \tag{14.4}$$

Again we see that although the applied loads produce only direct stresses on planes perpendicular and parallel to the axis of the cylinder, both direct and shear stresses exist on inclined planes. Furthermore, for given values of σ_x and σ_y (i.e. p) the shear stress τ is a maximum on planes inclined at 45° to the axis of the cylinder.

Example 14.1   A cylindrical pressure vessel has an internal diameter of 2 m and is fabricated from plates 20 mm thick. If the pressure inside the vessel is $1 \cdot 5 \ N/mm^2$ and, in addition, the vessel is subjected to an axial tensile load of 2500 kN, calculate the direct and shear stresses on a plane inclined at an angle of 60° to the axis of the vessel. Calculate also the maximum shear stress.

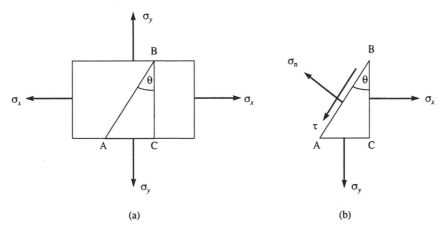

Fig. 14.5 Determination of stresses on an inclined plane in a biaxial stress system

From Eq. (7.77) the circumferential stress is

$$\frac{pd}{2t} = \frac{1 \cdot 5 \times 2 \times 10^3}{2 \times 20} = 75 \text{ N/mm}^2$$

From Eq. (7.76) the longitudinal stress is

$$\frac{pd}{4t} = 37 \cdot 5 \text{ N/mm}^2$$

The direct stress due to axial load is, from Eq. (7.1),

$$\frac{2500 \times 10^3}{\pi \times 2000 \times 20} = 19 \cdot 9 \text{ N/mm}^2$$

Thus on a rectangular element at any point in the wall of the vessel there is a biaxial stress system as shown in Fig. 14.6. Now considering the equilibrium of the triangular element ABC we have, resolving forces perpendicular to AB,

$$\sigma_n AB \times 20 = 57 \cdot 4 \, BC \times 20 \cos 30° + 75 AC \times 20 \cos 60°$$

Since the walls of the vessel are thin the thickness of the two-dimensional element may be taken as 20 mm. However, as can be seen, the thickness cancels out of the above equation so that it is simpler to assume unit thickness for two-dimensional elements in all cases. Thus

$$\sigma_n = 57 \cdot 4 \cos^2 30° + 75 \cos^2 60°$$

which gives

$$\sigma_n = 61 \cdot 8 \text{ N/mm}^2$$

Resolving parallel to AB

$$\tau AB = 57 \cdot 4 \, BC \cos 60° - 75 AC \sin 60°$$

or

$$\tau = 57 \cdot 4 \sin 60° \cos 60° - 75 \cos 60° \sin 60°$$

from which

$$\tau = -7 \cdot 6 \text{ N/mm}^2$$

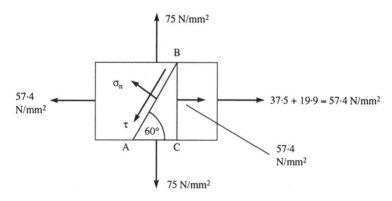

Fig. 14.6 Biaxial stress system of Ex. 14.1

The negative sign of τ indicates that τ acts in the direction AB and not, as was assumed, in the direction BA. From Eq. (14.4) it can be seen that the maximum shear stress occurs on planes inclined at $45°$ to the axis of the cylinder and is given by

$$\tau_{max} = \frac{57\cdot4 - 75}{2} = -8\cdot8 \text{ N/mm}^2$$

Again the negative sign of τ_{max} indicates that the direction of τ_{max} is opposite to that assumed.

General two-dimensional case

If we now apply a torque to the cylinder of Fig. 14.4 in an anticlockwise sense when viewed from the right-hand end, shear and complementary shear stresses are induced on the sides of the rectangular element in addition to the direct stresses already present. The value of these shear stresses is given by Eq. (11.21) since the cylinder is thin-walled. We now have a general two-dimensional stress system acting on the element as shown in Fig. 14.7(a). The suffixes employed in designating shear stress refer to the plane on which the stress acts and its direction. Thus τ_{xy} is a shear stress acting on an x plane in the y direction. Conversely τ_{yx} acts on a y plane in the x direction. However, since $\tau_{xy} = \tau_{yx}$ we label both shear and complementary shear stresses τ_{xy} as in Fig. 14.7(b).

Considering the equilibrium of the triangular element ABC in Fig. 14.7(b) and resolving forces in a direction perpendicular to AB

$$\sigma_n \text{AB} = \sigma_x \text{BC} \cos\theta + \sigma_y \text{AC} \sin\theta + \tau_{xy} \text{BC} \sin\theta + \tau_{xy} \text{AC} \cos\theta$$

Dividing through by AB and simplifying we obtain

$$\sigma_n = \sigma_x \cos^2\theta + \sigma_y \sin^2\theta + \tau_{xy} \sin 2\theta \qquad (14.5)$$

Now resolving forces parallel to BA

$$\tau \text{AB} = \sigma_x \text{BC} \sin\theta - \sigma_y \text{AC} \cos\theta - \tau_{xy} \text{BC} \cos\theta + \tau_{xy} \text{AC} \sin\theta$$

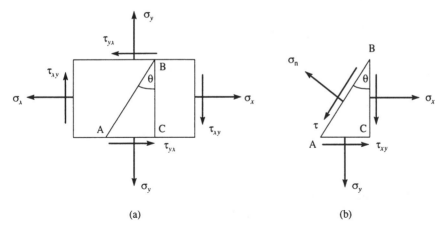

Fig. 14.7 General two-dimensional stress system

Again dividing through by AB and simplifying we have

$$\tau = \left(\frac{\sigma_x - \sigma_y}{2}\right)\sin 2\theta - \tau_{xy}\cos 2\theta \tag{14.6}$$

Example 14.2 A cantilever of solid, circular cross-section supports a compressive load of 50 000 N applied to its free end at a point 1·5 mm below a horizontal diameter in the vertical plane of symmetry together with a torque of 1200 N m (Fig. 14.8).

Calculate the direct and shear stresses on a plane inclined at 60° to the axis of the cantilever at a point on the lower edge of the vertical plane of symmetry.

The direct loading system is equivalent to an axial load of 50 000 N together with a bending moment of 50 000 × 1·5 = 75 000 N mm in a vertical plane. Thus at any point on the lower edge of the vertical plane of symmetry there are direct compressive stresses due to axial load and bending moment which act on planes perpendicular to the axis of the beam and are given, respectively, by Eqs (7.1) and (9.9). Therefore

$$\sigma_x(\text{axial load}) = \frac{50\,000}{\pi \times 60^2/4} = 17\cdot7 \text{ N/mm}^2$$

$$\sigma_x(\text{bending moment}) = \frac{75\,000 \times 30}{\pi \times 60^4/64} = 3\cdot5 \text{ N/mm}^2$$

The shear stress τ_{xy} at the same point due to the torque is obtained from Eq. (11.4) and is

$$\tau_{xy} = \frac{1200 \times 10^3 \times 30}{\pi \times 60^4/32} = 28\cdot3 \text{ N/mm}^2$$

The stress system acting on a two-dimensional rectangular element at the point is as

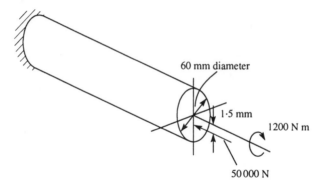

Fig. 14.8 Cantilever beam of Ex. 14.2

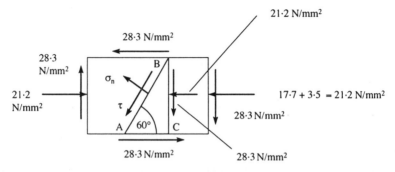

Fig. 14.9 Two-dimensional stress system in cantilever beam of Ex. 14.2

shown in Fig. 14.9. Thus considering the equilibrium of the triangular element ABC and resolving forces in a direction perpendicular to AB we have

$$\sigma_n \, AB = -21 \cdot 2 \, BC \cos 30° + 28 \cdot 3 \, BC \sin 30° + 28 \cdot 3 \, AC \cos 30°$$

Dividing through by AB we obtain

$$\sigma_n = -21 \cdot 2 \cos^2 30° + 28 \cdot 3 \cos 30° \sin 30° + 28 \cdot 3 \sin 30° \cos 30°$$

which gives $$\sigma_n = 8.6 \text{ N/mm}^2$$

Similarly resolving parallel to AB

$$\tau AB = -21 \cdot 2 \, BC \cos 60° - 28 \cdot 3 \, BC \sin 60° + 28 \cdot 3 \, AC \cos 60°$$

Thus $$\tau = -21 \cdot 2 \sin 60° \cos 60° - 28 \cdot 3 \sin^2 60° + 28 \cdot 3 \cos^2 60°$$

from which $$\tau = -23.3 \text{ N/mm}^2$$

acting in the direction AB.

14.3 Principal stresses

Equations (14.5) and (14.6) give the direct and shear stresses on an inclined plane at a point in a structural member subjected to a combination of loads which produces a general two-dimensional stress system at that point. Clearly for given values of σ_x, σ_y and τ_{xy}, in other words a given loading system, both σ_n and τ vary with the angle θ and will attain maximum or minimum values when $d\sigma_n/d\theta = 0$ and $d\tau/d\theta = 0$. From Eq. (14.5)

$$\frac{d\sigma_n}{d\theta} = -2\sigma_x \cos\theta \sin\theta + 2\sigma_y \sin\theta \cos\theta + 2\tau_{xy} \cos 2\theta = 0$$

Hence

$$-(\sigma_x - \sigma_y)\sin 2\theta + 2\tau_{xy}\cos 2\theta = 0$$

or

$$\tan 2\theta = \frac{2\tau_{xy}}{\sigma_x - \sigma_y} \qquad (14.7)$$

Two solutions, θ and $\theta + \pi/2$, satisfy Eq. (14.7) so that there are two mutually perpendicular planes on which the direct stress is either a maximum or a minimum. Furthermore, by comparison of Eqs (14.7) and (14.6) it can be seen that these planes correspond to those on which $\tau = 0$.

The direct stresses on these planes are called *principal stresses* and the planes are called *principal planes*.

From Eq. (14.7)

$$\sin 2\theta = \frac{2\tau_{xy}}{\sqrt{(\sigma_x - \sigma_y)^2 + 4\tau_{xy}^2}}, \qquad \cos 2\theta = \frac{\sigma_x - \sigma_y}{\sqrt{(\sigma_x - \sigma_y)^2 + 4\tau_{xy}^2}}$$

and

$$\sin 2(\theta + \pi/2) = \frac{-2\tau_{xy}}{\sqrt{(\sigma_x - \sigma_y)^2 + 4\tau_{xy}^2}}$$

$$\cos 2(\theta + \pi/2) = \frac{-(\sigma_x - \sigma_y)}{\sqrt{(\sigma_x - \sigma_y)^2 + 4\tau_{xy}^2}}$$

Rewriting Eq. (14.5) as

$$\sigma_n = \frac{\sigma_x}{2}(1 + \cos 2\theta) + \frac{\sigma_y}{2}(1 - \cos 2\theta) + \tau_{xy}\sin 2\theta$$

and substituting for $\{\sin 2\theta, \cos 2\theta\}$ and $\{\sin 2(\theta + \pi/2), \cos 2(\theta + \pi/2)\}$ in turn gives

$$\sigma_I = \frac{\sigma_x + \sigma_y}{2} + \frac{1}{2}\sqrt{(\sigma_x - \sigma_y)^2 + 4\tau_{xy}^2} \qquad (14.8)$$

$$\sigma_{II} = \frac{\sigma_x + \sigma_y}{2} - \frac{1}{2}\sqrt{(\sigma_x - \sigma_y)^2 + 4\tau_{xy}^2} \qquad (14.9)$$

where σ_I is the *maximum* or *major principal stress* and σ_{II} is the *minimum* or *minor principal stress*; σ_I is algebraically the greatest direct stress at the point while σ_{II} is algebraically the least. Thus, when σ_{II} is compressive, i.e. negative, it is possible for σ_{II} to be numerically greater than σ_I.

From Eq. (14.6)

$$\frac{d\tau}{d\theta} = (\sigma_x - \sigma_y) \cos 2\theta + 2\tau_{xy} \sin 2\theta = 0$$

giving

$$\tan 2\theta = -\frac{(\sigma_x - \sigma_y)}{2\tau_{xy}} \tag{14.10}$$

It follows that

$$\sin 2\theta = \frac{-(\sigma_x - \sigma_y)}{\sqrt{(\sigma_x - \sigma_y)^2 + 4\tau_{xy}^2}},$$

$$\cos 2\theta = \frac{2\tau_{xy}}{\sqrt{(\sigma_x - \sigma_y)^2 + 4\tau_{xy}^2}}$$

$$\sin 2(\theta + \pi/2) = \frac{(\sigma_x - \sigma_y)}{\sqrt{(\sigma_x - \sigma_y)^2 + 4\tau_{xy}^2}},$$

$$\cos 2(\theta + \pi/2) = \frac{-2\tau_{xy}}{\sqrt{(\sigma_x - \sigma_y)^2 + 4\tau_{xy}^2}}$$

Substituting these values in Eq. (14.6) gives

$$\tau_{max,min} = \pm\frac{1}{2}\sqrt{(\sigma_x - \sigma_y)^2 + 4\tau_{xy}^2} \tag{14.11}$$

Here, as in the case of the principal stresses, we take the maximum value as being the greater value algebraically.

Comparing Eq. (14.11) with Eqs (14.8) and (14.9) we see that

$$\tau_{max} = \frac{\sigma_I - \sigma_{II}}{2} \tag{14.12}$$

Equations (14.11) and (14.12) give alternative expressions for the maximum shear stress acting at the point *in the plane of the given stresses*. This is not necessarily the maximum shear stress in a three-dimensional element subjected to a two-dimensional stress system, as we shall see in Section 14.10.

Since Eq. (14.10) is the negative reciprocal of Eq. (14.7), the angles given by these two equations differ by 90° so that the planes of maximum shear stress are inclined at 45° to the principal planes.

We see now that the direct stresses, σ_x, σ_y, and shear stresses, τ_{xy}, are not, in a general case, the greatest values of direct and shear stress at the point. This fact is clearly important in designing structural members subjected to complex loading systems, as we shall see in Section 14.10. We can illustrate the stresses acting on the various planes at the point by considering a series of elements at the point as shown in Fig. 14.10. Note that generally there will be a direct stress on the planes on which τ_{max} acts.

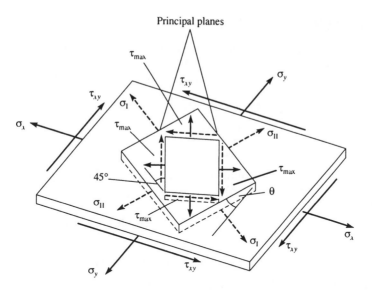

Fig. 14.10 Stresses acting on different planes at a point in a structural member

Example 14.3 A structural member supports loads which produce, at a particular point, a direct tensile stress of 80 N/mm² and a shear stress of 45 N/mm² on the same plane. Calculate the values and directions of the principal stresses at the point and also the maximum shear stress, stating on which planes this will act.

Suppose that the tensile stress of 80 N/mm² acts in the x direction. Then $\sigma_x = +80$ N/mm², $\sigma_y = 0$ and $\tau_{xy} = 45$ N/mm². Substituting these values in Eqs (14.8) and (14.9) in turn gives

$$\sigma_{\mathrm{I}} = \frac{80}{2} + \frac{1}{2}\sqrt{80^2 + 4 \times 45^2} = 100 \cdot 2 \text{ N/mm}^2$$

$$\sigma_{\mathrm{II}} = \frac{80}{2} - \frac{1}{2}\sqrt{80^2 + 4 \times 45^2} = -20 \cdot 2 \text{ N/mm}^2$$

From Eq. (14.7)

$$\tan 2\theta = \frac{2 \times 45}{80} = 1 \cdot 125$$

from which $\quad\quad\quad \theta = 24°11'$ (corresponding to σ_{I})

Also, the plane on which σ_{II} acts corresponds to $\theta = 24°11' + 90° = 114°11'$.

The maximum shear stress is most easily found from Eq. (14.12) and is given by

$$\tau_{\max} = \frac{100 \cdot 2 - (-20 \cdot 2)}{2} = 60 \cdot 2 \text{ N/mm}^2$$

The maximum shear stress acts on planes at 45° to the principal planes. Thus $\theta = 69°11'$ and $\theta = 159°11'$ give the planes of maximum shear stress.

14.4 Mohr's circle of stress

The state of stress at a point in a structural member may be conveniently represented graphically by *Mohr's circle of stress*. We have shown that the direct and shear stresses on an inclined plane are given, in terms of known applied stresses, by

$$\sigma_n = \sigma_x \cos^2 \theta + \sigma_y \sin^2 \theta + \tau_{xy} \sin 2\theta \qquad \text{(Eq. (14.5))}$$

and

$$\tau = \frac{(\sigma_x - \sigma_y)}{2} \sin 2\theta - \tau_{xy} \cos 2\theta \qquad \text{(Eq. (14.6))}$$

respectively. The positive directions of these stresses and the angle θ are defined in Fig. 14.7. We now write Eq. (14.5) in the form

$$\sigma_n = \frac{\sigma_x}{2} (1 + \cos 2\theta) + \frac{\sigma_y}{2} (1 - \cos 2\theta) + \tau_{xy} \sin 2\theta$$

or

$$\sigma_n - \tfrac{1}{2}(\sigma_x + \sigma_y) = \tfrac{1}{2}(\sigma_x - \sigma_y) \cos 2\theta + \tau_{xy} \sin 2\theta \qquad (14.13)$$

Now squaring and adding Eqs (14.6) and (14.13) we obtain

$$[\sigma_n - \tfrac{1}{2}(\sigma_x + \sigma_y)]^2 + \tau^2 = [\tfrac{1}{2}(\sigma_x - \sigma_y)]^2 + \tau_{xy}^2 \qquad (14.14)$$

Equation (14.14) represents the equation of a circle of radius

$$\pm \tfrac{1}{2}\sqrt{(\sigma_x - \sigma_y)^2 + 4\tau_{xy}^2}$$

and having its centre at the point $\left(\dfrac{\sigma_x + \sigma_y}{2}, 0\right)$.

The circle may be constructed by locating the points $Q_1(\sigma_x, \tau_{xy})$ and $Q_2(\sigma_y, -\tau_{xy})$ referred to axes $O\sigma\tau$ as shown in Fig. 14.11. The line Q_1Q_2 is then drawn and intersects the $O\sigma$ axis at C. From Fig. 14.11

$$OC = OP_1 - CP_1 = \sigma_x - (\sigma_x - \sigma_y)/2$$

so that

$$OC = (\sigma_x + \sigma_y)/2$$

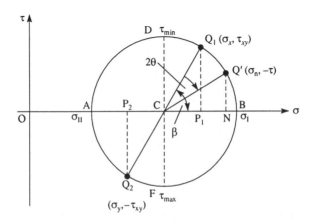

Fig. 14.11   Mohr's circle of stress

Thus the point C has coordinates $\left(\dfrac{\sigma_x + \sigma_y}{2}, 0\right)$ which, as we have seen, is the centre of the circle. Also

$$CQ_1 = \sqrt{CP_1^{\,2} + P_1 Q_1^{\,2}}$$

$$= \sqrt{[(\sigma_x - \sigma_y)/2]^2 + \tau_{xy}^{\,2}}$$

whence $\qquad CQ_1 = \tfrac{1}{2}\sqrt{(\sigma_x - \sigma_y)^2 + 4\tau_{xy}^{\,2}}$

which is the radius of the circle; the circle is then drawn as shown.

Now we set CQ′ at an angle 2θ (positive clockwise) to CQ_1; Q′ is then the point $(\sigma_n, -\tau)$ as demonstrated below.

From Fig. 14.11 we see that

$$ON = OC + CN$$

or, since $OC = (\sigma_x + \sigma_y)/2$, $CN = CQ'\cos(\beta - 2\theta)$ and $CQ' = CQ_1$, we have

$$\sigma_n = \frac{\sigma_x + \sigma_y}{2} + CQ_1(\cos \beta \cos 2\theta + \sin \beta \sin 2\theta)$$

But $\qquad CQ_1 = CP_1/\cos \beta$ and $CP_1 = (\sigma_x - \sigma_y)/2$

Hence $\qquad \sigma_n = \dfrac{\sigma_x + \sigma_y}{2} + \left(\dfrac{\sigma_x - \sigma_y}{2}\right)\cos 2\theta + CP_1 \tan \beta \sin 2\theta$

which, on rearranging, becomes

$$\sigma_n = \sigma_x \cos^2 \theta + \sigma_y \sin^2 \theta + \tau_{xy} \sin 2\theta$$

as in Eq. (14.5). Similarly it may be shown that

$$Q'N = \tau_{xy} \cos 2\theta - \left(\frac{\sigma_x - \sigma_y}{2}\right)\sin 2\theta = -\tau$$

as in Eq. (14.6). It must be remembered that the construction of Fig. 14.11 corresponds to the stress system of Fig. 14.7(b); any sign reversal must be allowed for. Also the $O\sigma$ and $O\tau$ axes must be constructed to the same scale otherwise the circle would not be that represented by Eq. (14.14).

The maximum and minimum values of the direct stress σ_n, that is the major and minor principal stresses σ_1 and σ_{11}, occur when N and Q′ coincide with B and A, respectively. Thus

$$\sigma_1 = OC + \text{radius of circle}$$

i.e. $\qquad \sigma_1 = \dfrac{\sigma_x + \sigma_y}{2} + \dfrac{1}{2}\sqrt{(\sigma_x - \sigma_y)^2 + 4\tau_{xy}^{\,2}} \qquad$ (as in Eq. (14.8))

and $\qquad \sigma_{11} = OC - \text{radius of circle}$

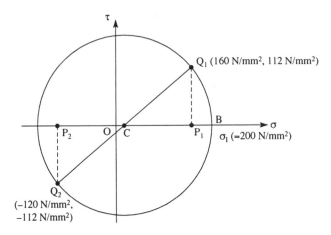

Fig. 14.12 Mohr's circle of stress for Ex. 14.4

whence $\qquad \sigma_{\mathrm{II}} = \dfrac{\sigma_x + \sigma_y}{2} - \dfrac{1}{2}\sqrt{(\sigma_x - \sigma_y)^2 + 4\tau_{xy}{}^2} \qquad$ (as in Eq. (14.9))

The principal planes are then given by $2\theta = \beta(\sigma_{\mathrm{I}})$ and $2\theta = \beta + \pi(\sigma_{\mathrm{II}})$.

The maximum and minimum values of the shear stress τ occur when Q' coincides with F and D at the lower and upper extremities of the circle. At these points $\tau_{\mathrm{max,min}}$ are clearly equal to the radius of the circle. Hence

$$\tau_{\mathrm{max,\,min}} = \pm\tfrac{1}{2}\sqrt{(\sigma_x - \sigma_y)^2 + 4\tau_{xy}{}^2} \qquad \text{(see Eq. (14.11))}$$

The minimum value of shear stress is the algebraic minimum. The planes of maximum and minimum shear stress are given by $2\theta = \beta + \pi/2$ and $2\theta = \beta + 3\pi/2$ and are inclined at 45° to the principal planes.

Example 14.4   Direct stresses of 160 N/mm², tension, and 120 N/mm², compression, are applied at a particular point in an elastic material on two mutually perpendicular planes. The maximum principal stress in the material is limited to 200 N/mm², tension. Use a graphical method to find the allowable value of shear stress at the point.

First, axes $O\sigma\tau$ are set up to a suitable scale. P_1 and P_2 are then located corresponding to $\sigma_x = 160$ N/mm² and $\sigma_y = -120$ N/mm², respectively; the centre C of the circle is mid-way between P_1 and P_2 (Fig. 14.12). The radius is obtained by locating $B(\sigma_1 = 200$ N/mm²$)$ and the circle then drawn. The maximum allowable applied shear stress, τ_{xy}, is then obtained by locating Q_1 or Q_2. The maximum shear stress at the point is equal to the radius of the circle and is 180 N/mm².

14.5   Stress trajectories

We have shown that direct and shear stresses at a point in a beam produced, say, by bending and shear and calculated by the methods of Chapters 9 and 10, respectively,

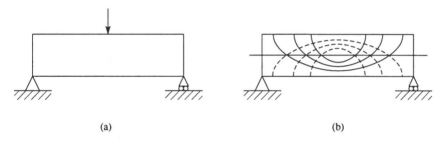

(a)   (b)

Fig. 14.13   Stress trajectories in a beam

are not necessarily the greatest values of direct and shear stress at the point. In order, therefore, to obtain a more complete picture of the distribution, magnitude and direction of the stresses in a beam we investigate the manner in which the principal stresses vary throughout a beam.

Consider the simply supported beam of rectangular section carrying a central concentrated load as shown in Fig. 14.13(a). Using Eqs (9.9) and (10.4) we can determine the direct and shear stress at any point in any section of the beam. Subsequently from Eqs (14.8), (14.9) and (14.7) we can find the principal stresses at the point and their directions. If this procedure is followed for very many points throughout the beam, curves, to which the principal stresses are tangential, may be drawn as shown in Fig. 14.13(b). These curves are known as *stress trajectories* and form two orthogonal systems; in Fig. 14.13(b) solid lines represent tensile principal stresses and dotted lines compressive principal stresses. The two sets of curves cross each other at right angles and all curves intersect the neutral axis at 45° where the direct stress (calculated from Eq. (9.9)) is zero. At the top and bottom surfaces of the beam where the shear stress (calculated from Eq. (10.4)) is zero the trajectories have either horizontal or vertical tangents.

Another type of curve that may be drawn from a knowledge of the distribution of principal stress is a *stress contour*. Such a curve connects points of equal principal stress.

14.6   Determination of strains on inclined planes

In Section 14.2 we investigated the two-dimensional state of stress at a point in a

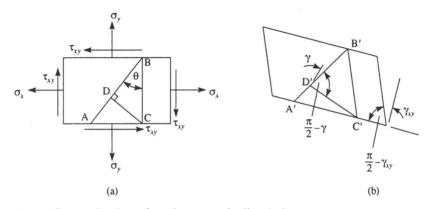

(a)   (b)

Fig. 14.14   Determination of strains on an inclined plane

structural member and determined direct and shear stresses on inclined planes; we shall now determine the accompanying strains.

Figure 14.14(a) shows a two-dimensional element subjected to a complex direct and shear stress system. The applied stresses will distort the rectangular element of Fig. 14.14(a) into the shape shown in Fig. 14.14(b). In particular, the triangular element ABC will suffer distortion to the shape A'B'C' with corresponding changes in the length CD and the angle BDC. The strains associated with the stresses σ_x, σ_y and τ_{xy} are ε_x, ε_y and γ_{xy}, respectively. We shall now determine the direct strain ε_n in a direction normal to the plane AB and the shear strain γ produced by the shear stress acting on the plane AB.

To a first order of approximation

$$\left. \begin{aligned} A'C' &= AC(1 + \varepsilon_x) \\ C'B' &= CB(1 + \varepsilon_y) \\ A'B' &= AB(1 + \varepsilon_{n+\pi/2}) \end{aligned} \right\} \tag{14.15}$$

where $\varepsilon_{n+\pi/2}$ is the direct strain in the direction AB. From the geometry of the triangle A'B'C' in which angle $B'C'A' = \pi/2 - \gamma_{xy}$

$$(A'B')^2 = (A'C')^2 + (C'B')^2 - 2(A'C')(C'B') \cos (\pi/2 - \gamma_{xy})$$

or, substituting from Eqs (14.15)

$$(AB)^2(1 + \varepsilon_{n+\pi/2})^2$$
$$= (AC)^2(1 + \varepsilon_x)^2 + (CB)^2(1 + \varepsilon_y)^2 - 2(AC)(CB)(1 + \varepsilon_x)(1 + \varepsilon_y) \sin \gamma_{xy}$$

Noting that $(AB)^2 = (AC)^2 + (CB)^2$ and neglecting squares and higher powers of small quantities, this equation may be rewritten

$$2(AB)^2\varepsilon_{n+\pi/2} = 2(AC)^2\varepsilon_x + 2(CB)^2\varepsilon_y - 2(AC)(CB)\gamma_{xy}$$

Dividing through by $2(AB)^2$ gives

$$\varepsilon_{n+\pi/2} = \varepsilon_x \sin^2\theta + \varepsilon_y \cos^2\theta - \sin \theta \cos \theta \gamma_{xy} \tag{14.16}$$

The strain ε_n in the direction normal to the plane AB is found by replacing the angle θ in Eq. (14.16) by $\theta - \pi/2$. Hence

$$\varepsilon_n = \varepsilon_x \cos^2 \theta + \varepsilon_y \sin^2 \theta + \frac{\gamma_{xy}}{2} \sin 2\theta \tag{14.17}$$

Now from triangle C'D'B' we have

$$(C'B')^2 = (C'D')^2 + (D'B')^2 - 2(C'D')(D'B') \cos (\pi/2 - \gamma) \tag{14.18}$$

in which
$$C'B' = CB(1 + \varepsilon_y)$$

$$C'D' = CD(1 + \varepsilon_n)$$

$$D'B' = DB(1 + \varepsilon_{n+\pi/2})$$

Substituting in Eq. (14.18) for C′B′, C′D′ and D′B′ and writing $\cos(\pi/2 - \gamma) = \sin\gamma$ we have

$$(CB)^2(1 + \varepsilon_y)^2 = (CD)^2(1 + \varepsilon_n)^2 + (DB)^2(1 + \varepsilon_{n+\pi/2})^2$$
$$- 2(CD)(DB)(1 + \varepsilon_n)(1 + \varepsilon_{n+\pi/2})\sin\gamma \quad (14.19)$$

Again ignoring squares and higher powers of strains and writing $\sin\gamma = \gamma$, Eq. (14.19) becomes

$$(CB)^2(1 + 2\varepsilon_y) = (CD)^2(1 + 2\varepsilon_n) + (DB)^2(1 + 2\varepsilon_{n+\pi/2}) - 2(CD)(DB)\gamma$$

From Fig. 14.14(a) we see that $(CB)^2 = (CD)^2 + (DB)^2$ and the above equation simplifies to

$$2(CB)^2\varepsilon_y = 2(CD)^2\varepsilon_n + 2(DB)^2\varepsilon_{n+\pi/2} - 2(CD)(DB)\gamma$$

Dividing through by $2(CB)^2$ and rearranging we obtain

$$\gamma = \frac{\varepsilon_n \sin^2\theta + \varepsilon_{n+\pi/2}\cos^2\theta - \varepsilon_y}{\sin\theta\cos\theta}$$

Substitution of ε_n and $\varepsilon_{n+\pi/2}$ from Eqs (14.17) and (14.16) yields

$$\frac{\gamma}{2} = \frac{(\varepsilon_x - \varepsilon_y)}{2}\sin 2\theta - \frac{\gamma_{xy}}{2}\cos 2\theta \quad (14.20)$$

14.7 Principal strains

From a comparison of Eqs (14.17) and (14.20) with Eqs (14.5) and (14.6) we observe that the former two equations may be obtained from Eqs (14.5) and (14.6) by replacing σ_n by ε_n, σ_x by ε_x, σ_y by ε_y, τ_{xy} by $\gamma_{xy}/2$ and τ by $\gamma/2$. It follows that for each deduction made from Eqs (14.5) and (14.6) concerning σ_n and τ there is a corresponding deduction from Eqs (14.17) and (14.20) regarding ε_n and $\gamma/2$. Thus at a point in a structural member there are two mutually perpendicular planes on which the shear strain γ is zero and normal to which the direct strain is the algebraic maximum or minimum direct strain at the point. These direct strains are the *principal strains* at the point and are given (from a comparison with Eqs (14.8) and (14.9)) by

$$\varepsilon_I = \frac{\varepsilon_x + \varepsilon_y}{2} + \frac{1}{2}\sqrt{(\varepsilon_x - \varepsilon_y)^2 + \gamma_{xy}^2} \quad (14.21)$$

and

$$\varepsilon_{II} = \frac{\varepsilon_x + \varepsilon_y}{2} - \frac{1}{2}\sqrt{(\varepsilon_x - \varepsilon_y)^2 + \gamma_{xy}^2} \quad (14.22)$$

Since the shear strain γ is zero on these planes it follows that the shear stress must also be zero and we deduce from Section 14.3 that the directions of the principal strains and principal stresses coincide. The related planes are then determined from Eq. (14.7) or from

$$\tan 2\theta = \frac{\gamma_{xy}}{\varepsilon_x - \varepsilon_y} \quad (14.23)$$

In addition the maximum shear strain at the point is given by

$$\left(\frac{\gamma}{2}\right)_{max} = \frac{1}{2}\sqrt{(\varepsilon_x - \varepsilon_y)^2 + \gamma_{xy}^2} \qquad (14.24)$$

or

$$\left(\frac{\gamma}{2}\right)_{max} = \frac{\varepsilon_1 - \varepsilon_{11}}{2} \qquad (14.25)$$

(cf. Eqs (14.11) and (14.12)).

14.8 Mohr's circle of strain

The argument of Section 14.7 may be applied to Mohr's circle of stress described in Section 14.4. A circle of strain, analogous to that shown in Fig. 14.11, may be drawn when σ_x, σ_y, etc., are replaced by ε_x, ε_y, etc., as specified in Section 14.7. The horizontal extremities of the circle represent the principal strains, the radius of the circle half the maximum shear strain, and so on.

Example 14.5   A structural member is loaded in such a way that at a particular point in the member a two-dimensional stress system exists consisting of $\sigma_x = +60 \text{ N/mm}^2$, $\sigma_y = -40 \text{ N/mm}^2$ and $\tau_{xy} = 50 \text{ N/mm}^2$.

(a) Calculate the direct strain in the x and y directions and the shear strain, γ_{xy}, at the point.

(b) Calculate the principal strains at the point and determine the position of the principal planes.

(c) Verify your answer using a graphical method. Take $E = 200\,000 \text{ N/mm}^2$ and Poisson's ratio, $\nu = 0.3$.

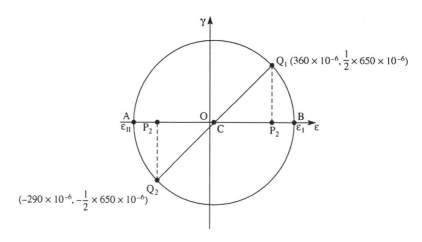

Fig. 14.15   Mohr's circle of strain for Ex. 14.5

(a) From Section 7.8

$$\varepsilon_x = \frac{1}{200\,000}\,(60 + 0\cdot3 \times 40) = 360 \times 10^{-6}$$

$$\varepsilon_y = \frac{1}{200\,000}\,(-40 - 0\cdot3 \times 60) = -290 \times 10^{-6}$$

The shear modulus, G, is obtained using Eq. (7.21); thus

$$G = \frac{E}{2(1+v)} = \frac{200\,000}{2(1+0\cdot3)} = 76\,923 \text{ N/mm}^2$$

Hence, from Eq. (7.9)

$$\gamma_{xy} = \frac{\tau_{xy}}{G} = \frac{50}{76\,923} = 650 \times 10^{-6}$$

(b) Now substituting in Eqs (14.21) and (14.22) for ε_x, ε_y and γ_{xy} we have

$$\varepsilon_1 = 10^{-6}\left[\frac{360-290}{2} + \frac{1}{2}\sqrt{(360+290)^2 + 650^2}\right]$$

which gives $\qquad\qquad \varepsilon_1 = 495 \times 10^{-6}$

Similarly $\qquad\qquad\qquad \varepsilon_{II} = -425 \times 10^{-6}$

From Eq. (14.23) we have

$$\tan 2\theta = \frac{650 \times 10^{-6}}{360 \times 10^{-6} + 290 \times 10^{-6}} = 1$$

Therefore $\qquad\qquad\qquad 2\theta = 45° \text{ or } 225°$

so that $\qquad\qquad\qquad \theta = 22\cdot5° \text{ or } 112\cdot5°$

(c) Axes $O\varepsilon$ and $O\gamma$ are set up and the points $Q_1(360 \times 10^{-6}, \frac{1}{2} \times 650 \times 10^{-6})$ and $Q_2(-290 \times 10^{-6}, -\frac{1}{2} \times 650 \times 10^{-6})$ located. The centre C of the circle is the intersection of Q_1Q_2 and the $O\varepsilon$ axis (Fig. 14.15). The circle is then drawn with radius equal to CQ_1 and the points $B(\varepsilon_1)$ and $A(\varepsilon_{II})$ located. Finally angle $Q_1CB = 2\theta$ and $Q_1CA = 2\theta + \pi$.

14.9 Experimental measurement of surface strains and stresses

Stresses at a point on the surface of a structural member may be determined by measuring the strains at the point, usually with electrical resistance strain gauges. These consist of a short length of fine wire sandwiched between two layers of impregnated paper, the whole being glued to the surface of the member. The resistance of the wire changes as the wire stretches or contracts so that as the surface of the member is strained the gauge indicates a change of resistance which is measurable on a Wheatstone bridge.

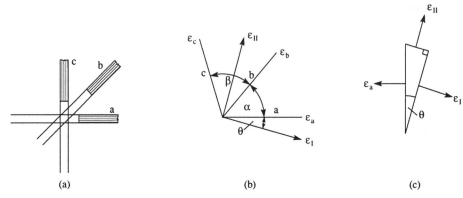

Fig. 14.16　Electrical resistance strain gauge measurement

Strain gauges measure direct strains only, but the state of stress at a point may be investigated in terms of principal stresses by using a strain gauge 'rosette'. This consists of three strain gauges inclined at a given angle to each other. Typical of these is the 45° or 'rectangular' strain gauge rosette illustrated in Fig. 14.16(a). An equiangular rosette has gauges inclined at 60°.

Suppose that a rosette consists of three arms, 'a', 'b' and 'c' inclined at angles α and β as shown in Fig. 14.16(b). Suppose also that ε_I and ε_{II} are the principal strains at the point and that ε_I is inclined at an unknown angle θ to the arm 'a'. Then if ε_a, ε_b and ε_c are the measured strains in the directions θ, $(\theta + \alpha)$ and $(\theta + \alpha + \beta)$ to ε_I we have, from Eq. (14.17)

$$\varepsilon_a = \varepsilon_I \cos^2 \theta + \varepsilon_{II} \sin^2 \theta \qquad (14.26)$$

in which ε_n has become ε_a, ε_x has become ε_I, ε_y has become ε_{II} and γ_{xy} is zero since the x and y directions have become principal directions. This situation is equivalent, as far as ε_a, ε_I and ε_{II} are concerned, to the strains acting on a triangular element as shown in Fig. 14.16(c). Rewriting Eq. (14.26) we have

$$\varepsilon_a = \frac{\varepsilon_I}{2}(1 + \cos 2\theta) + \frac{\varepsilon_{II}}{2}(1 - \cos 2\theta)$$

or

$$\varepsilon_a = \tfrac{1}{2}(\varepsilon_I + \varepsilon_{II}) + \tfrac{1}{2}(\varepsilon_I - \varepsilon_{II}) \cos 2\theta \qquad (14.27)$$

Similarly

$$\varepsilon_b = \tfrac{1}{2}(\varepsilon_I + \varepsilon_{II}) + \tfrac{1}{2}(\varepsilon_I - \varepsilon_{II}) \cos 2(\theta + \alpha) \qquad (14.28)$$

and

$$\varepsilon_c = \tfrac{1}{2}(\varepsilon_I + \varepsilon_{II}) + \tfrac{1}{2}(\varepsilon_I - \varepsilon_{II}) \cos 2(\theta + \alpha + \beta) \qquad (14.29)$$

Therefore if ε_a, ε_b and ε_c are measured in given directions, i.e. given angles α and β, then ε_I, ε_{II} and θ are the only unknowns in Eqs (14.27), (14.28) and (14.29).

Having determined the principal strains we obtain the principal stresses using relationships derived in Section 7.8. Thus

$$\varepsilon_I = \frac{1}{E}(\sigma_I - \nu \sigma_{II}) \qquad (14.30)$$

and

$$\varepsilon_{II} = \frac{1}{E}(\sigma_{II} - \nu \sigma_I) \qquad (14.31)$$

Solving Eqs (14.30) and (14.31) for σ_I and σ_{II} we have

$$\sigma_I = \frac{E}{1-v^2}(\varepsilon_I + v\varepsilon_{II}) \tag{14.32}$$

and

$$\sigma_{II} = \frac{E}{1-v^2}(\varepsilon_{II} + v\varepsilon_I) \tag{14.33}$$

For a 45° rosette $\alpha = \beta = 45°$ and the principal strains may be obtained using the geometry of Mohr's circle of strain. Suppose that the arm 'a' of the rosette is inclined at some unknown angle θ to the maximum principal strain as in Fig. 14.16(b). Then Mohr's circle of strain is as shown in Fig. 14.17; the shear strains γ_a, γ_b and γ_c do not feature in the discussion and are therefore ignored. From Fig. 14.17

$$OC = \tfrac{1}{2}(\varepsilon_a + \varepsilon_c)$$

$$CN = \varepsilon_a - OC = \tfrac{1}{2}(\varepsilon_a - \varepsilon_c)$$

$$QN = CM = \varepsilon_b - OC = \varepsilon_b - \tfrac{1}{2}(\varepsilon_a + \varepsilon_c)$$

The radius of the circle is CQ and

$$CQ = \sqrt{CN^2 + QN^2}$$

Hence

$$CQ = \sqrt{[\tfrac{1}{2}(\varepsilon_a - \varepsilon_c)]^2 + [\varepsilon_b - \tfrac{1}{2}(\varepsilon_a + \varepsilon_c)]^2}$$

which simplifies to

$$CQ = \frac{1}{\sqrt{2}}\sqrt{(\varepsilon_a - \varepsilon_b)^2 + (\varepsilon_c - \varepsilon_b)^2}$$

Therefore ε_I, which is given by

$$\varepsilon_I = OC + \text{radius of circle}$$

is

$$\varepsilon_I = \frac{1}{2}(\varepsilon_a + \varepsilon_c) + \frac{1}{\sqrt{2}}\sqrt{(\varepsilon_a - \varepsilon_b)^2 + (\varepsilon_c - \varepsilon_b)^2} \tag{14.34)}$$

Also

$$\varepsilon_{II} = OC - \text{radius of circle}$$

i.e.

$$\varepsilon_{II} = \frac{1}{2}(\varepsilon_a + \varepsilon_c) - \frac{1}{\sqrt{2}}\sqrt{(\varepsilon_a - \varepsilon_b)^2 + (\varepsilon_c - \varepsilon_b)^2} \tag{14.35}$$

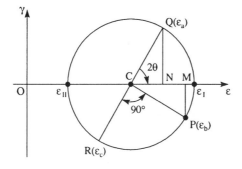

Fig. 14.17 Mohr's circle of strain for a 45° strain gauge rosette

Finally the angle θ is given by

$$\tan 2\theta = \frac{QN}{CN} = \frac{\varepsilon_b - \frac{1}{2}(\varepsilon_a + \varepsilon_c)}{\frac{1}{2}(\varepsilon_a - \varepsilon_c)}$$

i.e. $$\tan 2\theta = \frac{2\varepsilon_b - \varepsilon_a - \varepsilon_c}{\varepsilon_a - \varepsilon_c} \qquad (14.36)$$

A similar approach can be adopted for a 60° rosette.

Example 14.6   A shaft of solid circular cross-section has a diameter of 50 mm and is subjected to a torque, T, and axial load, P. A rectangular strain gauge rosette attached to the surface of the shaft recorded the following values of strain: $\varepsilon_a = 1000 \times 10^{-6}$, $\varepsilon_b = -200 \times 10^{-6}$ and $\varepsilon_c = -300 \times 10^{-6}$ where the gauges 'a' and 'c' are in line with and perpendicular to the axis of the shaft, respectively. If the material of the shaft has a Young's modulus of 70 000 N/mm² and a Poisson's ratio of 0·3, calculate the values of T and P.

Substituting the values of ε_a, ε_b and ε_c in Eq. (14.34) we have

$$\varepsilon_1 = \frac{10^{-6}}{2}(1000 - 300) + \frac{10^{-6}}{\sqrt{2}}\sqrt{(1000 + 200)^2 + (-200 + 300)^2}$$

which gives $$\varepsilon_1 = \frac{10^{-6}}{2}(700 + 1703) = 1202 \times 10^{-6}$$

It follows from Eq. (14.35) that

$$\varepsilon_{II} = \frac{10^{-6}}{2}(700 - 1703) = -502 \times 10^{-6}$$

Substituting for ε_1 and ε_{II} in Eq. (14.32) we have

$$\sigma_1 = \frac{70\,000 \times 10^{-6}}{1 - (0\cdot3)^2}(1202 - 0\cdot3 \times 502) = 80\cdot9\ \text{N/mm}^2$$

Similarly from Eq. (14.33)

$$\sigma_{II} = \frac{70\,000 \times 10^{-6}}{1 - (0\cdot3)^2}(-502 + 0\cdot3 \times 1202) = -10\cdot9\ \text{N/mm}^2$$

Since $\sigma_y = 0$ (note that the axial load produces σ_x only), Eqs (14.8) and (14.9) reduce to

$$\sigma_1 = \frac{\sigma_x}{2} + \frac{1}{2}\sqrt{\sigma_x^2 + 4\tau_{xy}^2} \qquad (i)$$

and $$\sigma_{II} = \frac{\sigma_x}{2} - \frac{1}{2}\sqrt{\sigma_x^2 + 4\tau_{xy}^2} \qquad (ii)$$

respectively. Adding Eqs (i) and (ii) we obtain

$$\sigma_{\mathrm{I}} + \sigma_{\mathrm{II}} = \sigma_x$$

Thus

$$\sigma_x = 80 \cdot 9 - 10 \cdot 9 = 70 \ \mathrm{N/mm^2}$$

Substituting for σ_x in either of Eqs (i) or (ii) gives

$$\tau_{xy} = 29 \cdot 7 \ \mathrm{N/mm^2}$$

For an axial load P

$$\sigma_x = 70 \ \mathrm{N/mm^2} = \frac{P}{A} = \frac{P}{(\pi/4) \times 50^2} \qquad \text{(Eq. (7.1))}$$

whence

$$P = 137 \cdot 4 \ \mathrm{kN}$$

Also for the torque T and using Eq. (11.4) we have

$$\tau_{xy} = 29 \cdot 7 \ \mathrm{N/mm^2} = \frac{Tr}{J} = \frac{T \times 25}{(\pi/32) \times 50^4}$$

which gives

$$T = 0 \cdot 7 \ \mathrm{kN\,m}$$

Note that P could have been found directly in this case from the axial strain ε_{a}. Thus from Eq. (7.8)

$$\sigma_x = E\varepsilon_{\mathrm{a}} = 70 \ 000 \times 1000 \times 10^{-6} = 70 \ \mathrm{N/mm^2}$$

as before.

14.10 Theories of elastic failure

The direct stress in a structural member subjected to simple tension or compression is directly proportional to strain up to the yield point of the material (Section 7.7). It is therefore a relatively simple matter to design such a member using the direct stress at yield as the design criterion. However, as we saw in Section 14.3, the direct and shear stresses at a point in a structural member subjected to a complex loading system are not necessarily the maximum values at the point. In such cases it is not clear how failure occurs, so that it is difficult to determine limiting values of load or alternatively to design a structural member for given loads. An obvious method, perhaps, would be to use direct experiment in which the structural member is loaded until deformations are no longer proportional to the applied load; clearly such an approach would be both time-wasting and uneconomical. Ideally a method is required that relates some parameter representing the applied stresses to, say, the yield stress in simple tension which is a constant for a given material.

In Section 14.3 we saw that a complex two-dimensional stress system comprising direct and shear stresses could be represented by a simpler system of direct stresses only, in other words, the principal stresses. The problem is therefore simplified to some extent since the applied loads are now being represented by a system of direct stresses only. Clearly this procedure could be extended to the three-dimensional case so that no matter how complex the loading and the resulting stress system, there would remain at the most just three principal stresses, σ_I, σ_{II} and σ_{III}, as shown, for a three-dimensional element, in Fig. 14.18.

It now remains to relate, in some manner, these principal stresses to the yield stress in simple tension, σ_Y, of the material.

Ductile materials

A number of theories of elastic failure have been proposed in the past for ductile materials but experience and experimental evidence have led to all but two being discarded.

Maximum shear stress theory

This theory is usually linked with the names of Tresca and Guest, although it is more widely associated with the former. The theory proposes that:

Failure (i.e. yielding) will occur when the maximum shear stress in the material is equal to the maximum shear stress at failure in simple tension.

For a two-dimensional stress system the maximum shear stress is given in terms of the principal stresses by Eq. (14.12). For a three-dimensional case the maximum shear stress is given by

$$\tau_{max} = \frac{\sigma_{max} - \sigma_{min}}{2} \tag{14.37}$$

where σ_{max} and σ_{min} are the algebraic maximum and minimum principal stresses. At failure in simple tension the yield stress σ_Y is in fact a principal stress and since there can be no direct stress perpendicular to the axis of loading, the maximum shear stress is, therefore, from either of Eqs (14.12) or (14.37),

$$\tau_{max} = \frac{\sigma_Y}{2} \tag{14.38}$$

Thus the theory proposes that failure in a complex system will occur when

$$\frac{\sigma_{max} - \sigma_{min}}{2} = \frac{\sigma_Y}{2}$$

or
$$\sigma_{max} - \sigma_{min} = \sigma_Y \tag{14.39}$$

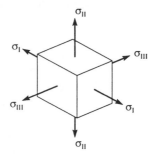

Fig. 14.18   Reduction of a complex three-dimensional stress system

Let us now examine stress systems having different relative values of σ_I, σ_{II} and σ_{III}. First suppose that $\sigma_I > \sigma_{II} > \sigma_{III} > 0$. From Eq. (14.39) failure occurs when

$$\sigma_I - \sigma_{III} = \sigma_Y \tag{14.40}$$

Second, suppose that $\sigma_I > \sigma_{II} > 0$ but $\sigma_{III} = 0$. In this case the three-dimensional stress system of Fig. 14.18 reduces to a two-dimensional stress system but *is still acting on a three-dimensional element*. Thus Eq. (14.39) becomes

$$\sigma_I - 0 = \sigma_Y$$

or

$$\sigma_I = \sigma_Y \tag{14.41}$$

Here we see an apparent contradiction of Eq. (14.12) where the maximum shear stress in a two-dimensional stress system is equal to half the difference of σ_I and σ_{II}. However, the maximum shear stress in that case occurs in the plane of the two-dimensional element, i.e. in the plane of σ_I and σ_{II}. In this case we have a three-dimensional element so that the maximum shear stress will lie in the plane of σ_I and σ_{III}.

Finally, let us suppose that $\sigma_I > 0$, $\sigma_{II} < 0$ and $\sigma_{III} = 0$. Again we have a two-dimensional stress system acting on a three-dimensional element but now σ_{II} is a compressive stress and algebraically less than σ_{III}. Thus Eq. (14.39) becomes

$$\sigma_I - \sigma_{II} = \sigma_Y \tag{14.42}$$

Shear strain energy theory

This particular theory of elastic failure was established independently by von Mises, Maxwell and Hencky but is now generally referred to as the von Mises criterion. The theory proposes that:

Failure will occur when the shear or distortion strain energy in the material reaches the equivalent value at yielding in simple tension.

In 1904 Huber proposed that the total strain energy, U_t, of an element of material could be regarded as comprising two separate parts: that due to change in volume and that due to change in shape. The former is termed the volumetric strain energy, U_v, the latter the distortion or shear strain energy, U_s. Thus

$$U_t = U_v + U_s \tag{14.43}$$

Since it is relatively simple to determine U_t and U_v, we obtain U_s by transposing Eq. (14.43). Hence

$$U_s = U_t - U_v \tag{14.44}$$

Initially, however, we shall demonstrate that the deformation of an element of material may be separated into change of volume and change in shape.

The principal stresses σ_I, σ_{II} and σ_{III} acting on the element of Fig. 14.18 may be written as

$$\sigma_I = \tfrac{1}{3}(\sigma_I + \sigma_{II} + \sigma_{III}) + \tfrac{1}{3}(2\sigma_I - \sigma_{II} - \sigma_{III})$$

$$\sigma_{II} = \tfrac{1}{3}(\sigma_I + \sigma_{II} + \sigma_{III}) + \tfrac{1}{3}(2\sigma_{II} - \sigma_I - \sigma_{III})$$

$$\sigma_{III} = \tfrac{1}{3}(\sigma_I + \sigma_{II} + \sigma_{III}) + \tfrac{1}{3}(2\sigma_{III} - \sigma_{II} - \sigma_I)$$

or

$$\left.\begin{array}{l} \sigma_1 = \bar{\sigma} + \sigma_1{}^1 \\[4pt] \sigma_{\mathrm{II}} = \bar{\sigma} + \sigma_{\mathrm{II}}{}^1 \\[4pt] \sigma_{\mathrm{III}} = \bar{\sigma} + \sigma_{\mathrm{III}}{}^1 \end{array}\right\} \tag{14.45}$$

Thus the stress system of Fig. 14.18 may be represented as the sum of two separate stress systems as shown in Fig. 14.19. The $\bar{\sigma}$ stress system is clearly equivalent to a hydrostatic or volumetric stress which will produce a change in volume but not a change in shape. The effect of the σ^1 stress system may be determined as follows. Adding together Eqs (14.45) we obtain

$$\sigma_1 + \sigma_{\mathrm{II}} + \sigma_{\mathrm{III}} = 3\bar{\sigma} + \sigma_1{}^1 + \sigma_{\mathrm{II}}{}^1 + \sigma_{\mathrm{III}}{}^1$$

but

$$\bar{\sigma} = \tfrac{1}{3}(\sigma_1 + \sigma_{\mathrm{II}} + \sigma_{\mathrm{III}})$$

so that

$$\sigma_1{}^1 + \sigma_{\mathrm{II}}{}^1 + \sigma_{\mathrm{III}}{}^1 = 0 \tag{14.46}$$

From the stress–strain relationships of Section 7.8 we have

$$\left.\begin{array}{l} \varepsilon_1{}^1 = \dfrac{\sigma_1{}^1}{E} - \dfrac{v}{E}(\sigma_{\mathrm{II}}{}^1 + \sigma_{\mathrm{III}}{}^1) \\[12pt] \varepsilon_{\mathrm{II}}{}^1 = \dfrac{\sigma_{\mathrm{II}}{}^1}{E} - \dfrac{v}{E}(\sigma_1{}^1 + \sigma_{\mathrm{III}}{}^1) \\[12pt] \varepsilon_{\mathrm{III}}{}^1 = \dfrac{\sigma_{\mathrm{III}}{}^1}{E} - \dfrac{v}{E}(\sigma_1{}^1 + \sigma_{\mathrm{II}}{}^1) \end{array}\right\} \tag{14.47}$$

The volumetric strain ε_{v} corresponding to $\sigma_1{}^1$, $\sigma_{\mathrm{II}}{}^1$ and $\sigma_{\mathrm{III}}{}^1$ is equal to the sum of the linear strains. Thus from Eqs (14.47)

$$\varepsilon_{\mathrm{v}} = \varepsilon_1{}^1 + \varepsilon_{\mathrm{II}}{}^1 + \varepsilon_{\mathrm{III}}{}^1 = \frac{(1 - 2v)}{E}(\sigma_1{}^1 + \sigma_{\mathrm{II}}{}^1 + \sigma_{\mathrm{III}}{}^1)$$

which, from Eq. (14.46), gives

$$\varepsilon_{\mathrm{v}} = 0$$

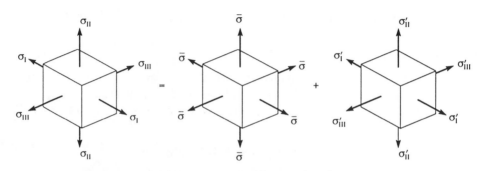

Fig. 14.19 Representation of principal stresses as volumetric and distortional stresses

It follows that σ_I', σ_{II}' and σ_{III}' produce no change in volume but only change in shape. We have therefore successfully divided the σ_I, σ_{II}, σ_{III} stress system into stresses ($\bar{\sigma}$) producing changes in volume and stresses (σ') producing changes in shape.

In Section 7.10 we derived an expression for the strain energy, U, of a member subjected to a direct stress, σ (Eq. (7.30)), i.e.

$$U = \frac{1}{2} \times \frac{\sigma^2}{E} \times \text{volume}$$

This equation may be rewritten

$$U = \tfrac{1}{2} \times \sigma \times \varepsilon \times \text{volume}$$

since $E = \sigma/\varepsilon$. The strain energy per unit volume is then $\sigma\varepsilon/2$. Thus for a three-dimensional element subjected to a stress $\bar{\sigma}$ on each of its six faces the strain energy in one direction is

$$\tfrac{1}{2}\bar{\sigma}\bar{\varepsilon}$$

where $\bar{\varepsilon}$ is the strain due to $\bar{\sigma}$ in each of the three directions. The total or volumetric strain energy per unit volume, U_v, of the element is then given by

$$U_v = 3(\tfrac{1}{2}\bar{\sigma}\bar{\varepsilon})$$

or, since

$$\bar{\varepsilon} = \frac{\bar{\sigma}}{E} - 2v\frac{\bar{\sigma}}{E} = \frac{\bar{\sigma}}{E}(1 - 2v)$$

$$U_v = \frac{1}{2}\bar{\sigma}\frac{3\bar{\sigma}}{E}(1 - 2v) \tag{14.48}$$

But

$$\bar{\sigma} = \tfrac{1}{3}(\sigma_I + \sigma_{II} + \sigma_{III})$$

so that Eq. (14.48) becomes

$$U_v = \frac{(1 - 2v)}{6E}(\sigma_I + \sigma_{II} + \sigma_{III})^2 \tag{14.49}$$

By a similar argument the total strain energy per unit volume, U_t, of an element subjected to stresses σ_I, σ_{II} and σ_{III} is

$$U_t = \tfrac{1}{2}\sigma_I\varepsilon_I + \tfrac{1}{2}\sigma_{II}\varepsilon_{II} + \tfrac{1}{2}\sigma_{III}\varepsilon_{III} \tag{14.50}$$

where

$$\varepsilon_I = \frac{\sigma_I}{E} - \frac{v}{E}(\sigma_{II} + \sigma_{III})$$

$$\varepsilon_{II} = \frac{\sigma_{II}}{E} - \frac{v}{E}(\sigma_I + \sigma_{III}) \tag{14.51}$$

and

$$\varepsilon_{III} = \frac{\sigma_{III}}{E} - \frac{v}{E}(\sigma_I + \sigma_{II})$$

(see Eqs (14.47))

Substituting for ε_1, etc. in Eq. (14.50) and then for U_v from Eq. (14.49) and U_t in Eq. (14.44) we have

$$U_s = \frac{1}{2E}$$

$$\times \left[\sigma_1^2 + \sigma_{11}^2 + \sigma_{111}^2 - 2v(\sigma_1\sigma_{11} + \sigma_{11}\sigma_{111} + \sigma_{111}\sigma_1) - \frac{(1-2v)}{6E}(\sigma_1 + \sigma_{11} + \sigma_{111})^2 \right]$$

which simplifies to

$$U_s = \frac{(1+v)}{6E}[(\sigma_1 - \sigma_{11})^2 + (\sigma_{11} - \sigma_{111})^2 + (\sigma_{111} - \sigma_1)^2]$$

per unit volume.

From Eq. (7.21) $E = 2G(1+v)$

Thus

$$U_s = \frac{1}{12G}[(\sigma_1 - \sigma_{11})^2 + (\sigma_{11} - \sigma_{111})^2 + (\sigma_{111} - \sigma_1)^2] \qquad (14.52)$$

The shear or distortion strain energy per unit volume at failure in simple tension corresponds to $\sigma_1 = \sigma_Y$, $\sigma_{11} = \sigma_{111} = 0$. Hence from Eq. (14.52)

$$U_s \text{ (at failure in simple tension)} = \frac{\sigma_Y^2}{6G} \qquad (14.53)$$

According to the von Mises criterion, failure occurs when U_s, given by Eq. (14.52), reaches the value of U_s, given by Eq. (14.53), i.e. when

$$(\sigma_1 - \sigma_{11})^2 + (\sigma_{11} - \sigma_{111})^2 + (\sigma_{111} - \sigma_1)^2 = 2\sigma_Y^2 \qquad (14.54)$$

For a two-dimensional stress system in which $\sigma_{111} = 0$, Eq. (14.54) becomes

$$\sigma_1^2 + \sigma_{11}^2 - \sigma_1\sigma_{11} = \sigma_Y^2 \qquad (14.55)$$

Design application

Codes of Practice for the use of structural steel in building use the von Mises criterion for a two-dimensional stress system (Eq. (14.55)) in determining an equivalent allowable stress for members subjected to bending and shear. Thus if σ_x and τ_{xy} are the direct and shear stresses, respectively, at a point in a member subjected to bending and shear, then the principal stresses at the point are, from Eqs (14.8) and (14.9)

$$\sigma_1 = \frac{\sigma_x}{2} + \frac{1}{2}\sqrt{\sigma_x^2 + 4\tau_{xy}^2}$$

$$\sigma_{11} = \frac{\sigma_x}{2} - \frac{1}{2}\sqrt{\sigma_x^2 + 4\tau_{xy}^2}$$

Substituting these expressions in Eq. (14.55) and simplifying we obtain

$$\sigma_Y = \sqrt{\sigma_x^2 + 3\tau_{xy}^2} \qquad (14.56)$$

In Codes of Practice σ_Y is termed an equivalent stress and allowable values are given for a series of different structural members.

Yield loci

Equations (14.39) and (14.54) may be plotted graphically for a two-dimensional stress system in which $\sigma_{III} = 0$ and in which it is assumed that the yield stress, σ_Y, is the same in tension and compression.

Figure 14.20 shows the yield locus for the maximum shear stress or Tresca theory of elastic failure. In the first and third quadrants, when σ_I and σ_{II} have the same sign, failure occurs when either $\sigma_I = \sigma_Y$ or $\sigma_{II} = \sigma_Y$ (see Eq. (14.41)) depending on which principal stress attains the value σ_Y first. For example, a structural member may be subjected to loads that produce a given value of σ_{II} ($<\sigma_Y$) and varying values of σ_I. If the loads were increased, failure would occur when σ_I reached the value σ_Y. Similarly for a fixed value of σ_I and varying σ_{II}. In the second and third quadrants where σ_I and σ_{II} have opposite signs, failure occurs when $\sigma_I - \sigma_{II} = \sigma_Y$ or $\sigma_{II} - \sigma_I = \sigma_Y$ (see Eq. (14.42)). Both these equations represent straight lines, each having a gradient of $45°$ and an intercept on the σ_{II} axis of σ_Y. Clearly all combinations of σ_I and σ_{II} that lie inside the locus will not cause failure, while all combinations of σ_I and σ_{II} on or outside the locus will. Thus the inside of the locus represents elastic conditions while the outside represents plastic conditions. Note that for the purposes of a yield locus, σ_I and σ_{II} are interchangeable.

The shear strain energy (von Mises) theory for a two-dimensional stress system is represented by Eq. (14.55). This equation may be shown to be that of an ellipse whose major and minor axes are inclined at $45°$ to the axes of σ_I and σ_{II} as shown in Fig. 14.21. It may also be shown that the ellipse passes through the six corners of the Tresca yield locus so that at these points the two theories give identical results. However, for other combinations of σ_I and σ_{II} the Tresca theory predicts failure

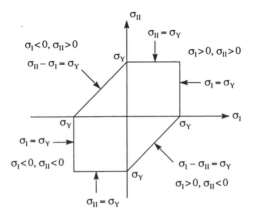

Fig. 14.20 Yield locus for the Tresca theory of elastic failure

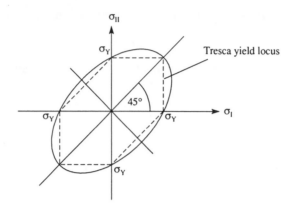

Fig. 14.21   Yield locus for the von Mises theory

where the von Mises theory does not so that the Tresca theory is the more conservative of the two.

The value of the yield loci lies in their use in experimental work on the validation of the different theories. Structural members fabricated from different materials may be subjected to a complete range of combinations of σ_I and σ_{II} each producing failure. The results are then plotted on the yield loci and the accuracy of each theory is determined for different materials.

Example 14.7   The state of stress at a point in a structural member is defined by a two-dimensional stress system as follows: $\sigma_x = +140$ N/mm², $\sigma_y = -70$ N/mm² and $\tau_{xy} = +60$ N/mm². If the material of the member has a yield stress in simple tension of 225 N/mm², determine whether or not yielding has occurred according to the Tresca and von Mises theories of elastic failure.

The first step is to determine the principal stresses σ_I and σ_{II}. From Eqs (14.8) and (14.9)

$$\sigma_I = \tfrac{1}{2}(140 - 70) + \tfrac{1}{2}\sqrt{(140 + 70)^2 + 4 \times 60^2}$$

i.e.
$$\sigma_I = 155 \cdot 9 \text{ N/mm}^2$$

and
$$\sigma_{II} = \tfrac{1}{2}(140 - 70) - \tfrac{1}{2}\sqrt{(140 + 70)^2 + 4 \times 60^2}$$

i.e.
$$\sigma_{II} = -85 \cdot 9 \text{ N/mm}^2$$

Since σ_{II} is algebraically less than $\sigma_{III}(=0)$, Eq. (14.42) applies.

Thus
$$\sigma_I - \sigma_{II} = 241 \cdot 8 \text{ N/mm}^2.$$

This value is greater than $\sigma_Y (=225$ N/mm²) so that according to the Tresca theory failure has, in fact, occurred.

Substituting the above values of σ_I and σ_{II} in Eq. (14.55) we have

$$(155 \cdot 9)^2 + (-85 \cdot 9)^2 - (155 \cdot 9)(-85 \cdot 9) = 45\ 075 \cdot 4$$

The square root of this expression is 212·3 N/mm² so that according to the von Mises theory the material has not failed.

Example 14.8 The rectangular cross-section of a thin-walled box girder (Fig. 14.22) is subjected to a bending moment of 250 kN m and a torque of 200 kN m. If the allowable equivalent stress for the material of the box girder is 180 N/mm², determine whether or not the design is satisfactory using the requirement of Eq. (14.56).

The maximum shear stress in the cross-section occurs in the vertical walls of the section and is given by Eq. (11.22), i.e.

$$\tau_{max} = \frac{T_{max}}{2At_{min}} = \frac{200 \times 10^6}{2 \times 500 \times 250 \times 10} = 80 \text{ N/mm}^2$$

The maximum stress due to bending occurs at the top and bottom of each vertical wall and is given by Eq. (9.9), i.e.

$$\sigma = \frac{My}{I}$$

where $I = 2 \times 12 \times 250 \times 250^2 + \dfrac{2 \times 10 \times 500^3}{12}$ (see Section 9.6)

i.e. $I = 583 \cdot 3 \times 10^6 \text{ mm}^4$

Thus $\sigma = \dfrac{250 \times 10^6 \times 250}{583 \cdot 3 \times 10^6} = 107 \cdot 1 \text{ N/mm}^2$

Substituting these values in Eq. (14.56) we have

$$\sqrt{\sigma_x^2 + 3\tau_{xy}^2} = \sqrt{107 \cdot 1^2 + 3 \times 80^2} = 175 \cdot 1 \text{ N/mm}^2$$

This equivalent stress is less than the allowable value of 180 N/mm² so that the box girder section is satisfactory.

Example 14.9 A beam of rectangular cross-section 60 mm × 100 mm is subjected to an axial tensile load of 60 000 N. If the material of the beam fails in simple

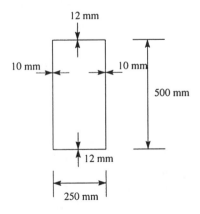

Fig. 14.22 Box girder beam section of Ex. 14.8

tension at a stress of 150 N/mm^2 determine the maximum shear force that can be applied to the beam section in a direction parallel to its longest side using the Tresca and von Mises theories of elastic failure.

The direct stress σ_x due to the axial load is uniform over the cross-section of the beam and is given by

$$\sigma_x = \frac{60\,000}{60 \times 100} = 10 \text{ N/mm}^2$$

The maximum shear stress τ_{max} occurs at the horizontal axis of symmetry of the beam section and is, from Eq. (10.7)

$$\tau_{max} = \frac{3}{2} \times \frac{S_y}{60 \times 100} \tag{i}$$

Thus from Eqs (14.8) and (14.9)

$$\sigma_1 = \frac{10}{2} + \frac{1}{2}\sqrt{10^2 + 4\tau_{max}^2}, \qquad \sigma_{II} = \frac{10}{2} - \frac{1}{2}\sqrt{10^2 + 4\tau_{max}^2}$$

or
$$\sigma_1 = 5 + \sqrt{25 + \tau_{max}^2}, \qquad \sigma_{II} = 5 - \sqrt{25 + \tau_{max}^2} \tag{ii}$$

It is clear from the second of Eqs (ii) that σ_{II} is negative since $|\sqrt{25 + \tau_{max}^2}| > 5$. Thus in the Tresca theory Eq. (14.42) applies and

$$\sigma_1 - \sigma_{II} = 2\sqrt{25 + \tau_{max}^2} = 150 \text{ N/mm}^2$$

from which
$$\tau_{max} = 74 \cdot 8 \text{ N/mm}^2$$
Thus from Eq. (i)

$$S_y = 299 \cdot 3 \text{ kN}$$

Now substituting for σ_1 and σ_{II} in Eq. (14.55) we have

$$(5 + \sqrt{25 + \tau_{max}^2})^2 + (5 - \sqrt{25 + \tau_{max}^2})^2$$
$$-(5 + \sqrt{25 + \tau_{max}^2})(5 - \sqrt{25 + \tau_{max}^2}) = 150^2$$

which gives
$$\tau_{max} = 86 \cdot 4 \text{ N/mm}^2$$

Again from Eq. (i)

$$S_y = 345 \cdot 6 \text{ kN}$$

Brittle materials

When subjected to tensile stresses brittle materials such as cast iron, concrete and ceramics fracture at a value of stress very close to the elastic limit with little or no permanent yielding on the planes of maximum shear stress. In fact the failure plane is generally flat and perpendicular to the axis of loading, unlike ductile materials which have failure planes inclined at approximately 45° to the axis of loading; in the latter case failure occurs on planes of maximum shear stress (see Sections 8.3 and

14.2). This would suggest, therefore, that shear stresses have no effect on the failure of brittle materials and that a direct relationship exists between the principal stresses at a point in a brittle material subjected to a complex loading system and the failure stress in simple tension or compression. This forms the basis for the most widely accepted theory of failure for brittle materials.

Maximum normal stress theory

This theory, frequently attributed to Rankine, states that:

Failure occurs when one of the principal stresses reaches the value of the yield stress in simple tension or compression.

For most brittle materials the yield stress in tension is very much less than the yield stress in compression, for example for concrete σ_Y (compression) is approximately $20\sigma_Y$ (tension). Thus it is essential in any particular problem to know which of the yield stresses is achieved first.

Suppose that a brittle material is subjected to a complex loading system which produces principal stresses σ_1, σ_{11} and σ_{111} as in Fig. 14.18. Thus for $\sigma_1 > \sigma_{11} > \sigma_{111} > 0$ failure occurs when

$$\sigma_1 = \sigma_Y \text{ (tension)} \tag{14.57}$$

Alternatively, for $\sigma_1 > \sigma_{11} > 0$, $\sigma_{111} < 0$ and $\sigma_1 < \sigma_Y$ (tension) failure occurs when

$$\sigma_{111} = \sigma_Y \text{ (compression)} \tag{14.58}$$

and so on.

A yield locus may be drawn for the two-dimensional case, as for the Tresca and von Mises theories of failure for ductile materials, and is shown in Fig. 14.23. Note that since the failure stress in tension, $\sigma_Y(T)$, is generally less than the failure stress in compression, $\sigma_Y(C)$, the yield locus is not symmetrically arranged about the σ_1 and σ_{11} axes. Again combinations of stress corresponding to points inside the locus will not cause failure, whereas combinations of σ_1 and σ_{11} on or outside the locus will.

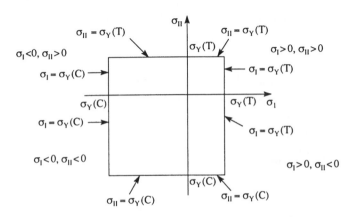

Fig. 14.23 Yield locus for a brittle material

Example 14.10   A concrete beam has a rectangular cross-section 250 mm × 500 mm and is simply supported over a span of 4 m. Determine the maximum mid-span concentrated load the beam can carry if the failure stress in simple tension of concrete is 1·5 N/mm². Neglect the self-weight of the beam.

If the central concentrated load is W N the maximum bending moment occurs at mid-span and is

$$\frac{4W}{4} = W \ \text{N m}   \text{(see Ex. 3.6)}$$

The maximum direct tensile stress due to bending occurs at the soffit of the beam and is

$$\sigma = \frac{W \times 10^3 \times 250 \times 12}{250 \times 500^3} = W \times 9\cdot6 \times 10^{-5} \ \text{N/mm}^2   \text{(Eq. 9.9)}$$

At this point the maximum principal stress is, from Eq. (14.8),

$$\sigma_1 = W \times 9\cdot6 \times 10^{-5} \ \text{N/mm}^2$$

Thus from Eq. (14.57) the maximum value of W is given by

$$\sigma_1 = W \times 9\cdot6 \times 10^{-5} = \sigma_Y(\text{tension}) = 1\cdot5 \ \text{N/mm}^2$$

from which $W = 15\cdot6$ kN.

The maximum shear stress occurs at the horizontal axis of symmetry of the beam section over each support and is, from Eq. (10.7),

$$\tau_{max} = \frac{3}{2} \times \frac{W/2}{250 \times 500}$$

i.e.

$$\tau_{max} = W \times 0\cdot6 \times 10^{-5} \ \text{N/mm}^2$$

Again, from Eq. (14.8), the maximum principal stress is

$$\sigma_1 = W \times 0\cdot6 \times 10^{-5} \ \text{N/mm}^2 = \sigma_Y(\text{tension}) = 1\cdot5 \ \text{N/mm}^2$$

from which $\qquad\qquad\qquad\qquad W = 250$ kN

Thus the maximum allowable value of W is 15·6 kN.

Problems

P.14.1   At a point in an elastic material there are two mutually perpendicular planes, one of which carries a direct tensile stress of 50 N/mm² and a shear stress of 40 N/mm² while the other plane is subjected to a direct compressive stress of 35 N/mm² and a complementary shear stress of 40 N/mm². Determine the principal stresses at the point, the position of the planes on which they act and the position of the planes on which there is no direct stress.

Ans. $\sigma_1 = 66$ N/mm², $\theta = 21°37'$; $\sigma_{11} = -51$ N/mm², $\theta = 111°37'$.

No direct stress on planes at 70°17′ and −26°48′ to the plane on which the 50 N/mm² stress acts.

P.14.2 One of the principal stresses in a two-dimensional stress system is 139 N/mm² acting on a plane A. On another plane B normal and shear stresses of 108 N/mm² and 62 N/mm², respectively, act. Determine

(a) the angle between the planes A and B,
(b) the other principal stress,
(c) the direct stress on the plane perpendicular to plane B.

Ans. (a) 26°34′, (b) −16 N/mm², (c) 15 N/mm².

P.14.3 The state of stress at a point in a structural member may be represented by a two-dimensional stress system in which $\sigma_x = 100$ N/mm², $\sigma_y = -80$ N/mm² and $\tau_{xy} = 45$ N/mm². Determine the direct stress on a plane inclined at 60° to the positive direction of σ_x and also the principal stresses. Calculate also the inclination of the principal planes to the plane on which σ_x acts. Verify your answers by a graphical method.

Ans. $\sigma_n = 94$ N/mm², $\sigma_1 = 110.5$ N/mm², $\sigma_{11} = -90.5$ N/mm², $\theta = 13°18′$ and $103°18′$.

P.14.4 Determine the normal and shear stress on the plane AB shown in Fig. P.14.4 when

(i) $\alpha = 60°$, $\sigma_x = 54$ N/mm², $\sigma_y = 30$ N/mm², $\tau_{xy} = 5$ N/mm²,
(ii) $\alpha = 120°$, $\sigma_x = -60$ N/mm², $\sigma_y = -36$ N/mm², $\tau_{xy} = 5$ N/mm².

Ans. (i) $\sigma_n = 43.7$ N/mm², $\tau = 12.9$ N/mm²,
　　 (ii) $\sigma_n = -49.7$ N/mm², $\tau = 12.9$ N/mm².

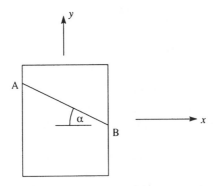

Fig. P.14.4

P.14.5 A shear stress τ_{xy} acts in a two-dimensional field in which the maximum allowable shear stress is denoted by τ_{max} and the major principal stress by σ_1. Derive, using the geometry of Mohr's circle of stress, expressions for the maximum values of direct stress which may be applied to the x and y planes in terms of the parameters given.

Ans. $\sigma_x = \sigma_1 - \tau_{max} + \sqrt{\tau_{max}^2 - \tau_{xy}^2}$, $\qquad \sigma_y = \sigma_1 - \tau_{max} - \sqrt{\tau_{max}^2 - \tau_{xy}^2}$.

P.14.6 In an experimental determination of principal stresses a cantilever of hollow circular cross-section is subjected to a varying bending moment and torque;

the internal and external diameters of the cantilever are 40 mm and 50 mm, respectively. For a given loading condition the bending moment and torque at a particular section of the cantilever are 100 Nm and 50 Nm, respectively. Calculate the maximum and minimum principal stresses at a point on the outer surface of the cantilever at this section where the direct stress produced by the bending moment is tensile. Determine also the maximum shear stress at the point and the inclination of the principal stresses to the axis of the cantilever.

The experimental values of principal stress are estimated from readings obtained from a 45° strain gauge rosette aligned so that one of its three arms is parallel to and another perpendicular to the axis of the cantilever. For the loading condition of zero torque and varying bending moment, comment on the ratio of these strain gauge readings.

Ans. $\sigma_1 = 14\cdot6$ N/mm^2, $\sigma_{II} = -0.8$ N/mm^2, $\tau_{max} = 7\cdot7$ N/mm^2,
$\theta = 12°53'$ and $102°53'$.

P.14.7   A thin-walled cylinder has an internal diameter of 1200 mm and has walls 1·2 mm thick. It is subjected to an internal pressure of 0·7 N/mm^2 and a torque, about its longitudinal axis, of 500 kN m. Determine the principal stresses at a point in the wall of the cylinder and also the maximum shear stress.

Ans.   466·5 N/mm^2, 58·5 N/mm^2, 204 N/mm^2.

P.14.8   A rectangular piece of material is subjected to tensile stresses of 83 N/mm^2 and 65 N/mm^2 on mutually perpendicular faces. Find the strain in the direction of each stress and in the direction perpendicular to both stresses. Determine also the maximum shear strain, the maximum shear stress and their directions. Take $E = 200\ 000$ N/mm^2 and $v = 0\cdot3$.

Ans. $3\cdot18 \times 10^{-4}$, $2\cdot01 \times 10^{-4}$, $-2\cdot22 \times 10^{-4}$, $\gamma_{max} = 1\cdot17 \times 10^{-4}$,
$\tau_{max} = 9\cdot0$ N/mm^2 at 45° to the directions of the given stresses.

P.14.9   A cantilever beam of length 2 m has a rectangular cross-section 100 mm wide and 200 mm deep. The beam is subjected to an axial tensile load, P, and a vertically downward uniformly distributed load of intensity w. A rectangular strain gauge rosette attached to a vertical side of the beam at the built-in end and in the neutral plane of the beam recorded the following values of strain: $\varepsilon_a = 1000 \times 10^{-6}$, $\varepsilon_b = 100 \times 10^{-6}$, $\varepsilon_c = -300 \times 10^{-6}$. The arm 'a' of the rosette is aligned with the longitudinal axis of the beam while the arm 'c' is perpendicular to the longitudinal axis.

Calculate the value of Poisson's ratio, the principal strains at the point and hence the values of P and w. Young's modulus, $E = 200\ 000$ N/mm^2.

Ans. $P = 4000$ kN, $w = 255\cdot3$ kN/m.

P.14.10   A beam has a rectangular thin-walled box section 50 mm wide by 100 mm deep and has walls 2 mm thick. At a particular section the beam carries a bending moment M and a torque T. A rectangular strain gauge rosette positioned on the top horizontal wall of the beam at this section recorded the following values of strain: $\varepsilon_a = 1000 \times 10^{-6}$, $\varepsilon_b = -200 \times 10^{-6}$, $\varepsilon_c = -300 \times 10^{-6}$. If the strain gauge 'a' is aligned with the longitudinal axis of the beam and the strain gauge 'c' is perpendicular to the longitudinal axis, calculate the values of M and T. Take $E = 200\ 000$ N/mm^2 and $v = 0\cdot3$.

Ans. $M = 3333$ N m, $T = 1692$ N m.

P.14.11 The simply supported beam shown in Fig. P.14.11 carries two symmetrically placed transverse loads, W. A rectangular strain gauge rosette positioned at the point P gave strain readings as follows: $\varepsilon_a = -222 \times 10^{-6}$, $\varepsilon_b = -213 \times 10^{-6}$, $\varepsilon_c = 45 \times 10^{-6}$. Also the direct stress at P due to an external axial compressive load is 7 N/mm^2. Calculate the magnitude of the transverse load. Take $E = 31\ 000$ N/mm^2, $v = 0\cdot2$.

Ans. $W = 98\cdot1$ kN.

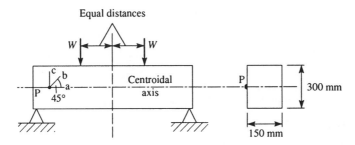

Fig. P.14.11

P.14.12 In a tensile test on a metal specimen having a cross-section 20 mm by 10 mm elastic breakdown occurred at a load of 70 000 N.

A thin plate made from the same material is to be subjected to loading such that at a certain point in the plate the stresses are $\sigma_y = -70$ N/mm^2, $\tau_{xy} = 60$ N/mm^2 and σ_x. Determine the maximum allowable values of σ_x using the Tresca and von Mises theories of elastic breakdown.

Ans. 259 N/mm^2 (Tresca),   294 N/mm^2 (von Mises).

P.14.13 A beam of circular cross-section is 3000 mm long and is attached at each end to supports which allow rotation of the ends of the beam in the longitudinal vertical plane of symmetry but prevent rotation of the ends in vertical planes perpendicular to the axis of the beam (Fig. P.14.13). The beam supports an offset load of 40 000 N at mid-span.

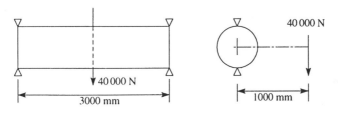

Fig. P.14.13

If the material of the beam suffers elastic breakdown in simple tension at a stress of 145 N/mm², calculate the minimum diameter of the beam on the basis of the Tresca and von Mises theories of elastic failure.

Ans.   136 mm (Tresca),   135 mm (von Mises).

P.14.14   A cantilever of circular cross-section has a diameter of 150 mm and is made from steel, which, when subjected to simple tension suffers elastic breakdown at a stress of 150 N/mm².

The cantilever supports a bending moment and a torque, the latter having a value numerically equal to twice that of the former. Calculate the maximum allowable values of the bending moment and torque on the basis of the Tresca and von Mises theories of elastic failure.

Ans. $M = 22 \cdot 3$ kNm, $T = 44 \cdot 4$ kNm (Tresca).

$M = 24 \cdot 9$ kNm, $T = 49 \cdot 8$ kNm (von Mises).

P.14.15   A certain material has a yield stress limit in simple tension of 387 N/mm². The yield limit in compression can be taken to be equal to that in tension. The material is subjected to three stresses in mutually perpendicular directions, the stresses being in the ratio $3 : 2 : -1 \cdot 8$. Determine the stresses that will cause failure according to the von Mises and Tresca theories of elastic failure.

Ans.   Tresca: $\sigma_I = 240$ N/mm², $\sigma_{II} = 160$ N/mm², $\sigma_{III} = -144$ N/mm².

von Mises: $\sigma_I = 263$ N/mm², $\sigma_{II} = 175$ N/mm², $\sigma_{III} = -158$ N/mm².

P.14.16   A column has the cross-section shown in Fig. P.14.16 and carries a compressive load P parallel to its longitudinal axis. If the failure stresses of the material of the column are 4 N/mm² and 22 N/mm² in simple tension and compression, respectively, determine the maximum allowable value of P using the maximum normal stress theory.

Ans.   640 kN.

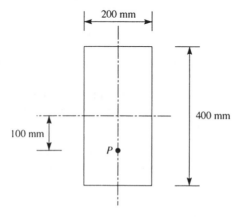

Fig. P.14.16

CHAPTER 15

Virtual Work and Energy Methods

The majority of the structural problems we have encountered so far have involved structures in which the support reactions and the internal force systems are statically determinate. Thus we have analysed beams, trusses, cables and three-pinned arches and, in the case of beams, calculated displacements. Some statically indeterminate structures have also been investigated. These include the simple truss and composite structural members in Section 7.14 and the circular section beams subjected to torsion and supported at each end in Section 11.1. These relatively simple problems were solved using a combination of statical equilibrium and compatibility of displacements. Further, in Section 13.7, a statically indeterminate propped cantilever was analysed using the principle of superposition (Section 3.7) while the support reactions for some cases of fixed beams were determined by combining the conditions of statical equilibrium with the moment-area method (Section 13.3). These methods are perfectly adequate for the comparatively simple problems to which they have been applied. However, other more powerful methods of analysis are required for more complex structures which may possess a high degree of statical indeterminacy. These methods will, in addition, be capable of providing rapid solutions for some statically determinate problems, particularly those involving the calculation of displacements.

The methods fall into two categories and are based on two important concepts; the first, *the principle of virtual work*, is the most fundamental and powerful tool available for the analysis of statically indeterminate structures and has the advantage of being able to deal with conditions other than those in the elastic range, while the second, based on *strain energy*, can provide approximate solutions of complex problems for which exact solutions may not exist. The two methods are, in fact, equivalent in some cases since, although the governing equations differ, the equations themselves are identical.

In modern structural analysis, computer-based techniques are widely used; these include the flexibility and stiffness methods. However, the formulation of, say, stiffness matrices for the elements of a complex structure is based on one of the above approaches, so that a knowledge and understanding of their application is advantageous. We shall briefly examine the flexibility and stiffness methods in Chapter 16 and their role in computer-based analysis.

Other specialist approaches have been developed for particular problems. Examples of these are the slope–deflection method for beams and the moment distribution method for beams and frames; these will also be described in Chapter 16 where we shall consider statically indeterminate structures. Initially, however, in this chapter, we shall examine the principle of virtual work, the different energy theorems and some of the applications of these two concepts.

15.1 Work

Before we consider the principle of virtual work in detail, it is important to clarify exactly what is meant by *work*. The basic definition of work in elementary mechanics is that 'work is done when a force moves its point of application'. However, we shall require a more exact definition since we shall be concerned with work done by both forces and moments and with the work done by a force when the body on which it acts is given a displacement which is not coincident with the line of action of the force.

Consider the force, F, acting on a particle, A, in Fig. 15.1(a). If the particle is given a displacement, Δ, by some external agency so that it moves to A′ in a direction at an angle α to the line of action of F, the work, W_F, done by F is given by

$$W_F = F(\Delta \cos \alpha) \tag{15.1}$$

or
$$W_F = (F \cos \alpha)\Delta \tag{15.2}$$

Thus we see that the work done by the force, F, as the particle moves from A to A′ may be regarded as either the product of F and the component of Δ in the direction of F (Eq. (15.1)) or as the product of the component of F in the direction of Δ and Δ (Eq. (15.2)).

Now consider the couple (pure moment) in Fig. 15.1(b) and suppose that the couple is given a small rotation of θ radians. The work done by each force F is then $F(a/2)\theta$ so that the total work done, W_C, by the couple is

$$W_C = F\,\frac{a}{2}\,\theta + F\,\frac{a}{2}\,\theta = Fa\theta$$

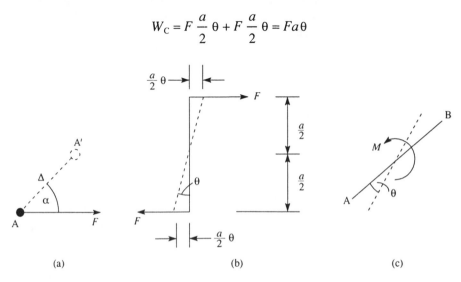

Fig. 15.1 Work done by a force and a moment

It follows that the work done, W_M, by the pure moment, M, acting on the bar AB in Fig. 15.1(c) as it is given a small rotation, θ, is

$$W_M = M\theta \tag{15.3}$$

Note that in the above the force, F, and moment, M, are in position before the displacements take place and are not the cause of them. Also, in Fig. 15.1(a), the component of Δ parallel to the direction of F is in the same direction as F; if it had been in the opposite direction the work done would have been negative. The same argument applies to the work done by the moment, M, where we see in Fig. 15.1(c) that the rotation, θ, is in the same sense as M. Note also that if the displacement, Δ, had been perpendicular to the force, F, no work would have been done by F.

Finally it should be remembered that work is a scalar quantity since it is not associated with direction (in Fig. 15.1(a) the force F does work if the particle is moved in any direction). Thus the work done by a series of forces is the algebraic sum of the work done by each force.

15.2 Principle of virtual work

The establishment of the principle will be carried out in stages. First we shall consider a particle, then a rigid body and finally a deformable body, which is the practical application we require when analysing structures.

Principle of virtual work for a particle

In Fig. 15.2 a particle, A, is acted upon by a number of concurrent forces, $F_1, F_2, \ldots, F_k, \ldots, F_r$; the resultant of these forces is R. Suppose that the particle is given a small arbitrary displacement, Δ_v, to A' in some specified direction; Δ_v is an imaginary or *virtual* displacement and is sufficiently small so that the directions of F_1, F_2, etc., are unchanged. Let θ_R be the angle that the resultant, R, of the forces makes with the direction of Δ_v and $\theta_1, \theta_2, \ldots, \theta_k, \ldots, \theta_r$ the angles that $F_1, F_2, \ldots, F_k, \ldots, F_r$ make with the direction of Δ_v, respectively. Then, from either

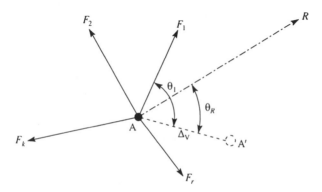

Fig. 15.2 Virtual work for a system of forces acting on a particle

of Eqs (15.1) or (15.2) the total virtual work, W_F, done by the forces F as the particle moves through the virtual displacement, Δ_v, is given by

$$W_F = F_1 \Delta_v \cos \theta_1 + F_2 \Delta_v \cos \theta_2 + \cdots + F_k \Delta_v \cos \theta_k + \cdots + F_r \Delta_v \cos \theta_r$$

Thus
$$W_F = \sum_{k=1}^{r} F_k \Delta_v \cos \theta_k$$

or, since Δ_v is a fixed, although imaginary displacement,

$$W_F = \Delta_v \sum_{k=1}^{r} F_k \cos \theta_k \tag{15.4}$$

In Eq. (15.4)

$$\sum_{k=1}^{r} F_k \cos \theta_k$$

is the sum of all the components of the forces, F, in the direction of Δ_v and therefore must be equal to the component of the resultant, R, of the forces, F, in the direction of Δ_v, i.e.

$$W_F = \Delta_v \sum_{k=1}^{r} F_k \cos \theta_k = \Delta_v R \cos \theta_R \tag{15.5}$$

If the particle, A, is in equilibrium under the action of the forces, $F_1, F_2, \ldots, F_k, \ldots, F_r$, the resultant, R, of the forces is zero (Chapter 2). It follows from Eq. (15.5) that the virtual work done by the forces, F, during the virtual displacement, Δ_v, is zero.

We can therefore state the *principle of virtual work* for a particle as follows:

If a particle is in equilibrium under the action of a number of forces the total work done by the forces for a small arbitrary displacement of the particle is zero.

It is possible for the total work done by the forces to be zero even though the particle is not in equilibrium if the virtual displacement is taken to be in a direction perpendicular to their resultant, R. We cannot, therefore, state the converse of the above principle unless we specify that the total work done must be zero for *any* arbitrary displacement. Thus:

A particle is in equilibrium under the action of a system of forces if the total work done by the forces is zero for any virtual displacement of the particle.

Note that in the above, Δ_v is a purely imaginary displacement and is not related in any way to the possible displacement of the particle under the action of the forces, F. Δ_v has been introduced purely as a device for setting up the work–equilibrium relationship of Eq. (15.5). The forces, F, therefore remain unchanged in magnitude and direction during this imaginary displacement; this would not be the case if the displacement were real.

Principle of virtual work for a rigid body

Consider the rigid body shown in Fig. 15.3, which is acted upon by a system of external forces, $F_1, F_2, \ldots, F_k, \ldots, F_r$. These external forces will induce internal forces in the body, which may be regarded as comprising an infinite number of particles; on adjacent particles, such as A_1 and A_2, these internal forces will be equal and opposite, in other words self-equilibrating. Suppose now that the rigid body is given a small, imaginary, that is virtual, displacement, Δ_v (or a rotation or a combination of both), in some specified direction. The external and internal forces then do virtual work and the total virtual work done, W_t, is the sum of the virtual work, W_e, done by the external forces and the virtual work, W_i, done by the internal forces. Thus

$$W_t = W_e + W_i \tag{15.6}$$

Since the body is rigid, all the particles in the body move through the same displacement, Δ_v, so that the virtual work done on all the particles is numerically the same. However, for a pair of adjacent particles, such as A_1 and A_2 in Fig. 15.3, the self-equilibrating forces are in opposite directions, which means that the work done on A_1 is opposite in sign to the work done on A_2. Thus the sum of the virtual work done on A_1 and A_2 is zero. The argument can be extended to the infinite number of pairs of particles in the body from which we conclude that the internal virtual work produced by a virtual displacement in a rigid body is zero. Equation (15.6) then reduces to

$$W_t = W_e \tag{15.7}$$

Since the body is rigid and the internal virtual work is therefore zero, we may regard the body as a large particle. It follows that if the body is in equilibrium under the action of a set of forces, $F_1, F_2, \ldots, F_k, \ldots, F_r$, the total virtual work done by the external forces during an arbitrary virtual displacement of the body is zero.

The principle of virtual work is, in fact, an alternative to Eqs (2.10) for specifying the necessary conditions for a system of coplanar forces to be in equilibrium. To illustrate the truth of this we shall consider the calculation of the support reactions in a simple beam.

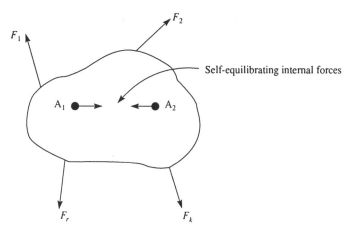

Fig. 15.3 Virtual work for a rigid body

Example 15.1   Calculate the support reactions in the simply supported beam shown in Fig. 15.4.

Only a vertical load is applied to the beam so that only vertical reactions, R_A and R_C, are produced.

Suppose that the beam at C is given a small imaginary, that is a virtual, displacement, $\Delta_{v,C}$, in the direction of R_C as shown in Fig. 15.4(b). Since we are concerned here solely with the *external* forces acting on the beam we may regard the beam as a rigid body. The beam therefore rotates about A so that C moves to C' and B moves to B'. From similar triangles we see that

$$\Delta_{v,B} = \frac{a}{a+b}\,\Delta_{v,C} = \frac{a}{L}\,\Delta_{v,C} \tag{i}$$

The total virtual work, W_t, done by all the forces acting on the beam is then given by

$$W_t = R_C\Delta_{v,C} - W\Delta_{v,B} \tag{ii}$$

Note that the work done by the load, W, is negative since $\Delta_{v,B}$ is in the opposite direction to its line of action. Note also that the support reaction, R_A, does no work since the beam only rotates about A. Now substituting for $\Delta_{v,B}$ in Eq. (ii) from Eq. (i) we have

$$W_t = R_C\,\Delta_{v,C} - W\,\frac{a}{L}\,\Delta_{v,C} \tag{iii}$$

Since the beam is in equilibrium, W_t is zero from the principal of virtual work. Hence, from Eq. (iii)

$$R_C\,\Delta_{v,C} - W\,\frac{a}{L}\,\Delta_{v,C} = 0$$

which gives

$$R_C = W\,\frac{a}{L}$$

which is the result that would have been obtained from a consideration of the moment equilibrium of the beam about A. R_A follows in a similar manner. Suppose now that instead of the single displacement $\Delta_{v,C}$ the complete beam is given a vertical virtual displacement, Δ_v, together with a virtual rotation, θ_v, about A as shown in Fig. 15.4(c). The total virtual work, W_t, done by the forces acting on the beam is now given by

$$W_t = R_A\,\Delta_v - W(\Delta_v + a\theta_v) + R_C(\Delta_v + L\theta_v) = 0 \tag{iv}$$

since the beam is in equilibrium. Rearranging Eq. (iv)

$$(R_A + R_C - W)\Delta_v + (R_CL - Wa)\theta_v = 0 \tag{v}$$

Equation (v) is valid for all values of Δ_v and θ_v so that

$$R_A + R_C - W = 0 \qquad \text{and} \qquad R_CL - Wa = 0$$

which are the equations of equilibrium we would have obtained by resolving forces vertically and taking moments about A.

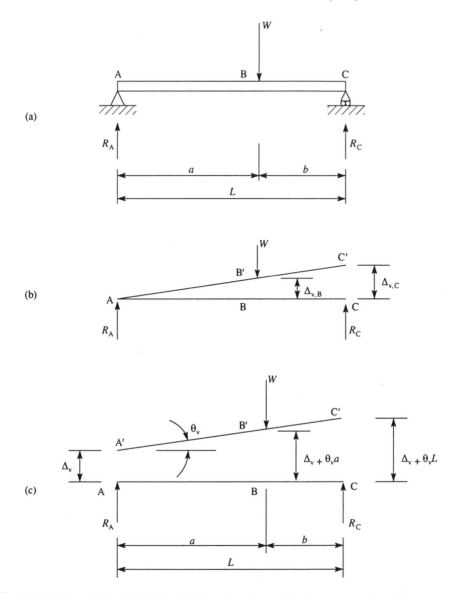

Fig. 15.4 Use of the principle of virtual work to calculate support reactions

It is not being suggested here that the application of Eqs (2.10) should be abandoned in favour of the principle of virtual work. The purpose of Ex. 15.1 is to illustrate the application of a virtual displacement and the manner in which the principle is used.

Virtual work in a deformable body

In structural analysis we are not generally concerned with forces acting on a rigid body. Structures and structural members deform under load, which means that if we assign a virtual displacement to a particular point in a structure, not all points in the structure will suffer the same virtual displacement as would be the case if the structure were rigid. Thus the virtual work produced by the internal forces is not zero as it is in the rigid body case, since the virtual work produced by the self-equilibrating forces on adjacent particles does not cancel out. The total virtual work produced by applying a virtual displacement to a deformable body acted upon by a system of external forces is therefore given by Eq. (15.6).

If the body is in equilibrium under the action of the external force system then every particle in the body is also in equilibrium. Thus, from the principle of virtual work, the virtual work done by the forces acting on the particle is zero irrespective of whether the forces are external or internal. Therefore, since the virtual work is zero for all particles in the body, it is zero for the complete body and Eq. (15.6) becomes

$$W_e + W_i = 0 \tag{15.8}$$

Note that in the above argument only the conditions of equilibrium and the concept of work are employed. Thus Eq. (15.8) does not require the deformable body to be linearly elastic (i.e. it need not obey Hooke's law) so that the principle of virtual work may be applied to any body or structure that is rigid, elastic or plastic. The principle does require that displacements, whether real or imaginary, must be small, so that we may assume that external and internal forces are unchanged in magnitude and direction during the displacements. In addition the virtual displacements must be compatible with the geometry of the structure and the constraints that are applied, such as those at a support. The exception is the situation we have in Ex. 15.1 where we apply a virtual displacement at a support. This approach is valid since we include the work done by the support reactions in the total virtual work equation.

Work done by internal force systems

The calculation of the work done by an external force is straightforward in that it is the product of the force and the displacement of its point of application in its own line of action (Eqs (15.1), (15.2) or (15.3)) whereas the calculation of the work done by an internal force system during a displacement is much more complicated. In Chapter 3 we saw that no matter how complex a loading system is, it may be simplified to a combination of up to four load types: axial load, shear force, bending moment and torsion; these in turn produce corresponding internal force systems. We shall now consider the work done by these internal force systems during arbitrary virtual displacements.

Axial force

Consider the elemental length, δz, of a structural member as shown in Fig. 15.5 and suppose that it is subjected to a positive internal force system comprising a normal force (i.e. axial force), N, a shear force, S, a bending moment, M, and a torque, T, produced by some external loading system acting on the structure of which the member is part. The stress distributions corresponding to these internal forces have been related in previous chapters to an axis system whose origin coincides with the

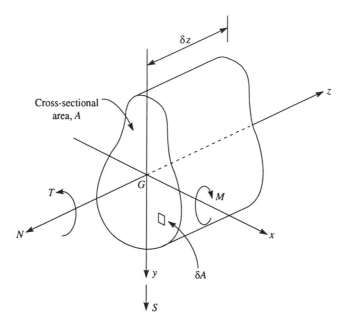

Fig. 15.5 Virtual work due to internal force system

centroid of area of the cross-section. We shall, in fact, be using these stress distributions in the derivation of expressions for internal virtual work in linearly elastic structures so that it is logical to assume the same origin of axes here; we shall also assume that the y axis is an axis of symmetry. Initially we shall consider the normal force, N.

The direct stress, σ, at any point in the cross-section of the member is given by $\sigma = N/A$ (Eq. (7.1)). Therefore the normal force on the element δA at the point (x, y) is

$$\delta N = \sigma \delta A = \frac{N}{A} \delta A$$

Suppose now that the structure is given an arbitrary virtual displacement which produces a virtual axial strain, ε_v, in the element. Thus the internal virtual work, $\delta w_{\mathrm{i},N}$, done by the axial force on the elemental length of the member is given by

$$\delta w_{\mathrm{i},N} = \int_A \frac{N}{A} \, dA \varepsilon_\mathrm{v} \, \delta z$$

which, since $\int_A dA = A$, reduces to

$$\delta w_{\mathrm{i},N} = N \varepsilon_\mathrm{v} \, \delta z \tag{15.10}$$

In other words, the virtual work done by N is the product of N and the virtual axial displacement of the element of the member. For a member of length L, the virtual work, $w_{\mathrm{i},N}$, done during the arbitrary virtual strain is then

$$w_{\mathrm{i},N} = \int_L N \varepsilon_\mathrm{v} \, dz \tag{15.11}$$

For a structure comprising a number of members, the total internal virtual work, $W_{i,N}$, done by axial force is the sum of the virtual work of each of the members. Thus

$$W_{i,N} = \sum \int_L N \varepsilon_v \, dz \tag{15.12}$$

Note that in the derivation of Eq. (15.12) we have made no assumption regarding the material properties of the structure so that the relationship holds for non-elastic as well as elastic materials. However, for a linearly elastic material, i.e. one that obeys Hooke's law (Section 7.7), we can express the virtual strain in terms of an equivalent virtual normal force. Thus

$$\varepsilon_v = \frac{\sigma_v}{E} = \frac{N_v}{EA}$$

Therefore, if we designate the *actual* normal force in a member by N_A, Eq. (15.12) may be expressed in the form

$$W_{i,N} = \sum \int_L \frac{N_A N_v}{EA} \, dz \tag{15.13}$$

Shear force

The shear force, S, acting on the member section in Fig. 15.5 produces a distribution of vertical shear stress which, as we saw in Section 10.2, depends upon the geometry of the cross-section. However, since the element, δA, is infinitesimally small, we may regard the shear stress, τ, as constant over the element. The shear force, δS, on the element is then

$$\delta S = \tau \, \delta A \tag{15.14}$$

Suppose that the structure is given an arbitrary virtual displacement which produces a virtual shear strain, γ_v, at the element. This shear strain represents the angular rotation in a vertical plane of the element $\delta A \times \delta z$ relative to the longitudinal centroidal axis of the member. The vertical displacement at the section being considered is therefore $\gamma_v \, \delta z$. The internal virtual work, $\delta w_{i,S}$, done by the shear force, S, on the elemental length of the member is given by

$$\delta w_{i,S} = \int_A \tau \, dA \gamma_v \, \delta z$$

We saw in Section 13.6 that we could assume a uniform shear stress through the cross-section of a beam if we allowed for the actual variation by including a form factor, β. Thus the expression for the internal virtual work in the member may be written

$$\delta w_{i,S} = \int_A \beta \left(\frac{S}{A} \right) dA \gamma_v \, \delta z$$

or

$$\delta w_{i,S} = \beta S \gamma_v \, \delta z \tag{15.15}$$

Hence the virtual work done by the shear force during the arbitrary virtual strain in a member of length L is

$$w_{i,S} = \beta \int_L S \gamma_v \, dz \tag{15.16}$$

For a linearly elastic member, as in the case of axial force, we may express the virtual shear strain, γ_v, in terms of an equivalent virtual shear force, S_v. Thus, from Section 7.7

$$\gamma_v = \frac{\tau_v}{G} = \frac{S_v}{GA}$$

so that from Eq. (15.16)

$$w_{i,S} = \beta \int_L \frac{S_A S_v}{GA} \, dz \tag{15.17}$$

For a structure comprising a number of linearly elastic members the total internal work, $W_{i,S}$, done by the shear forces is

$$W_{i,S} = \sum \beta \int_L \frac{S_A S_v}{GA} \, dz \tag{15.18}$$

Bending moment

The bending moment, M, acting on the member section in Fig. 15.5 produces a distribution of direct stress, σ, through the depth of the member cross-section. The normal force on the element, δA, corresponding to this stress is therefore $\sigma \, \delta A$. Again we shall suppose that the structure is given a small arbitrary virtual displacement which produces a virtual direct strain, ε_v, in the element $\delta A \times \delta z$. Thus the virtual work done by the normal force acting on the element δA is $\sigma \, \delta A \, \varepsilon_v \, \delta z$. Hence, integrating over the complete cross-section of the member we obtain the internal virtual work, $\delta w_{i,M}$, done by the bending moment, M, on the elemental length of member, i.e.

$$\delta w_{i,M} = \int_A \sigma \, dA \varepsilon_v \, \delta z \tag{15.19}$$

The virtual strain, ε_v, in the element $\delta A \times \delta z$ is, from Eq. (9.1), given by

$$\varepsilon_v = \frac{y}{R_v}$$

where R_v is the radius of curvature of the member produced by the virtual displacement. Thus, substituting for ε_v in Eq. (15.19), we obtain

$$\delta w_{i,M} = \int_A \sigma \frac{y}{R_v} \, dA \, \delta z$$

or, since $\sigma y \, \delta A$ is the moment of the normal force on the element, δA, about the x axis,

$$\delta w_{i,M} = \frac{M}{R_v} \, \delta z$$

Therefore, for a member of length L, the internal virtual work done by an actual bending moment, M_A, is given by

$$w_{i,M} = \int_L \frac{M_A}{R_v} \, dz \tag{15.20}$$

In the derivation of Eq. (15.20) no specific stress–strain relationship has been assumed, so that it is applicable to a non-linear system. For the particular case of a linearly elastic system, the virtual curvature $1/R_v$ may be expressed in terms of an equivalent virtual bending moment, M_v, using the relationship of Eq. (9.11). Thus

$$\frac{1}{R_v} = \frac{M_v}{EI}$$

Substituting for $1/R_v$ in Eq. (15.20) we have

$$w_{i,M} = \int_L \frac{M_A M_v}{EI} \, dz \tag{15.21}$$

so that for a structure comprising a number of members the total internal virtual work, $W_{i,M}$, produced by bending is

$$W_{i,M} = \sum \int_L \frac{M_A M_v}{EI} \, dz \tag{15.22}$$

In Chapter 9 we used the suffix 'x' to denote a bending moment in a vertical plane about the x axis (M_x) and the second moment of area of the member section about the x axis (I_x). Clearly the bending moments in Eq. (15.22) need not be restricted to those in a vertical plane; the suffixes are therefore omitted.

Torsion

The internal virtual work, $w_{i,T}$, due to torsion in the particular case of a linearly elastic circular section bar may be found in a similar manner and is given by

$$w_{i,T} = \int_L \frac{T_A T_v}{GI_o} \, dz \tag{15.23}$$

in which I_o is the polar second moment of area of the cross-section of the bar (see Section 11.1). For beams of non-circular cross-section, I_o is replaced by a torsion constant, J, which, for many practical beam sections is determined empirically (Section 11.5).

Hinges

In some cases it is convenient to impose a virtual rotation, θ_v, at some point in a structural member where, say, the actual bending moment is M_A. The internal virtual work done by M_A is then $M_A \theta_v$ (see Eq. (15.3)); physically this situation is equivalent to inserting a hinge at the point.

Sign of internal virtual work

So far we have derived expressions for internal work without considering whether it is positive or negative in relation to external virtual work.

Suppose that the structural member, AB, in Fig. 15.6(a) is, say, a member of a truss and that it is in equilibrium under the action of two externally applied axial tensile loads, *P*; clearly the internal axial, that is normal, force at any section of the member is *P*. Suppose now that the member is given a virtual extension, δ_v, such that B moves to B'. Then the virtual work done by the applied load, *P*, is positive since the displacement, δ_v, is in the same direction as its line of action. However, the virtual work done by the internal force, *N* (=*P*), is negative since the displacement of B is in the opposite direction to its line of action; in other words work is done *on* the member. Thus, from Eq. (15.8), we see that in this case

$$W_e = W_i \tag{15.24}$$

Equation (15.24) would apply if the virtual displacement had been a contraction and not an extension, in which case the signs of the external and internal virtual work in Eq. (15.8) would have been reversed. Clearly the above applies equally if *P* is a compressive load. The above arguments may be extended to structural members subjected to shear, bending and torsional loads, so that Eq. (15.24) is generally applicable.

Virtual work due to external force systems

So far in our discussion we have only considered the virtual work produced by externally applied concentrated loads. For completeness we must also consider the virtual work produced by moments, torques and distributed loads.

In Fig. 15.7 a structural member carries a distributed load, $w(z)$, and at a particular point a concentrated load, *W*, a moment, *M*, and a torque, *T*. Suppose that at the point a virtual displacement is imposed that has translational components, $\Delta_{v,y}$ and $\Delta_{v,z}$, parallel to the *y* and *z* axes, respectively, and rotational components, θ_v and ϕ_v, in the *yz* and *xy* planes, respectively.

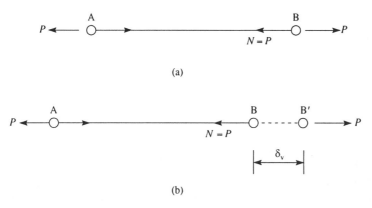

(a)

(b)

Fig. 15.6 Sign of the internal virtual work in an axially loaded member

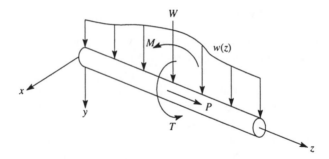

Fig. 15.7   Virtual work due to externally applied loads

If we consider a small element, δz, of the member at the point, the distributed load may be regarded as constant over the length δz and acting, in effect, as a concentrated load $w(z)\,\delta z$. Thus the virtual work, w_e, done by the complete external force system is given by

$$w_e = W\Delta_{v,y} + P\Delta_{v,z} + M\theta_v + T\phi_v + \int_L w(z)\Delta_{v,y}\,dz$$

For a structure comprising a number of load positions, the total external virtual work done is then

$$W_e = \sum \left[W\Delta_{v,y} + P\Delta_{v,z} + M\theta_v + T\phi_v + \int_L w(z)\Delta_{v,y}\,dz \right] \qquad (15.25)$$

In Eq. (15.25) there need not be a complete set of external loads applied at every loading point so, in fact, the summation is for the appropriate number of loads. Further, the virtual displacements in the above are related to forces and moments applied in a vertical plane. We could, of course, have forces and moments and components of the virtual displacement in a horizontal plane, in which case Eq. (15.25) would be extended to include their contribution.

The internal virtual work equivalent of Eq. (15.25) for a linear system is, from Eqs (15.13), (15.18), (15.22) and (15.23)

$$W_i = \sum \left[\int_L \frac{N_A N_v}{EA}\,dz + \beta \int_L \frac{S_A S_v}{GA}\,dz + \int_L \frac{M_A M_v}{EI}\,dz + \int_L \frac{T_A T_v}{GJ}\,dz + M_A\theta_v \right] \qquad (15.26)$$

in which the last term on the right-hand side is the virtual work produced by an actual internal moment at a hinge (see above). Note that the summation in Eq. (15.26) is taken over all the *members* of the structure.

Use of virtual force systems

So far, in all the structural systems we have considered, virtual work has been produced by actual forces moving through imposed virtual displacements. However, the actual forces are not related to the virtual displacements in any way since, as we have seen, the magnitudes and directions of the actual forces are unchanged by the virtual displacements so long as the displacements are small. Thus the principle of virtual work applies for *any* set of forces in equilibrium and *any* set of displacements. Equally, therefore, we could specify that the forces are a set of virtual forces *in equilibrium* and that the displacements are actual displacements. Thus, instead of relating actual external and internal force systems through virtual displacements, we can relate actual external and internal displacements through virtual forces.

If we apply a virtual force system to a deformable body it will induce an internal virtual force system which will move through the actual displacements; thus, internal virtual work will be produced. In this case, for example, Eq. (15.11) becomes

$$w_{i,N} = \int_L N_v \varepsilon_A \, dz$$

in which N_v is the internal virtual normal force and ε_A is the actual strain. Thus, for a linear system, in which the actual internal normal force is N_A, $\varepsilon_A = N_A/EA$, so that for a structure comprising a number of members the total internal virtual work due to a virtual normal force is

$$W_{i,N} = \sum \int_L \frac{N_v N_A}{EA} \, dz$$

which is identical to Eq. (15.13). Equations (15.18), (15.22) and (15.23) may be shown to apply to virtual force systems in a similar manner.

Applications of the principal of virtual work

We have now seen that the principle of virtual work may be used either in the form of imposed virtual displacements or in the form of imposed virtual forces. Generally the former approach, as we saw in Ex. 15.1, is used to determine forces, while the latter is used to obtain displacements.

For statically determinate structures the use of virtual displacements to determine force systems is a relatively trivial use of the principle although problems of this type provide a useful illustration of the method. The real power of this approach lies in its application to the solution of statically indeterminate structures, as we shall see in Chapter 16. However, the use of virtual forces is particularly useful in determining actual displacements of structures. We shall illustrate both approaches by examples.

Example 15.2 Determine the bending moment at the point B in the simply supported beam ABC shown in Fig. 15.8(a).

We determined the support reactions for this particular beam in Ex. 15.1. In this example, however, we are interested in the actual internal moment, M_B, at the point of application of the load. We must therefore impose a virtual displacement which will relate the internal moment at B to the applied load and which will exclude other unknown external forces such as the support reactions, and unknown internal force systems such as the bending moment distribution along the length of the beam. Thus, if we imagine that the beam is hinged at B and that the lengths AB and BC are rigid, a virtual displacement, $\Delta_{v,B}$, at B will result in the displaced shape shown in Fig. 15.8(b).

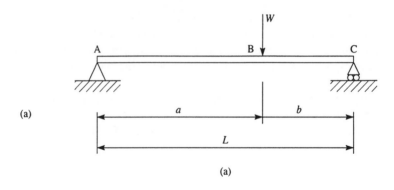

(a)

(a)

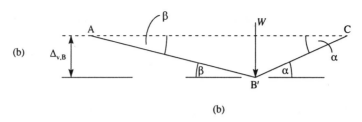

(b)

(b)

Fig. 15.8 Determination of bending moment at a point in the beam of Ex. 15.2 using virtual work

Note that the support reactions at A and C do no work and that the internal moments in AB and BC do no work because AB and BC are rigid links. From Fig. 15.8(b)

$$\Delta_{v,B} = a\beta = b\alpha \qquad (i)$$

Hence

$$\alpha = \frac{a}{b}\beta$$

and the angle of rotation of BC relative to AB is then

$$\theta_B = \beta + \alpha = \beta\left(1 + \frac{a}{b}\right) = \frac{L}{b}\beta \qquad (ii)$$

Now equating the external virtual work done by W to the internal virtual work done by M_B (see Eq. (15.24)) we have

$$W\,\Delta_{v,B} = M_B\theta_B \qquad (iii)$$

Substituting in Eq. (iii) for $\Delta_{v,B}$ from Eq. (i) and for θ_B from Eq. (ii) we have

$$Wa\beta = M_B\,\frac{L}{b}\,\beta$$

whence

$$M_B = \frac{Wab}{L}$$

which is the result we would have obtained by calculating the moment of R_C $(=Wa/L$ from Eq. 15.1) about B.

Example 15.3 Determine the force in the member AB in the truss shown in Fig. 15.9(a).

We are required to calculate the force in the member AB, so that again we need to relate this internal force to the externally applied loads without involving the internal forces in the remaining members of the truss. We therefore impose a virtual extension, $\Delta_{v,B}$, at B in the member AB, such that B moves to B'. If we assume that the remaining members are rigid, the forces in them will do no work. Further, the triangle BCD will rotate as a rigid body about D to B'C'D as shown in Fig. 15.9(b). The horizontal displacement of C, Δ_C, is then given by

$$\Delta_C = 4\alpha$$

while
$$\Delta_{v,B} = 3\alpha$$

Hence
$$\Delta_C = 4\,\Delta_{v,B}/3 \tag{i}$$

Thus, equating the external virtual work done by the 30 kN load to the internal virtual work done by the force, F_{BA}, in the member, AB, we have (see Eq. (15.24) and Fig. 15.6)

$$30\,\Delta_C = F_{BA}\Delta_{v,B} \tag{ii}$$

Substituting for Δ_C from Eq. (i) in Eq. (ii),

$$30 \times \tfrac{4}{3}\,\Delta_{v,B} = F_{BA}\Delta_{v,B}$$

Whence
$$F_{BA} = +40 \text{ kN (i.e. } F_{BA} \text{ is tensile)}$$

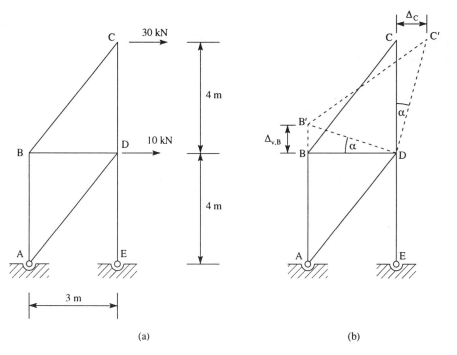

(a)

(b)

Fig. 15.9 Determination of the internal force in a member of a truss using virtual work

In the above we are, in effect, assigning a positive (i.e. tensile) sign to F_{BA} by imposing a virtual extension on the member AB.

The actual sign of F_{BA} is then governed by the sign of the external virtual work. Thus, if the 30 kN load had been in the opposite direction to Δ_C the external work done would have been negative, so that F_{BA} would be negative and therefore compressive. This situation can be verified by inspection. Alternatively, for the loading as shown in Fig. 15.9(a), a contraction in AB would have implied that F_{BA} was compressive. In this case DC would have rotated in an anticlockwise sense, Δ_C would have been in the opposite direction to the 30 kN load so that the external virtual work done would be negative, resulting in a negative value for the compressive force F_{BA}; F_{BA} would therefore be tensile as before. Note also that the 10 kN load at D does no work since D remains undisplaced.

We shall now consider problems involving the use of virtual forces. Generally we shall require the displacement of a particular point in a structure, so that if we apply a virtual force to the structure at the point and in the direction of the required displacement the external virtual work done will be the product of the virtual force and the actual displacement, which may then be equated to the internal virtual work produced by the internal virtual force system moving through actual displacements. Since the choice of the virtual force is arbitrary, we may give it any convenient value; the simplest type of virtual force is therefore a unit load and the method then becomes the *unit load method*.

Example 15.4   Determine the vertical deflection of the free end of the cantilever beam shown in Fig. 15.10(a).

Let us suppose that the actual deflection of the cantilever at B produced by the uniformly distributed load is v_B and that a vertically downward virtual unit load was applied at B before the actual deflection took place. The external virtual work done by

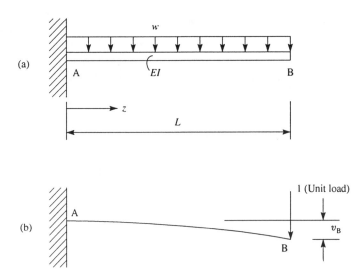

Fig. 15.10   Deflection of the free end of a cantilever beam using the unit load method

the unit load is, from Fig. 15.10(b), $1v_B$. The deflection, v_B, is assumed to be caused by bending only, i.e. we are ignoring any deflections due to shear. The internal virtual work is given by Eq. (15.22) which, since only one member is involved, becomes

$$W_{i,M} = \int_0^L \frac{M_A M_v}{EI} \, dz \tag{i}$$

The virtual moments, M_v, are produced by a unit load so that we shall replace M_v by M_1. Thus

$$W_{i,M} = \int_0^L \frac{M_A M_1}{EI} \, dz \tag{ii}$$

At any section of the beam a distance z from the built-in end

$$M_A = -\frac{w}{2}(L - z)^2, \quad M_1 = -1(L - z)$$

Substituting for M_A and M_1 in Eq. (ii) and equating the external virtual work done by the unit load to the internal virtual work we have

$$1v_B = \int_0^L \frac{w}{2EI}(L - z)^3 \, dz$$

which gives

$$v_B = -\frac{w}{2EI}\left[\frac{1}{4}(L - z)^4\right]_0^L$$

so that

$$v_B = \frac{wL^4}{8EI} \text{ (as in Ex. 13.2)}$$

Example 15.5 Determine the rotation, i.e. the slope, of the beam ABC shown in Fig. 15.11(a) at A.

The actual rotation of the beam at A produced by the actual concentrated load, W, is θ_A. Let us suppose that a virtual unit moment is applied at A before the actual rotation takes place, as shown in Fig. 15.11(b). The virtual unit moment induces virtual support reactions of $R_{v,A}$ ($=1/L$) acting downwards and $R_{v,C}$ ($=1/L$) acting upwards. The actual internal bending moments are

$$M_A = +\frac{W}{2}z \qquad 0 \leqslant z \leqslant L/2$$

$$M_A = +\frac{W}{2}(L - z) \qquad L/2 \leqslant z \leqslant L$$

The internal virtual bending moment is

$$M_v = 1 - \frac{1}{L}z \qquad 0 \leqslant z \leqslant L$$

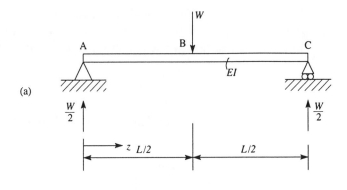

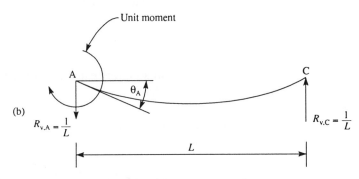

Fig. 15.11   Determination of the rotation of a simply supported beam at a support using the unit load method

The external virtual work done is $1\theta_A$ (the virtual support reactions do no work as there is no vertical displacement of the beam at the supports) and the internal virtual work done is given by Eq. (15.22). Hence

$$1\theta_A = \frac{1}{EI}\left[\int_0^{L/2} \frac{W}{2} z\left(1 - \frac{z}{L}\right)dz + \int_{L/2}^{L} \frac{W}{2}(L - z)\left(1 - \frac{z}{L}\right)dz\right] \qquad \text{(i)}$$

Simplifying Eq. (i) we have

$$\theta_A = \frac{W}{2EIL}\left[\int_0^{L/2}(Lz - z^2)\,dz + \int_{L/2}^{L}(L - z)^2\,dz\right] \qquad \text{(ii)}$$

Hence

$$\theta_A = \frac{W}{2EIL}\left\{\left[L\frac{z^2}{2} - \frac{z^3}{3}\right]_0^{L/2} - \frac{1}{3}\left[(L - z)^3\right]_{L/2}^{L}\right\}$$

from which

$$\theta_A = \frac{WL^2}{16EI}$$

which is the result that may be obtained from Eq. (iii) of Ex. 13.5.

Example 15.6 Calculate the vertical deflection of the joint B and the horizontal movement of the support D in the truss shown in Fig. 15.12(a). The cross-sectional area of each member is 1800 mm² and Young's modulus, E, for the material of the members is 200 000 N/mm².

The virtual force systems, i.e. unit loads, required to determine the vertical deflection of B and the horizontal deflection of D are shown in Fig. 15.12(b) and (c), respectively. Thus, if the actual vertical deflection at B is $\delta_{B,v}$ and the horizontal deflection at D is $\delta_{D,h}$ the external virtual work done by the unit loads is $1\delta_{B,v}$ and $1\delta_{D,h}$, respectively. The internal actual and virtual force systems comprise axial forces in all the members. These axial forces are constant along the length of each member so that for a truss comprising n members, Eq. (15.13) reduces to

$$W_{i,N} = \sum_{j=1}^{n} \frac{F_{A,j} F_{v,j} L_j}{E_j A_j} \tag{i}$$

in which $F_{A,j}$ and $F_{v,j}$ are the actual and virtual forces in the jth member which has a length L_j, an area of cross-section A_j and a Young's modulus E_j.

Since the forces $F_{v,j}$ are due to a unit load, we shall write Eq. (i) in the form

$$W_{i,N} = \sum_{j=1}^{n} \frac{F_{A,j} F_{1,j} L_j}{E_j A_j} \tag{ii}$$

Also, in this particular example, the area of cross-section, A, and Young's modulus, E, are the same for all members so that it is sufficient to calculate $\sum_{j=1}^{n} F_{A,j} F_{1,j} L_j$ and then divide by EA to obtain $W_{i,N}$.

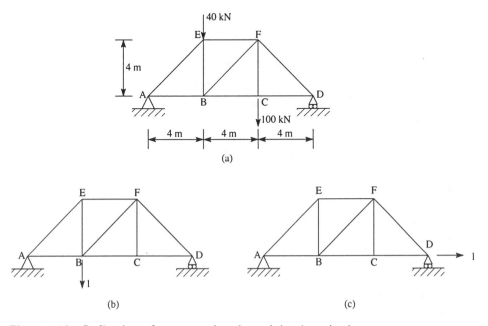

Fig. 15.12 Deflection of a truss using the unit load method

Table 15.1

Member	L(m)	F_A(kN)	$F_{1,B}$	$F_{1,D}$	$F_A F_{1,B} L$(kN m)	$F_A F_{1,D} L$(kN m)
AE	5·7	−84·9	−0·94	0	+451·4	0
AB	4·0	+60·0	+0·67	+1·0	+160·8	+240·0
EF	4·0	−60·0	−0·67	0	+160·8	0
EB	4·0	+20·0	+0·67	0	+53·6	0
BF	5·7	−28·3	+0·47	0	−75·2	0
BC	4·0	+80·0	+0·33	+1·0	+105·6	+320·0
CD	4·0	+80·0	+0·33	+1·0	+105·6	+320·0
CF	4·0	+100·0	0	0	0	0
DF	5·7	−113·1	−0·47	0	+301·0	0
					$\Sigma = +1263·6$	$\Sigma = +880·0$

The forces in the members, whether actual or virtual, may be calculated by the method of joints (Section 4.3). Note that the support reactions corresponding to the three sets of applied loads (one actual, two virtual) must be calculated before the internal force systems can be determined. However, in Fig. 15.12(c), it is clear from inspection that $F_{1,AB} = F_{1,BC} = F_{1,CD} = +1$ while the forces in all other members are zero. The calculations are presented in Table 15.1; note that positive signs indicate tension and negative signs compression.

Thus equating internal and external virtual work done (Eq. (15.24)) we have

$$1\delta_{B,v} = \frac{1263·6 \times 10^6}{200\,000 \times 1800}$$

whence

$$\delta_{B,v} = 3·51 \text{ mm}$$

and

$$1\delta_{D,h} = \frac{880 \times 10^6}{200\,000 \times 1800}$$

which gives

$$\delta_{D,h} = 2·44 \text{ mm}$$

Both deflections are positive which indicates that the deflections are in the directions of the applied unit loads. Note that in the above it is unnecessary to specify units for the unit load since the unit load appears, in effect, on both sides of the virtual work equation (the internal F_1 forces are directly proportional to the unit load).

Examples 15.2–15.6 illustrate the application of the principle of virtual work to the solution of problems involving statically determinate linearly elastic structures. We have also previously seen its application in the plastic bending of beams (Fig. 9.42), thereby demonstrating that the method is not restricted to elastic systems. We shall now examine the alternative energy methods but we shall return to the use of virtual work in Chapter 16 when we consider statically indeterminate structures.

15.3 Energy methods

Although it is generally accepted that energy methods are not as powerful as the principle of virtual work in that they are limited to elastic analysis, they possibly find their greatest use in providing rapid approximate solutions of problems for which exact solutions do not exist. Also, many statically indeterminate structures may be conveniently analysed using energy methods while, in addition, they are capable of providing comparatively simple solutions for deflection problems which are not readily solved by more elementary means.

Energy methods involve the use of either the *total complementary energy* or the *total potential energy* of a structural system. Either method may be employed to solve a particular problem, although as a general rule displacements are more easily found using complementary energy while forces are more easily found using potential energy.

Strain energy and complementary energy

In Section 7.10 we investigated strain energy in a linearly elastic member subjected to an axial load. Subsequently in Sections 9.4, 10.3 and 11.2 we derived expressions for the strain energy in a linearly elastic member subjected to bending, shear and

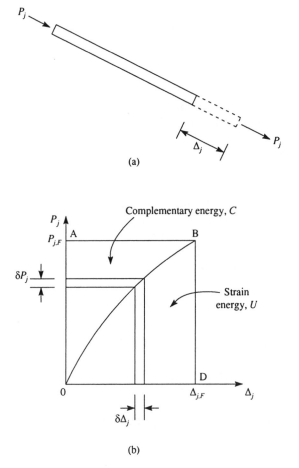

(a)

(b)

Fig. 15.13 Load–deflection curve for a non-linearly elastic member

torsional loads, respectively. We shall now examine the more general case of a member that is not linearly elastic.

Figure 15.13(a) shows the jth member of a structure comprising n members. The member is subjected to a gradually increasing load, P_j, which produces a gradually increasing displacement, Δ_j. If the member possesses non-linear elastic characteristics, the load–deflection curve will take the form shown in Fig. 15.13(b). Let us suppose that the final values of P_j and Δ_j are $P_{j,\mathrm{F}}$ and $\Delta_{j,\mathrm{F}}$.

As the member extends (or contracts if P_j is a compressive load) P_j does work which, as we saw in Section 7.10, is stored in the member as strain energy. The work done by P_j as the member extends by a small amount $\delta\Delta_j$ is given by

$$\delta W_j = P_j\, \delta\Delta_j$$

Therefore the total work done by P_j, and therefore the strain energy stored in the member, as P_j increases from zero to $P_{j,\mathrm{F}}$ is given by

$$u_j = \int_0^{\Delta_{j,\mathrm{F}}} P_j\, d\Delta_j \tag{15.27}$$

which is clearly the area OBD under the load–deflection curve in Fig. 15.13(b). Similarly the area OAB, which we shall denote by c_j, above the load–deflection curve is given by

$$c_j = \int_0^{P_{j,\mathrm{F}}} \Delta_j\, dP_j \tag{15.28}$$

It may be seen from Fig. 15.13(b) that the area OABD represents the work done by a constant force $P_{j,\mathrm{F}}$ moving through the displacement $\Delta_{j,\mathrm{F}}$. Thus from Eqs (15.27) and (15.28).

$$u_j + c_j = P_{j,\mathrm{F}}\, \Delta_{j,\mathrm{F}} \tag{15.29}$$

It follows that since u_j has the dimensions of work, c_j also has the dimensions of work but otherwise c_j has no physical meaning. It can, however, be regarded as the complement of the work done by P_j in producing the displacement Δ_j and is therefore called the *complementary energy*.

The total strain energy, U, of the structure is the sum of the individual strain energies of the members. Thus

$$U = \sum_{j=1}^{n} u_j$$

which becomes, when substituting for u_j from Eq. (15.27)

$$U = \sum_{j=1}^{n} \int_0^{\Delta_{j,\mathrm{F}}} P_j\, d\Delta_j \tag{15.30}$$

Similarly, the total complementary energy, C, of the structure is given by

$$C = \sum_{j=1}^{n} c_j$$

whence, from Eq. (15.28)

$$C = \sum_{j=1}^{n} \int_{0}^{P_{j,F}} \Delta_j \, dP_j \qquad (15.31)$$

Equation (15.30) may be written in expanded form as

$$U = \int_{0}^{\Delta_{1,F}} P_1 \, d\Delta_1 + \int_{0}^{\Delta_{2,F}} P_2 \, d\Delta_2 + \cdots + \int_{0}^{\Delta_{j,F}} P_j \, d\Delta_j + \cdots + \int_{0}^{\Delta_{n,F}} P_n \, d\Delta_n \qquad (15.32)$$

Partially differentiating Eq. (15.32) with respect to a particular displacement, say Δ_j, gives

$$\frac{\partial U}{\partial \Delta_j} = P_j \qquad (15.33)$$

Equation (15.33) states that the partial derivative of the strain energy in an elastic structure with respect to a displacement Δ_j is equal to the corresponding force P_j; clearly U must be expressed as a function of the displacements. This equation is generally known as *Castigliano's first theorem (Part I)* after the Italian engineer who derived and published it in 1879. One of its primary uses is in the analysis of non-linearly elastic structures, which is outside the scope of this book.

Now writing Eq. (15.31) in expanded form we have

$$C = \int_{0}^{P_{1,F}} \Delta_1 \, dP_1 + \int_{0}^{P_{2,F}} \Delta_2 \, dP_2 + \cdots + \int_{0}^{P_{j,F}} \Delta_j \, dP_j + \cdots + \int_{0}^{P_{n,F}} \Delta_n \, dP_n \qquad (15.34)$$

The partial derivative of Eq. (15.34) with respect to one of the loads, say P_j, is then

$$\frac{\partial C}{\partial P_j} = \Delta_j \qquad (15.35)$$

Equation (15.35) states that the partial derivative of the complementary energy of an elastic structure with respect to an applied load, P_j, gives the displacement of that load in its own line of action; C in this case is expressed as a function of the loads. Equation (15.35) is sometimes called the *Crotti–Engesser theorem* after the two engineers, one Italian, one German, who derived the relationship independently, Crotti in 1879 and Engesser in 1889.

Now consider the situation that arises when the load–deflection curve is linear, as shown in Fig. 15.14. In this case the areas OBD and OAB are equal so that the strain and complementary energies are equal. Thus we may replace the complementary energy, C, in Eq. (15.35) by the strain energy, U. Hence

$$\frac{\partial U}{\partial P_j} = \Delta_j \qquad (15.36)$$

Equation (15.36) states that, for a linearly elastic structure, the partial derivative of the strain energy of a structure with respect to a load gives the displacement of the load in its own line of action. This is generally know as *Castigliano's first theorem (Part II)*. Its direct use is limited in that it enables the displacement at a particular point in a structure to be determined *only* if there is a load applied at the point and

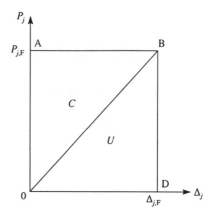

Fig. 15.14 Load–deflection curve for a linearly elastic member

only in the direction of the load. It could not therefore be used to solve for the required displacements at B and D in the truss in Ex. 15.6.

The principle of the stationary value of the total complementary energy

Suppose that an elastic structure comprising n members is in equilibrium under the action of a number of forces, $P_1, P_2, \ldots, P_k, \ldots, P_r$, which produce corresponding actual displacements, $\Delta_1, \Delta_2, \ldots, \Delta_k, \ldots, \Delta_r$, and actual internal forces, $F_1, F_2, \ldots, F_j, \ldots, F_n$. Now let us suppose that a system of elemental virtual forces, $\delta P_1, \delta P_2, \ldots, \delta P_k, \ldots, \delta P_r$, are imposed on the structure and act through the actual displacements. The external virtual work, δW_e, done by these elemental virtual forces is, from Section 15.2,

$$\delta W_e = \delta P_1 \Delta_1 + \delta P_2 \Delta_2 + \cdots + \delta P_k \Delta_k + \cdots + \delta P_r \Delta_r$$

or
$$\delta W_e = \sum_{k=1}^{r} \Delta_k \, \delta P_k \tag{15.37}$$

At the same time the elemental external virtual forces are in equilibrium with an elemental internal virtual force system, $\delta F_1, \delta F_2, \ldots, \delta F_j, \ldots, \delta F_n$, which moves through actual internal deformations, $\delta_1, \delta_2, \ldots, \delta_j, \ldots, \delta_n$. Hence the internal elemental virtual work done is

$$\delta W_i = \sum_{j=1}^{n} \delta_j \, \delta F_j \tag{15.38}$$

From Eq. (15.24)
$$\sum_{k=1}^{r} \Delta_k \, \delta P_k = \sum_{j=1}^{n} \delta_j \, \delta F_j$$

so that
$$\sum_{j=1}^{n} \delta_j \, \delta F_j - \sum_{k=1}^{r} \Delta_k \, \delta P_k = 0 \tag{15.39}$$

Equation (15.39) may be written

$$\delta\left(\sum_{j=1}^{n}\int_{0}^{F_j}\delta_j\,\mathrm{d}F_j-\sum_{k=1}^{r}\Delta_k P_k\right)=0 \qquad (15.40)$$

From Eq. (15.31) we see that the first term in Eq. (15.40) represents the complementary energy, C_i, of the actual internal force system, while the second term represents the complementary energy, C_e, of the external force system. C_i and C_e are opposite in sign since C_e is the complement of the work done *by* the external force system while C_i is the complement of the work done *on* the structure. Rewriting Eq. (15.40), we have

$$\delta(C_i+C_e)=0 \qquad (15.41)$$

In Eq. (15.40) the displacements, Δ_k, and the deformations, δ_j, are the actual displacements and deformations of the elastic structure. They therefore obey the condition of compatibility of displacement so that Eqs (15.41) and (15.40) are equations of geometrical compatibility. Also Eq. (15.41) establishes the *principle of the stationary value of the total complementary energy* which may be stated as:

For an elastic body in equilibrium under the action of applied forces the true internal forces (or stresses) and reactions are those for which the total complementary energy has a stationary value.

In other words the true internal forces (or stresses) and reactions are those that satisfy the condition of compatibility of displacement. This property of the total complementary energy of an elastic structure is particularly useful in the solution of statically indeterminate structures in which an infinite number of stress distributions and reactive forces may be found to satisfy the requirements of equilibrium so that, as we have already seen, equilibrium conditions are insufficient for a solution.

We shall examine the application of the principle in the solution of statically indeterminate structures in Chapter 16. Meanwhile we shall illustrate its application to the calculation of displacements in statically determinate structures.

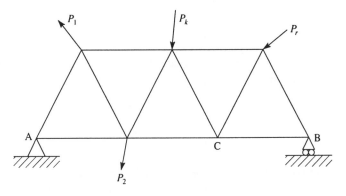

Fig. 15.15 Deflection of a truss using complementary energy

Example 15.7   The calculation of deflections in a truss.

Suppose that we wish to calculate the deflection, Δ_2, in the direction of the load, P_2, and at the joint at which P_2 is applied in a truss comprising n members and carrying a system of loads $P_1, P_2, \ldots, P_k, \ldots, P_r$, as shown in Fig. 15.15. From Eq. (15.40) the total complementary energy, C, of the truss is given by

$$C = \sum_{j=1}^{n} \int_{0}^{F_j} \delta_j \, \mathrm{d} F_j - \sum_{k=1}^{r} \Delta_k P_k \tag{i}$$

From the principle of the stationary value of the total complementary energy with respect to the load P_2, we have

$$\frac{\partial C}{\partial P_2} = \sum_{j=1}^{n} \delta_j \frac{\partial F_j}{\partial P_2} - \Delta_2 = 0 \tag{ii}$$

from which

$$\Delta_2 = \sum_{j=1}^{n} \delta_j \frac{\partial F_j}{\partial P_2} \tag{iii}$$

Note that the partial derivatives with respect to P_2 of the fixed loads, $P_1, P_3, \ldots, P_k, \ldots, P_r$, vanish.

To complete the solution we require the load–displacement characteristics of the structure. For a non-linear system in which, say,

$$F_j = b(\delta_j)^c$$

where b and c are known, Eq. (iii) becomes

$$\Delta_2 = \sum_{j=1}^{n} \left(\frac{F_j}{b} \right)^{1/c} \frac{\partial F_j}{\partial P_2} \tag{iv}$$

In Eq. (iv) F_j may be obtained from basic equilibrium conditions, e.g. the method of joints, and expressed in terms of P_2; hence $\partial F_j/\partial P_2$ is found. The actual value of P_2 is then substituted in the expression for F_j and the product $(F_j/b)^{1/c} \partial F_j/\partial P_2$ calculated for each member. Summation then gives Δ_2.

In the case of a linearly elastic structure δ_j is, from Sections 7.4 and 7.7, given by

$$\delta_j = \frac{F_j}{E_j A_j} L_j$$

in which E_j, A_j and L_j are Young's modulus, the area of cross-section and the length of the jth member. Substituting for δ_j in Eq. (iii) we obtain

$$\Delta_2 = \sum_{j=1}^{n} \frac{F_j L_j}{E_j A_j} \frac{\partial F_j}{\partial P_2} \tag{v}$$

Equation (v) could have been derived directly from Castigliano's first theorem (Part II) which is expressed in Eq. (15.36) since, for a linearly elastic system, the complementary and strain energies are identical; in this case the strain energy of the jth member is $F_j^2 L_j / 2 A_j E_j$ from Eq. (7.29). Other aspects of the solution merit discussion.

We note that the support reactions at A and B do not appear in Eq. (i). This convenient absence derives from the fact that the displacements, $\Delta_1, \Delta_2, ..., \Delta_k, ..., \Delta_r$, are the actual displacements of the truss and fulfil the conditions of geometrical compatibility and boundary restraint. The complementary energy of the reactions at A and B is therefore zero since both of their corresponding displacements are zero.

In Eq. (v) the term $\partial F_j/\partial P_2$ represents the rate of change of the actual forces in the members of the truss with P_2. This may be found, as described in the non-linear case, by calculating the forces, F_j, in the members in terms of P_2 and then differentiating these expressions with respect to P_2. Subsequently the actual value of P_2 would be substituted in the expressions for F_j and thus, using Eq. (v), Δ_2 obtained. This approach is rather clumsy. A simpler alternative would be to calculate the forces, F_j, in the members produced by the applied loads including P_2, then remove all the loads and apply P_2 only as an unknown force and recalculate the forces F_j as functions of P_2; $\partial F_j/\partial P_2$ is then obtained by differentiating these functions.

This procedure indicates a method for calculating the displacement of a point in the truss in a direction not coincident with the line of action of a load or, in fact, of a point such as C which carries no load at all. Initially the forces F_j in the members due to $P_1, P_2, ..., P_k, ..., P_r$ are calculated. These loads are then removed and a *dummy* or *fictitious* load, P_f, applied at the point and in the direction of the required displacement. A new set of forces, F_j, are calculated in terms of the dummy load, P_f, and thus $\partial F_j/\partial P_f$ is obtained. The required displacement, say Δ_C of C, is then given by

$$\Delta_C = \sum_{j=1}^{n} \frac{F_j L_j}{E_j A_j} \frac{\partial F_j}{\partial P_f} \tag{vi}$$

The simplification may be taken a stage further. The force F_j in a member due to the dummy load may be expressed, since the system is linearly elastic, in terms of the dummy load as

$$F_j = \frac{\partial F_j}{\partial P_f} P_f \tag{vii}$$

Suppose now that $P_f = 1$, i.e. a *unit load*. Equation (vii) then becomes

$$F_j = \frac{\partial F_j}{\partial P_f} 1$$

so that $\partial F_j/\partial P_f = F_{1,j}$, the load in the jth member due to a unit load applied at the point and in the direction of the required displacement. Thus, Eq. (vi) may be written

$$\Delta_C = \sum_{j=1}^{n} \frac{F_j F_{1,j} L_j}{E_j A_j} \tag{viii}$$

in which a unit load has been applied at C in the direction of the required displacement. Note that Eq. (viii) is identical in form to Eq. (ii) of Ex. 15.6.

In the above we have concentrated on members subjected to axial loads. The arguments apply in cases where structural members carry bending moments that produce rotations, shear loads that cause shear deflections and torques that produce angles of twist. We shall now demonstrate the application of the method to structures subjected to other than axial loads.

Example 15.8   Calculate the deflection, v_B, at the free end of the cantilever beam shown in Fig. 15.16(a).

We shall assume that deflections due to shear are negligible so that v_B is due entirely to bending action in the beam. In this case the total complementary energy of the beam is, from Eq. (15.40)

$$C = \int_0^L \int_0^M d\theta \, dM - W v_B \tag{i}$$

in which M is the bending moment acting on an element, δz, of the beam; δz subtends a small angle, $\delta\theta$, at the centre of curvature of the beam. The radius of curvature of the beam at the section is R as shown in Fig. 15.16(b) where, for clarity, we represent the beam by its neutral plane. From the principle of the stationary value of the total complementary energy of the beam

$$\frac{\partial C}{\partial W} = \int_0^L \frac{\partial M}{\partial W} d\theta - v_B = 0$$

whence

$$v_B = \int_0^L \frac{\partial M}{\partial W} d\theta \tag{ii}$$

In Eq. (ii)

$$\delta\theta = \frac{\delta z}{R}$$

and from Eq. (9.11)

$$\frac{1}{R} = \frac{M}{EI}$$

so that

$$\delta\theta = \frac{M}{EI} \delta z$$

Substituting in Eq. (ii) for $\delta\theta$ we have

$$v_B = \int_0^L \frac{M}{EI} \frac{\partial M}{\partial W} dz \tag{iii}$$

so that

$$\frac{\partial U}{\partial W} = \int_0^L \frac{M}{EI} \frac{\partial M}{\partial W} dz = v_B$$

From Fig. 15.16(a) we see that

$$M = -W(L - z)$$

Hence

$$\frac{\partial M}{\partial W} = -(L - z)$$

Note: Eq. (iii) could have been obtained directly from Eq. (9.21) by using Castigliano's first theorem (Part II).

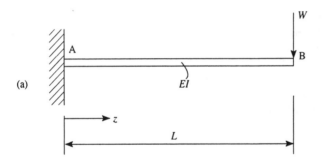

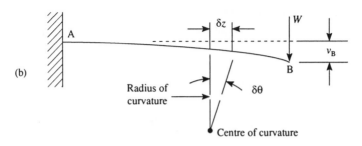

Fig. 15.16 Deflection of a cantilever beam using complementary energy

Equation (iii) then becomes

$$v_B = \int_0^L \frac{W}{EI} (L - z)^2 \, dz$$

whence

$$v_B = \frac{WL^3}{3EI} \quad \text{(as in Ex 13.1)}$$

Example 15.9 Determine the deflection, v_B, of the free end of a cantilever beam carrying a uniformly distributed load of intensity w. The beam is represented in Fig. 15.17 by its neutral plane; the flexural rigidity of the beam is EI.

For this example we use the dummy load method to determine v_B since we require the deflection at a point which does not coincide with the position of a concentrated

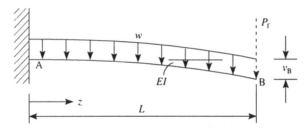

Fig. 15.17 Deflection of a cantilever beam using the dummy load method

load; thus we apply a dummy load, P_f, at B as shown. The total complementary energy, C, of the beam includes that produced by the uniformly distributed load; thus

$$C = \int_0^L \int_0^M d\theta \, dM - P_f v_B - \int_0^L vw \, dz \qquad \text{(i)}$$

in which v is the displacement of an elemental length, δz, of the beam at any distance z from the built-in end. Then

$$\frac{\partial C}{\partial P_f} = \int_0^L d\theta \, \frac{\partial M}{\partial P_f} - v_B = 0$$

so that

$$v_B = \int_0^L d\theta \, \frac{\partial M}{\partial P_f} \qquad \text{(ii)}$$

Note that in Eq. (i) v is an actual displacement and w an actual load, so that the last term disappears when C is partially differentiated with respect to P_f. As in Ex. 15.8

$$\delta\theta = \frac{M}{EI} \, \delta z$$

Also

$$M = -P_f(L - z) - \frac{w}{2}(L - z)^2$$

in which P_f is imaginary and therefore disappears when we substitute for M in Eq. (ii). Then

$$\frac{\partial M}{\partial P_f} = -(L - z)$$

so that

$$v_B = \int_0^L \frac{w}{2EI}(L - z)^3 \, dz$$

whence

$$v_B = \frac{wL^4}{8EI} \qquad \text{(see Ex. 13.2)}$$

For a linearly elastic system the bending moment, M_f, produced by a dummy load, P_f, may be written as

$$M_f = \frac{\partial M}{\partial P_f} P_f$$

If $P_f = 1$, i.e. a *unit load*

$$M_f = \frac{\partial M}{\partial P_f} 1$$

so that $\partial M/\partial P_f = M_1$, the bending moment due to a unit load applied at the point and in the direction of the required deflection. Thus we could write an equation for deflection, such as Eq. (ii), in the form

$$v = \int_0^L \frac{M_A M_1}{EI} \, dz \tag{iii}$$

in which M_A is the actual bending moment at any section of the beam and M_1 is the bending moment at any section of the beam due to a unit load applied at the point and in the direction of the required deflection. Thus, in this example

$$M_A = -\frac{w}{2}(L-z)^2, \qquad M_1 = -1(L-z)$$

so that

$$v_B = \int_0^L \frac{w}{2EI}(L-z)^3 \, dz$$

as before.

Temperature effects

The principle of the stationary value of the total complementary energy in conjunction with the unit load method may be used to determine the effect of a temperature gradient through the depth of a beam.

Normally, if a structural member is subjected to a uniform temperature rise, t, it will expand as shown in Fig. 15.18. However, a variation in temperature through the depth of the member such as the linear variation shown in Fig. 15.19(b) causes the

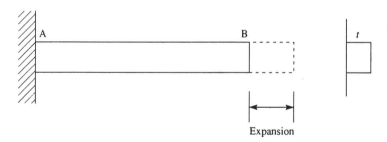

Expansion

Fig. 15.18 Expansion of a member due to a uniform temperature rise

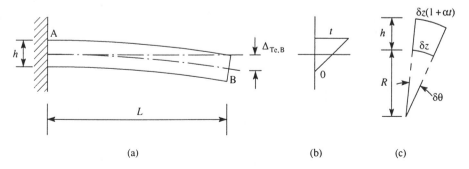

(a)   (b)   (c)

Fig. 15.19 Bending of a beam due to a linear temperature gradient

upper fibres to expand more than the lower ones so that bending strains, without bending stresses, are induced as shown in Fig. 15.19(a). Note that the undersurface of the member is unstrained since the change in temperature in this region is zero.

Consider an element, δz, of the member. The upper surface will increase in length to $\delta z(1 + \alpha t)$, while the length of the lower surface remains equal to δz as shown in Fig. 15.19(c); α is the coefficient of linear expansion of the material of the member. Thus, from Fig. 15.19(c),

$$\frac{R}{\delta z} = \frac{R+h}{\delta z(1 + \alpha t)}$$

so that

$$R = \frac{h}{\alpha t}$$

Also

$$\delta\theta = \frac{\delta z}{R}$$

whence

$$\delta\theta = \frac{\alpha t \, \delta z}{h} \tag{15.42}$$

If we require the deflection, $\Delta_{\text{Te,B}}$, of the free end of the member due to the temperature rise, we can employ the unit load method as in Ex. 15.9. Thus, by comparison with Eq. (ii) in Ex. 15.9.

$$\Delta_{\text{Te,B}} = \int_0^L d\theta \, \frac{\partial M}{\partial P_f} \tag{15.43}$$

in which, as we have seen, $\partial M/\partial P_f = M_1$, the bending moment at any section of the member produced by a unit load acting vertically downwards at B. Now substituting for $\delta\theta$ in Eq. (15.43) from Eq. (15.42)

$$\Delta_{\text{Te,B}} = -\int_0^L M_1 \frac{\alpha t}{h} \, dz \tag{15.44}$$

In the case of a beam carrying actual external loads the total deflection is, from the principle of superposition (Section 3.7), the sum of the bending, shear (unless neglected) and temperature deflections. Note that in Eq. (15.44) t can vary arbitrarily along the length of the beam but only linearly with depth. Note also that the temperature gradient shown in Fig. 15.19(b) produces a hogging deflected shape for the member. Thus, strictly speaking, the radius of curvature, R, in the derivation of Eq. (15.42) is negative (compare with Fig. 9.4) so that we must insert a minus sign in Eq. (15.44) as shown.

Example 15.10   Determine the deflection of the free end of the cantilever beam in Fig. 15.20 when subjected to the temperature gradients shown.

The temperature, t, at any section z of the beam is given by

$$t = \frac{z}{L} t_0$$

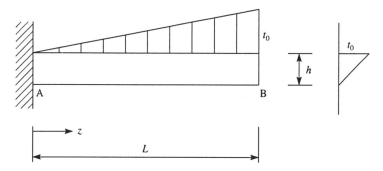

Fig. 15.20 Deflection of a cantilever beam having linear lengthwise and depthwise temperature gradients

Thus, substituting for t in Eq. (15.44), which applies since the variation of temperature through the depth of the beam is identical to that in Fig. 15.19(b), and noting that $M_1 = -1(L - z)$ we have

$$\Delta_{\text{Te, B}} = -\int_0^L [-1(L - z)] \frac{\alpha}{h} \frac{z}{L} t_0 \, dz$$

which simplifies to

$$\Delta_{\text{Te, B}} = \frac{\alpha t_0}{hL} \int_0^L (Lz - z^2) \, dz$$

whence

$$\Delta_{\text{Te, B}} = \frac{\alpha t_o L^2}{6h}$$

Potential energy

In the spring–mass system shown in its unstrained position in Fig. 15.21(a) the *potential energy* of the mass, m, is defined as the product of its weight and its height, h, above some arbitrary fixed datum. In other words, it possesses energy by virtue of its position. If the mass is allowed to deflect to the equilibrium position shown in Fig. 15.21(b) it has lost an amount of potential energy $mg \, \Delta_F$. Thus deflection is associated with a loss of potential energy or, alternatively, we could say that the loss of potential energy of the mass represents a *negative gain* in potential energy. Thus, if we define the potential energy of the mass as zero in its undeflected position in Fig. 15.21(a), which is the same as taking the position of the datum such that $h = 0$, its actual potential energy in its deflected state in Fig. 15.21(b) is $-mgh$. Thus, in the deflected state, the total energy of the spring–mass system is the sum of the potential energy of the mass ($-mgh$) and the strain energy of the spring.

Applying the above argument to the elastic member in Fig. 15.13(a) and defining the *total potential energy* (TPE) of the member as the sum of the strain energy, U, of the member and the potential energy, V, of the load, we have

$$\text{TPE} = U + V = \int_0^{\Delta_{j,F}} P_j \, d\Delta_j - P_{j,F} \Delta_{j,F} \quad \text{(see Eq. (15.25))} \qquad (15.45)$$

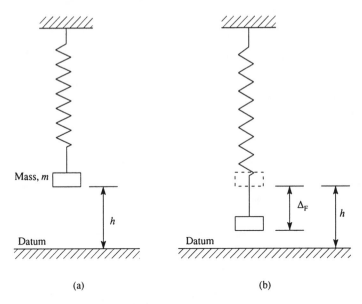

Fig. 15.21   Potential energy of a spring–mass system

Thus, for a structure comprising n members and subjected to a system of loads, $P_1, P_2, ..., P_k, ..., P_r$, the total potential energy is given by

$$\text{TPE} = U + V = \sum_{j=1}^{n} \int_{0}^{\Delta_{j,\text{F}}} P_j \, d\Delta_j - \sum_{k=1}^{r} P_k \Delta_k \tag{15.46}$$

in which P_j is the internal force in the jth member, $\Delta_{j,\text{F}}$ is its extension or contraction and Δ_k is the displacement of the load, P_k, in its line of action.

The principle of the stationary value of the total potential energy

Let us now consider an elastic body in equilibrium under a series of loads, $P_1, P_2, ..., P_k, ..., P_r$, and let us suppose that we impose infinitesimally small virtual displacements, $\delta\Delta_1, \delta\Delta_2, ..., \delta\Delta_k, ..., \delta\Delta_r$, at the points of application and in the directions of the loads. The virtual work done by the loads is then

$$\delta W_e = \sum_{k=1}^{r} P_k \, \delta\Delta_k \tag{15.47}$$

This virtual work will be accompanied by an increment of virtual strain energy, δU, or internal virtual work since, by imposing virtual displacements at the points of application of the loads we induce accompanying virtual strains in the body itself. Thus from the principle of virtual work (Eq. (15.24)) we have

$$\delta W_e = \delta U$$

or

$$\delta U - \delta W_e = 0$$

Substituting for δW_e from Eq. (15.47) we obtain

$$\delta U - \sum_{k=1}^{r} P_k \, \delta\Delta_k = 0 \qquad (15.48)$$

which may be written in the form

$$\delta \left(U - \sum_{k=1}^{r} P_k \Delta_k \right) = 0$$

in which we see that the second term is the potential energy, V, of the applied loads. Hence the equation becomes

$$\delta(U + V) = 0 \qquad (15.49)$$

and we see that the total potential energy of an elastic system has a stationary value for all small displacements if the system is in equilibrium.

It may also be shown that if the stationary value is a minimum, the equilibrium is stable. This may be demonstrated by examining the states of equilibrium of the particle at the positions A, B and C in Fig. 15.22. The total potential energy of the particle is proportional to its height, h, above some arbitrary datum, u; note that a single particle does not possess strain energy, so that in this case TPE $= V$. Clearly, at each position of the particle, the first-order variation, $\partial(U + V)/\partial u$, is zero (indicating equilibrium) but only at B, where the total potential energy is a minimum, is the equilibrium stable; at A the equilibrium is unstable while at C the equilibrium is neutral.

The *principle of the stationary value of the total potential energy* may therefore be stated as:

The total potential energy of an elastic system has a stationary value for all small displacements when the system is in equilibrium; further, the equilibrium is stable if the stationary value is a minimum.

Potential energy can often be used in the approximate analysis of structures in cases where an exact analysis does not exist. We shall illustrate such an application for a simple beam in Ex. 15.11 below and in Chapter 18 in the case of a buckled column;

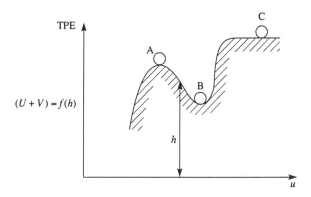

Fig. 15.22 States of equilibrium of a particle

in both cases we shall suppose that the deflected form is unknown and has to be initially assumed (this approach is called the *Rayleigh–Ritz method*). For a linearly elastic system, of course, the methods of complementary energy and potential energy are identical.

Example 15.11   Determine the deflection of the mid-span point of the linearly elastic, simply supported beam ABC shown in Fig. 15.23(a).

We shall suppose that the deflected shape of the beam is unknown. Initially, therefore, we shall assume a deflected shape that satisfies the boundary conditions for the beam. Generally, trigonometric or polynomial functions have been found to be the most convenient where the simpler the function the less accurate the solution. Let us suppose that the displaced shape of the beam is given by

$$v = v_B \sin \frac{\pi z}{L} \tag{i}$$

in which v_B is the deflection at the mid-span point. From Eq. (i) we see that when $z = 0$ and $z = L$, $v = 0$ and that when $z = L/2$, $v = v_B$. Furthermore, $dv/dz = (\pi/L)v_B \cos (\pi z/L)$ which is zero when $z = L/2$. Thus the displacement function satisfies the boundary conditions of the beam.

The strain energy due to bending of the beam is given by Eq. (9.21), i.e.

$$U = \int_0^L \frac{M^2}{2EI} \, dz \tag{ii}$$

Also, from Eq. (13.3)

$$M = -EI \frac{d^2 v}{dz^2} \tag{iii}$$

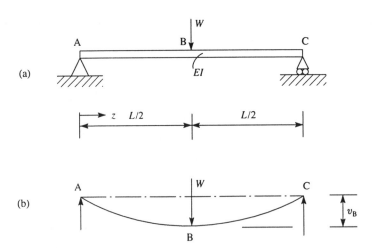

Fig. 15.23 Approximate value for beam deflection using total potential energy

Substituting in Eq. (iii) for v from Eq. (i), and for M in Eq. (ii) from Eq. (iii), we have

$$U = \frac{EI}{2} \int_0^L \frac{v_B^2 \pi^4}{L^4} \sin^2 \frac{\pi z}{L} \, dz$$

which gives

$$U = \frac{\pi^4 EI v_B^2}{4L^3}$$

The TPE of the beam is then given by

$$\text{TPE} = U + V = \frac{\pi^4 EI v_B^2}{4L^3} - W v_B$$

Hence, from the principle of the stationary value of the total potential energy

$$\frac{\partial(U+V)}{\partial v_B} = \frac{\pi^4 EI v_B}{2L^3} - W = 0$$

whence

$$v_B = \frac{2WL^3}{\pi^4 EI} = 0 \cdot 02053 \, \frac{WL^3}{EI} \tag{iv}$$

The exact expression for the deflection at the mid-span point was found in Ex. 13.5 and is

$$v_B = \frac{WL^3}{48EI} = 0 \cdot 02083 \, \frac{WL^3}{EI} \tag{v}$$

Comparing the exact and approximate results we see that the difference is less than two percent. Furthermore, the approximate deflection is less than the exact deflection because, by assuming a deflected shape, we have, in effect, forced the beam into that shape by imposing restraints; the beam is therefore stiffer.

15.4 Reciprocal theorems

There are two reciprocal theorems: one, attributed to Maxwell, is the *theorem of reciprocal displacements* (often referred to as Maxwell's reciprocal theorem) and the other, derived by Betti and Rayleigh, is the *theorem of reciprocal work*. We shall see, in fact, that the former is a special case of the latter. We shall also see that their proofs rely on the principle of superposition (Section 3.7) so that their application is limited to linearly elastic structures.

Theorem of reciprocal displacements

In a linearly elastic body a load, P_1, applied at a point 1 will produce a displacement, Δ_1, at the point and in its own line of action given by

$$\Delta_1 = a_{11} P_1$$

in which a_{11} is a *flexibility coefficient* which is defined as the displacement at the point 1 in the direction of P_1 produced by a unit load at the point 1 in the direction of P_1. It follows that if the elastic body is subjected to a series of loads, $P_1, P_2, ..., P_k, ..., P_r$, each of the loads will contribute to the displacement of point 1. Thus the corresponding displacement, Δ_1, at the point 1 (i.e. the total displacement in the direction of P_1 produced by all the loads) is then

$$\Delta_1 = a_{11}P_1 + a_{12}P_2 + \cdots + a_{1k}P_k + \cdots + a_{1r}P_r$$

in which a_{12} is the displacement at the point 1 in the direction of P_1 produced by a unit load at 2 in the direction of P_2, and so on. The corresponding displacements at the points of application of the loads are then

$$\left.\begin{aligned}
\Delta_1 &= a_{11}P_1 + a_{12}P_2 + \cdots + a_{1k}P_k + \cdots + a_{1r}P_r \\
\Delta_2 &= a_{21}P_1 + a_{22}P_2 + \cdots + a_{2k}P_k + \cdots + a_{2r}P_r \\
&\qquad\qquad\qquad \vdots \\
\Delta_k &= a_{k1}P_1 + a_{k2}P_2 + \cdots + a_{kk}P_k + \cdots + a_{kr}P_r \\
&\qquad\qquad\qquad \vdots \\
\Delta_r &= a_{r1}P_1 + a_{r2}P_2 + \cdots + a_{rk}P_k + \cdots + a_{rr}P_r
\end{aligned}\right\} \tag{15.50}$$

or, in matrix form

$$
\begin{Bmatrix} \Delta_1 \\ \Delta_2 \\ \vdots \\ \Delta_k \\ \vdots \\ \Delta_r \end{Bmatrix}
=
\begin{bmatrix}
a_{11} & a_{12} & \cdots & a_{1k} & \cdots & a_{1r} \\
a_{21} & a_{22} & \cdots & a_{2k} & \cdots & a_{2r} \\
& & \vdots & & & \\
a_{k1} & a_{k2} & \cdots & a_{kk} & \cdots & a_{kr} \\
& & \vdots & & & \\
a_{r1} & a_{r2} & \cdots & a_{rk} & \cdots & a_{rr}
\end{bmatrix}
\begin{Bmatrix} P_1 \\ P_2 \\ \vdots \\ P_k \\ \vdots \\ P_r \end{Bmatrix}
\tag{15.51}
$$

which may be written in matrix shorthand notation as

$$\{\Delta\} = [A]\{P\}$$

Suppose now that a linearly elastic body is subjected to a gradually applied load, P_1, at a point 1 and then, while P_1 remains in position, a load P_2 is gradually applied at another point 2. The total strain energy, U_1, of the body is equal to the external work done by the loads; thus

$$U_1 = \frac{P_1}{2}(a_{11}P_1) + \frac{P_2}{2}(a_{22}P_2) + P_1(a_{12}P_2) \tag{15.52}$$

The third term on the right-hand side of Eq. (15.52) results from the additional work done by P_1 as it is displaced through a further distance $a_{12}P_2$ by the action of P_2. If we now remove the loads and then apply P_2 followed by P_1, the strain energy, U_2, is given by

$$U_2 = \frac{P_2}{2}(a_{22}P_2) + \frac{P_1}{2}(a_{11}P_1) + P_2(a_{21}P_1) \tag{15.53}$$

By the principle of superposition the strain energy of the body is independent of the order in which the loads are applied. Hence

$$U_1 = U_2$$

so that

$$a_{12} = a_{21} \qquad\qquad (15.54)$$

Thus, in its simplest form, the theorem of reciprocal displacements states that:

The displacement at a point 1 in a given direction due to a unit load at a point 2 in a second direction is equal to the displacement at the point 2 in the second direction due to a unit load at the point 1 in the given direction.

The theorem of reciprocal displacements may also be expressed in terms of moments and rotations. Thus:

The rotation at a point 1 due to a unit moment at a point 2 is equal to the rotation at the point 2 produced by a unit moment at the point 1.

Finally we have:

The rotation in radians at a point 1 due to a unit load at a point 2 is numerically equal to the displacement at the point 2 in the direction of the unit load due to a unit moment at the point 1.

Example 15.12   A cantilever 800 mm long with a prop 500 mm from its built-in end deflects in accordance with the following observations when a concentrated load of 40 kN is applied at its free end:

Table 15.2

Distance from fixed end (mm)	0	100	200	300	400	500	600	700	800
Deflection (mm)	0	−0·3	−1·4	−2·5	−1·9	0	2·3	4·8	10·6

What will be the angular rotation of the beam at the prop due to a 30 kN load applied 200 mm from the built-in end together with a 10 kN load applied 350 mm from the built-in end?

The initial deflected shape of the cantilever is plotted to a suitable scale from the above observations and is shown in Fig. 15.24(a). Thus, from Fig. 15.24(a) we see that the deflection at D due to a 40 kN load at C is −1·4 mm. Hence the deflection at C due to a 40 kN load at D is, from the reciprocal theorem, −1·4 mm. It follows that the deflection at C due to a 30 kN load at D is equal to $(3/4) \times (-1 \cdot 4) = -1 \cdot 05$ mm. Again, from Fig. 15.24(a), the deflection at E due to a 40 kN load at C is −2·4 mm. Thus the deflection at C due to a 10 kN load at E is equal to $(1/4) \times (-2 \cdot 4) = -0 \cdot 6$ mm. Therefore the total deflection at C due to a 30 kN load at D and a 10 kN load at E is $-1 \cdot 05 - 0 \cdot 6 = -1 \cdot 65$ mm. From Fig. 15.24(b) we see that the rotation of the beam at B is given by

$$\theta_B = \tan^{-1}\frac{1 \cdot 65}{300} = \tan^{-1} 0 \cdot 0055$$

or

$$\theta_B = 0°19'$$

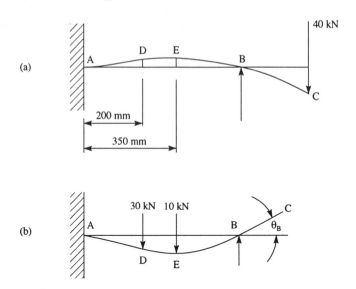

Fig. 15.24   Deflection of a propped cantilever using the reciprocal theorem

Example 15.13   An elastic member is pinned to a drawing board at its ends A and B. When a moment, M, is applied at A, A rotates by θ_A, B rotates by θ_B and the centre deflects by δ_1. The same moment, M, applied at B rotates B by θ_C and deflects the centre through δ_2. Find the moment induced at A when a load, W, is applied to the centre in the direction of the measured deflections and A and B are restrained against rotation.

The three load conditions and the relevant displacements are shown in Fig. 15.25. Thus, from Figs 15.25(a) and (b) the rotation at A due to M at B is, from the reciprocal theorem, equal to the rotation at B due to M at A.

Thus
$$\theta_{A(b)} = \theta_B$$

It follows that the rotation at A due to M_B at B is

$$\theta_{A(c),1} = \frac{M_B}{M}\,\theta_B \qquad\qquad (i)$$

where (b) and (c) refer to (b) and (c) in Fig. 15.25.

Also, the rotation at A due to a unit load at C is equal to the deflection at C due to a unit moment at A. Therefore

$$\frac{\theta_{A(c),2}}{W} = \frac{\delta_1}{M}$$

or
$$\theta_{A(c),2} = \frac{W}{M}\,\delta_1 \qquad\qquad (ii)$$

in which $\theta_{A(c),2}$ is the rotation at A due to W at C. Finally the rotation at A due to M_A

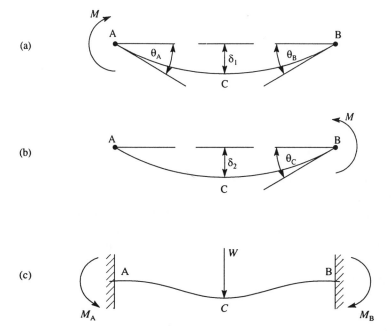

Fig. 15.25 Model analysis of a fixed beam

at A is, from Figs 15.25(a) and (c)

$$\theta_{A(c),3} = \frac{M_A}{M}\,\theta_A \tag{iii}$$

The total rotation at A produced by M_A at A, W at C and M_B at B is, from Eqs (i), (ii) and (iii)

$$\theta_{A(c),1} + \theta_{A(c),2} + \theta_{A(c),3} = \frac{M_B}{M}\,\theta_B + \frac{W}{M}\,\delta_1 + \frac{M_A}{M}\,\theta_A = 0 \tag{iv}$$

since the end A is restrained against rotation. In a similar manner the rotation at B is given by

$$\frac{M_B}{M}\,\theta_C + \frac{W}{M}\,\delta_2 + \frac{M_A}{M}\,\theta_B = 0 \tag{v}$$

Solving Eqs (iv) and (v) for M_A gives

$$M_A = W\left(\frac{\delta_2\theta_B - \delta_1\theta_C}{\theta_A\theta_C - \theta_B^2}\right)$$

The fact that the arbitrary moment, M, does not appear in the expression for the restraining moment at A (similarly it does not appear in M_B) produced by the load W indicates an extremely useful application of the reciprocal theorem, namely the model analysis of statically indeterminate structures. For example, the fixed beam of

Fig. 15.25(c) could possibly be a full-scale bridge girder. It is then only necessary to construct a model, say, of perspex, having the same flexural rigidity, *EI*, as the full-scale beam and measure rotations and displacements produced by an arbitrary moment, *M*, to obtain the fixed-end moments in the full-scale beam supporting a full-scale load.

Theorem of reciprocal work

Let us suppose that a linearly elastic body is to be subjected to two systems of loads, $P_1, P_2, \ldots, P_k, \ldots, P_r$, and, $Q_1, Q_2, \ldots, Q_i, \ldots, Q_m$, which may be applied simultaneously or separately. Let us also suppose that corresponding displacements are $\Delta_{P,1}, \Delta_{P,2}, \ldots, \Delta_{P,k}, \ldots, \Delta_{P,r}$ due to the loading system, P, and $\Delta_{Q,1}, \Delta_{Q,2}, \ldots, \Delta_{Q,i}, \ldots, \Delta_{Q,m}$ due to the loading system, Q. Finally, let us suppose that the loads, P, produce displacements $\Delta'_{Q,1}, \Delta'_{Q,2}, \ldots, \Delta'_{Q,i}, \ldots, \Delta'_{Q,m}$ at the points of application and in the direction of the loads, Q, while the loads, Q, produce displacements $\Delta'_{P,1}, \Delta'_{P,2}, \ldots, \Delta'_{P,k}, \ldots, \Delta'_{P,r}$ at the points of application and in the directions of the loads, P.

Now suppose that the loads P and Q are applied to the elastic body gradually and simultaneously. The total work done, and hence the strain energy stored, is then given by

$$U_1 = \tfrac{1}{2}P_1(\Delta_{P,1} + \Delta'_{P,1}) + \tfrac{1}{2}P_2(\Delta_{P,2} + \Delta'_{P,2}) + \cdots + \tfrac{1}{2}P_k(\Delta_{P,k} + \Delta'_{P,k})$$

$$+ \cdots + \tfrac{1}{2}P_r(\Delta_{P,r} + \Delta'_{P,r}) + \tfrac{1}{2}Q_1(\Delta_{Q,1} + \Delta'_{Q,1}) + \tfrac{1}{2}Q_2(\Delta_{Q,2} + \Delta'_{Q,2})$$

$$+ \cdots + \tfrac{1}{2}Q_i(\Delta_{Q,i} + \Delta'_{Q,i}) + \cdots + \tfrac{1}{2}Q_m(\Delta_{Q,m} + \Delta'_{Q,m}) \qquad (15.55)$$

If now we apply the P loading system followed by the Q loading system, the total strain energy stored is

$$U_2 = \tfrac{1}{2}P_1\Delta_{P,1} + \tfrac{1}{2}P_2\Delta_{P,2} + \cdots + \tfrac{1}{2}P_k\Delta_{P,k} + \cdots + \tfrac{1}{2}P_r\Delta_{P,r} + \tfrac{1}{2}Q_1\Delta_{Q,1} + \tfrac{1}{2}Q_2\Delta_{Q,2}$$

$$+ \cdots + \tfrac{1}{2}Q_i\Delta_{Q,i} + \cdots + \tfrac{1}{2}Q_m\Delta_{Q,m} + P_1\Delta'_{P,1} + P_2\Delta'_{P,2} + P_k\Delta'_{P,k} + \cdots + P_r\Delta'_{P,r} \quad (15.56)$$

Since, by the principle of superposition, the total strain energies, U_1 and U_2, must be the same, we have from Eqs (15.55) and (15.56)

$$-\tfrac{1}{2}P_1\Delta'_{P,1} - \tfrac{1}{2}P_2\Delta'_{P,2} - \cdots - \tfrac{1}{2}P_k\Delta'_{P,k} - \cdots - \tfrac{1}{2}P_r\Delta'_{P,r}$$

$$= -\tfrac{1}{2}Q_1\Delta'_{Q,1} - \tfrac{1}{2}Q_2\Delta'_{Q,2} - \cdots - \tfrac{1}{2}Q_i\Delta'_{Q,i} - \cdots - \tfrac{1}{2}Q_m\Delta'_{Q,m}$$

In other words

$$\sum_{k=1}^{r} P_k\Delta'_{P,k} = \sum_{i=1}^{m} Q_m\Delta'_{Q,m} \qquad (15.57)$$

The expression on the left-hand side of Eq. (15.57) is the sum of the products of the P loads and their corresponding displacements produced by the Q loads. The right-hand side of Eq. (15.57) is the sum of the products of the Q loads and their corresponding displacements produced by the P loads. Thus the *theorem of reciprocal work* may be stated as:

The work done by a first loading system when moving through the corresponding displacements produced by a second loading system is equal to the work done by the second loading system when moving through the corresponding displacements produced by the first loading system.

Again, as in the theorem of reciprocal displacements, the loading systems may be either forces or moments and the displacements may be deflections or rotations.

If, in the above, the P and Q loading systems comprise just two loads, say P_1 and Q_2, then, from Eq. (15.57), we see that

$$P_1(a_{12}Q_2) = Q_2(a_{21}P_1)$$

so that

$$a_{12} = a_{21}$$

as in the theorem of reciprocal displacements. Therefore, as stated initially, we see that the theorem of reciprocal displacements is a special case of the theorem of reciprocal work.

In addition to the use of the reciprocal theorems in the model analysis of structures as described in Ex. 15.13, they are used to establish the symmetry of, say, the stiffness matrix in the matrix analysis of some structural systems. We shall examine this procedure in Chapter 16.

Problems

P.15.1   Use the principle of virtual work to determine the support reactions in the beam ABCD shown in Fig. P.15.1.

Ans. $R_A = 1 \cdot 25W$, $R_D = 1 \cdot 75W$.

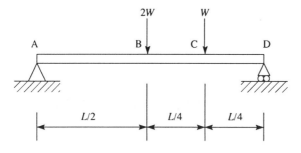

Fig. P.15.1

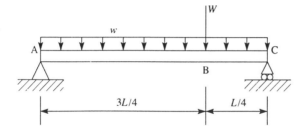

Fig. P.15.2

P.15.2 Find the support reactions in the beam ABC shown in Fig. P.15.2 using the principle of virtual work.

Ans. $R_A = (W + 2wL)/4$, $R_C = (3W + 2wL)/4$.

P.15.3 Determine the reactions at the built-in end of the cantilever beam ABC shown in Fig. P.15.3 using the principle of virtual work.

Ans. $R_A = 3W$, $M_A = 2 \cdot 5WL$.

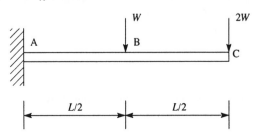

Fig. P.15.3

P.15.4 Find the bending moment at the three-quarter-span point in the beam shown in Fig. P.15.4. Use the principle of virtual work.

Ans. $3wL^2/32$.

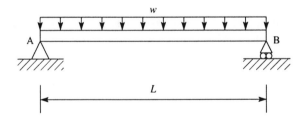

Fig. P.15.4

P.15.5 Calculate the forces in the members FG, GD and CD of the truss shown in Fig. P.15.5 using the principle of virtual work. All horizontal and vertical members are 1 m long.

Ans. FG = +20 kN,   GD = +28·3 kN,   CD = −20 kN.

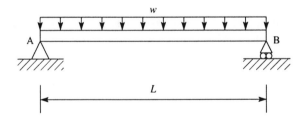

Fig. P.15.5

P.15.6 Use the unit load method to calculate the vertical displacements at the quarter- and mid-span points in the beam shown in Fig. P.15.6.

Ans. $119wL^4/24576\,EI$, $5wL^4/384\,EI$.

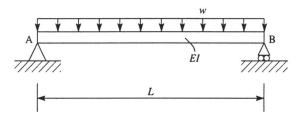

Fig. P.15.6

P.15.7 Calculate the deflection of the free end C of the cantilever beam ABC shown in Fig. P.15.7 using the unit load method.

Ans. $wa^3(4L-a)/24EI$.

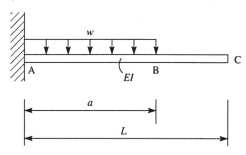

Fig. P.15.7

P.15.8 Use the unit load method to calculate the deflection at the free end of the cantilever beam ABC shown in Fig. P.15.8.

Ans. $3WL^3/8EI$.

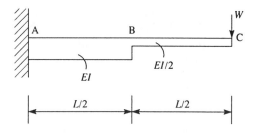

Fig. P.15.8

P.15.9 Use the unit load method to find the magnitude and direction of the deflection of the joint C in the truss shown in Fig. P.15.9. All members have a cross-sectional area of 500 mm^2 and a Young's modulus of 200 000 N/mm^2.

Ans. 23·4 mm, 9·8° to left of vertical.

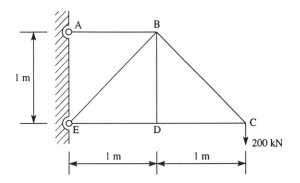

Fig. P.15.9

P.15.10 Calculate the magnitude and direction of the deflection of the joint A in the truss shown in Fig. P.15.10. The cross-sectional area of the compression members is 1000 mm² while that of the tension members is 750 mm². Young's modulus is 200 000 N/mm².

Ans. 15·03 mm, 9·6° to right of vertical.

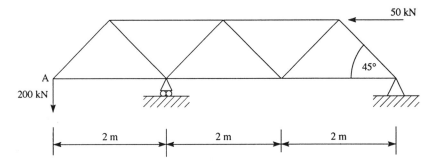

Fig. P.15.10

P.15.11 A rigid triangular plate is suspended from a horizontal plane by three vertical wires attached to its corners. The wires are each 1 mm diameter, 1440 mm long with a modulus of elasticity of 196 000 N/mm². The ratio of the lengths of the sides of the plate is 3 : 4 : 5. Calculate the deflection at the point of application of a load of 100 N placed at a point equidistant from the three sides of the plate.

Ans. 0·33 mm.

P.15.12 The pin-jointed space frame shown in Fig. P.15.12 is pinned to supports 0, 4, 5 and 9 and is loaded by a force P in the x direction and a force $3P$ in the negative y direction at the point 7. Find the rotation of the member 27 about the z axis due to this loading. All members have the same cross-sectional area, A, and Young's modulus, E. (Hint. Calculate the deflections in the x direction of joints 2 and 7.)

Ans. $382P/9AE$.

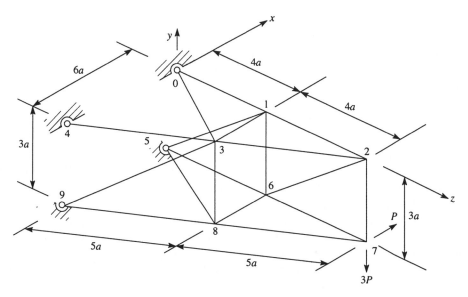

Fig. P.15.12

P.15.13 The tubular steel post shown in Fig. P.15.13 carries a load of 250 N at the free end C. The outside diameter of the tube is 100 mm and its wall thickness is 3 mm. If the modulus of elasticity of the steel is 206 000 N/mm², calculate the horizontal movement of C.

Ans. 53·5 mm.

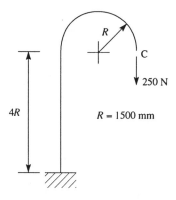

Fig. P.15.13

P.15.14 A cantilever beam of length L and depth h is subjected to a uniform temperature rise along its length. At any section, however, the temperature increases linearly from t_1 on the undersurface of the beam to t_2 on its upper surface. If the coefficient of linear expansion of the material of the beam is α, calculate the deflection at its free end.

Ans. $\alpha(t_2 - t_1)L^2/2h$.

P.15.15   A simply supported beam of span L is subjected to a temperature gradient which increases linearly from zero at the left-hand support to t_o at the right-hand support. If the temperature gradient also varies linearly through the depth, h, of the beam and is zero on its undersurface, calculate the deflection of the beam at its mid-span point. The coefficient of linear expansion of the material of the beam is α.

Ans. $-\alpha t_o L^2/48h$.

P.15.16   Figure P.15.16 shows a frame pinned to supports at A and B. The frame centre-line is a circular arc and its section is uniform, of bending stiffness EI and depth d. Find the maximum stress in the frame produced by a uniform temperature gradient through the depth, the temperatures on the outer and inner surfaces being raised and lowered by an amount T. The coefficient of linear expansion of the material of the frame is α. (Hint. Treat half the frame as a curved cantilever built-in on its axis of symmetry and determine the horizontal reaction at a support by equating the horizontal deflection produced by the temperature gradient to the horizontal deflection produced by the reaction).

Ans. $1 \cdot 29 E T \alpha$.

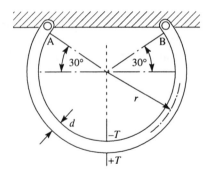

Fig. P.15.16

P.15.17   Calculate the deflection at the mid-span point of the beam of Ex. 15.11 by assuming a deflected shape function of the form

$$v = v_1 \sin \frac{\pi z}{L} + v_3 \sin \frac{3\pi z}{L}$$

in which v_1 and v_3 are unknown displacement parameters. Note:

$$\int_0^L \sin^2(n\pi z/L)\, \mathrm{d}z = L/2, \qquad \int_0^L \sin(m\pi z/L)\sin(n\pi z/L)\, \mathrm{d}z = 0$$

Ans. $0 \cdot 02078\ WL^3/EI$.

P.15.18   A beam is supported at both ends and has the central half of its span reinforced such that its flexural rigidity is $2EI$; the flexural rigidity of the remaining parts of the beam is EI. The beam has a span L and carries a vertically downward concentrated load, W, at its mid-span point. Assuming a deflected shape function of the form

$$v = \frac{4v_m z^2}{L^3} (3L - 4z) \quad (0 \leqslant z \leqslant L/2)$$

in which v_m is the deflection at the mid-span point, determine the value of v_m.

Ans. 0·00358 WL^3/EI.

P.15.19 Figure P.15.19 shows two cantilevers, the end of one being vertically above the end of the other and connected to it by a spring AB. Initially the system is unstrained. A weight, W, placed at A causes a vertical deflection at A of δ_1 and a vertical deflection at B of δ_2. When the spring is removed the weight W at A causes a deflection at A of δ_3. Find the extension of the spring when it is replaced and the weight, W, is transferred to B.

Ans. $\delta_2(\delta_1 - \delta_2)/(\delta_3 - \delta_1)$

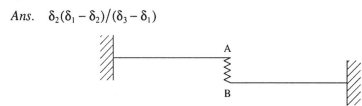

Fig. P.15.19

P.15.20 A beam 2·4 m long is simply supported at two points A and B which are 1·44 m apart; point A is 0·36 m from the left-hand end of the beam and point B is 0·6 m from the right-hand end; the value of EI for the beam is 240×10^8 N mm². Find the slope at the supports due to a load of 2 kN applied at the mid-point of AB.

Use the reciprocal theorem in conjunction with the above result to find the deflection at the mid-point of AB due to loads of 3 kN applied at each end of the beam.

Ans. 0·011, 15·8 mm.

CHAPTER 16

Analysis of Statically Indeterminate Structures

Statically indeterminate structures occur more frequently in practice than those that are statically determinate and are generally more economical in that they are stiffer and stronger. For example, a fixed beam carrying a concentrated load at mid-span has a central displacement that is one quarter of that of a simply supported beam of the same span and carrying the same load, while the maximum bending moment is reduced by half. It follows that a smaller beam section would be required in the fixed beam case, resulting in savings in material. There are, however, disadvantages in the use of this type of beam for, as we saw in Section 13.7, the settling of a support in a fixed beam causes bending moments that are additional to those produced by the loads, a serious problem in areas prone to subsidence. Another disadvantage of statically indeterminate structures is that their analysis requires the calculation of displacements so that their cross-sectional dimensions are required at the outset. The design of such structures therefore becomes a matter of trial and error, whereas the forces in the members of a statically determinate structure are independent of member size. On the other hand, failure of, say, a member in a statically indeterminate frame would not necessarily be catastrophic since alternative load paths would be available, at least temporarily. However, the failure of a member in, say, a statically determinate truss would lead, almost certainly, to a rapid collapse.

The choice between statically determinate and statically indeterminate structures depends to a large extent upon the purpose for which a particular structure is required. As we have seen, fixed or continuous beams are adversely affected by support settlement so that the insertion of hinges at, say, points of contraflexure would reduce the structure to a statically determinate state and eliminate the problem. This procedure would not be practical in the construction of skeletal structures for high-rise buildings so that these structures are statically indeterminate. Clearly, both types of structure exist in practice so that methods of analysis are required for both statically indeterminate and statically determinate structures.

In this chapter we shall examine methods of analysis of different forms of statically indeterminate structures; as a preliminary we shall discuss the basis of the different methods, and investigate methods of determining the degree of statical and kinematic indeterminacy, an essential part of the analytical procedure.

16.1 Flexibility and stiffness methods

In Section 4.4 we briefly discussed the statical indeterminacy of trusses and established a condition, not always applicable, for a truss to be stable and statically determinate. This condition, which related the number of members and the number of joints, did not involve the support reactions which themselves could be either statically determinate or indeterminate. The condition was therefore one of *internal statical determinacy*; clearly the determinacy, or otherwise, of the support reactions is one of *external statical determinacy*.

Consider the portal frame shown in Fig. 16.1. The frame carries loads, *P* and *W*, in its own plane so that the system is two-dimensional. Since the vertical members AB and FD of the frame are fixed at A and F, the applied loads will generate a total of six reactions of force and moment as shown. For a two-dimensional system there are three possible equations of statical equilibrium (Eqs (2.10) so that the frame is externally statically determinate to the *third degree*. The situation is not improved by taking a section through one of the members since this procedure, although eliminating one of the sets of reactive forces, would introduce three internal stress resultants. If, however, three of the support reactions were known or, alternatively, if the three internal stress resultants were known, the remaining three unknowns could be determined from the equations of statical equilibrium and the solution completed.

A different situation arises in the simple truss shown in Fig. 4.7(b) where, as we saw, the additional diagonal results in the truss becoming internally statically indeterminate to the *first degree*; note that the support reactions are statically determinate.

In the analysis of statically indeterminate structures two basic methods are employed. In one the structure is reduced to a statically determinate state by employing *releases*, i.e. by eliminating a sufficient number of unknowns to enable the support reactions and/or the internal stress resultants to be found from a consideration of statical equilibrium. In the frame in Fig. 16.1, for example, the number of support reactions would be reduced to three if one of the supports was

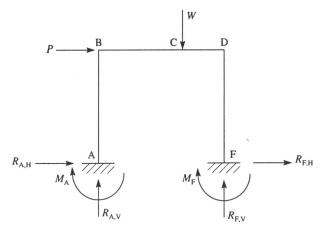

Fig. 16.1 Statical indeterminacy of a portal frame

pinned and the other was a pinned roller support. The same result would be achieved if one support remained fixed and the other support was removed entirely. Also, in the truss in Fig. 4.7.(b), removing a diagonal, vertical or horizontal member would result in the truss becoming statically determinate. Releasing a structure in this way would produce displacements that would not otherwise be present. These displacements may be calculated by analysing the released statically determinate structure; the force system required to eliminate them is then obtained, i.e. we are employing a compatibility of displacement condition. This method is generally termed the *flexibility* or *force method*; in effect this method was used in the solution of the propped cantilever in Fig. 13.22.

The alternative procedure, known as the *stiffness* or *displacement method* is analogous to the flexibility method, the major difference being that the unknowns are the displacements at specific points in the structure. Generally the procedure requires a structure to be divided into a number of elements for each of which load–displacement relationships are known. Equations of equilibrium are then written down in terms of the displacements at the element junctions and are solved from the required displacements; the complete solution follows.

Both the flexibility and stiffness methods generally result, for practical structures having a high degree of statical indeterminacy, in a large number of simultaneous equations which are most readily solved by computer-based techniques. However, the flexibility method requires the structure to be reduced to a statically determinate state by inserting releases, a procedure requiring some judgement on the part of the analyst. The stiffness method, on the other hand, requires no such judgement to be made and is therefore particularly suitable for automatic computation.

Although the practical application of the flexibility and stiffness methods is generally computer-based, they are fundamental to 'hand' methods of analysis as we shall see. Before investigating these hand methods we shall examine in greater detail the indeterminacy of structures since we shall require the degree of indeterminacy of a structure before, in the case of the flexibility method, the appropriate number of releases can be determined. At the same time the *kinematic indeterminacy* of a structure is needed to determine the number of constraints that must be applied to render the structure kinematically determinate in the stiffness method.

16.2 Degree of statical indeterminacy

In some cases the degree of statical indeterminacy of a structure is obvious from inspection. The portal frame in Fig. 16.1, for example, has a degree of external statical indeterminacy of 3, while the truss of Fig. 4.7(b) has a degree of internal statical indeterminacy of 1. However, in many cases, the degree is not obvious and in other cases the internal and external indeterminacies may not be independent so that we need to consider the complete structure, including the support system. A more formal and methodical approach is therefore required.

Rings

The simplest approach is to insert constraints in a structure until it becomes a series of completely stiff *rings*. The statical indeterminacy of a ring is known and hence that of the completely stiff structure. Thus by inserting the number of releases required to return the completely stiff structure to its original state, the degree of indeterminacy of the actual structure is found.

Consider the single ring shown in Fig. 16.2(a); the ring is in equilibrium in space under the action of a number of forces that are not coplanar. If, say, the ring is cut at some point, X, the cut ends of the ring will be displaced relative to each other as shown in Fig. 16.2(b) since, in effect, the internal forces equilibrating the external forces have been removed. The cut ends of the ring will move relative to each other in up to six possible ways until a new equilibrium position is found, i.e. translationally along the *x*, *y* and *z* axes and rotationally about the *x*, *y* and *z* axes, as shown in Fig. 16.2(c). The ring is now statically determinate and the internal force system at any section may be obtained from simple equilibrium considerations. To rejoin the ends of the ring we require forces and moments proportional to the displacements, i.e. three forces and three moments. Thus at any section in a complete ring subjected to an arbitrary external loading system there are three internal forces and three internal moments, none of which may be obtained by statics. Thus a ring is six times statically indeterminate. For a two-dimensional system in which the forces are applied in the plane of the ring, the internal force system is reduced to an axial force, a shear force and a moment, so that a two-dimensional ring is three times statically indeterminate.

The above arguments apply to any closed loop so that a ring may be of any shape. Furthermore, a ring may be regarded as comprising any number of members which form a closed loop and which are joined at *nodes*, a node being defined as a point at the end of a member. Examples of rings are shown in Fig. 16.3 where the number

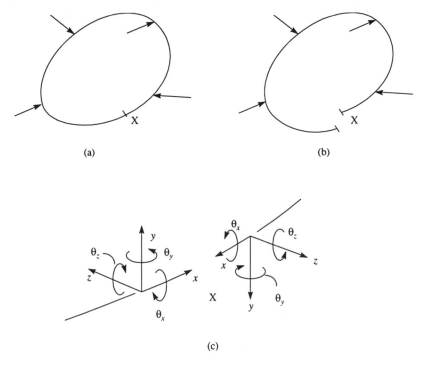

(a)

(b)

(c)

Fig. 16.2 Statical indeterminacy of a ring

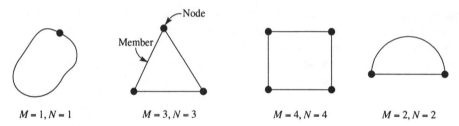

Fig. 16.3 Examples of rings

of members, M, and the number of nodes, N, are given. Note that the number of members is equal to the number of nodes in every case. However, when a ring is cut we introduce an additional member and two additional nodes, as shown in Fig. 16.4.

The entire structure

Since we shall require the number of rings in a structure, and since it is generally necessary to include the support system, we must decide what constitutes the structure. In Fig. 16.5, for example, the members AB and BC are pinned to the foundation at A and C. The foundation therefore acts as a member of very high stiffness. In this simple illustration it is obvious that the members AB and BC, with the foundation, form a ring if the pinned joints are replaced by rigid joints. In more complex structures we must ensure that just sufficient of the foundation is included so that superfluous indeterminacies are not introduced; the structure is then termed the *entire structure*. This condition requires that the points of support are *singly connected* such that for any two points A and B in the foundation system there is only one path from A to B that does involve retracing any part of the path. In

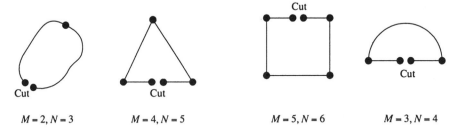

Fig. 16.4 Effect on members and nodes of cutting a ring

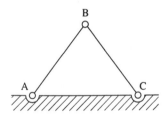

Fig. 16.5 Foundation acting as a structural member

Figs 16.6(a) and (b), for example, there is only one path between A and B which does not involve retracing part of the path. In Fig. 16.6(c), however, there are two possible paths from A to B, one via G and one via F and E. Thus the support points in Fig. 16.6(a) and (b) are singly connected, while those in Fig. 16.6(c) are multiply connected. We note from the above that there may be a number of ways of singly connecting the support points in a foundation system and that each support point in the entire structure is attached to at least one foundation 'member'. Including the foundation members increases the number of members, but the number of nodes is unchanged.

The completely stiff structure

Having established the entire structure we now require the *completely stiff structure* in which there is no point or member where any stress resultant is always zero for any possible loading system. Thus the completely stiff structure (Fig. 16.7(b)) corresponding to the simple truss in Fig. 16.7(a) has rigid joints (nodes), members that are capable of resisting shear loads as well as axial loads and a single foundation member. Note that the completely stiff structure comprises two rings, is two-dimensional and therefore six times statically indeterminate. We shall consider how such a structure is 'released' to return it to its original state (Fig. 16.7(a)) after considering the degree of indeterminacy of a three-dimensional system.

Degree of statical indeterminacy

Consider the frame structure shown in Fig. 16.8(a). It is three-dimensional and comprises three portal frames that are rigidly built-in at the foundation. Its completely stiff equivalent is shown in Fig. 16.8(b) where we observe by inspection that it consists of three rings, each of which is six times statically indeterminate so that the completely stiff structure is $3 \times 6 = 18$ times statically indeterminate. Although the number of rings in simple cases such as this is easily found by inspection, more complex cases require a more methodical approach.

Suppose that the members are disconnected until the structure becomes singly connected as shown in Fig. 16.8(c). (A singly connected structure is defined in the same way as a singly connected foundation.) Each time a member is disconnected, the number of nodes increases by one, while the number of rings is reduced by one;

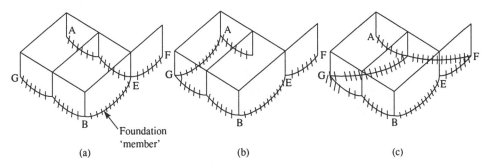

Fig. 16.6 Determination of the entire structure

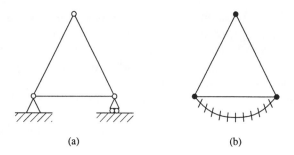

Fig. 16.7 A completely stiff structure

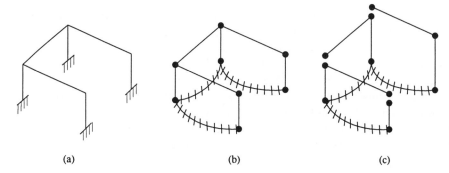

Fig. 16.8 Determination of the degree of statical indeterminacy of a structure

the number of members remains the same. The final number of nodes, N', in the singly connected structure is therefore given by

$$N' = M + 1 \quad (M = \text{number of members})$$

Suppose now that the members are reconnected to form the original completely stiff structure. Each reconnection forms a ring, i.e. each time a node disappears a ring is formed so that the number of rings, R, is equal to the number of nodes lost during the reconnection. Thus

$$R = N' - N$$

where N is the number of nodes in the completely stiff structure. Substituting for N' from the above we have

$$R = M - N + 1$$

In Fig. 16.8(b), $M = 10$ and $N = 8$ so that $R = 3$ as deduced by inspection. Therefore, since each ring is six times statically indeterminate, the degree of statical indeterminacy, n'_s, of the completely stiff structure is given by

$$n'_s = 6(M - N + 1) \tag{16.1}$$

For an actual entire structure, releases must be inserted to return the completely stiff structure to its original state. Each release will reduce the statical indeterminacy by one, so that if r is the total number of releases required, the degree of statical indeterminacy, n_s, of the actual structure is

$$n_s = n_s' - r$$

or, substituting for n_s' from Eq. (16.1)

$$n_s = 6(M - N + 1) - r \tag{16.2}$$

Note that in Fig. 16.8 no releases are required to return the completely stiff structure of Fig. 16.8(b) to its original state in Fig. 16.8(a) so that its degree of indeterminacy is 18.

In the case of two-dimensional structures in which a ring is three times statically indeterminate, Eq. (16.2) becomes

$$n_s = 3(M - N + 1) - r \tag{16.3}$$

Pin-jointed frames

A difficulty arises in determining the number of releases required to return the completely stiff equivalent of a pin-jointed frame to its original state.

Consider the completely stiff equivalent of a plane truss shown in Fig. 16.9(a); we are not concerned here with the indeterminacy or otherwise of the support system which is therefore omitted. In the actual truss each member is assumed to be capable of resisting axial load only so that there are two releases for each member, one of shear and one of moment, a total of $2M$ releases. Thus, if we insert a hinge at the end of each member as shown in Fig. 16.9(b) we have achieved the required number, $2M$, of releases. However, in this configuration, each joint would be free to rotate as a mechanism through an infinitesimally small angle, independently of the members; the truss is then excessively pin-jointed. This situation can be prevented by removing one hinge at each joint as shown, for example, at joint B in Fig. 16.9(c). The member BC then prevents rotation of the joint at B. Furthermore, the presence of a hinge at B in BA and at B in BE ensures that there is no moment at B in BC so that the conditions for a truss are satisfied.

From the above we see that the total number, $2M$, of releases is reduced by one for each node. Thus the required number of releases in a plane truss is

$$r = 2M - N \tag{16.4}$$

so that Eq. (16.3) becomes

$$n_s = 3(M - N + 1) - (2M - N)$$

or
$$n_s = M - 2N + 3 \tag{16.5}$$

Equation (16.5) refers only to the internal indeterminacy of a truss so that the degree of indeterminacy of the support system is additional. Also, returning to the simple triangular truss of Fig. 16.7(a) we see that its degree of internal indeterminacy is, from Eq. (16.5), given by

$$n_s = 3 - 2 \times 3 + 3 = 0$$

as expected.

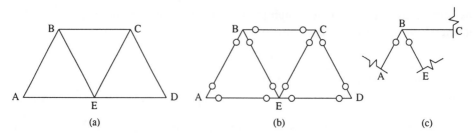

Fig. 16.9 Number of releases for a plane truss

A similar situation arises in a pin-jointed space frame where, again, each member is required to resist axial load only so that there are $5M$ releases for the complete frame. This could be achieved by inserting ball joints at the ends of each member. However, we would then be in the same kind of position as the plane truss of Fig. 16.9(b) in that each joint would be free to rotate through infinitesimally small angles about each of the three axes (the members in the plane truss can only rotate about one axis) so that three constraints are required at each node, a total of $3N$ constraints. Therefore the number of releases is given by

$$r = 5M - 3N$$

so that Eq. (16.2) becomes

$$n_s = 6(M - N + 1) - (5M - 3N)$$

or

$$n_s = M - 3N + 6 \tag{16.6}$$

For statically determinate plane trusses and pin-jointed space frames, i.e. $n_s = 0$, Eqs (16.5) and (16.6) become, respectively,

$$M = 2N - 3, \qquad M = 3N - 6 \tag{16.7}$$

which are the results deduced in Section 4.4 (Eqs (4.1) and (4.2)).

16.3 Kinematic indeterminacy

We have seen that the degree of statical indeterminacy of a structure is, in fact, the number of forces or stress resultants, which cannot be determined using the equations of statical equilibrium. Another form of the indeterminacy of a structure is expressed in terms of its *degrees of freedom*; this is known as the *kinematic indeterminacy, n_k*, of a structure and is of particular relevance in the stiffness method of analysis where the unknowns are the displacements.

A simple approach to calculating the kinematic indeterminacy of a structure is to sum the degrees of freedom of the nodes and then subtract those degrees of freedom that are prevented by constraints such as support points. It is therefore important to remember that in three-dimensional structures each node possesses six degrees of freedom while in plane structures each node possess three degrees of freedom.

Example 16.1 Determine the degrees of statical and kinematic indeterminacy of the beam ABC shown in Fig. 16.10(a).

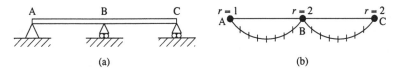

Fig. 16.10 Determination of the statical and kinematic indeterminacies of the beam of Ex. 16.1

The completely stiff structure is shown in Fig. 16.10(b) where we see that $M = 4$ and $N = 3$. The number of releases, r, required to return the completely stiff structure to its original state is five, as indicated in Fig. 16.10(b); these comprise a moment release at each of the three supports and a translational release at each of the supports B and C. Therefore, from Eq. (16.3)

$$n_s = 3(4 - 3 + 1) - 5 = 1$$

so that the degree of statical indeterminacy of the beam is one.

Each of the three nodes possesses three degrees of freedom, a total of nine. There are four constraints so that the degree of kinematic indeterminacy is given by

$$n_k = 9 - 4 = 5$$

Example 16.2 Determine the degree of statical and kinematic indeterminacy of the pin-jointed frame shown in Fig. 16.11(a).

The completely stiff structure is shown in Fig. 16.11(b) in which we see that $M = 17$ and $N = 8$. However, since the frame is pin-jointed, we can obtain the internal statical indeterminacy directly from Eq. (16.5) in which $M = 16$, the actual number of frame members. Thus

$$n_s = 16 - 16 + 3 = 3$$

and since, as can be seen from inspection, the support system is statically determinate, the complete structure is three times statically indeterminate.

Alternatively, considering the completely stiff structure in Fig. 16.11(b) in which $M = 17$ and $N = 8$, we can use Eq. (16.3). The number of internal releases is found from Eq. (16.4) and is $r = 2 \times 16 - 8 = 24$. There are three additional releases from the support system giving a total of twenty-seven releases. Thus, from Eq. (16.3)

$$n_s = 3(17 - 8 + 1) - 27 = 3$$

as before.

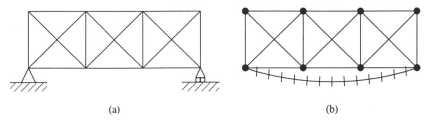

Fig. 16.11 Determinacy of the pin-jointed frame of Ex. 16.2

The kinematic indeterminacy is found as before by examining the total degrees of freedom of the nodes and the constraints, which in this case are provided solely by the support system. There are eight nodes each having two translational degrees of freedom. The rotation at a node does not result in a stress resultant and is therefore irrelevant. There are therefore two degrees of freedom at a node in a pin-jointed plane frame and three in a pin-jointed space frame. In this example there are then $8 \times 2 = 16$ degrees of freedom and three translational constraints from the support system. Thus

$$n_k = 16 - 3 = 13$$

Example 16.3   Calculate the degree of statical and kinematic indeterminacy of the frame shown in Fig. 16.12(a).

In the completely stiff structure shown in Fig. 16.12(a), $M = 7$ and $N = 6$. The number of releases, r, required to return the completely stiff structure to its original state is 3. Thus, from Eq. (16.3)

$$n_s = 3(7 - 6 + 1) - 3 = 3$$

The number of nodes is six, each having three degrees of freedom, a total of eighteen. The number of constraints is three so that the kinematic indeterminacy of the frame is given by

$$n_k = 18 - 3 = 15$$

Example 16.4   Determine the degree of statical and kinematic indeterminacy in the space frame shown in Fig. 16.13(a).

In the completely stiff structure shown in Fig. 16.13(b), $M = 19$, $N = 13$ and $r = 0$. Therefore from Eq. (16.2)

$$n_s = 6(19 - 13 + 1) - 0 = 42$$

There are thirteen nodes each having six degrees of freedom, a total of seventy-eight. There are six constraints at each of the four supports, a total of twenty-four. Thus

$$n_k = 78 - 24 = 54$$

We shall now consider different types of statically indeterminate structure and the methods that may be used to analyse them; the methods are based on the work and energy methods described in Chapter 15.

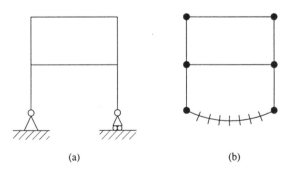

(a)   (b)

Fig. 16.12   Statical and kinematic indeterminacies of the frame of Ex. 16.3

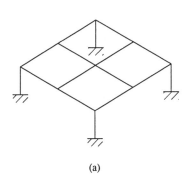

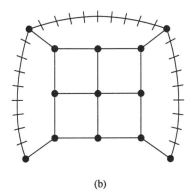

(a)   (b)

Fig. 16.13 Space frame of Ex. 16.4

16.4 Statically indeterminate beams

Beams are statically indeterminate generally because of their support systems. In this category are propped cantilevers, fixed beams and continuous beams. A propped cantilever and some fixed beams were analysed in Section 13.7 using either the principle of superposition or moment-area methods. We shall now apply the methods described in Chapter 15 to some examples of statically indeterminate beams.

Example 16.5 Calculate the support reaction at B in the propped cantilever shown in Fig. 16.14.

In this example it is unnecessary to employ the procedures described in Section 16.2 to calculate the degree of statical indeterminacy since this is obvious by inspection. Thus the removal of the vertical support at B would result in a statically determinate cantilever beam so that we deduce that the degree of statical indeterminacy is one. Furthermore, it is immaterial whether we use the principle of virtual work or complementary energy in the solution since, for linearly elastic systems, they result in the same equations (see Chapter 15). First, we shall adopt the complementary energy approach.

The total complementary energy, C, of the beam is given, from Eq. (i) of Ex. 15.8, by

$$C = \int_0^L \int_0^M d\theta \, dM - R_B v_B \qquad (i)$$

in which v_B is the vertical displacement of the cantilever at B (in this case $v_B = 0$ since the beam is supported at B).

From the principle of the stationary value of the total complementary energy we have

$$\frac{\partial C}{\partial R_B} = \int_0^L \frac{\partial M}{\partial R_B} d\theta - v_B = 0 \qquad (ii)$$

which, by comparison with Eq. (iii) of Ex. 15.8, becomes

$$v_B = \int_0^L \frac{M}{EI} \frac{\partial M}{\partial R_B} dz = 0 \qquad (iii)$$

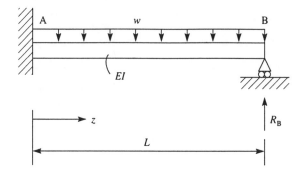

Fig. 16.14   Propped cantilever of Ex. 16.5

The bending moment, M, at any section of the beam is given by

$$M = R_B(L - z) - \frac{w}{2}(L - z)^2$$

Hence

$$\frac{\partial M}{\partial R_B} = L - z$$

Substituting in Eq. (iii) for M and

$$\frac{\partial M}{\partial R_B}$$

we have

$$\int_0^L \left\{ R_B(L - z)^2 - \frac{w}{2}(L - z)^3 \right\} dz = 0 \qquad \text{(iv)}$$

from which

$$R_B = \frac{3wL}{8}$$

which is the result obtained in Ex. 13.19.

The algebra in the above solution would have been slightly simplified if we had assumed an origin for z at the end B of the beam. Equation (iv) would then become

$$\int_0^L \left(R_B z^2 - \frac{w}{2} z^3 \right) dz = 0$$

which again gives

$$R_B = \frac{3wL}{8}$$

Having obtained R_B, the remaining support reactions follow from statics.

An alternative approach is to release the structure so that it becomes statically determinate by removing the support at B (one release only is required in this case) and then to calculate the vertical displacement at B due to the applied load using, say, the unit load method. We then calculate the vertical displacement at B produced by R_B acting alone, again, say, by the unit load method. The sum of the two displacements must be zero since the beam at B is supported, so that we obtain an equation in which R_B is the unknown.

It is not essential to select the support reaction at B as the release. We could, in fact, choose the fixing moment at A in which case the beam would become a simply supported beam which, of course, is statically determinate. We would then determine the moment at A required to restore the slope of the beam at A to zero.

In the above, the released structure is frequently termed the *primary structure*.

Suppose that the vertical displacement at the free end of the released cantilever due to the uniformly distributed load is $v_{B,0}$. Then, from Eq. (iii) of Ex. 15.9 (noting that M_A in that equation has been replaced by M_a here to avoid confusion with the bending moment at A)

$$v_{B,0} = \int_0^L \frac{M_a M_1}{EI} \, dz \tag{v}$$

in which

$$M_a = -\frac{w}{2}(L-z)^2, \qquad M_1 = -1(L-z)$$

Hence, substituting for M_a and M_1 in Eq. (v), we have

$$v_{B,0} = \int_0^L \frac{w}{2EI}(L-z)^3 \, dz$$

which gives

$$v_{B,0} = \frac{wL^4}{8EI} \tag{vi}$$

We now apply a vertically downward unit load at the B end of the cantilever from which the distributed load has been removed. The displacement, $v_{B,1}$, due to this unit load is then, from Eq. (v)

$$v_{B,1} = \int_0^L \frac{1}{EI}(L-z)^2 \, dz$$

from which

$$v_{B,1} = \frac{L^3}{3EI} \tag{vii}$$

The displacement due to R_B at B is $-R_B v_{B,1}$ (R_B acts in the opposite direction to the unit load) so that the total displacement, v_B, at B due to the uniformly distributed load and R_B is, using the principle of superposition,

$$v_B = v_{B,0} - R_B v_{B,1} = 0 \tag{viii}$$

Substituting for $v_{B,0}$ and $v_{B,1}$ from Eqs (vi) and (vii) we have

$$\frac{wL^4}{8EI} - R_B \frac{L^3}{3EI} = 0$$

whence

$$R_B = \frac{3wL}{8}$$

as before. This approach is the flexibility method described in Section 16.1 and is, in effect, identical to the method used in Ex. 13.9.

In Eq. (viii) $v_{B,1}$ is the displacement at B in the direction of R_B due to a unit load at B applied in the direction of R_B (either in the same or opposite directions). For a beam that has a degree of statical indeterminacy greater than one there will be a series of equations of the same form as Eq. (viii) but which will contain the displacements at a specific point produced by the redundant forces. We shall therefore employ the *flexibility coefficient* a_{kj} $(k = 1, 2, ..., r; j = 1, 2, ..., r)$ which we defined in Section 15.4 as the displacement at a point k in a given direction produced by a unit load at a point j in a second direction. Thus, in the above, $v_{B,1} = a_{11}$ so that Eq. (viii) becomes

$$v_{B,0} - a_{11}R_B = 0 \qquad (ix)$$

It is also convenient, since the flexibility coefficients are specified by numerical subscripts, to redesignate R_B as R_1. Thus Eq. (ix) becomes

$$v_{B,0} - a_{11}R_1 = 0 \qquad (x)$$

Example 16.6   Determine the support reaction at B in the propped cantilever shown in Fig. 16.15(a).

As in Ex. 16.5, the cantilever in Fig. 16.15(a) has a degree of statical indeterminacy equal to one. Again we shall choose the support reaction at B, R_1, as the indeterminacy; the released or primary structure is shown in Fig. 16.15(b). Initially we require the displacement, $v_{B,0}$, at B due to the applied load, W, at C. This may readily be found using the unit load method. Thus from Eq. (iii) of Ex. 15.9

$$v_{B,0} = \int_0^L \left\{ -\frac{W}{EI} \left(\frac{3L}{2} - z \right) \right\} \{ -1(L - z) \} \, dz$$

which gives

$$v_{B,0} = \frac{7WL^3}{12EI} \qquad (i)$$

Similarly, the displacement at B due to the unit load at B in the direction of R_1 (Fig. 16.15(c)) is

$$a_{11} = \frac{L^3}{3EI} \quad \text{(use Eq. (vii) of Ex. 16.5)}$$

Hence, since,

$$v_{B,0} - a_{11}R_1 = 0 \qquad (ii)$$

we have

$$\frac{7WL^3}{12EI} - \frac{L^3}{3EI} R_1 = 0$$

whence

$$R_1 = \frac{7W}{4}$$

Alternatively, we could select the fixing moment, M_A $(= M_2)$, at A as the release. The primary structure is then the simply supported beam shown in Fig. 16.16(a) where $R_A = -W/2$ and $R_B = 3W/2$. The rotation at A may be found by any of the methods previously described. They include the integration of the second-order

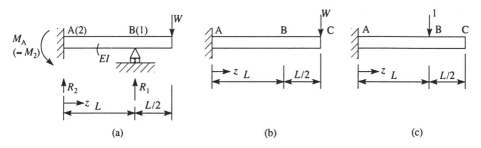

Fig. 16.15 Propped cantilever of Ex. 16.6

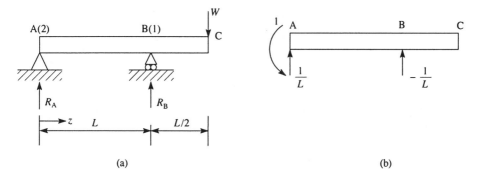

Fig. 16.16 Alternative solution for Ex. 16.6

differential equation of bending (Eq. (13.3)), the moment-area method described in Section 13.3 and the unit load method (in this case it would be a unit moment). Thus, using the unit load method and applying a unit moment at A as shown in Fig. 16.16(b) we have, from the principle of virtual work (see Ex. 15.5),

$$1\theta_{A,0} = \int_0^L \frac{M_a M_v}{EI}\, dz + \int_L^{3L/2} \frac{M_a M_v}{EI}\, dz \qquad\text{(iii)}$$

In Eq. (iii)

$$M_a = -\frac{W}{2}\, z, \qquad M_v = \frac{1}{L}\, z - 1 \quad (0 \leqslant z \leqslant L)$$

$$M_a = Wz - \frac{3WL}{2}, \qquad M_v = 0 \quad \left(L \leqslant z \leqslant \frac{3L}{2}\right)$$

Substituting in Eq. (iii) we have

$$\theta_{A,0} = \frac{W}{2EIL} \int_0^L (Lz - z^2)\, dz$$

from which

$$\theta_{A,0} = \frac{WL^2}{12EI} \quad \text{(anticlockwise)}$$

The flexibility coefficient, θ_{22}, i.e. the rotation at A (point 2), due to a unit moment at A is obtained from Fig. 16.16(b). Thus

$$\theta_{22} = \int_0^L \frac{1}{EI}\left(\frac{z}{L} - 1\right)^2 dz$$

from which $\qquad\qquad \theta_{22} = \dfrac{L}{3EI} \quad$ (anticlockwise)

Therefore, since the rotation at A in the actual structure is zero,

$$\theta_{A,0} + \theta_{22}M_2 = 0$$

or $\qquad\qquad \dfrac{WL^2}{12EI} + \dfrac{L}{3EI}M_2 = 0$

which gives $\qquad\qquad M_2 = -\dfrac{WL}{4} \quad$ (clockwise)

Considering now the statical equilibrium of the beam in Fig. 16.15(a) we have, taking moments about A

$$R_1 L - W\frac{3L}{2} - \frac{WL}{4} = 0$$

whence $\qquad\qquad R_1 = \dfrac{7WL}{4}$

as before.

Example 16.7   Determine the support reactions in the three-span continuous beam ABCD shown in Fig. 16.17(a).

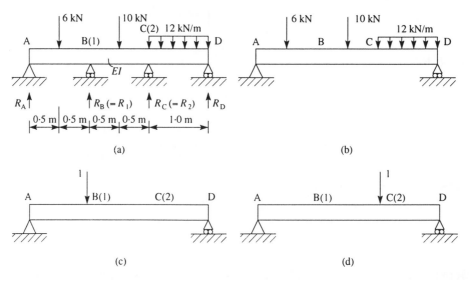

(a)

(b)

(c)

(d)

Fig. 16.17   Analysis of a three-span continuous beam

It is clear from inspection that the degree of statical indeterminacy of the beam is two. Therefore, if we choose the supports at B and C as the releases, the primary structure is that shown in Fig. 16.17(b). We therefore require the vertical displacements, $v_{B,0}$ and $v_{C,0}$, at the points B and C. These may readily be found using any of the methods previously described (unit load method, moment–area method, Macauley's method (Section 13.2)) and are

$$v_{B,0} = \frac{8\cdot88}{EI}, \qquad v_{C,0} = \frac{9\cdot08}{EI}$$

We now require the flexibility coefficients, a_{11}, a_{12}, a_{22} and a_{21}. The coefficients a_{11} and a_{21} are found by placing a unit load at B (point 1) as shown in Fig. 16.17(c) and then determining the displacements at B and C (point 2). Similarly, the coefficients a_{22} and a_{12} are found by placing a unit load at C and calculating the displacements at C and B; again, any of the methods listed above may be used. However, from the reciprocal theorem (Section 15.4) $a_{12} = a_{21}$ and from symmetry $a_{11} = a_{22}$. Therefore it is only necessary to calculate the displacements a_{11} and a_{21} from Fig. 16.17(c). These are

$$a_{11} = a_{22} = \frac{0\cdot45}{EI}, \qquad a_{21} = a_{12} = \frac{0\cdot39}{EI}$$

The total displacements at the support points B and C are zero so that

$$v_{B,0} - a_{11}R_1 - a_{12}R_2 = 0 \tag{i}$$

$$v_{C,0} - a_{21}R_1 - a_{22}R_2 = 0 \tag{ii}$$

or, substituting the calculated values of $v_{B,0}$, a_{11}, etc, in Eqs (i) and (ii) and multiplying through by EI,

$$8\cdot88 - 0\cdot45R_1 - 0\cdot39R_2 = 0 \tag{iii}$$

$$9\cdot08 - 0\cdot39R_1 - 0\cdot45R_2 = 0 \tag{iv}$$

Note that the negative signs in the terms involving R_1 and R_2 in Eqs (i) and (ii) are due to the fact that the unit loads were applied in the opposite directions to R_1 and R_2. Solving Eqs (iii) and (iv) we obtain

$$R_1(=R_B) = 8\cdot7 \text{ kN}, \qquad R_2(=R_C) = 12\cdot68 \text{ kN}$$

The remaining reactions are determined by considering the statical equilibrium of the beam and are

$$R_A = 1\cdot97 \text{ kN}, \qquad R_B = 4\cdot65 \text{ kN}$$

In Exs 16.5–16.7 we have assumed that the beam supports are not subjected to a vertical displacement themselves. It is possible, as we have previously noted, that a support may sink, so that the right-hand side of the compatibility equations, Eqs (viii), (ix) and (x) in Ex. 16.5, Eq. (ii) in Ex. 16.6 and Eqs (i) and (ii) in Ex. 16.7, would not be zero but equal to the actual displacement of the support. In such a situation one of the releases should coincide with the displaced support.

It is clear from Ex. 16.7 that the number of simultaneous equations of the form of Eqs (i) and (ii) requiring solution is equal to the degree of statical indeterminacy of the structure. For structures possessing a high degree of statical indeterminacy the solution, by hand, of a large number of simultaneous equations is not practicable. The equations would then be expressed in matrix form and solved using a computer-based approach. Thus for a structure having a degree of statical indeterminacy equal to n there would be n compatibility equations of the form

$$v_{1,0} + a_{11}R_1 + a_{12}R_2 + \cdots + a_{1n}R_n = 0$$
$$\vdots$$
$$v_{n,0} + a_{n1}R_1 + a_{n2}R_2 + \cdots + a_{nn}R_n = 0$$

or, in matrix form

$$\left\{ \begin{matrix} v_{1,0} \\ \vdots \\ v_{n,0} \end{matrix} \right\} = - \begin{bmatrix} a_{11} & a_{12} & \cdots & a_{1n} \\ & \vdots & & \\ a_{n1} & a_{n2} & \cdots & a_{nn} \end{bmatrix} \left\{ \begin{matrix} R_1 \\ \vdots \\ R_n \end{matrix} \right\}$$

Note that here n is n_s, the degree of statical indeterminacy; the subscript 's' has been omitted for convenience.

Alternative methods of solution of continuous beams are the slope–deflection method described in Section 16.9 and the iterative moment distribution method described in Section 16.10. The latter method is capable of producing relatively rapid solutions for beams having several spans.

16.5   Statically indeterminate trusses

A truss may be internally and/or externally statically indeterminate. For a truss that is externally statically indeterminate, the support reactions may be found by the methods described in Section 16.4. For a truss that is internally statically indeterminate the flexibility method may be employed as illustrated in the following examples.

Example 16.8   Determine the forces in the members of the truss shown in Fig. 16.18(a); the cross-sectional area, A, and Young's modulus, E, are the same for all members.

The truss in Fig. 16.18(a) is clearly externally statically determinate but, from Eq. (16.5), has a degree of internal statical indeterminacy equal to one ($M = 6$, $N = 4$). We therefore release the truss so that it becomes statically determinate by 'cutting' one of the members, say BD, as shown in Fig. 16.18(b). Due to the actual loads (P in this case) the cut ends of the member BD will separate or come together, depending on whether the force in the member (before it was cut) was tensile or compressive; we shall assume that it was tensile.

We are assuming that the truss is linearly elastic so that the relative displacement of the cut ends of the member BD (in effect the movement of B and D away from or towards each other along the diagonal BD) may be found using, say, the unit load method as illustrated in Exs 15.6 and 15.7. Thus we determine the forces $F_{a,j}$, in the members produced by the actual loads. We then apply equal and opposite unit loads

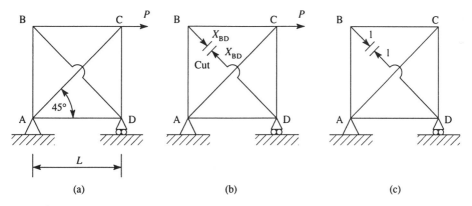

Fig. 16.18 Analysis of a statically indeterminate truss

to the cut ends of the member BD as shown in Fig. 16.18(c) and calculate the forces, $F_{1,j}$ in the members. The displacement of B relative to D, Δ_{BD}, is then given by

$$\Delta_{BD} = \sum_{j=1}^{n} \frac{F_{a,j} F_{1,j} L_j}{AE} \quad \text{(see Eq. (viii) in Ex. 15.7)}$$

The forces, $F_{a,j}$, are the forces in the members of the released truss due to the actual loads and are not, therefore, the actual forces in the members of the complete truss. We shall therefore redesignate the forces in the members of the released truss as $F_{0,j}$. The expression for Δ_{BD} then becomes

$$\Delta_{BD} = \sum_{j=1}^{n} \frac{F_{0,j} F_{1,j} L_j}{AE} \tag{i}$$

In the actual structures this displacement is prevented by the force, X_{BD}, in the redundant member BD. If, therefore, we calculate the displacement, a_{BD}, in the direction of BD produced by a unit value of X_{BD}, the displacement due to X_{BD} will be $X_{BD} a_{BD}$. Clearly, from compatibility

$$\Delta_{BD} + X_{BD} a_{BD} = 0 \tag{ii}$$

from which X_{BD} is found. Again, as in the case of statically indeterminate beams, a_{BD} is a flexibility coefficient. Having determined X_{BD}, the actual forces in the members of the complete truss may be calculated by, say, the method of joints or the method of sections.

In Eq. (ii), a_{BD} is the displacement of the released truss in the direction of BD produced by a unit load. Thus, in using the unit load method to calculate this displacement, the actual member forces ($F_{0,j}$) and the member forces produced by the unit load ($F_{1,j}$) are the same. Therefore, from Eq (i).

$$a_{BD} = \sum_{j=1}^{n} \frac{F_{1,j}^2 L_j}{AE} \tag{iii}$$

The solution is completed in Table 16.1.

Table 16.1

Member	L_j (m)	$F_{0,j}$	$F_{1,j}$	$F_{0,j}F_{1,j}L_j$	$F_{1,j}^2 L_j$	$F_{a,j}$
AB	L	0	$-0\cdot71$	0	$0\cdot5L$	$+0\cdot40P$
BC	L	0	$-0\cdot71$	0	$0\cdot5L$	$+0\cdot40P$
CD	L	$-P$	$-0\cdot71$	$0\cdot71\,PL$	$0\cdot5L$	$-0\cdot60P$
BD	$1\cdot41L$	–	$1\cdot0$	–	$1\cdot41L$	$-0\cdot56P$
AC	$1\cdot41L$	$1\cdot41P$	$1\cdot0$	$2\cdot0PL$	$1\cdot41L$	$+0\cdot85P$
AD	L	0	$-0\cdot71$	0	$0\cdot5L$	$+0\cdot40\,P$
				$\Sigma = 2\cdot71\,PL$	$\Sigma = 4\cdot82L$	

From Table 16.1

$$\Delta_{BD} = \frac{2\cdot71\,PL}{AE}, \qquad a_{BD} = \frac{4\cdot82\,L}{AE}$$

Substituting these values in Eq. (i) we have

$$\frac{2\cdot71\,PL}{AE} + X_{BD}\,\frac{4\cdot82\,L}{AE} = 0$$

whence $\qquad\qquad X_{BD} = -0\cdot56P \quad$ (i.e. compression)

The actual forces, $F_{a,j}$, in the members of the complete truss of Fig. 16.18(a) are now calculated using the method of joints and are listed in the final column of Table 16.1.

We note in the above that Δ_{BD} is positive, which means that Δ_{BD} is in the direction of the unit loads, i.e. B approaches D and the diagonal BD in the released structure decreases in length. Therefore in the complete structure the member BD, which prevents this shortening, must be in compression as shown; also a_{BD} will always be positive since it contains the term $F_{1,j}^2$. Finally, we note that the cut member BD is included in the calculation of the displacements in the released structure since its deformation, under a unit load, contributes to a_{BD}.

Example 16.9   Calculate the forces in the members of the truss shown in Fig. 16.19(a). All members have the same cross-sectional area, A, and Young's modulus, E.

By inspection we see that the truss is both internally and externally statically indeterminate since it would remain stable and in equilibrium if one of the diagonals, AD or BD, and the support at C were removed; the degree of indeterminacy is therefore 2. Unlike the truss in Ex. 16.18, we could not remove *any* member since, if BC or CD were removed, the outer half of the truss would become a mechanism while the portion ABDE would remain statically indeterminate. Therefore we select AD and the support at C as the releases, giving the statically determinate truss shown in Fig. 16.19(b); we shall designate the force in the member AD as X_1 and the vertical reaction at C as R_2.

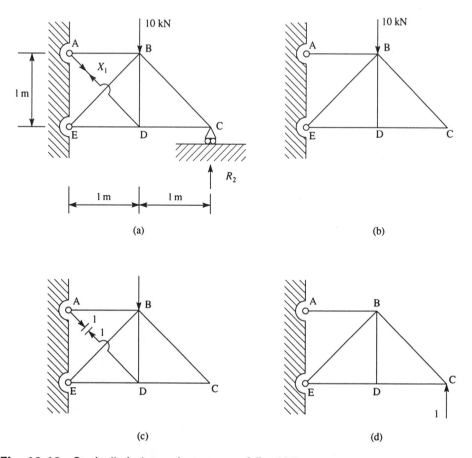

Fig. 16.19 Statically indeterminate truss of Ex. 16.9

In this case we shall have two compatibility conditions, one for the diagonal AD and one for the support at C. We therefore need to investigate three loading cases: one in which the actual loads are applied to the released statically determinate truss in Fig. 16.19(b), a second in which unit loads are applied to the cut member AD (Fig. 16.19(c)) and a third in which a unit load is applied at C in the direction of R_2 (Fig. 16.19(d)). By comparison with the previous example, the compatibility conditions are

$$\Delta_{AD} + a_{11}X_1 + a_{12}R_2 = 0 \tag{i}$$

$$v_C + a_{21}X_1 + a_{22}R_2 = 0 \tag{ii}$$

in which Δ_{AD} and v_C are, respectively, the change in length of the diagonal AD and the vertical displacement of C due to the actual loads acting on the released truss, while a_{11}, a_{12}, etc., are flexibility coefficients, which we have previously defined (see Ex. 16.7). The calculations are similar to those carried out in Ex. 16.8 and are shown in Table 16.2.

Table 16.2

Member	L_j	$F_{0,j}$	$F_{1,j}(X_1)$	$F_{1,j}(R_2)$	$F_{0,j}F_{1,j}(X_1)L_j$	$F_{0,j}F_{1,j}(R_2)L_j$	$F_{1,j}^2(X_1)L_j$	$F_{1,j}^2(R_2)L_j$	$F_{1,j}(X_1)F_{1,j}(R_2)L_j$	$F_{a,j}$
AB	1	10·0	-0·71	-2·0	-7·1	-20·0	0·5	4·0	1·41	0·67
BC	1·41	0	0	-1·41	0	0	0	2·81	0	-4·45
CD	1	0	0	1·0	0	0	0	1·0	0	3·15
DE	1	0	-0·71	1·0	0	0	0·5	1·0	-0·71	0·12
AD	1·41	0	1·0	0	0	0	1·41	0	0	4·28
BE	1·41	-14·14	1·0	1·41	-20·0	-28·11	1·41	2·81	2·0	-5·4
BD	1	0	-0·71	0	0	0	0·5	0	0	-3·03
					$\Sigma = -27·1$	$\Sigma = -48·11$	$\Sigma = 4·32$	$\Sigma = 11·62$	$\Sigma = 2·7$	

From Table 16.2

$$\Delta_{AD} = \sum_{j=1}^{n} \frac{F_{0,j} F_{1,j}(X_1) L_j}{AE} = \frac{-27 \cdot 1}{AE} \quad \text{(i.e. AD increases in length)}$$

$$v_C = \sum_{j=1}^{n} \frac{F_{0,j} F_{1,j}(R_2) L_j}{AE} = \frac{-48 \cdot 11}{AE} \quad \text{(i.e. C displaced downwards)}$$

$$a_{11} = \sum_{j=1}^{n} \frac{F_{1,j}^2(X_1) L_j}{AE} = \frac{4 \cdot 32}{AE}$$

$$a_{22} = \sum_{j=1}^{n} \frac{F_{1,j}^2(R_2) L_j}{AE} = \frac{11 \cdot 62}{AE}$$

$$a_{12} = a_{21} \sum_{j=1}^{n} \frac{F_{1,j}(X_1) F_{1,j}(R_2) L_j}{AE} = \frac{2 \cdot 7}{AE}$$

Substituting in Eqs (i) and (ii) and multiplying through by AE we have

$$-27 \cdot 1 + 4 \cdot 32 X_1 + 2 \cdot 7 R_2 = 0 \tag{iii}$$

$$-48 \cdot 11 + 2 \cdot 7 X_1 + 11 \cdot 62 R_2 = 0 \tag{iv}$$

Solving Eqs (iii) and (iv) we obtain

$$X_1 = 4 \cdot 28 \text{ kN}, \ R_2 = 3 \cdot 15 \text{ kN}$$

The actual forces, $F_{a,j}$, in the members of the complete truss are now calculated by the method of joints and are listed in the final column of Table 16.2.

Self-straining trusses

Statically indeterminate trusses, unlike the statically determinate type, may be subjected to self-straining in which internal forces are present before external loads are applied. Such a situation may be caused by a local temperature change or by an initial lack of fit of a member. In cases such as these, the term on the right-hand side of the compatibility equations, Eq. (ii) in Ex. 16.8 and Eqs (i) and (ii) in Ex. 16.9, would not be zero.

Example 16.10 The truss shown in Fig. 16.20(a) is unstressed when the temperature of each member is the same, but due to local conditions the temperature in the member BC is increased by 30°C. If the cross-sectional area of each member is 200 mm^2 and the coefficient of linear expansion of the members is 7×10^{-6}/°C, calculate the resulting forces in the members; Young's modulus $E = 200\ 000$ N/mm^2.

Due to the temperature rise, the increase in length of the member BC is $3 \times 10^3 \times 30 \times 7 \times 10^{-6} = 0 \cdot 63$ mm. The truss has a degree of internal statical indeterminacy equal to 1 (by inspection). We therefore release the truss by cutting the member BC, which has experienced the temperature rise, as shown in

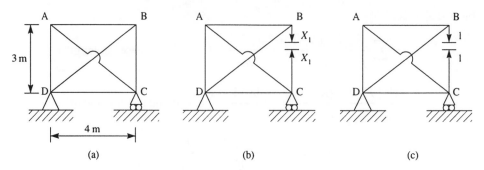

Fig. 16.20   Self-straining due to a temperature change

Fig. 16.20(b); we shall suppose that the force in BC is X_1. Since there are no external loads on the truss, Δ_{BC} is zero and the compatibility condition becomes

$$a_{11}X_1 = -0.63 \text{ mm} \tag{i}$$

in which, as before,

$$a_{11} = \sum_{j=1}^{n} \frac{F_{1,j}^2 L_j}{AE}$$

Note that the extension of BC is negative since it is opposite in direction to X_1. The solution is now completed in Table 16.3.

$$\text{Hence} \quad a_{11} = \frac{48\,000}{200 \times 200\,000} = 1.2 \times 10^{-3}$$

Thus, from Eq. (i)

$$X_1 = -525 \text{ N}$$

The forces, $F_{a,j}$, in the members of the complete truss are given in the final column of Table 16.3.

An alternative approach to the solution of statically indeterminate trusses, both self-straining and otherwise, is to use the principle of the stationary value of the

Table 16.3

Member	L_j (mm)	$F_{1,j}$	$F_{1,j}^2 L_j$	$F_{a,j}$ (N)
AB	4000	1·33	7111·1	−700
BC	3000	1·0	3000·0	−525
CD	4000	1·33	7111·1	−700
DA	3000	1·0	3000·0	−525
AC	5000	−1·67	13888·9	875
DB	5000	−1·67	13888·9	875
			$\Sigma = 48000\cdot0$	

total complementary energy. Thus, for the truss of Ex. 16.8, the total complementary energy, C, is, from Eq. (15·40), given by

$$C = \sum_{j=1}^{n} \int_{0}^{F_j} \delta_j \, dF_j - P\Delta_C$$

in which Δ_C is the displacement of the joint C in the direction of P. Let us suppose that the member BD is short by an amount λ_{BD} (i.e. the lack of fit of BD), then

$$C = \sum_{j=1}^{n} \int_{0}^{F_j} \delta_j \, dF_j - P\Delta_C X_1 \lambda_{BD}$$

From the principle of the stationary value of the total complementary energy we have

$$\frac{\partial C}{\partial X_1} = \sum_{j=1}^{n} \delta_j \frac{\partial F_j}{\partial X_1} - \lambda_{BD} = 0 \tag{16.8}$$

Assuming that the truss is linearly elastic, Eq. (16.8) may be written

$$\frac{\partial C}{\partial X_1} = \sum_{j=1}^{n} \frac{F_j L_j}{A_j E_j} \frac{\partial F_j}{\partial X_1} - \lambda_{BD} = 0 \tag{16.9}$$

or since, for linearly elastic systems, the complementary energy, C, and the strain energy, U, are interchangeable,

$$\frac{\partial U}{\partial X_1} = \sum_{j=1}^{n} \frac{F_j L_j}{A_j E_j} \frac{\partial F_j}{\partial X_1} = \lambda_{BD} \tag{16.10}$$

Equation (16.10) expresses mathematically what is generally referred to as Castigliano's second theorem which states that

For a linearly elastic structure the partial differential coefficient of the total strain energy of the structure with respect to the force in a redundant member is equal to the initial lack of fit of that member.

The application of complementary energy to the solution of statically indeterminate trusses is very similar to the method illustrated in Exs 16.8–16.10. For example, the solution of Ex. 16.8 would proceed as follows.

Again we select BD as the redundant member and suppose that the force in BD is X_1. The forces, $F_{a,j}$, in the complete truss are calculated in terms of P and X_1, and hence $\partial F_{a,j}/\partial X_1$ obtained for each member. The term $(F_{a,j} L_j/A_j E_j) \partial F_{a,j}/\partial X_1$ is calculated for each member and then summed for the complete truss. Equation (16.9) (or (16.10)) in which $\lambda_{BD} = 0$ then gives X_1 in terms of P. The solution is illustrated in Table 16.4. Thus from Eq. (16.9)

$$\frac{1}{AE}(2\cdot71PL + 4\cdot82X_1 L) = 0$$

whence

$$X_1 = -0\cdot56P$$

as before.

Table 16.4

Member	L_j	$F_{a,j}$	$\partial F_{a,j}/\partial X_1$	$F_{a,j}L_j(\partial F_{a,j}/\partial X_1)$
AB	L	$-0.71X_1$	-0.71	$0.5LX_1$
BC	L	$-0.71X_1$	-0.71	$0.5LX_1$
CD	L	$-P-0.71X_1$	-0.71	$(0.71P+0.5X_1)L$
DA	L	$-0.71X_1$	-0.71	$0.5LX_1$
AC	$1.41L$	$1.41P+X_1$	1.0	$(2P+1.41X_1)L$
BD	$1.41L$	X_1	1.0	$1.41X_1L$

$$\Sigma = 2.71PL + 4.82X_1L$$

Of the two approaches illustrated by the two solutions of Ex. 16.8, it can be seen that the use of the principle of the stationary value of the total complementary energy results in a slightly more algebraically clumsy solution. This will be even more the case when the degree of indeterminacy of a structure is greater than 1 and the forces $F_{a,j}$ are expressed in terms of the applied loads and all the redundant forces. There will, of course, be as many equations of the form of Eq. (16.9) as there are redundancies.

16.6 Braced beams

Some structures consist of beams that are stiffened by trusses in which the beam portion of the structure is capable of resisting shear forces and bending moments in addition to axial forces. Generally, however, displacements produced by shear forces are negligibly small and may be ignored. Therefore, in such structures we shall assume that the members of the truss portion of the structure resist axial forces only while the beam portion resists bending moments and axial forces; in some cases the axial forces in the beam are also ignored since their effect, due to the larger area of cross-section, is small.

Example 16.11   The beam ABC shown in Fig. 16.21(a) is simply supported and stiffened by a truss whose members are capable of resisting axial forces only. The beam has a cross-sectional area of 6000 mm^2 and a second moment of area of $7.2 \times 10^6 \text{ mm}^4$. If the cross-sectional area of the members of the truss is 400 mm^2, calculate the forces in the members of the truss and the maximum value of the bending moment in the beam. Young's modulus, E, is the same for all members.

We observe that if the beam were capable of resisting axial forces only, the structure would be a relatively simple statically determinate truss. However, the beam, in addition to axial forces, resists bending moments (we are ignoring the effect of shear) so that the structure is statically indeterminate with a degree of indeterminacy equal to 1, the bending moment at any section of the beam. Therefore we require just one release to produce a statically determinate structure; it does not necessarily have to be the bending moment in the beam, so we shall choose the truss member ED as shown in Fig. 16.21(b) since this will produce benefits from symmetry when we consider the unit load application in Fig. 16.21(c).

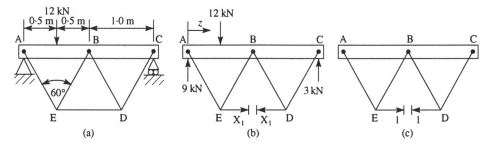

Fig. 16.21 Braced beam of Ex. 16.11

In this example displacements are produced by the bending of the beam as well as by the axial forces in the beam and truss members. Thus, in the released structure of Fig. 16.21(b), the relative displacement, Δ_{ED}, of the cut ends of the member ED is, from the unit load method (see Eq. (iii) of Ex. 15.9 and Exs 16.8–16.10), given by

$$\Delta_{ED} = \int_{ABC} \frac{M_0 M_1}{EI} \, dz + \sum_{j=1}^{n} \frac{F_{0,j} F_{1,j} L_j}{A_j E} \tag{i}$$

in which M_0 is the bending moment at any section of the beam ABC in the released structure. Further, the flexibility coefficient, a_{11}, of the member ED is given by

$$a_{11} = \int_{ABC} \frac{M_1^2}{EI} \, dz + \sum_{j=1}^{n} \frac{F_{1,j}^2 L_j}{A_j E} \tag{ii}$$

In Eqs (i) and (ii) the length, L_j, is constant, as is Young's modulus, E. These may therefore be omitted in the calculation of the summation terms in Table 16.5.

Examination of Table 16.5 shows that the displacement, Δ_{ED}, in the released structure is due solely to the bending of the beam, i.e. the second term on the right-hand side of Eq. (i) is zero; this could have been deduced by inspection of the released structure. Also the contribution to displacement of the axial forces in the beam may be seen, from the first two terms in the penultimate column of Table 16.5, to be negligibly small.

Table 16.5

Member	A_j (mm²)	$F_{0,j}$ (kN)	$F_{1,j}$	$F_{0,j}F_{1,j}/A_j$	$F_{1,j}^2/A_j$	$F_{a,j}$ (kN)
AB	6000	0	−0·5	0	$4·17 \times 10^{-5}$	−2·01
BC	6000	0	−0·5	0	$4·17 \times 10^{-5}$	−2·01
CD	400	0	1·0	0	$2·5 \times 10^{-3}$	4·02
ED	400	0	1·0	0	$2·5 \times 10^{-3}$	4·02
BD	400	0	−1·0	0	$2·5 \times 10^{-3}$	−4·02
EB	400	0	−1·0	0	$2·5 \times 10^{-3}$	−4·02
AE	400	0	1·0	0	$2·5 \times 10^{-3}$	4·02
				$\Sigma = 0$	$\Sigma = 0·0126$	

The contribution to Δ_{ED} of the bending of the beam will now be calculated. Thus from Fig. 16.21(b)

$$M_0 = 9z \quad (0 \leqslant z \leqslant 0.5 \text{ m})$$

$$M_0 = 9z - 12(z - 0.5) = 6 - 3z \quad (0.5 \leqslant z \leqslant 2.0 \text{ m})$$

$$M_1 = -0.87z \quad (0 \leqslant z \leqslant 1.0 \text{ m})$$

$$M_1 = -0.87z + 1.74(z - 1.0) = 0.87z - 1.74 \quad (1.0 \leqslant z \leqslant 2.0 \text{ m})$$

Substituting from M_0 and M_1 in Eq. (i) we have

$$\int_{ABC} \frac{M_0 M_1}{EI} \, dz$$

$$= \frac{1}{EI} \left[-\int_0^{0.5} 9 \times 0.87z^2 \, dz - \int_{0.5}^{1.0} (6 - 3z)0.87z \, dz + \int_{1.0}^{2.0} (6 - 3z)(0.87z - 1.74) \, dz \right]$$

from which
$$\int_{ABC} \frac{M_0 M_1}{EI} \, dz = -\frac{0.33 \times 10^6}{E} \text{ mm}$$

Similarly
$$\int_{ABC} \frac{M_1^2}{EI} \, dz = \frac{1}{EI} \left[\int_0^{1.0} 0.87^2 z^2 \, dz + \int_{1.0}^{2.0} (0.87z - 1.74)^2 \, dz \right]$$

from which
$$\int_{ABC} \frac{M_1^2}{EI} \, dz = \frac{0.083 \times 10^3}{EI} \text{ mm/N}$$

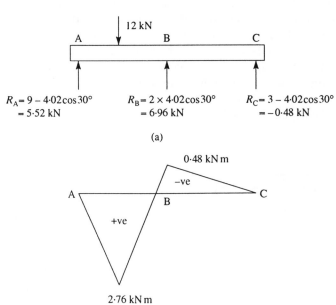

(a)

(b)

Fig. 16.22   Bending moment distribution in the beam of Ex. 16.11

The compatibility condition gives

$$\Delta_{ED} + a_{11}X_1 = 0$$

whence

$$-\frac{0.33 \times 10^6}{E} + \frac{0.083 \times 10^3}{E} X_1 = 0$$

which gives $\quad X_1 = 4018.1 \text{ N} \quad$ or $\quad X_1 = 4.02 \text{ kN}$

The axial forces in the beam and truss may now be calculated using the method of joints and are given in the final column of Table 16.5. The forces acting on the beam in the complete structure are shown in Fig. 16.22(a) together with the bending moment diagram in Fig. 16.22(b), from which we see that the maximum bending moment in the beam is 2.76 kN m.

16.7   Portal frames

The flexibility method may be applied to the analysis of portal frames although, as we shall see, in all but simple cases the degree of statical indeterminacy is high so that the number of compatibility equations requiring solution becomes too large for hand computation.

Consider the portal frame shown in Fig. 16.23(a). From Section 16.2 we see that the frame, together with its foundation, forms a single two-dimensional ring and is therefore three times statically indeterminate. Therefore we require three releases to obtain the statically determinate primary structure. These may be obtained by removing the foundation at the foot of one of the vertical legs as shown in Fig. 16.23(b); we then have two releases of force and one of moment and the primary structure is, in effect, a cranked cantilever. In this example there would be three compatibility equations requiring solution, two of translation and one of rotation. Clearly, for a plane, two-bay portal frame we would have six compatibility equations so that the solution would then become laborious; further additions to the frame would make a hand method of solution impracticable. Furthermore, as we shall see in Section 16.10, the moment distribution method produces a rapid solution for frames although it should be noted that using this method requires that the sway of the frame, that is its lateral movement, is considered separately whereas, in the flexibility method, sway is automatically included.

Example 16.12   Determine the distribution of bending moment in the frame ABCD shown in Fig. 16.24(a); the flexural rigidity of all the members of the frame is EI.

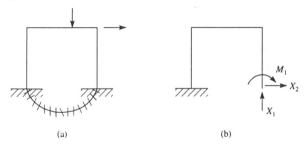

(a)   (b)

Fig. 16.23   Indeterminacy of a portal frame

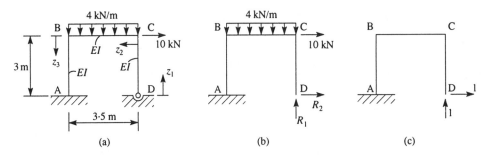

Fig. 16.24 Portal frame of Ex. 16.12

Comparison with Fig. 16.23(a) shows that the frame has a degree of statical indeterminacy equal to two since the vertical leg CD is pinned to the foundation at D. We therefore require just two releases of reaction, as shown in Fig. 16.24(b), to obtain the statically determinate primary structure. For frames of this type it is usual to neglect the displacements produced by axial force and to assume that they are caused solely by bending.

The point D in the primary structure will suffer vertical and horizontal displacements, $\Delta_{D,V}$ and $\Delta_{D,H}$. Thus if we designate the redundant reactions as R_1 and R_2, the equations of compatibility are

$$\Delta_{D,V} + a_{11}R_1 + a_{12}R_2 = 0 \tag{i}$$

$$\Delta_{D,H} + a_{21}R_1 + a_{22}R_2 = 0 \tag{ii}$$

in which the flexibility coefficients have their usual meaning. Again, as in the preceding examples, we employ the unit load method to calculate the displacements and flexibility coefficients. Thus

$$\Delta_{D,V} = \sum \int_L \frac{M_0 M_{1,V}}{EI} \, dz$$

in which $M_{1,V}$ is the bending moment at any point in the frame due to a unit load applied vertically at D.

Similarly
$$\Delta_{D,H} = \sum \int_L \frac{M_0 M_{1,H}}{EI} \, dz$$

and
$$a_{11} = \sum \int_L \frac{M_{1,V}^2}{EI} \, dz, \quad a_{22} = \sum \int_L \frac{M_{1,H}^2}{EI} \, dz, \quad a_{12} = a_{21} = \sum \int_L \frac{M_{1,V} M_{1,H}}{EI} \, dz$$

We shall now write down expressions for bending moment in the members of the frame; we shall designate a bending moment as positive when it causes tension on the outside of the frame. Thus in DC

$$M_0 = 0, \quad M_{1,V} = 0, \quad M_{1,H} = -1z_1$$

In CB
$$M_0 = 4z_2 \frac{z_2}{2} = 2z_2^2, \quad M_{1,V} = -1z_2, \quad M_{1,H} = -3$$

In BA $M_0 = 4 \times 3 \cdot 5 \times 1 \cdot 75 + 10z_3 = 24 \cdot 5 + 10z_3, \, M_{1,V} = -3 \cdot 5, \, M_{1,H} = -1(3-z_3)$

Hence

$$\Delta_{D,V} = \frac{1}{EI} \left[\int_0^{3.5} (-2z_2^3)\, dz_2 + \int_0^3 -(24\cdot5 + 10z_3)3\cdot5\, dz_3 \right] = -\frac{489\cdot8}{EI}$$

$$\Delta_{D,H} = \frac{1}{EI} \left[\int_0^{3.5} (-6z_2^2)\, dz_2 + \int_0^3 -(24\cdot5 + 10z_3)(3 - z_3)\, dz_3 \right] = -\frac{241\cdot0}{EI}$$

$$a_{11} = \frac{1}{EI} \left[\int_0^{3.5} z_2^2\, dz_2 + \int_0^3 3\cdot5^2\, dz_3 \right] = \frac{51\cdot0}{EI}$$

$$a_{22} = \frac{1}{EI} \left[\int_0^3 z_1^2\, dz_1 + \int_0^{3.5} 3^2\, dz_2 + \int_0^3 (3 - z_3)^2\, dz_3 \right] = \frac{49\cdot5}{EI}$$

$$a_{12} = a_{21} = \frac{1}{EI} \left[\int_0^{3.5} 3z_2\, dz_2 + \int_0^3 3\cdot5(3 - z_3)\, dz_3 \right] = \frac{34\cdot1}{EI}$$

Substituting for $\Delta_{D,V}$, $\Delta_{D,H}$, a_{11}, etc., in Eqs (i) and (ii) we obtain

$$-\frac{489\cdot8}{EI} + \frac{51\cdot0}{EI} R_1 + \frac{34\cdot1}{EI} R_2 = 0 \tag{iii}$$

and

$$-\frac{241\cdot0}{EI} + \frac{34\cdot1}{EI} R_1 + \frac{49\cdot5}{EI} R_2 = 0 \tag{iv}$$

Solving Eqs (iii) and (iv) we have

$$R_1 = 11\cdot8 \text{ kN}, \quad R_2 = -3\cdot3 \text{ kN}$$

The bending moment diagram is then drawn as shown in Fig. 16.25.

It can be seen that the amount of computation for even the relatively simple frame of Ex. 16.12 is quite considerable. Generally, therefore, as stated previously, the moment distribution method or a computer-based analysis would be employed.

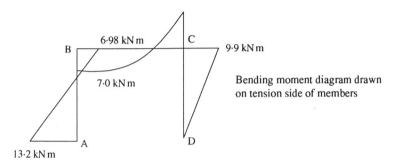

Fig. 16.25 Bending moment diagram for the frame of Ex. 16.12 (diagram drawn on tension side of members)

16.8 Two-pinned arches

In Chapter 6 we saw that a three-pinned arch is statically determinate due to the presence of the third pin or hinge at which the internal bending moment is zero; in effect the presence of the third pin provides a release. Therefore a two-pinned arch such as that shown in Fig. 16.26(a) has a degree of statical indeterminacy equal to 1. This is also obvious from inspection since, as in the three-pinned arch, there are two reactions at each of the supports.

The analysis of two-pinned arches, i.e. the determination of the support reactions, may be carried out using the flexibility method; again, as in the case of portal frames, it is usual to ignore the effect of axial force on displacements and to assume that they are caused by bending action only.

The arch in Fig. 16.26(a) has a profile whose equation may be expressed in terms of the reference axes x and y. The second moment of area of the cross-section of the arch is I and we shall designate the distance round the profile from A as s.

Initially we choose a release, say the horizontal reaction, R_1, at B, to obtain the statically determinate primary structure shown in Fig. 16.26(b). We then employ the unit load method to determine the horizontal displacement, $\Delta_{B,H}$, of B in the primary structure and the flexibility coefficient, a_{11}. Then, from compatibility

$$\Delta_{B,H} - a_{11}R_1 = 0 \tag{16.11}$$

in which the term containing R_1 is negative since R_1 is opposite in direction to the unit load (see Fig. 16.26(c)).

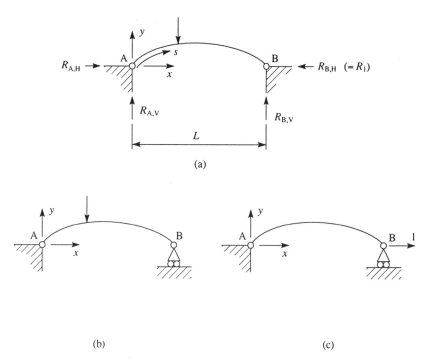

(a)

(b)   (c)

Fig. 16.26 Solution of a two-pinned arch

Then, with the usual notation

$$\Delta_{B,H} = \int_{Profile} \frac{M_0 M_1}{EI} \, ds \tag{16.12}$$

in which M_0 depends upon the applied loading and $M_1 = 1y$ (a moment is positive if it produces tension on the undersurface of the arch). Also

$$a_{11} = \int_{Profile} \frac{M_1^2}{EI} \, ds = \int_{Profile} \frac{y^2}{EI} \, ds \tag{16.13}$$

Substituting for M_1 in Eq. (16.12) and then for $\Delta_{B,H}$ and a_{11} in Eq. (16.11) we obtain

$$R_1 = \frac{\displaystyle\int_{Profile} \frac{M_0 y}{EI} \, ds}{\displaystyle\int_{Profile} \frac{y^2}{EI} \, ds} \tag{16.14}$$

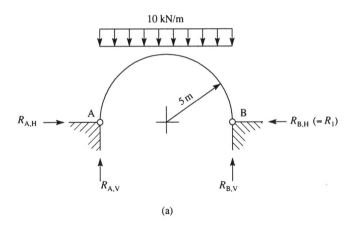

(a)

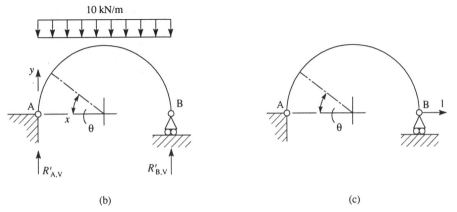

(b)   (c)

Fig. 16.27 Semicircular arch of Ex. 16.13

Example 16.13 Determine the support reactions in the semicircular two-pinned arch shown in Fig. 16.27(a). The flexural rigidity, EI, of the arch is constant throughout.

Again we shall choose the horizontal reaction at the support B as the release so that $R_{B,H}$ $(=R_1)$ is given directly by Eq. (16.14) in which M_0 and s are functions of x and y. The computation will therefore be simplified if we use an angular coordinate system so that, from the primary structure shown in Fig. 16.27(b)

$$M_0 = R'_{B,V}(5 + 5 \cos \theta) - \tfrac{10}{2}(5 + 5 \cos \theta)^2 \tag{i}$$

in which $R'_{B,V}$ is the vertical reaction at B in the primary structure. From Fig. 16.27(b) in which, from symmetry, $R'_{B,V} = R'_{A,V}$, we have $R'_{B,V} = 50$ kN. Substituting for $R'_{B,V}$ in Eq. (i) we obtain

$$M_0 = 125 \sin^2 \theta \tag{ii}$$

Also $y = 5 \sin \theta$ and $ds = 5 \, d\theta$, so that from Eq. (16.14) we have

$$R_1 = \frac{\int_0^\pi 125 \sin^2\theta \; 5\sin\theta \; 5 d\theta}{\int_0^\pi 25 \sin^2\theta \; 5 d\theta}$$

or
$$R_1 = \frac{\int_0^\pi 25 \sin^3 \theta \; d\theta}{\int_0^\pi \sin^2 \theta \; d\theta} \tag{iii}$$

which gives
$$R_1 = 21 \cdot 2 \text{ kN } (=R_{B,H})$$

The remaining reactions follow from a consideration of the statical equilibrium of the arch and are

$$R_{A,H} = 21 \cdot 2 \text{ kN}, \qquad R_{A,V} = R_{B,V} = 50 \text{ kN}$$

The integrals in Eq. (iii) of Ex. 16.13 are relatively straightforward to evaluate; the numerator may be found by integration by parts, while the denominator is found by replacing $\sin^2 \theta$ by $(1 - \cos 2\theta)/2$. Furthermore, in an arch having a semicircular profile, M_0, y and ds are simply expressed in terms of an angular coordinate system. However, in a two-pinned arch having a parabolic profile this approach cannot be used and complex integrals result. Such cases may be simplified by specifying that the second moment of area of the cross-section of the arch varies round the profile; one such variation is known as the secant assumption and is described below.

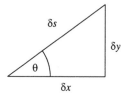

Fig. 16.28   Elemental length of arch

Secant assumption

In Eq. (16.14) the term ds/I appears. If this term could be replaced by a term that is a function of either x or y, the solution would be simplified.

Consider the elemental length, δs, of the arch shown in Fig. 16.28 and its projections, δx and δy, on the x and y axes. From the elemental triangle

$$\delta x = \delta s \cos \theta$$

or, in the limit as $\delta s \to 0$

$$ds = dx/\cos \theta = dx \sec \theta$$

Thus

$$\frac{ds}{I} = \frac{dx \sec \theta}{I}$$

Let us suppose that I varies round the profile of the arch such that $I = I_0 \sec \theta$ where I_0 is the second moment of area at the crown of the arch (i.e. where $\theta = 0$). Then

$$\frac{ds}{I} = \frac{dx \sec \theta}{I_0 \sec \theta} = \frac{dx}{I_0}$$

Thus substituting in Eq. (16.14) for ds/I we have

$$R_1 = \frac{\displaystyle\int_{\text{Profile}} \frac{M_0 y}{EI_0} \, dx}{\displaystyle\int_{\text{Profile}} \frac{y^2}{EI_0} \, dx}$$

or

$$R_1 = \frac{\displaystyle\int_{\text{Profile}} M_0 y \, dx}{\displaystyle\int_{\text{Profile}} y^2 \, dx} \tag{16.15}$$

Example 16.14 Determine the support reactions in the parabolic arch shown in Fig. 16.29 assuming that the second moment of area of the cross-section of the arch varies in accordance with the secant assumption.

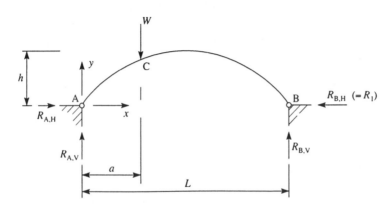

Fig. 16.29 Parabolic arch of Ex. 16.14

The equation of the arch may be shown to be

$$y = \frac{4h}{L^2}(Lx - x^2) \tag{i}$$

Again we shall release the arch at B as in Fig. 16.26(b). Then

$$M_0 = R'_{A,V}x \quad (0 \leqslant x \leqslant a)$$

$$M_0 = R'_{A,V}x - W(x-a) \quad (a \leqslant x \leqslant L)$$

in which $R'_{A,V}$ is the vertical reaction at A in the released structure. Now taking moments about B we have

$$R'_{A,V}L - W(L-a) = 0$$

whence

$$R'_{A,V} = \frac{W}{L}(L-a)$$

Substituting in the expressions for M_0 gives

$$M_0 = \frac{W}{L}(L-a)x \quad (0 \leqslant x \leqslant a) \tag{ii}$$

$$M_0 = \frac{Wa}{L}(L-x) \quad (a \leqslant x \leqslant L) \tag{iii}$$

The denominator in Eq. (16.15) may be evaluated separately. Thus, from Eq. (i)

$$\int_{\text{Profile}} y^2 \, \mathrm{d}x = \int_0^L \left(\frac{4h}{L^2}\right)^2 (Lx - x^2)^2 \, \mathrm{d}x = \frac{8h^2L}{15}$$

Then, from Eq. (16.15) and Eqs (ii) and (iii)

$$R_1 = \frac{15}{8h^2L}\left[\int_0^a \frac{W}{L}(L-a)x \frac{4h}{L^2}(Lx - x^2) \, \mathrm{d}x + \int_a^L \frac{Wa}{L}(L-x)\frac{4h}{L^2}(Lx - x^2)\, \mathrm{d}x\right]$$

which gives

$$R_1 = \frac{5Wa}{8hL^3}(L^3 + a^3 - 2La^2) \tag{iv}$$

The remaining support reactions follow from a consideration of the statical equilibrium of the arch.

If, in Ex. 16.14, we had expressed the load position in terms of the span of the arch, say $a = kL$, Eq. (iv) in Ex. 16.14 becomes

$$R_1 = \frac{5WL}{8h}(k + k^4 - 2k^3) \tag{16.16}$$

Therefore, for a series of concentrated loads positioned at distances k_1L, k_2L, k_3L, etc., from A, the reaction, R_1, may be calculated for each load acting separately using Eq. (16.16) and the total reaction due to all the loads obtained by superposition.

The result expressed in Eq. (16.16) may be used to determine the reaction, R_1, due to a part-span uniformly distributed load. Consider the arch shown in Fig 16.30. The arch profile is parabolic and its second moment of area varies as the secant assumption. An elemental length, δx, of the load produces a load $w\,\delta x$ on the arch. Thus, since δx is very small, we may regard this load as a concentrated load. This will then produce an increment, δR_1, in the horizontal support reaction which, from Eq. (16.16), is given by

$$\delta R_1 = \frac{5}{8}\, w\, \delta x\, \frac{L}{h}\,(k + k^4 - 2k^3)$$

in which $k = x/L$. Therefore, substituting for k in the expression for δR_1 and then integrating over the length of the load we obtain

$$R_1 = \frac{5wL}{8h} \int_{x_1}^{x_2} \left(\frac{x}{L} + \frac{x^4}{L^4} - \frac{2x^3}{L^3} \right) dx$$

which gives

$$R_1 = \frac{5wL}{8h} \left[\frac{x^2}{2L} + \frac{x^5}{5L^4} - \frac{x^4}{2L^3} \right]_{x_1}^{x_2}$$

For a uniformly distributed load covering the complete span, i.e. $x_1 = 0$, $x_2 = L$, we have

$$R_1 = \frac{5wL}{8h} \left(\frac{L^2}{2L} + \frac{L^5}{5L^4} - \frac{L^4}{2L^3} \right) = \frac{wL^2}{8h}$$

The bending moment at any point (x, y) in the arch is then

$$M = \frac{wL}{2} x - \frac{wx^2}{2} - \frac{wL^2}{8h} \left[\frac{4h}{L^2}(Lx - x^2) \right]$$

i.e.

$$M = \frac{wL}{2} x - \frac{wx^2}{2} - \frac{wL}{2} x + \frac{wx^2}{2} = 0$$

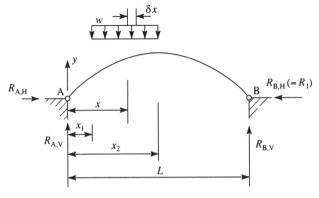

Fig. 16.30 Parabolic arch carrying a part-span uniformly distributed load

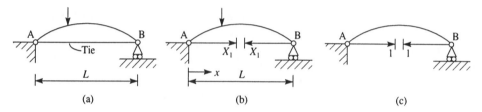

Fig. 16.31 Solution for a tied two-pinned arch

Therefore, for a parabolic two-pinned arch carrying a uniformity distributed load over its complete span, the bending moment in the arch is everywhere zero; the same result was obtained for the three-pinned arch in Chapter 6.

Although the secant assumption appears to be an artificial simplification in the solution of parabolic arches it would not, in fact, produce a great variation in second moment of area in, say, large-span shallow arches. The assumption would therefore provide reasonably accurate solutions for some practical cases.

Tied arches

In some cases the horizontal support reactions are replaced by a tie which connects the ends of the arch as shown in Fig. 16.31(a). In this case we select the axial force, X_1, in the tie as the release. The primary structure is then as shown in Fig. 16.31(b) with the tie cut. The unit load method, Fig. 16.31(c), is then used to determine the horizontal displacement of B in the primary structure. This displacement will receive contributions from the bending of the arch and the axial force in the tie. Thus, with the usual notation

$$\Delta_{B,H} = \int_{\text{Profile}} \frac{M_0 M_1}{EI}\, ds + \int_0^L \frac{F_0 F_1 L}{AE}\, dx$$

and

$$a_{11} = \int_{\text{Profile}} \frac{M_1^2}{EI}\, ds + \int_0^L \frac{F_1^2 L}{AE}\, dx$$

The compatibility condition is then

$$\Delta_{B,H} + a_{11} X_1 = 0$$

Segmental arches

A segmental arch is one comprising segments having different curvatures or different equations describing their profiles. The analysis of such arches is best carried out using a computer-based approach such as the stiffness method in which the stiffness of an individual segment may be found by determining the force–displacement relationships using an energy approach. Such considerations are, however, outside the scope of this book.

16.9 Slope–deflection method

An essential part of the computer-based stiffness method of analysis and also of the moment distribution method are the slope–deflection relationships for beam elements. In these, the shear forces and moments at the ends of a beam element are related to the end displacements and rotations. In addition these relationships provide a method of solution for the determination of end moments in statically indeterminate beams and frames; this method is known as the *slope–deflection method*.

Consider the beam, AB, shown in Fig. 16.32. The beam has flexural rigidity EI and is subjected to moments, M_{AB} and M_{BA}, and shear forces, S_{AB} and S_{BA}, at its ends. The shear forces and moments produce displacements v_A and v_B and rotations θ_A and θ_B as shown. Here we are concerned with moments at the ends of a beam. The usual sagging/hogging sign convention is therefore insufficient to describe these moments since a clockwise moment at the left-hand end of a beam coupled with an anticlockwise moment at the right-hand end would induce a positive bending moment at all sections of the beam. We shall therefore adopt a sign convention such that the moment at a point is positive when it is applied in a clockwise sense and negative when in an anticlockwise sense; thus in Fig. 16.32 both moments M_{AB} and M_{BA} are positive. We shall see in the solution of a particular problem how these end moments are interpreted in terms of the bending moment distribution along the length of a beam. In the analysis we shall ignore axial force effects since these would have a negligible effect in the equation for moment equilibrium. Also, the moments M_{AB} and M_{BA} are independent of each other but the shear forces, which in the absence of lateral loads are equal and opposite, depend upon the end moments.

From Eq. (13.3) and Fig. 16.32

$$EI\frac{d^2v}{dz^2} = -M_{AB} + S_{AB}z$$

Hence
$$EI\frac{dv}{dz} = -M_{AB}z + S_{AB}\frac{z^2}{2} + C_1 \tag{16.17}$$

and
$$EIv = -M_{AB}\frac{z^2}{2} + S_{AB}\frac{z^3}{6} + C_1z + C_2 \tag{16.18}$$

When $z = 0$, $\quad \dfrac{dv}{dz} = \theta_A \quad$ and $\quad v = v_A$

Therefore, from Eq. (16.17) $C_1 = EI\theta_A$ and from Eq. (16.18), $C_2 = EIv_A$. Equations (16.17) and (16.18) then become, respectively,

$$EI\frac{dv}{dz} = -M_{AB}z + S_{AB}\frac{z^2}{2} + EI\theta_A \tag{16.19}$$

and
$$EIv = -M_{AB}\frac{z^2}{2} + S_{AB}\frac{z^3}{6} + EI\theta_Az + EIv_A \tag{16.20}$$

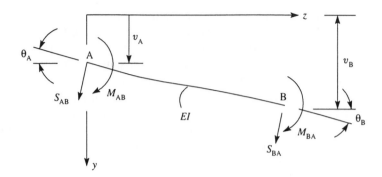

Fig. 16.32   Slope and deflection of a beam

Also, at $z = L$, $dv/dz = \theta_B$ and $v = v_B$. Thus, from Eqs (16.19) and (16.20) we have

$$EI\theta_B = -M_{AB}L + S_{AB}\frac{L}{2} + EI\theta_A \qquad (16.21)$$

and

$$EIv_B = -M_{AB}\frac{L^2}{2} + S_{AB}\frac{L^3}{6} + EI\theta_A L + EIv_A \qquad (16.22)$$

Solving Eqs (16.21) and (16.22) for M_{AB} and S_{AB} gives

$$M_{AB} = \frac{2EI}{L}\left[2\theta_A + \theta_B + \frac{3}{L}(v_A - v_B)\right] \qquad (16.23)$$

and

$$S_{AB} = \frac{6EI}{L^2}\left[\theta_A + \theta_B + \frac{2}{L}(v_A - v_B)\right] \qquad (16.24)$$

Now, from the moment equilibrium of the beam about B, we have

$$M_{BA} - S_{AB}L + M_{AB} = 0$$

or

$$M_{BA} = S_{AB}L - M_{AB}$$

Substituting for S_{AB} and M_{AB} in this expression from Eqs (16.24) and (16.23) we obtain

$$M_{BA} = \frac{2EI}{L}\left[2\theta_B + \theta_A + \frac{3}{L}(v_A - v_B)\right] \qquad (16.25)$$

Further, since $S_{BA} = -S_{AB}$

$$S_{BA} = \frac{-6EI}{L^2}\left[\theta_A + \theta_B + \frac{2}{L}(v_A - v_B)\right] \qquad (16.26)$$

Equations (16.23)–(16.26) are usually written in the form

$$M_{AB} = \frac{6EI}{L^2} v_A + \frac{4EI}{L} \theta_A - \frac{6EI}{L^2} v_B + \frac{2EI}{L} \theta_B$$

$$S_{AB} = \frac{12EI}{L^3} v_A + \frac{6EI}{L^2} \theta_A - \frac{12EI}{L^3} v_B + \frac{6EI}{L^2} \theta_B$$

$$M_{BA} = \frac{6EI}{L^2} v_A + \frac{2EI}{L} \theta_A - \frac{6EI}{L^2} v_B + \frac{4EI}{L} \theta_B$$

$$S_{BA} = -\frac{12EI}{L^3} v_A - \frac{6EI}{L^2} \theta_A + \frac{12EI}{L^3} v_B - \frac{6EI}{L^2} \theta_B$$

(16.27)

Equations (16.27) are known as the slope–deflection equations and establish force–displacement relationships for the beam as opposed to the displacement–force relationships of the flexibility method. The coefficients that pre-multiply the components of displacement in Eqs (16.27) are known as *stiffness coefficients*.

The beam in Fig. 16.32 is not subject to lateral loads. Clearly, in practical cases, unless we are interested solely in the effect of a sinking support, lateral loads will be present. These will cause additional moments and shear forces at the ends of the beam. Equations (16.23)–(16.26) may then be written as

$$M_{AB} = \frac{2EI}{L} \left[2\theta_A + \theta_B + \frac{3}{L} (v_A - v_B) \right] + M_{AB}^F$$

(16.28)

$$S_{AB} = \frac{6EI}{L^2} \left[\theta_A + \theta_B + \frac{2}{L} (v_A - v_B) \right] + S_{AB}^F$$

(16.29)

$$M_{BA} = \frac{2EI}{L} \left[2\theta_B + \theta_A + \frac{3}{L} (v_A - v_B) \right] + M_{BA}^F$$

(16.30)

$$S_{BA} = \frac{-6EI}{L^2} \left[\theta_A + \theta_B + \frac{2}{L} (v_A - v_B) \right] + S_{BA}^F$$

(16.31)

in which M_{AB}^F and M_{BA}^F are the moments at the ends of the beam caused by the applied loads and correspond to $\theta_A = \theta_B = 0$ and $v_A = v_B = 0$, i.e. they are *fixed-end moments*. Similarly the shear forces S_{AB}^F and S_{BA}^F correspond to the fixed-end case.

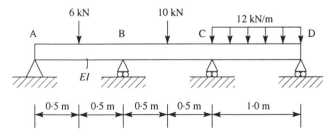

Fig. 16.33 Continuous beam of Ex. 16.15

Example 16.15   Find the support reactions in the three-span continuous beam shown in Fig. 16.33.

The beam in Fig. 16.33 is the beam that was solved using the flexibility method in Ex. 16.7, so that this example provides a comparison between the two methods.

Initially we consider the beam as comprising three separate fixed beams AB, BC and CD and calculate the values of the fixed-end moments, M_{AB}^F, M_{BA}^F, M_{BC}^F, etc. Thus, using the results of Exs 13.20 and 13.22 and remembering that clockwise moments are positive and anticlockwise moments negative

$$M_{AB}^F = -M_{BA}^F = -\frac{6 \times 1 \cdot 0}{8} = -0 \cdot 75 \text{ kN m}$$

$$M_{BC}^F = -M_{CB}^F = -\frac{10 \times 1 \cdot 0}{8} = -1 \cdot 25 \text{ kN m}$$

$$M_{CD}^F = -M_{DC}^F = -\frac{12 \times 1 \cdot 0^2}{12} = -1 \cdot 0 \text{ kN m}$$

In the beam of Fig. 16.33 the vertical displacements at all the supports are zero, i.e. v_A, v_B, v_C and v_D are zero. Therefore, from Eqs (16.28) and (16.30) we have

$$M_{AB} = \frac{2EI}{1 \cdot 0} (2\theta_A + \theta_B) - 0 \cdot 75 \tag{i}$$

$$M_{BA} = \frac{2EI}{1 \cdot 0} (2\theta_B + \theta_A) + 0 \cdot 75 \tag{ii}$$

$$M_{BC} = \frac{2EI}{1 \cdot 0} (2\theta_B + \theta_C) - 1 \cdot 25 \tag{iii}$$

$$M_{CB} = \frac{2EI}{1 \cdot 0} (2\theta_C + \theta_B) + 1 \cdot 25 \tag{iv}$$

$$M_{CD} = \frac{2EI}{1 \cdot 0} (2\theta_C + \theta_D) - 1 \cdot 0 \tag{v}$$

$$M_{DC} = \frac{2EI}{1 \cdot 0} (2\theta_D + \theta_C) + 1 \cdot 0 \tag{vi}$$

From the equilibrium of moments at the supports

$$M_{AB} = 0, \qquad M_{BA} + M_{BC} = 0, \qquad M_{CB} + M_{CD} = 0, \qquad M_{DC} = 0$$

Substituting for M_{AB}, etc., from Eqs (i)–(vi) in these expressions we obtain

$$4EI\theta_A + 2EI\theta_B - 0 \cdot 75 = 0 \tag{vii}$$

$$2EI\theta_A + 8EI\theta_B + 2EI\theta_C - 0 \cdot 5 = 0 \tag{viii}$$

$$2EI\theta_B + 8EI\theta_C + 2EI\theta_D + 0 \cdot 25 = 0 \tag{ix}$$

$$4EI\theta_D + 2EI\theta_C + 1 \cdot 0 = 0 \tag{x}$$

The solution of Eqs (vii)–(x) gives

$$EI\theta_A = 0.183, \qquad EI\theta_B = 0.008, \qquad EI\theta_C = 0.033, \qquad EI\theta_D = -0.267$$

Substituting these values in Eqs (i)–(vi) gives

$$M_{AB} = 0, \quad M_{BA} = 1.15, \quad M_{BC} = -1.15, \quad M_{CB} = 1.4, \quad M_{CD} = -1.4, \quad M_{DC} = 0$$

The end moments acting on the three spans of the beam are now shown in Fig. 16.34. They produce reactions R_{AB}, R_{BA}, etc., at the supports; thus

$$R_{AB} = -R_{BA} = -\frac{1.15}{1.0} = -1.15 \text{ kN}$$

$$R_{BC} = -R_{CB} = -\frac{(1.4 - 1.15)}{1.0} = -0.25 \text{ kN}$$

$$R_{CD} = -R_{DC} = \frac{1.4}{1.0} = 1.40 \text{ kN}$$

Therefore, due to the end moments *only*, the support reactions are

$$R_{A,M} = -1.15 \text{ kN}, \qquad R_{B,M} = 1.15 - 0.25 = 0.9 \text{ kN},$$
$$R_{C,M} = 0.25 + 1.4 = 1.65 \text{ kN}, \qquad R_{D,M} = -1.4 \text{ kN}$$

In addition to these reactions there are the reactions due to the actual loading, which may be obtained by analysing each span as a simply supported beam (the effects of the end moments have been calculated above). In this example these reactions may be obtained by inspection. Thus

$$R_{A,S} = 3.0 \text{ kN}, \qquad R_{B,S} = 3.0 + 5.0 = 8.0 \text{ kN}, \qquad R_{C,S} = 5.0 + 6.0 = 11.0 \text{ kN},$$
$$R_{D,S} = 6.0 \text{ kN}$$

The final reactions at the supports are then

$$R_A = R_{A,M} + R_{A,S} = -1.15 + 3.0 = 1.85 \text{ kN}$$

$$R_B = R_{B,M} + R_{B,S} = 0.9 + 8.0 = 8.9 \text{ kN}$$

$$R_C = R_{C,M} + R_{C,S} = 1.65 + 11.0 = 12.65 \text{ kN}$$

$$R_D = R_{D,M} + R_{D,S} = -1.4 + 6.0 = 4.6 \text{ kN}$$

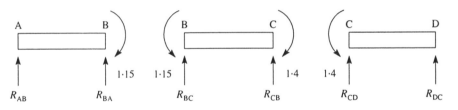

Fig. 16.34 Moments and reactions at the ends of the spans of the continuous beam of Ex. 16.15

Alternatively, we could have obtained these reactions by the slightly lengthier procedure of substituting for θ_A, θ_B, etc., in Eqs (16.29) and (16.31). Thus, for example,

$$S_{AB} = -R_A = \frac{6EI}{L^2}(\theta_A + \theta_B) - 3\cdot0 \quad (v_A = v_B = 0)$$

which gives $R_A = 1\cdot85$ kN as before.

Comparing the above solution with that of Ex. 16.7 we see that there are small discrepancies; these are caused by rounding-off errors.

Having obtained the support reactions, the bending moment distribution (reverting to the sagging (positive) and hogging (negative) sign convention) is obtained in the usual way and is shown in Fig. 16.35.

Example 16.16   Determine the end moments in the members of the portal frame shown in Fig. 16.36; the second moment of area of the vertical members is $2\cdot5I$ while that of the horizontal member is I.

In this particular problem the approach is very similar to that for the continuous beam of Ex. 16.15. However, due to the unsymmetrical geometry of the frame and also to the application of the 10 kN load, the frame will sway such that there will be

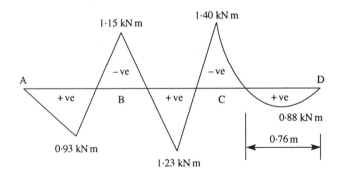

Fig. 16.35   Bending moment diagram for the beam of Ex. 16.15

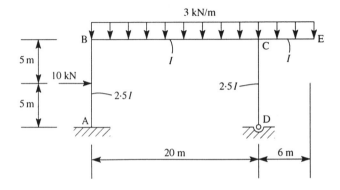

Fig. 16.36   Portal frame of Ex. 16.16

horizontal displacements, v_B and v_C, at B and C in the members BA and CD. Since we are ignoring displacements produced by axial forces then $v_B = v_C = v_1$, say. We would, in fact, have a similar situation in a continuous beam if one or more of the supports experienced settlement. Also we note that the rotation, θ_A, at A must be zero since the end A of the member AB is fixed.

Initially, as in Ex. 16.15, we calculate the fixed-end moments in the members of the frame, again using the results of Exs 13.20 and 13.22. The effect of the cantilever CE may be included by replacing it by its end moment, thereby reducing the number of equations to be solved. Thus, from Fig. 16.36 we have

$$M_{CE}^F = -\frac{3 \times 6^2}{2} = -54 \text{ kN m}$$

$$M_{AB}^F = -M_{BA}^F = -\frac{10 \times 10}{8} = -12 \cdot 5 \text{ kN m}$$

$$M_{BC}^F = -M_{CB}^F = -\frac{3 \times 20^2}{12} = -100 \text{ kN m}, \qquad M_{CD}^F = M_{DC}^F = 0$$

Now, from Eqs (16.28) and (16.30)

$$M_{AB} = \frac{2 \times 2 \cdot 5EI}{10}\left(\theta_B - \frac{3}{10}v_1\right) - 12 \cdot 5 \tag{i}$$

$$M_{BA} = \frac{2 \times 2 \cdot 5EI}{10}\left(2\theta_B - \frac{3}{10}v_1\right) + 12 \cdot 5 \tag{ii}$$

In Eqs (i) and (ii) we are assuming that the displacement, v_1, is to the right. Furthermore

$$M_{BC} = \frac{2EI}{20}(2\theta_B + \theta_C) - 100 \tag{iii}$$

$$M_{CB} = \frac{2EI}{20}(2\theta_C + \theta_B) + 100 \tag{iv}$$

$$M_{CD} = \frac{2 \times 2 \cdot 5EI}{10}\left(2\theta_C + \theta_D + \frac{3}{10}v_1\right) \tag{v}$$

$$M_{DC} = \frac{2 \times 2 \cdot 5EI}{10}\left(2\theta_D + \theta_C + \frac{3}{10}v_1\right) \tag{vi}$$

From the equilibrium of the member end moments at the joints

$$M_{BA} + M_{BC} = 0, \qquad M_{CB} + M_{CD} - 54 = 0, \qquad M_{DC} = 0$$

Substituting in the equilibrium equations for M_{BA}, M_{BC}, etc., from Eqs (i)–(vi) we obtain

$$1 \cdot 25EI\theta_B + 0 \cdot 1EI\theta_C - 0 \cdot 15EIv_1 - 87 \cdot 5 = 0 \qquad \text{(vii)}$$

$$1 \cdot 2EI\theta_C + 0 \cdot 1EI\theta_B + 0 \cdot 5EI\theta_C + 0 \cdot 15EIv_1 + 46 = 0 \qquad \text{(viii)}$$

$$EI\theta_D + 0 \cdot 5EI\theta_C + 0 \cdot 15EIv_1 = 0 \qquad \text{(ix)}$$

Since there are four unknown displacements we require a further equation for a solution. This may be obtained by considering the overall horizontal equilibrium of the frame. Thus

$$S_{AB} + S_{DC} + 10 = 0$$

in which, from Eq. (16.29)

$$S_{AB} = \frac{6 \times 2 \cdot 5EI}{10^2} \theta_B - \frac{12 \times 2 \cdot 5EI}{10^3} v_1 - 5$$

where the last term on the right-hand side is $S_{AB}^F (= -5 \text{ kN})$, the contribution of the 10 kN horizontal load to S_{AB}. Also

$$S_{DC} = \frac{6 \times 2 \cdot 5EI}{10^2} (\theta_D + \theta_C) - \frac{12 \times 2 \cdot 5EI}{10^3} v_1$$

Hence, substituting for S_{AB} and S_{DC} in the equilibrium equations, we have

$$EI\theta_B + EI\theta_D + EI\theta_C - 0 \cdot 4EIv_1 + 33 \cdot 3 = 0 \qquad \text{(x)}$$

Solving Eqs (vii)–(x) we obtain

$$EI\theta_B = 101 \cdot 5, \qquad EI\theta_C = -73 \cdot 2, \qquad EI\theta_D = 9 \cdot 8, \qquad EIv_1 = 178 \cdot 6$$

Substituting these values in Eqs (i)–(vi) yields

$$M_{AB} = 11 \cdot 5 \text{ kN m}, \qquad M_{BA} = 87 \cdot 2 \text{ kN m}, \qquad M_{BC} = -87 \cdot 2 \text{ kN m},$$

$$M_{CB} = 95 \cdot 5 \text{ kN m}, \qquad M_{CD} = -41 \cdot 5 \text{ kN m},$$

$$M_{DC} = 0 \qquad \text{and} \qquad M_{CE} = -54 \text{ kN m}$$

16.10   Moment distribution

Examples 16.15 and 16.16 show that the greater the complexity of a structure, the greater the number of unknowns and therefore the greater the number of simultaneous equations requiring solution; hand methods of analysis then become extremely tedious if not impracticable so that alternatives are desirable. One obvious alternative is to employ computer-based techniques but another, quite powerful hand method is an iterative procedure known as the *moment distribution method*. The method was derived by Professor Hardy Cross and presented in a paper to the ASCE in 1932.

Principle

Consider the three-span continuous beam shown in Fig. 16.37(a). The beam carries loads that, as we have previously seen, will cause rotations, θ_A, θ_B, θ_C and θ_D at the

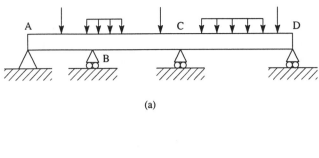

(a)

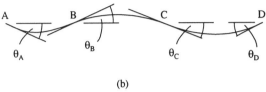

(b)

Fig. 16.37   Principle of the moment distribution method

supports as shown in Fig. 16.37(b). In Fig. 16.37(b), θ_A and θ_C are positive (corresponding to positive moments) and θ_B and θ_D are negative.

Suppose that the beam is clamped at the supports before the loads are applied, thereby preventing these rotations. Each span then becomes a fixed beam with moments at each end, i.e. fixed-end moments (FEMs). Using the same notation as in the slope–deflection method these moments are M_{AB}^F, M_{BA}^F, M_{BC}^F, M_{CB}^F, M_{CD}^F and M_{DC}^F. If we now release the beam at the support B, say, the resultant moment at B, $M_{BA}^F + M_{BC}^F$, will cause rotation of the beam at B until equilibrium is restored; $M_{BA}^F + M_{BC}^F$ is the *out of balance* moment at B. Note that, at this stage, the rotation of the beam at B *is not* θ_B. By allowing the beam to rotate to an equilibrium position at B we are, in effect, applying a balancing moment at B equal to $-(M_{BA}^F + M_{BC}^F)$. Part of this balancing moment will cause rotation in the span BA and part will cause rotation in the span BC. In other words the balancing moment at B has been *distributed* into the spans BA and BC, the relative amounts depending upon the *stiffness*, or the resistance to rotation, of BA and BC. This procedure will affect the fixed-end moments at A and C so that they will no longer be equal to M_{AB}^F and M_{CB}^F. We shall see later how they are modified.

We now clamp the beam at B in its new equilibrium position and release the beam at, say, C. This will produce an out of balance moment at C which will cause the beam to rotate to a new equilibrium position at C. The fixed-end moment at D will then be modified and there will now be an out of balance moment at B. The beam is now clamped at C and released in turn at A and D, thereby modifying the moments at B and C.

The beam is now in a position in which it is clamped at each support but in which it has rotated at the supports through angles that are not yet equal to θ_A, θ_B, θ_C and θ_D. Clearly the out of balance moment at each support will not be as great as it was initially since some rotation has taken place; the beam is now therefore closer to the equilibrium state of Fig. 16.37(b). The release/clamping procedure is repeated until the difference between the angle of rotation at each support and the equilibrium state of Fig. 16.37(b) is negligibly small. Fortunately this occurs after relatively few release/clamping operations.

In applying the moment distribution method we shall require the fixed-end moments in the different members of a beam or frame. We shall also need to determine the distribution of the balancing moment at a support into the adjacent spans and also the fraction of the distributed moment which is *carried over* to each adjacent support.

Table 16.6

Load case	Fixed-end moments (FEMs)	
	M_{AB}^{F}	M_{BA}^{F}
	$-\dfrac{WL}{8}$	$+\dfrac{WL}{8}$
	$-\dfrac{Wab^2}{L^2}$	$+\dfrac{Wa^2 b}{L^2}$
	$-\dfrac{wL^2}{12}$	$+\dfrac{wL^2}{12}$
	$-\dfrac{w}{L^2}\left[\dfrac{L^2}{2}(b^2-a^2)-\dfrac{2}{3}L(b^3-a^3)+\dfrac{1}{4}(b^4-a^4)\right]$	$+\dfrac{wb^3}{L^2}\left(\dfrac{L}{3}-\dfrac{b}{4}\right)$
	$+\dfrac{M_0 b}{L^2}(2a-b)$	$+\dfrac{M_0 a}{L^2}(2b-a)$
	$-\dfrac{6EI\delta}{L^2}$	$-\dfrac{6EI\delta}{L^2}$
	0	$-\dfrac{3EI\delta}{L^2}$

The sign convention we shall adopt for the fixed-end moments is identical to that for the end moments in the slope–deflection method; thus clockwise moments are positive, anticlockwise are negative.

Fixed-end moments

We shall require values of fixed-end moments for a variety of loading cases. It will be useful, therefore, to list them for the more common loading causes; others may be found using the moment-area method described in Section 13.3. Included in Table 16.6 are the results for the fixed beams analysed in Section 13.7.

Stiffness coefficient

A moment applied at a point on a beam causes a rotation of the beam at that point, the angle of rotation being directly proportional to the applied moment (see Eq. (9.19)). Thus for a beam AB and a moment M_{BA} applied at the end B

$$M_{BA} = K_{AB}\theta_B \tag{16.32}$$

in which $K_{AB}(=K_{BA})$ is the rotational stiffness of the beam AB. The value of K_{AB} depends, as we shall see, upon the support conditions at the ends of the beam.

Distribution factor

Suppose that in Fig. 16.38 the out of balance moment at the support B in the beam ABC to be distributed into the spans BA and BC is $M_B(=M_{BA}^F+M_{BC}^F)$ at the first release). Let M_{BA}' be the fraction of M_B to be distributed into BA and M_{BC}' be the fraction of M_B to be distributed into BC. Suppose also that the angle of rotation at B due to M_B is θ_B'. Then, from Eq. (16.32)

$$M_{BA}' = K_{BA}\theta_B' \tag{16.33}$$

and

$$M_{BC}' = K_{BC}\theta_B' \tag{16.34}$$

but

$$M_{BA}' + M_{BC}' + M_B = 0$$

Note that M_{BA}' and M_{BC}' are fractions of the balancing moment while M_B is the out of balance moment. Substituting in this equation for M_{BA}' and M_{BC}' from Eqs (16.33) and (16.34)

$$\theta_B'(K_{BA} + K_{BC}) = -M_B$$

so that

$$\theta_B' = -\frac{M_B}{K_{BA} + K_{BC}} \tag{16.35}$$

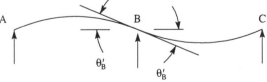

Fig. 16.38 Determination of distribution factor

Substituting in Eqs (16.33) and (16.34) for θ'_B from Eq. (16.35) we have

$$M'_{BA} = \frac{K_{BA}}{K_{BA} + K_{BC}} (-M_B), \qquad M'_{BC} = \frac{K_{BC}}{K_{BA} + K_{BC}} (-M_B) \qquad (16.36)$$

The terms $K_{BA}/(K_{BA} + K_{BC})$ and $K_{BC}/(K_{BA} + K_{BC})$ are the *distribution factors* (DFs) at the support B.

Stiffness coefficients and carry over factors

We shall now derive values of stiffness coefficient (K) and carry over factor (COF) for a number of support and loading conditions. These will be of use in the solution of a variety of problems. For this purpose we use the slope–deflection equations, Eqs (16.28) and (16.30). Thus for a span AB of a beam

$$M_{AB} = \frac{2EI}{L} \left[2\theta_A + \theta_B + \frac{3}{L} (v_A - v_B) \right]$$

and

$$M_{BA} = \frac{2EI}{L} \left[2\theta_B + \theta_A + \frac{3}{L} (v_A - v_B) \right]$$

In some problems we shall be interested in the displacement of one end of a beam span relative to the other, i.e. the effect of a sinking support. Thus for, say, $v_A = 0$ and $v_B = \delta$ (the final two load cases in Table 16.6) the above equations become

$$M_{AB} = \frac{2EI}{L} \left(2\theta_A + \theta_B + \frac{3}{L} \delta \right) \qquad (16.37)$$

and

$$M_{BA} = \frac{2EI}{L} \left(2\theta_B + \theta_A - \frac{3}{L} \delta \right) \qquad (16.38)$$

Rearranging Eqs (16.37) and (16.38) we have

$$2\theta_A + \theta_B - \frac{3}{L} \delta = \frac{L}{2EI} M_{AB} \qquad (16.39)$$

and

$$2\theta_B + \theta_A - \frac{3}{L} \delta = \frac{L}{2EI} M_{BA} \qquad (16.40)$$

Equations (16.39) and (16.40) may be expressed in terms of various combinations of θ_A, θ_B and δ. Thus subtracting Eq. (16.39) from Eq. (16.40) and rearranging we obtain

$$\theta_B - \theta_A = \frac{L}{2EI} (M_{BA} - M_{AB}) \qquad (16.41)$$

Multiplying Eq. (16.39) by 2 and subtracting from Eq. (16.40) gives

$$\frac{\delta}{L} - \theta_A = \frac{L}{6EI} (M_{BA} - 2M_{AB}) \qquad (16.42)$$

Now eliminating θ_A between Eqs (16.39) and (16.40) we have

$$\theta_B - \frac{\delta}{L} = \frac{L}{6EI}(2M_{BA} - M_{AB}) \tag{16.43}$$

We shall now use Eqs (16.41)–(16.43) to determine stiffness coefficients and carry over factors for a variety of support and loading conditions at A and B.

Case 1: A fixed, B simply supported, moment M_{BA} applied at B

This is the situation arising when a beam has been released at a support (B) and we require the stiffness coefficient of the span BA so that we can determine the distribution factor; we also require the fraction of the moment, M_{BA}, which is carried over to the support at A.

In this case $\theta_A = \delta = 0$ so that, from Eq. (16.42)

$$M_{AB} = \tfrac{1}{2}M_{BA}$$

Therefore one-half of the applied moment, M_{BA}, is carried over to A so that the carry over factor (COF) = 1/2. Now from Eq. (16.43) we have

$$\theta_B = \frac{L}{6EI}\left(2M_{BA} - \frac{M_{BA}}{2}\right)$$

so that

$$M_{BA} = \frac{4EI}{L}\theta_B$$

from which (see Eq. (16.32)) $K_{BA} = \dfrac{4EI}{L}\,(= K_{AB})$

Case 2: A simply supported, B simply supported, moment M_{BA} applied at B

This situation arises when we release the beam at an internal support (B) and the adjacent support (A) is an outside support which is pinned and therefore free to rotate. In this case the moment, M_{BA}, does not affect the moment at A, which is always zero; there is, therefore, no carry over from B to A.

From Eq. (16.43)

$$\theta_B = \frac{L}{6EI}\,2M_{BA} \quad (M_{AB} = 0)$$

whence

$$M_{BA} = \frac{3EI}{L}\theta_B$$

so that

$$K_{BA} = \frac{3EI}{L}\,(= K_{AB})$$

Case 3: A and B simply supported, equal moments M_{BA} and $-M_{AB}$ applied at B and A

This case is of use in a symmetrical beam that is symmetrically loaded and would apply to the central span. Thus identical operations will be carried out at each end of the central span so that there will be no carry over of moment from B to A or A to B. Also $\theta_B = -\theta_A$ so that from Eq. (16.41)

$$M_{BA} = \frac{2EI}{L}\theta_B$$

and

$$K_{BA} = \frac{2EI}{L}(=K_{AB})$$

Case 4: A and B simply supported, the beam antisymmetrically loaded such that $M_{BA} = M_{AB}$

This case uses the antisymmetry of the beam and loading in the same way that Case 3 used symmetry. There is therefore no carry over of moment from B to A or A to B and $\theta_A = \theta_B$. Therefore, from Eq. (16.43).

$$M_{BA} = \frac{6EI}{L}\theta_B$$

$$\text{so that} \quad K_{BA} = \frac{6EI}{L}(=K_{AB})$$

We are now in a position to apply the moment distribution method to beams and frames. Note that the successive releasing and clamping of supports is, in effect, carried out simultaneously in the analysis.

First we shall consider continuous beams.

Continuous beams

Example 16.17 Determine the support reactions in the continuous beam ABCD shown in Fig. 16.39; its flexural rigidity EI is constant throughout.

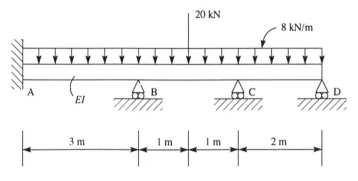

Fig. 16.39 Beam of Ex. 16.17

Initially we calculate the fixed-end moments (FEMs) for each of the three spans using the results presented in Table 16.6. Thus

$$M^F_{AB} = -M^F_{BA} = -\frac{8 \times 3^2}{12} = -6 \cdot 0 \text{ kN m}$$

$$M^F_{BC} = -M^F_{CB} = -\frac{8 \times 2^2}{12} - \frac{20 \times 2}{8} = -7 \cdot 67 \text{ kN m}$$

$$M^F_{CD} = -M^F_{DC} = -\frac{8 \times 2^2}{12} = -2 \cdot 67 \text{ kN m}$$

In this particular example certain features should be noted. First, the support at A is a fixed support so that it will not be released and clamped in turn. In other words, the moment at A will always be balanced (by the fixed support) but will be continually modified as the beam at B is released and clamped. Secondly, the support at D is an outside pinned support so that the final moment at D must be zero. We can therefore reduce the amount of computation by balancing the beam at D initially and then leaving the support at D pinned so that there will be no carry over of moment from C to D in the subsequent moment distribution. However, the stiffness coefficient of CD must be modified to allow for this since the span CD will then correspond to Case 2 as the beam is released at C and is free to rotate at D. Thus $K_{CD} = K_{DC} = 3EI/L$. All other spans correspond to Case 1 where, as we release the beam at a support, that support is a pinned support while the beam at the adjacent support is fixed. Therefore, for the spans AB and BC, the stiffness coefficients are $4EI/L$ and the carry over factors are equal to $1/2$.

The distribution factions (DFs) are obtained from Eqs (16.36). Thus

$$\text{DF}_{BA} = \frac{K_{BA}}{K_{BA} + K_{BC}} = \frac{4EI/3}{4EI/3 + 4EI/2} = 0 \cdot 4$$

$$\text{DF}_{BC} = \frac{K_{BC}}{K_{BA} + K_{BC}} = \frac{4EI/2}{4EI/3 + 4EI/2} = 0 \cdot 6$$

$$\text{DF}_{CB} = \frac{K_{CB}}{K_{CB} + K_{CD}} = \frac{4EI/2}{4EI/2 + 3EI/2} = 0 \cdot 57$$

$$\text{DF}_{CD} = \frac{K_{CD}}{K_{CB} + K_{CD}} = \frac{3EI/2}{4EI/2 + 3EI/2} = 0 \cdot 43$$

Note that the sum of the distribution factors at a support must always be equal to unity since they represent the fraction of the out of balance moment which is distributed into the spans meeting at that support. The solution is now completed as shown in Table 16.7.

Note that there is a rapid convergence in the moment distribution. As a general rule it is sufficient to stop the procedure when the distributed moments are of the order of 2% of the original fixed-end moments. In the table the last moment at C in

Table 16.7

	A	B		C		D
DFs	–	0·4	0·6	0·57	0·43	1·0
FEMs	−6·0	+6·0	−7·67	+7·67	−2·67	+2·67
Balance D						−2·67
Carry over					−1·34	
Balance		+0·67	+1·0	−2·09	−1·58	
Carry over	+0·34		−1·05	+0·5		
Balance		+0·42	+0·63	−0·29	−0·21	
Carry over	+0·21		−0·15	+0·32		
Balance		+0·06	+0·09	−0·18	−0·14	
Carry over	+0·03		−0·09	+0·05		
Balance		+0·04	+0·05	−0·03	−0·02	
Final moments	−5·42	+7·19	−7·19	+5·95	−5·95	0

CD is −0·02 which is 0·75% of the original fixed-end moment, while the last moment at B in BC is +0·05 which is 0·65% of the original fixed-end moment. We could, therefore, have stopped the procedure at least one step earlier and still have retained sufficient accuracy.

The final reactions at the supports are now calculated from the final support moments and the reactions corresponding to the actual loads, i.e. the free reactions; these are calculated as though each span were simply supported. The procedure is identical to that in Ex. 16.15.

Table 16.8

	A	B		C		D
Free reactions	↑12·0	12·0↑	↑18·0	18·0↑	↑ 8·0	8·0 ↑
Final moment reactions	↓0·6	0·6↑	↑ 0·6	0·6↓	↑ 2·98	2·98↓
Total reactions (kN)	↑11·4	12·6↑	↑18·6	17·4↑	↑10·98	5·02↑

In Table 16.8 the final moment reactions in AB, for example, form a couple to balance the clockwise moment of $7·19 − 5·42 = 1·77$ kN m acting on AB. Thus at A the reaction is $1·77/3·0 = 0·6$ kN acting downwards while at B in AB the reaction is 0·6 kN acting upwards. The remaining final moment reactions are calculated in the same way.

Finally the complete reactions at each of the supports are

$$R_A = 11·4 \text{ kN}, \qquad R_B = 12·6 + 18·6 = 31·2 \text{ kN},$$
$$R_C = 17·4 + 10·98 = 28·38 \text{ kN}, \qquad R_D = 5·02 \text{ kN}$$

Example 16.18   Calculate the support reactions in the beam shown in Fig. 16.40; the flexural rigidity, *EI*, of the beam is constant throughout.

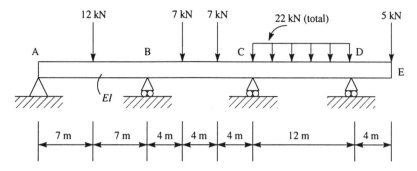

Fig. 16.40 Beam of Ex. 16.18

This example differs slightly from Ex. 16.17 in that there is no fixed support and there is a cantilever overhang at the right-hand end of the beam. We therefore treat the support at A in exactly the same way as the support at D in the previous example. The effect of the cantilever overhang may be treated in a similar manner since we know that the final value of moment at D is $-5 \times 4 = -20$ kN m. We therefore calculate the fixed-end moments M_{DE}^F ($=-20$ kN m) and M_{DC}^F, balance the beam at D, carry over to C and then leave the beam at D balanced and pinned; again the stiffness coefficient, K_{DC}, is modified to allow for this (Case 2).

The fixed-end moments are again calculated using the appropriate results from Table 16.6. Thus

$$M_{AB}^F = -M_{BA}^F = -\frac{12 \times 14}{8} = -21 \text{ kN m}$$

$$M_{BC}^F = -M_{CB}^F = -\frac{7 \times 4 \times 8^2}{12^2} - \frac{7 \times 8 \times 4^2}{12^2} = -18 \cdot 67 \text{ kN m}$$

$$M_{CD}^F = -M_{DC}^F = -\frac{22 \times 12}{12} = -22 \text{ kN m}$$

$$M_{DE}^F = -5 \times 4 = -20 \text{ kN m}$$

The distribution factors are calculated as follows.

$$DF_{BA} = \frac{K_{BA}}{K_{BA} + K_{BC}} = \frac{3EI/14}{3EI/14 + 4EI/12} = 0 \cdot 39$$

Hence

$$DF_{BC} = 1 - 0 \cdot 39 = 0 \cdot 61$$

$$DF_{CB} = \frac{K_{CB}}{K_{CB} + K_{CD}} = \frac{4EI/12}{4EI/12 + 3EI/12} = 0 \cdot 57$$

Hence

$$DF_{CD} = 1 - 0 \cdot 57 = 0 \cdot 43$$

The solution is completed as follows.

	A	B		C		D	E
DFs	1	0·39	0·61	0·57	0·43	1·0 0	–
FEMs	−21·0	+21·0	−18·67	+18·67	−22·0	+22·0 −20·0	0
Balance A and D	+21·0					−2·0	
Carry over		+10·5		−1·0			
Balance		−5·0	−7·83	+2·47	+1·86		
Carry over			+1·24	−3·92			
Balance		−0·48	−0·76	+2·23	+1·69		
Carry over			+1·12	−0·38			
Balance		−0·44	−0·68	+0·22	+0·16		
Carry over			+0·11	−0·34			
Balance		−0·04	−0·07	+0·19	+0·15		
Final moments	0	+25·54	−25·54	+19·14	−19·14	+20·0 −20·0	0

The support reactions are now calculated in an identical manner to that in Ex. 16.17 and are

$$R_A = 4·18 \text{ kN}, \qquad R_B = 15·35 \text{ kN}, \qquad R_C = 17·4 \text{ kN}, \qquad R_D = 16·07 \text{ kN}$$

Example 16.19 Calculate the reactions at the supports in the beam ABCD shown in Fig. 16.41. The flexural rigidity of the beam is constant throughout.

The beam in Fig. 16.41 is symmetrically supported and loaded about its centre-line; we may therefore use this symmetry to reduce the amount of computation.

In the centre span, BC, $M_{BC}^F = -M_{CB}^F$ and will remain so during the distribution. This situation corresponds to Case 3, so that if we reduce the stiffness (K_{BC}) of BC to $2EI/L$ there will be no carry over of moment from B to C (or C to B) and we can consider just half the beam. The outside pinned support at A is treated in exactly the same way as the outside pinned supports in Exs 16.17 and 16.18.

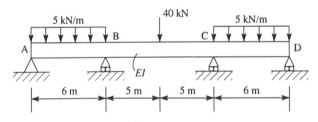

Fig. 16.41 Symmetrical beam of Ex. 16.19

The fixed-end moments are

$$M^F_{AB} = -M^F_{BA} = -\frac{5 \times 6^2}{12} = -15 \text{ kN m}$$

$$M^F_{BC} = -M^F_{CB} = -\frac{40 \times 5}{8} = -25 \text{ kN m}$$

The distribution factors are

$$DF_{AB} = \frac{K_{BA}}{K_{BA} + K_{BC}} = \frac{3EI/6}{3EI/6 + 2EI/10} = 0\cdot71$$

Hence

$$DF_{BC} = 1 - 0\cdot71 = 0\cdot29$$

The solution is completed as follows:

	A	B	
DFs	1	0·71	0·29
FEMs	−15·0	+15·0	−25·0
Balance A	+15·0		
Carry over		+7·5	
Balance B		+1·78	+0·72
Final moments	0	+24·28	−24·28

Note that we only need to balance the beam at B once. The use of symmetry therefore leads to a significant reduction in the amount of computation.

Example 16.20 Calculate the end moments at the supports in the beam shown in Fig. 16.42 if the support at B is subjected to a settlement of 12 mm. Furthermore, the second moment of area of the cross-section of the beam is $9 \times 10^6 \text{ mm}^4$ in the span AB and $12 \times 10^6 \text{ mm}^4$ in the span BC; Young's modulus, E, is 200 000 N/mm^2.

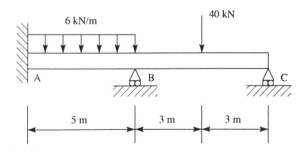

Fig. 16.42 Beam of Ex. 16.20

In this example the fixed-end moments produced by the applied loads are modified by additional moments produced by the sinking support. Thus, using Table 16.6

$$M^F_{AB} = -\frac{6 \times 5^2}{12} - \frac{6 \times 200\,000 \times 9 \times 10^6 \times 12}{(5 \times 10^3)^2 \times 10^6} = -17 \cdot 7 \text{ kN m}$$

$$M^F_{BA} = +\frac{6 \times 5^2}{12} - \frac{6 \times 200\,000 \times 9 \times 10^6 \times 12}{(5 \times 10^3)^2 \times 10^6} = +7 \cdot 3 \text{ kN m}$$

Since the support at C is an outside pinned support, the effect on the fixed-end moments in BC of the settlement of B is reduced (see the last case in Table 16.6). Thus

$$M^F_{BC} = -\frac{40 \times 6}{8} + \frac{3 \times 200\,000 \times 12 \times 10^6 \times 12}{(6 \times 10^3)^2 \times 10^6} = -27 \cdot 6 \text{ kN m}$$

$$M^F_{CB} = +\frac{40 \times 6}{8} = +30 \cdot 0 \text{ kN m}$$

The distribution factors are

$$DF_{BA} = \frac{K_{BA}}{K_{BA} + K_{BC}} = \frac{4E \times 9 \times 10^6/5}{(4E \times 9 \times 10^6)/5 + (3E \times 12 \times 10^6)/6} = 0 \cdot 55$$

Hence

$$DF_{BC} = 1 - 0 \cdot 55 = 0 \cdot 45$$

	A	B		C
DFs	–	0·55	0·45	1·0
FEMs	−17·7	+7·3	−27·6	+30·0
Balance C				−30·0
Carry over			−15·0	
Balance B		+19·41	+15·89	
Carry over	+9·71			
Final moments	−7·99	+26·71	−26·71	0

Note that in this example balancing the beam at B has a significant effect on the fixing moment at A; we therefore complete the distribution after a carry over to A.

Portal frames

Portal frames fall into two distinct categories. In the first the frames, such as that shown in Fig. 16.43(a), are symmetrical in geometry and symmetrically loaded, while in the second (Fig. 16.43(b)) the frames are unsymmetrical due either to their

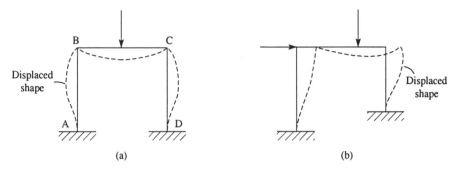

Fig. 16.43 Symmetrical and unsymmetrical portal frames

geometry, the loading or a combination of both. The displacements in the symmetrical frame of Fig. 16.43(a) are such that the joints at B and C remain in their original positions (we are ignoring axial and shear displacements and we assume that the joints remain rigid so that the angle between adjacent members at a joint is unchanged by the loading). In the unsymmetrical frame there are additional displacements due to side sway or *sway* as it is called. This sway causes additional moments at the ends of the members which must be allowed for in the analysis.

Initially we shall consider frames in which there is no sway. The analysis is then virtually identical to that for continuous beams with only, in some cases, the added complication of more than two members meeting at a joint.

Example 16.21 Obtain the bending moment diagram for the frame shown in Fig. 16.44; the flexural rigidity EI is the same for all members.

In this example the frame is unsymmetrical but sway is prevented by the member BC which is fixed at C. Also, the member DA is fixed at D while the member EB is pinned at E.

The fixed-end moments are calculated using the results of Table 16.6 and are

$$M_{AD}^F = M_{DA}^F = 0, \qquad M_{BE}^F = M_{EB}^F = 0$$

$$M_{AB}^F = -M_{BA}^F = -\frac{12 \times 4 \times 8^2}{12^2} - \frac{12 \times 8 \times 4^2}{12^2} = -32 \text{ kN m}$$

$$M_{BC}^F = -M_{CB}^F = -\frac{1 \times 16^2}{12} = -21 \cdot 3 \text{ kN m}$$

Since the vertical member EB is pinned at E, the final moment at E is zero. We may therefore treat E as an outside pinned support, balance E initially and reduce the stiffness coefficient, K_{BE}, as before. However, there is no fixed-end moment at E so that the question of balancing E initially does not arise. The distribution factors are now calculated.

$$DF_{AD} = \frac{K_{AD}}{K_{AD} + K_{AB}} = \frac{4EI/12}{4EI/12 + 4EI/12} = 0 \cdot 5$$

Hence
$$DF_{AB} = 1 - 0 \cdot 5 = 0 \cdot 5$$

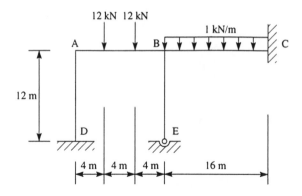

Fig. 16.44   Beam of Ex. 16.21

$$DF_{BA} = \frac{K_{BA}}{K_{BA} + K_{BC} + K_{BE}} = \frac{4EI/12}{4EI/12 + 4EI/16 + 3EI/12} = 0{\cdot}4$$

$$DF_{BC} = \frac{K_{BC}}{K_{BA} + K_{BC} + K_{BE}} = \frac{4EI/16}{4EI/12 + 4EI/16 + 3EI/12} = 0{\cdot}3$$

Hence $DF_{BE} = 1 - 0{\cdot}4 - 0{\cdot}3 = 0{\cdot}3$

The solution is now completed below.

Joint	D	A		B			C	E
Member	DA	AD	AB	BA	BE	BC	CB	EB
DFs	–	0·5	0·5	0·4	0·3	0·3	–	1·0
FEMs	0	0	−32·0	+32·0	0	−21·3	+21·3	0
Balance A & B		+16·0	+16·0	−4·3	−3·2	−3·2		
Carry over	+8·0		−2·15	+8·0			−1·6	
Balance		+1·08	+1·08	−3·2	−2·4	−2·4		
Carry over	+0·54		−1·6	+0·54			−1·2	
Balance		+0·8	+0·8	−0·22	−0·16	−0·16		
Carry over	+0·4		−0·11	+0·4			−0·08	
Balance		+0·05	+0·06	−0·16	−0·12	−0·12		
Final moments	+8·94	+17·93	−17·93	+33·08	−5·88	−27·18	+18·42	0

The bending moment diagram is shown in Fig. 16.45 and is drawn on the tension side of each member. The bending moment distributions in the members AB and BC are determined by superimposing the fixing moment diagram on the free bending moment diagram, i.e. the bending moment diagram obtained by supposing that AB and BC are simply supported.

We shall now consider frames that are subject to sway. For example, the frame shown in Fig. 16.46(a), although symmetrical itself, is unsymmetrically loaded and will therefore sway. Let us suppose that the final end moments in the members of the

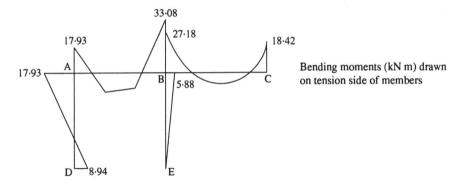

Fig. 16.45 Bending moment diagram for the frame of Ex. 16.21 (bending moments [kN m] drawn on tension side of members)

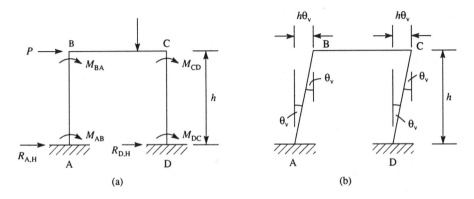

Fig. 16.46 Calculation of sway effect in a portal frame

frame are M_{AB}, M_{BA}, M_{BC}, etc. Since we are assuming a linearly elastic system we may calculate the end moments produced by the applied loads assuming that the frame does not sway, then calculate the end moments due solely to sway and superimpose the two cases. Thus

$$M_{AB} = M_{AB}^{NS} + M_{AB}^{S}, \quad M_{BA} = M_{BA}^{NS} + M_{BA}^{S}, \text{ etc.}$$

in which M_{AB}^{NS} is the end moment at A in the member AB due to the applied loads, assuming that sway is prevented, while M_{AB}^{S} is the end moment at A in the member AB produced by sway only, and so on for M_{BA}, M_{BC}, etc.

We shall now use the principle of virtual work (Section 15.2) to establish a relationship between the final end moments in the member and the applied loads. Thus we impose a small virtual displacement on the frame comprising a rotation, θ_v, of the members AB and DC as shown in Fig. 16.46(b). This displacement *should not be confused with the sway of the frame* which may, or may not, have the same form depending on the loads that are applied. In Fig. 16.46(b) the members are rotating as rigid links so that the internal moments in the members do no work. Therefore the total virtual work comprises external virtual work only (the end moments M_{AB}, M_{BA}, etc. are externally applied moments as far as each frame member is concerned) so that, from the principle of virtual work

$$M_{AB}\theta_v + M_{BA}\theta_v + M_{CD}\theta_v + M_{DC}\theta_v + Ph\theta_v = 0$$

Hence
$$M_{AB} + M_{BA} + M_{CD} + M_{DC} + Ph = 0 \qquad (16.44)$$

Note that, in this case, the member BC does not rotate so that the end moments M_{BC} and M_{CB} do no virtual work. Now substituting for M_{AB}, M_{BA}, etc. in Eq. (16.44) we have

$$M_{AB}^{NS} + M_{AB}^{S} + M_{BA}^{NS} + M_{BA}^{S} + M_{CD}^{NS} + M_{CD}^{S} + M_{DC}^{NS} + M_{DC}^{S} + Ph = 0 \qquad (16.45)$$

in which the no-sway end moments, M_{AB}^{NS}, etc. are found in an identical manner to those in the frame of Ex. 16.21.

Let us now impose an arbitrary sway on the frame; this can be of any convenient magnitude. The arbitrary sway and moments, M_{AB}^{AS}, M_{BA}^{AS}, etc., are calculated using the moment distribution method in the usual way except that the fixed-end moments will be caused solely by the displacement of one end of a member relative to the other. Since the system is linear the member end moments will be directly proportional to the sway so that the end moments corresponding to the actual sway will be directly proportional to the end moments produced by the arbitrary sway. Thus, $M_{AB}^{S} = kM_{AB}^{AS}$, $M_{BA}^{S} = kM_{BA}^{AS}$, etc. in which k is a constant. Substituting in Eq. (16.45) for M_{AB}^{S}, M_{BA}^{S}, etc. we obtain

$$M_{AB}^{NS} + M_{BA}^{NS} + M_{CD}^{NS} + M_{DC}^{NS} + k(M_{AB}^{AS} + M_{BA}^{AS} + M_{CD}^{AS} + M_{DC}^{AS}) + Ph = 0 \qquad (16.46)$$

Substituting the calculated values of M_{AB}^{AS}, M_{AB}^{AS}, etc. in Eq. (16.46) gives k. The actual sway moments M_{AB}^{S}, etc., follow as do the final end moments, $M_{AB}(=M_{AB}^{NS} + M_{AB}^{S})$, etc.

An alternative method of establishing Eq. (16.44) is to consider the equilibrium of the members AB and DC. Thus, from Fig. 16.46(a) in which we consider the moment equilibrium of the member AB about B we have

$$R_{A,H}h - M_{AB} - M_{BA} = 0$$

which gives
$$R_{A,H} = \frac{M_{AB} + M_{BA}}{h}$$

Similarly, by considering the moment equilibrium of DC about C

$$R_{D,H} = \frac{M_{DC} + M_{CD}}{h}$$

Now, from the horizontal equilibrium of the frame

$$R_{A,H} + R_{D,H} + P = 0$$

so that, substituting for $R_{A,H}$ and $R_{D,H}$ we obtain

$$M_{AB} + M_{BA} + M_{DC} + M_{CD} + Ph = 0$$

which is Eq. (16.44).

Example 16.22 Obtain the bending moment diagram for the portal frame shown in Fig. 16.47(a). The flexural rigidity of the horizontal member BC is $2EI$ while that of the vertical members AB and CD is EI.

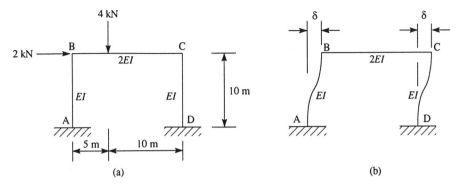

Fig. 16.47 Portal frame of Ex. 16.22

First we shall determine the end moments in the members assuming that the frame does not sway. The corresponding fixed-end moments are found using the results in Table 16.6 and are as follows:

$$M^F_{AB} = M^F_{BA} = 0, \qquad M^F_{CD} = M^F_{DC} = 0$$

$$M^F_{BC} = -\frac{4 \times 5 \times 10^2}{15^2} = -8.89 \text{ kN m}$$

$$M^F_{CB} = +\frac{4 \times 10 \times 5^2}{15^2} = +4.44 \text{ kN m}$$

The distribution factors are

$$DF_{BA} = \frac{K_{BA}}{K_{BA} + K_{BC}} = \frac{4EI/10}{4EI/10 + 4 \times 2EI/15} = 0.43$$

Hence
$$DF_{BC} = 1 - 0.43 = 0.57$$

From the symmetry of the frame, $DF_{CB} = 0.57$, $DF_{CD} = 0.43$.

The no-sway moments are determined in the table overleaf. We now assume that the frame sways by an arbitrary amount, δ, as shown in Fig. 16.47(b). Since we are ignoring the effect of axial strains, the horizontal movements of B and C are both δ. The fixed-end moments corresponding to this sway are then (see Table 16.6).

$$M^F_{AB} = M^F_{BA} = -\frac{6EI\delta}{10^2} = M^F_{DC} = M^F_{CD}$$

$$M^F_{BC} = M^F_{CB} = 0$$

Suppose that $\delta = 100 \times 10^2/6EI$. Then
$$M^F_{AB} = M^F_{BA} = M^F_{DC} = M^F_{CD} = -100 \text{ kN m} \quad \text{(a convenient value)}$$

The distribution factors for the members are the same as those in the no-sway case since they are functions of the member stiffness. We now obtain the member end moments corresponding to the arbitrary sway.

No-sway case

	A	B		C		D
DFs	–	0·43	0·57	0·57	0·43	–
FEMs	0	0	−8·89	+4·44	0	0
Balance		+3·82	+5·07	−2·53	−1·91	
Carry over	+1·91		−1·26	+2·53		−0·95
Balance		+0·54	+0·72	−1·44	−1·09	
Carry over	+0·27		−0·72	+0·36		−0·55
Balance		+0·31	+0·41	−0·21	−0·15	
Carry over	+0·15		−0·11	+0·21		−0·08
Balance		+0·05	+0·06	−0·12	−0·09	
Carry over	+0·03		−0·06	+0·03		−0·05
Balance		+0·03	+0·03	−0·02	−0·01	
Final moments (M^{NS})	+2·36	+4·75	−4·75	+3·25	−3·25	−1·63

Sway case

	A	B		C		D
DFs	–	0·43	0·57	0·57	0·43	–
FEMs	−100	−100	0	0	−100	−100
Balance		+43	+57	+57	+43	
Carry over	+21·5		+28·5	+28·5		+21·5
Balance		−12·3	−16·2	−16·2	−12·3	
Carry over	−6·2		−8·1	−8·1		−6·2
Balance		+3·5	+4·6	+4·6	+3·5	
Carry over	+1·8		+2·3	+2·3		+1·8
Balance		−1·0	−1·3	−1·3	−1·0	
Final arbitrary sway moments (M^{AS})	−82·9	−66·8	+66·8	+66·8	−66·8	−82·9

Comparing the frames shown in Figs 16.47 and 16.46 we see that they are virtually identical. We may therefore use Eq. (16.46) directly. Thus, substituting for the no-sway and arbitrary-sway end moments we have

$$2·36 + 4·75 − 3·25 − 1·63 + k(−82·9 − 66·8 − 66·8 − 82·9) + 2 \times 10 = 0$$

which gives

$$k = 0·074$$

The actual sway moments are then

$$M_{AB}^S = kM_{AB}^{AS} = 0.074 \times (-82.9) = -6.14 \text{ kN m}$$

Similarly $M_{BA}^S = -4.94 \text{ kN m},$ $\qquad M_{BC}^S = 4.94 \text{ kN m},$ $\qquad M_{CB}^S = 4.94 \text{ kN m}$

$$M_{CD}^S = -4.94 \text{ kN m}, \qquad M_{DC}^S = -6.14 \text{ kN m}$$

Thus the final end moments are

$$M_{AB} = M_{AB}^{NS} + M_{AB}^S = 2.36 - 6.14 = -3.78 \text{ kN m}$$

Similarly $M_{BA} = -0.19 \text{ kN m},$ $\qquad M_{BC} = 0.19 \text{ kN m},$ $\qquad M_{CB} = 8.19 \text{ kN m}$

$$M_{CD} = -8.19 \text{ kN m}, \qquad M_{DC} = -7.77 \text{ kN m}$$

The bending moment diagram is shown in Fig. 16.48 and is drawn on the tension side of the members.

Example 16.23 Calculate the end moments in the members of the frame shown in Fig. 16.49. All members have the same flexural rigidity, EI; note that the member CD is pinned to the foundation at D.

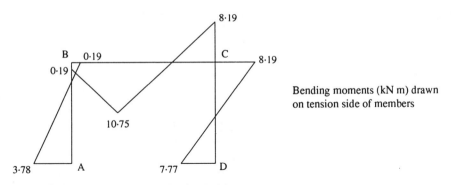

Bending moments (kN m) drawn on tension side of members

Fig. 16.48 Bending moment diagram for the portal frame of Ex. 16.22

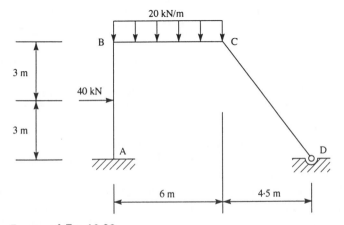

Fig. 16.49 Frame of Ex. 16.23

Initially, the fixed-end moments produced by the applied loads are calculated. Thus, from Table 16.6

$$M^F_{AB} = -M^F_{BA} = -\frac{40 \times 6}{8} = -30 \text{ kN m}$$

$$M^F_{BC} = -M^F_{CB} = -\frac{20 \times 6^2}{12} = -60 \text{ kN m}$$

$$M^F_{CD} = M^F_{DC} = 0$$

The distribution factors are calculated as before. Note that the length of the member $CD = \sqrt{6^2 + 4 \cdot 5^2} = 7 \cdot 5$ m.

$$DF_{BA} = \frac{K_{BA}}{K_{BA} + K_{BC}} = \frac{4EI/6}{4EI/6 + 4EI/6} = 0 \cdot 5$$

Hence
$$DF_{BC} = 1 - 0 \cdot 5 = 0 \cdot 5$$

$$DF_{CB} = \frac{K_{CB}}{K_{CB} + K_{CD}} = \frac{4EI/6}{4EI/6 + 3EI/7 \cdot 5} = 0 \cdot 625$$

Therefore
$$DF_{CD} = 1 - 0 \cdot 625 = 0 \cdot 375$$

No-sway case

	A	B		C		D
DFs	–	0·5	0·5	0·625	0·375	1·0
FEMs	−30·0	+30·0	−60·0	+60·0	0	0
Balance		+15·0	+15·0	−37·5	−22·5	
Carry over	+7·5		−18·8	+7·5		
Balance		+9·4	+9·4	−4·7	−2·8	
Carry over	+4·7		−2·4	+4·7		
Balance		+1·2	+1·2	−2·9	−1·8	
Carry over	+0·6		−1·5	+0·6		
Balance		+0·75	+0·75	−0·38	−0·22	
Final moments (M^{NS})	−17·2	+56·35	−56·35	+27·32	−27·32	0

Unlike the frame in Ex. 16.22 the frame itself in this case is unsymmetrical. Therefore the geometry of the frame, after an imposed arbitrary sway, will not have the simple form shown in Fig. 16.47(b). Furthermore, since the member CD is inclined, an arbitrary sway will cause a displacement of the joint C relative to the joint B. This also means that in the application of the principle of virtual work a virtual rotation of the member AB will result in a rotation of the member BC, so that the end moments M_{BC} and M_{CB} will do work; Eq. (16.46) cannot, therefore, be used in its existing form. In this situation we can make use of the geometry of the frame after an arbitrary virtual displacement to deduce the relative displacements of the joints produced by an imposed arbitrary sway; the fixed-end moments due to the arbitrary sway may then be calculated.

Figure 16.50 shows the displaced shape of the frame after a rotation, θ, of the member AB. This diagram will serve, as stated above, to deduce the fixed-end moments due to sway and also to establish a virtual work equation similar to Eq. (16.46). It is helpful, when calculating the rotations of the different members, to employ an instantaneous centre, I. This is the point about which the triangle IBC rotates as a rigid body to IB′C′; thus all sides of the triangle rotate through the same angle which, since BI = 8 m (obtained from similar triangles AID and BIC), is 3θ/4. The relative displacements of the joints are then as shown.

The fixed-end moments due to the arbitrary sway are, from Table 16.6 and Fig. 16.50

$$M_{AB}^{F} = M_{BA}^{F} = -\frac{6EI(6\theta)}{6^2} = -EI\theta$$

$$M_{BC}^{F} = M_{CB}^{F} = +6EI(4 \cdot 5\theta)/6^2 = +0 \cdot 75EI\theta$$

$$M_{CD}^{F} = -3EI(7 \cdot 5\theta)/7 \cdot 5^2 = -0 \cdot 4EI\theta$$

If we impose an arbitrary sway such that $EI\theta = 100$ we have

$$M_{AB}^{F} = M_{BA}^{F} = -100 \text{ kN m}, \quad M_{BC}^{F} = M_{CB}^{F} = +75 \text{ kN m}, \quad M_{CD}^{F} = -40 \text{ kN m}$$

Sway case

	A	B		C		D
DFs	–	0·5	0·5	0·625	0·375	1·0
FEMs	−100	−100	+75	+75	−40	0
Balance		+12·5	+12·5	−21·9	−13·1	
Carry over	+6·3		−10·9	+6·3		
Balance		+5·45	+5·45	−3·9	−2·4	
Carry over	+2·72		−1·95	+2·72		
Balance		−0·97	+0·97	−1·7	−1·02	
Carry over	+0·49		−0·85	+0·49		
Balance		+0·43	+0·43	−0·31	−0·18	
Final arbitrary sway moments (M^{AS})	−90·49	−80·65	+80·65	+56·7	−56·7	

Now using the principle of virtual work and referring to Fig. 16.50 we have

$$M_{AB}\theta + M_{BA}\theta + M_{BC}(-3\theta/4) + M_{CB}(-3\theta/4)$$
$$+ M_{CD}\theta + 40(6\theta/2) + 20 \times 6(-4 \cdot 5\theta/2) = 0$$

Hence
$$4(M_{AB} + M_{BA} + M_{CD}) - 3(M_{BC} + M_{CB}) - 600 = 0 \qquad \text{(i)}$$

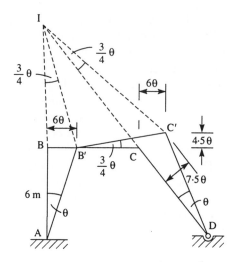

Fig. 16.50 Arbitrary sway and virtual displacement geometry of frame of Ex. 16.23

Now replacing M_{AB}, etc., by $M_{AB}^{NS} + kM_{AB}^{AS}$, etc., Eq. (i) becomes

$$4(M_{AB}^{NS} + M_{BA}^{NS} + M_{CD}^{NS}) - 3(M_{BC}^{NS} + M_{CB}^{NS})$$
$$+ k[4(M_{AB}^{AS} + M_{BA}^{AS} + M_{CD}^{AS}) - 3(M_{BC}^{AS} + M_{CB}^{AS})] - 600 = 0$$

Substituting the values of M_{AB}^{NS} and M_{AB}^{AS}, etc., we have

$$4(-17\cdot2 + 56\cdot35 - 27\cdot32) - 3(-56\cdot35 + 27\cdot32)$$
$$+ k[4(-90\cdot49 - 80\cdot65 - 56\cdot7) - 3(80\cdot65 + 56\cdot7)] - 600 = 0$$

from which $k = -0\cdot352$. The final end moments are calculated from $M_{AB} = M_{AB}^{NS} - 0\cdot352M_{AB}^{AS}$, etc., and are given below.

	AB	BA	BC	CB	CD	DC
No-sway moments	−17·2	+56·4	−56·4	+27·3	−27·3	0
Sway moments	+31·9	+28·4	−28·4	−20·0	+20·0	0
Final moments	+14·7	+84·8	−84·8	+7·3	−7·3	0

16.11 Introduction to matrix methods

In Section 16.1 we discussed the flexibility and stiffness methods of analysis of statically indeterminate structure and saw that the flexibility method involved releasing the structures, determining the displacements in the released structure and then finding the forces required to fulfil the compatibility of displacement condition in the complete structure. The method was applied to statically indeterminate beams, trusses, braced beams, portal frames and two-pinned arches in Sections 16.4–16.8. It is clear from the analysis of these types of structure that the greater the degree of indeterminacy the higher the number of simultaneous equations requiring solution; for large numbers of equations a computer approach then becomes necessary. Furthermore, the flexibility method requires judgements to be made in terms of the release selected, so that a more automatic procedure is desirable so long, of course, as the fundamental behaviour of the structure is understood.

In Section 16.9 we examined the slope–deflection method for the solution of statically indeterminate beams and frames; the slope–deflection equations also form the basis of the moment distribution method described in Section 16.10. These equations are, in fact, force–displacement relationships as opposed to the displacement–force relationships of the flexibility method. The slope–deflection and moment distribution methods are therefore *stiffness* or *displacement* methods.

The stiffness method basically requires that a structure, which has a degree of *kinematic indeterminacy* equal to n_k, is initially rendered determinate by imposing a system of n_k constraints. Thus, for example, in the slope–deflection analysis of a continuous beam (e.g. Ex. 16.15) the beam is initially fixed at each support and the fixed-end moments calculated. This generally gives rise to an unbalanced system of forces at each node. Then by allowing displacements to occur at each node we obtain a series of force–displacement states (Eqs (i)–(vi) in Ex. 16.15). The n_k equilibrium conditions at the nodes are then expressed in terms of the displacements, giving n_k equations (Eqs (vii)–(x) in Ex. 16.15), the solution of which gives the true values of the displacements at the nodes. The internal stress resultants follow from the known force–displacement relationships for each member of the structure (Eqs (i)–(vi) in Ex. 16.15) and the complete solution is then the sum of the determinate solution and the set of n_k indeterminate systems.

Again, as in the flexibility method, we see that the greater the degree of indeterminacy (kinematic in this case) the greater the number of equations requiring solution, so that a computer-based approach is necessary when the degree of interdeterminacy is high. Generally this requires that the force–displacement relationships in a structure are expressed in matrix form. We therefore need to establish force–displacement relationships for structural members and to examine the way in which these individual force–displacement relationships are combined to produce a force–displacement relationship for the complete structure. Initially we shall investigate members that are subjected to axial force only.

Axially loaded members

Consider the axially loaded member, AB, shown in Fig. 16.51(a) and suppose that it is subjected to axial forces, F_A and F_B, and that the corresponding displacements are w_A and w_B; the member has a cross-sectional area, A, and Young's modulus, E. An elemental length, δz, of the member is subjected to forces and displacements as shown in Fig. 16.51(b) so that its change in length from its unloaded state is $w + \delta w - w = \delta w$. Thus, from Eq. (7.4), the strain, ε, in the element is given by

$$\varepsilon = \frac{\mathrm{d}w}{\mathrm{d}z}$$

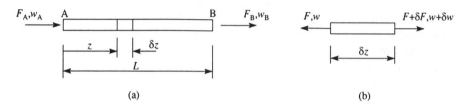

Fig. 16.51   Axially loaded member

Further, from Eq. (7.8)

$$\frac{F}{A} = E\,\frac{dw}{dz}$$

so that

$$dw = \frac{F}{AE}\,dz$$

Therefore the axial displacement at the section a distance z from A is given by

$$w = \int_0^z \frac{F}{AE}\,dz$$

which gives

$$w = \frac{F}{AE}\,z + C_1$$

in which C_1 is a constant of integration. When $z = 0$, $w = w_A$ so that $C_1 = w_A$ and the expression for w may be written as

$$w_B = \frac{F}{AE}\,z + w_A \tag{16.47}$$

In the absence of any loads applied between A and B, $F = F_B = -F_A$ and Eq. (16.47) may be written as

$$w = \frac{F_B}{AE}\,z + w_A \tag{16.48}$$

Thus, when $z = L$, $w = w_B$ so that from Eq. (16.48)

$$w_B = \frac{F_B L}{AE} + w_A$$

or

$$F_B = \frac{AE}{L}\,(w_B - w_A) \tag{16.49}$$

Furthermore, since $F_B = -F_A$ we have, from Eq. (16.49)

$$-F_A = \frac{AE}{L}\,(w_B - w_A)$$

or

$$F_A = -\frac{AE}{L}\,(w_B - w_A) \tag{16.50}$$

Eqs (16.49) and (16.50) may be expressed in matrix form as follows

$$\begin{Bmatrix} F_A \\ F_B \end{Bmatrix} = \begin{bmatrix} AE/L & -AE/L \\ -AE/L & AE/L \end{bmatrix} \begin{Bmatrix} w_A \\ w_B \end{Bmatrix}$$

or
$$\begin{Bmatrix} F_A \\ F_B \end{Bmatrix} = \frac{AE}{L} \begin{bmatrix} 1 & -1 \\ -1 & 1 \end{bmatrix} \begin{Bmatrix} w_A \\ w_B \end{Bmatrix} \tag{16.51}$$

Eq. (16.51) may be written in the general form

$$\{F\} = [K_{AB}]\{w\} \tag{16.52}$$

in which $\{F\}$ and $\{w\}$ are generalized force and displacement matrices and $[K_{AB}]$ is the *stiffness matrix* of the member AB.

Suppose now that we have two axially loaded members, AB and BC, in line and connected at their common node B as shown in Fig. 16.52.

In Fig. 16.52 the force, F_B, comprises two components: $F_{B,AB}$ due to the change in length of AB, and $F_{B,BC}$ due to the change in length of BC. Thus, using the results of Eqs (16.49) and (16.50)

$$F_A = \frac{A_{AB}E_{AB}}{L_{AB}}(w_A - w_B) \tag{16.53}$$

$$F_B = F_{B,AB} + F_{B,BC} = \frac{A_{AB}E_{AB}}{L_{AB}}(w_B - w_A) + \frac{A_{BC}E_{BC}}{L_{BC}}(w_B - w_C) \tag{16.54}$$

$$F_C = \frac{A_{BC}E_{BC}}{L_{BC}}(w_C - w_B) \tag{16.55}$$

in which A_{AB}, E_{AB} and L_{AB} are the cross-sectional area, Young's modulus and length of the member AB; similarly for the member BC. The term AE/L is a measure of the stiffness of a member, this we shall designate by k. Thus, Eqs (16.53)–(16.55) become

$$F_A = k_{AB}(w_A - w_B) \tag{16.56}$$

$$F_B = -k_{AB}w_A + (k_{AB} + k_{BC})w_B - k_{BC}w_C \tag{16.57}$$

$$F_C = k_{BC}(w_C - w_B) \tag{16.58}$$

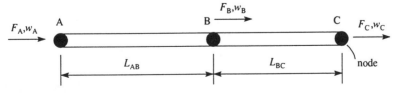

Fig. 16.52 Two axially loaded members in line

Equations (16.56)–(16.58) are expressed in matrix form as

$$\begin{Bmatrix} F_A \\ F_B \\ F_C \end{Bmatrix} = \begin{bmatrix} k_{AB} & -k_{AB} & 0 \\ -k_{AB} & k_{AB} + k_{BC} & -k_{BC} \\ 0 & -k_{BC} & k_{BC} \end{bmatrix} \begin{Bmatrix} w_A \\ w_B \\ w_C \end{Bmatrix} \tag{16.59}$$

Note that in Eq. (16.59) the stiffness matrix is a symmetric matrix of order 3×3, which, as can be seen, connects *three* nodal forces to *three* nodal displacements. Also, in Eq. (16.51), the stiffness matrix is a 2×2 matrix connecting *two* nodal forces to *two* nodal displacements. We deduce, therefore, that a stiffness matrix for a structure in which n nodal forces relate to n nodal displacements will be a symmetric matrix of the order $n \times n$.

In more general terms the matrix in Eq. (16.59) may be written in the form

$$[K] = \begin{bmatrix} k_{11} & k_{12} & k_{13} \\ k_{21} & k_{22} & k_{23} \\ k_{31} & k_{32} & k_{33} \end{bmatrix} \tag{16.60}$$

in which the element k_{11} relates the force at node 1 to the displacement at node 1, k_{12} relates the force at node 1 to the displacement at node 2, and so on. Now, for the member connecting nodes 1 and 2

$$[K_{12}] = \begin{bmatrix} k_{11} & k_{12} \\ k_{21} & k_{22} \end{bmatrix}$$

and for the member connecting nodes 2 and 3

$$[K_{23}] = \begin{bmatrix} k_{22} & k_{23} \\ k_{32} & k_{33} \end{bmatrix}$$

Therefore we may assemble a stiffness matrix for a complete structure, not by the procedure used in establishing Eqs (16.56)–(16.58) but by writing down the matrices for the individual members and then inserting them into the overall stiffness matrix such as that in Eq. (16.60). The element k_{22} appears in both $[K_{12}]$ and $[K_{23}]$ and will therefore receive contributions from both matrices. Hence, from Eq. (16.51)

$$[K_{AB}] = \begin{bmatrix} k_{AB} & -k_{AB} \\ -k_{AB} & k_{AB} \end{bmatrix}$$

and

$$[K_{BC}] = \begin{bmatrix} k_{BC} & -k_{BC} \\ -k_{BC} & k_{BC} \end{bmatrix}$$

Inserting these matrices into Eq. (16.60) we obtain

$$[K_{ABC}] = \begin{bmatrix} k_{AB} & -k_{AB} & 0 \\ -k_{AB} & k_{AB} + k_{BC} & -k_{BC} \\ 0 & -k_{BC} & k_{BC} \end{bmatrix}$$

as before. We see that only the k_{22} term (linking the force at node 2 (B) to the displacement at node 2) receives contributions from both members AB and BC. This results from the fact that node 2(B) is directly connected to both nodes 1(A) and 3(C) while nodes 1 and 3 are connected directly to node 2. Nodes 1 and 3 are not directly connected so that the terms k_{13} and k_{31} are both zero, i.e. they are not affected by each other's displacement.

To summarize, the formation of the stiffness matrix for a complete structure is carried out as follows: terms of the form k_{ii} on the main diagonal consist of the sum of the stiffnesses of all the structural elements meeting at node i, while the off-diagonal terms of the form k_{ij} consist of the sum of the stiffnesses of all the elements connecting node i to node j.

Equation (16.59) may be solved for a specific case in which certain boundary conditions are specified. Thus, for example, the member AB may be fixed at A and loads F_B and F_C applied. Then $w_A = 0$ and F_A is a reaction force. Inversion of the resulting matrix enables w_B and w_C to be found.

In a practical situation a member subjected to an axial load could be part of a truss which would comprise several members set at various angles to one another. Therefore, to assemble a stiffness matrix for a complete structure, we need to refer axial forces and displacements to a common, or *global*, axis system.

Consider the member shown in Fig. 16.53 . It is inclined at an angle θ to a global axis system denoted by zy. The member connects node i to node j, and has *member* or *local* axes $\bar{z}, \bar{y}$. Thus nodal forces and displacements referred to local axes are written as $\bar{F}, \bar{w}$, etc., so that, by comparison with Eq. (16.51), we see that

$$\left\{\begin{array}{c} \bar{F}_{z,i} \\ \bar{F}_{z,j} \end{array}\right\} = \frac{AE}{L}\left[\begin{array}{cc} 1 & -1 \\ -1 & 1 \end{array}\right]\left\{\begin{array}{c} \bar{w}_i \\ \bar{w}_j \end{array}\right\} \tag{16.61}$$

where the member stiffness matrix is written as $[\bar{K}_{ij}]$.

In Fig. 16.53 external forces $\bar{F}_{z,i}$, and $\bar{F}_{z,j}$ are applied to i and j. It should be noted that $\bar{F}_{y,i}$ and $\bar{F}_{y,j}$ do not exist since the member can only support axial forces. However, $\bar{F}_{z,i}$ and $\bar{F}_{z,j}$ have components $F_{z,i}$, $F_{y,i}$ and $F_{z,j}$, $F_{y,j}$ respectively, so that whereas only two force components appear for the member in local coordinates, four components are present when global coordinates are used. Therefore, if we are to transfer from local to global coordinates, Eq. (16.61) must be expanded to an order consistent with the use of global coordinates. Thus,

$$\left\{\begin{array}{c} \bar{F}_{z,i} \\ \bar{F}_{y,i} \\ \bar{F}_{z,i} \\ \bar{F}_{y,j} \end{array}\right\} = \frac{AE}{L}\left[\begin{array}{cccc} 1 & 0 & -1 & 0 \\ 0 & 0 & 0 & 0 \\ -1 & 0 & 1 & 0 \\ 0 & 0 & 0 & 0 \end{array}\right]\left\{\begin{array}{c} \bar{w}_i \\ \bar{v}_i \\ \bar{w}_j \\ \bar{v}_j \end{array}\right\} \tag{16.62}$$

Expansion of Eq. (16.62) shows that the basic relationship between $\bar{F}_{z,i}$, $\bar{F}_{z,j}$ and $\bar{w}_i, \bar{w}_j$ as defined in Eq. (16.61) is unchanged.

From Fig. 16.53 we see that

$$\bar{F}_{z,i} = F_{z,i} \cos \theta + F_{y,i} \sin \theta$$
$$\bar{F}_{y,i} = -F_{z,i} \sin \theta + F_{y,i} \cos \theta$$
and
$$\bar{F}_{z,j} = F_{z,j} \cos \theta + F_{y,j} \sin \theta$$
$$\bar{F}_{y,j} = -F_{z,j} \sin \theta + F_{y,j} \cos \theta$$

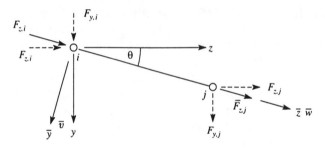

Fig. 16.53   Local and global axes systems for an axially loaded member

Writing λ for $\cos\theta$ and μ for $\sin\theta$ we express the above equations in matrix form as

$$
\begin{Bmatrix} \bar{F}_{z,i} \\ \bar{F}_{y,i} \\ \bar{F}_{z,j} \\ \bar{F}_{y,j} \end{Bmatrix}
=
\begin{bmatrix} \lambda & \mu & 0 & 0 \\ -\mu & \lambda & 0 & 0 \\ 0 & 0 & \lambda & \mu \\ 0 & 0 & -\mu & \lambda \end{bmatrix}
\begin{Bmatrix} F_{z,i} \\ F_{y,i} \\ F_{z,j} \\ F_{y,j} \end{Bmatrix}
\tag{16.63}
$$

or, in abbreviated form

$$
\{\bar{F}\} = [T]\{F\}
\tag{16.64}
$$

where $[T]$ is known as the *transformation matrix*. A similar relationship exists between the sets of nodal displacements. Thus,

$$
\{\bar{\delta}\} = [T]\{\delta\}
\tag{16.65}
$$

in which $\{\bar{\delta}\}$ and $\{\delta\}$ are generalized displacements referred to the local and global axes, respectively. Substituting now for $\{\bar{F}\}$ and $\{\bar{\delta}\}$ in Eq. (16.62) from Eqs (16.64) and (16.65) we have

$$
[T]\{F\} = [\bar{K}_{ij}][T]\{\delta\}
$$

Hence
$$
\{F\} = [T^{-1}][\bar{K}_{ij}][T]\{\delta\}
\tag{16.66}
$$

It may be shown that the inverse of the transformation matrix is its transpose, i.e.

$$
[T^{-1}] = [T]^{\mathrm{T}}
$$

Thus we rewrite Eq. (16.66) as

$$
\{F\} = [T]^{\mathrm{T}}[\bar{K}_{ij}][T]\{\delta\}
\tag{16.67}
$$

The nodal force system referred to the global axes, $\{F\}$, is related to the corresponding nodal displacements by

$$
\{F\} = [K_{ij}]\{\delta\}
\tag{16.68}
$$

in which $[K_{ij}]$ is the member stiffness matrix referred to global coordinates. Comparison of Eqs (16.67) and (16.68) shows that

$$
[K_{ij}] = [T]^{\mathrm{T}}[\bar{K}_{ij}][T]
$$

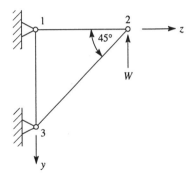

Fig. 16.54 Truss of Ex. 16.24

Substituting for $[T]$ from Eq. (16.63) and $[\bar{K}_{ij}]$ from Eq. (16.62) we obtain

$$[K_{ij}] = \frac{AE}{L} \begin{bmatrix} \lambda^2 & \lambda\mu & -\lambda^2 & -\lambda\mu \\ \lambda\mu & \mu^2 & -\lambda\mu & -\mu^2 \\ -\lambda^2 & -\lambda\mu & \lambda^2 & \lambda\mu \\ -\lambda\mu & -\mu^2 & \lambda\mu & \mu^2 \end{bmatrix} \qquad (16.69)$$

Evaluating $\lambda(=\cos\theta)$ and $\mu(=\sin\theta)$ for each member and substituting in Eq. (16.69) we obtain the stiffness matrix, referred to global axes, for each member of the framework.

Example 16.24 Determine the horizontal and vertical components of the deflection of node 2 and the forces in the members of the truss shown in Fig. 16.54. The product AE is constant for all members.

We see from Fig. 16.54 that the nodes 1 and 3 are pinned to the foundation and are therefore not displaced. Hence, referring to the global coordinate system shown,

$$w_1 = v_1 = w_3 = v_3 = 0$$

The external forces are applied at node 2 such that $F_{z,2} = 0$, $F_{y,2} = -W$; the nodal forces at 1 and 3 are then unknown reactions.

The first step in the solution is to assemble the stiffness matrix for the complete framework by writing down the member stiffness matrices referred to the global axes using Eq. (16.69). The direction cosines λ and μ take different values for each of the three members; therefore, remembering that the angle θ is measured clockwise from the positive direction of the z axis we have the following:

Member	θ (deg)	λ	μ
12	0	1	0
13	90	0	1
23	135	−0·707	0·707

The member stiffness matrices are therefore

$$[K_{12}] = \frac{AE}{L}\begin{bmatrix} 1 & 0 & -1 & 0 \\ 0 & 0 & 0 & 0 \\ -1 & 0 & 1 & 0 \\ 0 & 0 & 0 & 0 \end{bmatrix} \qquad [K_{13}] = \frac{AE}{L}\begin{bmatrix} 0 & 0 & 0 & 0 \\ 0 & 1 & 0 & -1 \\ 0 & 0 & 0 & 0 \\ 0 & -1 & 0 & 1 \end{bmatrix}$$

$$[K_{23}] = \frac{AE}{1\cdot414L}\begin{bmatrix} 0\cdot5 & -0\cdot5 & -0\cdot5 & 0\cdot5 \\ -0\cdot5 & 0\cdot5 & 0\cdot5 & -0\cdot5 \\ -0\cdot5 & 0\cdot5 & 0\cdot5 & -0\cdot5 \\ 0\cdot5 & -0\cdot5 & -0\cdot5 & 0\cdot5 \end{bmatrix} \tag{i}$$

The complete stiffness matrix is now assembled using the method suggested in the discussion of Eq. (16.60). The matrix will be a 6×6 matrix since there are six nodal forces connected to six nodal displacements; thus

$$\begin{Bmatrix} F_{z,1} \\ F_{y,1} \\ F_{z,2} \\ F_{y,2} \\ F_{z,3} \\ F_{y,3} \end{Bmatrix} = \frac{AE}{L}\begin{bmatrix} 1 & 0 & -1 & 0 & 0 & 0 \\ 0 & 1 & 0 & 0 & 0 & -1 \\ -1 & 0 & 1\cdot354 & -0\cdot354 & -0\cdot354 & 0\cdot354 \\ 0 & 0 & -0\cdot354 & 0\cdot354 & 0\cdot354 & -0\cdot354 \\ 0 & 0 & -0\cdot354 & 0\cdot354 & 0\cdot354 & -0\cdot354 \\ 0 & -1 & 0\cdot354 & -0\cdot354 & -0\cdot354 & 1\cdot354 \end{bmatrix}\begin{Bmatrix} w_1 = 0 \\ v_1 = 0 \\ w_2 \\ v_2 \\ w_3 = 0 \\ v_3 = 0 \end{Bmatrix} \tag{ii}$$

If we now delete rows and columns in the stiffness matrix corresponding to zero displacements, we obtain the unknown nodal displacements w_2 and v_2 in terms of the applied loads $F_{z,2}(=0)$ and $F_{y,2}(= -W)$. Thus

$$\begin{Bmatrix} F_{z,2} \\ F_{y,2} \end{Bmatrix} = \frac{AE}{L}\begin{bmatrix} 1\cdot354 & -0\cdot354 \\ -0\cdot354 & 0\cdot354 \end{bmatrix}\begin{Bmatrix} w_2 \\ v_2 \end{Bmatrix} \tag{iii}$$

Inverting Eq. (iii) gives

$$\begin{Bmatrix} w_2 \\ v_2 \end{Bmatrix} = \frac{L}{AE}\begin{bmatrix} 1 & 1 \\ 1 & 3\cdot828 \end{bmatrix}\begin{Bmatrix} F_{z,2} \\ F_{y,2} \end{Bmatrix} \tag{iv}$$

from which
$$w_2 = \frac{L}{AE}(F_{z,2} + F_{y,2}) = \frac{-WL}{AE}$$

$$v_2 = \frac{L}{AE}(F_{z,2} + 3\cdot828F_{y,2}) = \frac{-3\cdot828WL}{AE}$$

The reactions at nodes 1 and 3 are now obtained by substituting for w_2 and v_2 from Eq. (iv) into Eq. (ii). Hence

$$\begin{Bmatrix} F_{z,1} \\ F_{y,1} \\ F_{z,3} \\ F_{y,3} \end{Bmatrix} = \begin{bmatrix} -1 & 0 \\ 0 & 0 \\ -0\cdot354 & 0\cdot354 \\ 0\cdot354 & -0\cdot354 \end{bmatrix}\begin{bmatrix} 1 & 1 \\ 1 & 3\cdot828 \end{bmatrix}\begin{Bmatrix} F_{z,2} \\ F_{y,2} \end{Bmatrix} = \begin{bmatrix} -1 & -1 \\ 0 & 0 \\ 0 & 1 \\ 0 & -1 \end{bmatrix}\begin{Bmatrix} F_{z,2} \\ F_{y,2} \end{Bmatrix}$$

giving
$$F_{z,1} = -F_{z,2} - F_{y,2} = W$$
$$F_{y,1} = 0$$
$$F_{z,3} = F_{y,2} = -W$$
$$F_{y,3} = W$$

The internal forces in the members may be found from the axial displacements of the nodes. Thus, for a member ij, the internal force F_{ij} is given by

$$F_{ij} = \frac{AE}{L}(\bar{w}_j - \bar{w}_i) \tag{v}$$

But
$$\bar{w}_j = \lambda w_j + \mu v_j$$
$$\bar{w}_i = \lambda w_i + \mu v_i$$

Hence
$$\bar{w}_j - \bar{w}_i = \lambda(w_j - w_i) + \mu(v_j - v_i)$$

Substituting in Eq. (v) and rewriting in matrix form,

$$F_{ij} = \frac{AE}{L}\begin{bmatrix} \lambda & \mu \\ & ij \end{bmatrix}\begin{Bmatrix} w_j - w_i \\ v_j - v_i \end{Bmatrix} \tag{vi}$$

Thus, for the members of the framework

$$F_{12} = \frac{AE}{L}[1 \ \ 0]\begin{Bmatrix} \dfrac{-WL}{AE} - 0 \\[2mm] \dfrac{-3\cdot828WL}{AE} - 0 \end{Bmatrix} = -W \text{ (compression)}$$

$$F_{13} = \frac{AE}{L}[0 \ \ 1]\begin{Bmatrix} 0 - 0 \\ 0 - 0 \end{Bmatrix} = 0 \text{ (obvious from inspection)}$$

$$F_{23} = \frac{AE}{1\cdot414L}[-0\cdot707 \ \ 0\cdot707]\begin{Bmatrix} 0 + \dfrac{WL}{AE} \\[2mm] 0 + \dfrac{3\cdot828WL}{AE} \end{Bmatrix} = 1\cdot414W \text{ (tension)}$$

The matrix method of solution for the statically determinate truss of Ex. 16.24 is completely general and therefore applicable to any structural problem. We observe from the solution that the question of statical determinacy of the truss did not arise. Statically indeterminate trusses are therefore solved in an identical manner with the stiffness matrix for each redundant member being included in the complete stiffness matrix as described above. Clearly, the greater the number of members the greater the size of the stiffness matrix, so that a computer-based approach is essential.

Pin-jointed space frames may be analysed in a similar manner to plane trusses. In this case a member stiffness matrix is of the order 6×6 as is the transformation matrix. The analysis of these structures is, however, outside the scope of this book.

Beam elements

The matrix analysis of members subjected to axial forces may be extended to beams that carry bending moments and shear forces. These beams may be structures in their own right or, in fact, be elements of other structural forms such as portal frames. However, even in the case of a simple beam, matrix analysis requires the beam to be idealized into a number of elements where the end of an element, i.e. a node, coincides with a loading or structural discontinuity.

In Section 16.9 we derived the slope–deflection relationships for a beam AB (Eqs (16.27)). Rewriting these equations in matrix form for a beam connecting nodes i (A) and j (B) we have

$$\begin{Bmatrix} F_{y,i} \\ M_i \\ F_{y,j} \\ M_j \end{Bmatrix} = EI \begin{bmatrix} 12/L^3 & 6/L^2 & -12/L^3 & 6/L^2 \\ 6/L^2 & 4/L & -6/L^2 & 2/L \\ -12/L^3 & -6/L^2 & 12/L^3 & -6/L^2 \\ -6/L^2 & 2/L & -6/L^2 & 4/L \end{bmatrix} \begin{Bmatrix} v_i \\ \theta_i \\ v_j \\ \theta_j \end{Bmatrix} \tag{16.70}$$

in which $F_{y,i}$ has replaced S_{AB}, M_i has replaced M_{AB}, and so on. Equation (16.70) is of the form

$$\{F\} = [K_{ij}]\{\delta\}$$

where $[K_{ij}]$ is the stiffness matrix for the beam. This stiffness matrix applies to a beam where the axis is aligned with the z axis, so that it is actually $[\bar{K}_{ij}]$, the stiffness matrix referred to local or member axes. If the beam is positioned in the zy plane with its axis inclined to the z axis, then the zy axes are global axes and Eq. (16.70) must be transformed to allow for this. The procedure is similar to that for an axially loaded member except that $[\bar{K}_{ij}]$ must be expanded to allow for the fact that nodal displacements $\bar{w}_i$ and $\bar{w}_j$ which are irrelevant for the beam in local axes (we are not considering axial effects here) have components w_i, v_i and w_j, v_j referred to global axes. Thus

$$[\bar{K}_{ij}] = EI \begin{bmatrix} 0 & 0 & 0 & 0 & 0 & 0 \\ 0 & 12/L^3 & 6/L^2 & 0 & -12/L^3 & 6/L^2 \\ 0 & 6/L^2 & 4/L & 0 & -6/L^2 & 2/L \\ 0 & 0 & 0 & 0 & 0 & 0 \\ 0 & -12/L^3 & -6/L^2 & 0 & 12/L^3 & -6/L^2 \\ 0 & -6/L^2 & 2/L & 0 & -6/L^2 & 4/L \end{bmatrix} \tag{16.71}$$

The transformation matrix $[T]$ may be deduced from Eq. (16.63) if it is remembered that although w and v will transform in exactly the same way as in the case of the axially loaded member, the rotations θ remain the same in local and global axes. Hence

$$[T] = \begin{bmatrix} \lambda & \mu & 0 & 0 & 0 & 0 \\ -\mu & \lambda & 0 & 0 & 0 & 0 \\ 0 & 0 & 1 & 0 & 0 & 0 \\ 0 & 0 & 0 & \lambda & \mu & 0 \\ 0 & 0 & 0 & -\mu & \lambda & 0 \\ 0 & 0 & 0 & 0 & 0 & 1 \end{bmatrix} \tag{16.72}$$

where λ and μ have previously been defined. Therefore, since

$$[K_{ij}] = [T]^{\mathrm{T}}[\bar{K}_{ij}][T]$$

we have, from Eqs (16.71) and (16.72)

$$[K_{ij}] = EI \begin{bmatrix} 12\mu^2/L^3 \\ -12\lambda\mu/L^3 & 12\lambda^2/L^3 \\ -6\mu/L^2 & 6\lambda/L^2 & 4/L \\ -12\mu^2/L^3 & 12\lambda\mu/L^3 & 6\mu/L^2 & 12\mu^2/L^3 \\ 12\lambda\mu/L^3 & -12\lambda^2/L^3 & -6\lambda/L^2 & -12\lambda\mu/L^3 & 12\lambda^2/L^3 \\ -6\mu/L^2 & 6\lambda/L^2 & 2/L & -6\mu/L^2 & -6\lambda/L^2 & 4\lambda/L \end{bmatrix} \qquad (16.73)$$

Again, the stiffness matrix for the complete structure is assembled from the member stiffness matrices, the boundary conditions are applied and the resulting equations solved for the unknown nodal displacements and forces.

The matrix analysis of beams presented above is based on the condition that no external forces are applied between the nodes. In a practical situation a beam supports a variety of loads along its length and therefore such beams must be idealized into a number of *beam-elements* for which the above condition holds. Thus nodes are specified at points along the beam such that any element lying between adjacent nodes carries at the most a uniform shear force and a linearly varying bending moment. Beams carrying a distributed load require the load to be replaced by a series of statically equivalent point loads at a selected number of nodes. Clearly the greater the number of nodes chosen, the more accurate but more complex will be the analysis.

The discussion in this section is intended as an introduction to the matrix analysis of structures. The subject is extensive and complete texts are devoted to its presentation, which ranges from the relatively simple case of an axially loaded member to the sophisticated finite element method for the analysis of continuum structures. Such topics are advanced and fall outside the scope of this book.

Problems

P.16.1 Determine the degrees of static and kinematic indeterminacy in the plane structures shown in Fig. P.16.1.

Ans. (a) $n_s = 3$, $n_k = 6$, (b) $n_s = 1$, $n_k = 2$, (c) $n_s = 2$, $n_k = 4$,
(d) $n_s = 6$, $n_k = 15$, (e) $n_s = 2$, $n_k = 7$.

P.16.2 Determine the degrees of static and kinematic indeterminacy in the space frames shown in Figs P.16.2.

Ans. (a) $n_s = 6$, $n_k = 24$, (b) $n_s = 42$, $n_k = 36$, (c) $n_s = 18$, $n_k = 6$.

P.16.3 Calculate the support reactions in the beam shown in Fig. P.16.3 using a flexibility method.

Ans. $R_A = 3{\cdot}4$ kN, $R_B = 14{\cdot}5$ kN, $R_C = 4{\cdot}1$ kN, $M_A = 2{\cdot}4$ kN m (hogging).

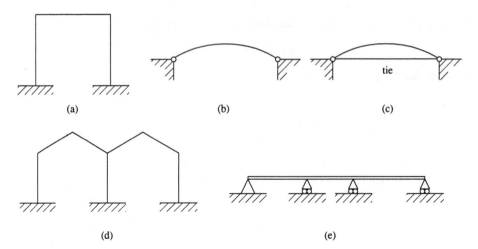

Fig. P.16.1

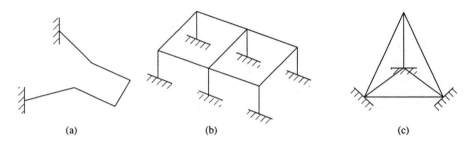

Fig. P.16.2

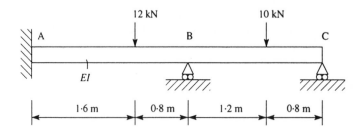

Fig. P.16.3

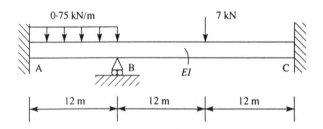

Fig. P.16.4

P.16.4 Determine the support reactions in the beam shown in Fig. P.16.4 using a flexibility method.

Ans. $R_A = 3.5$ kN, $R_B = 8.75$ kN, $R_C = 3.75$ kN, $M_A = 5$ kN m (hogging), $M_C = -23$ kN m (hogging).

P.16.5 Use a flexibility method to determine the support reactions in the beam shown in Fig. P.16.5. The flexural rigidity EI of the beam is constant throughout.

Ans. $R_A = 4.2$ k N, $R_B = 15.4$ kN, $R_C = 17.4$ kN, $R_D = 16.1$ kN.

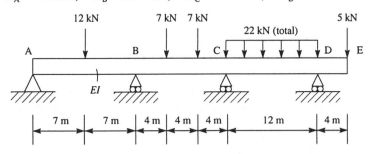

Fig. P.16.5

P.16.6 Calculate the forces in the members of the truss shown in Fig. P.16.6. The members AC and BD are 30 mm² in cross-section, all the other members are 20 mm² in cross-section. The members AD, BC and DC are each 800 mm long; $E = 200\,000$ N/mm².

Ans. AC = 48·2 N, BC = 87·6 N, BD = −1·8 N, CD = 2·1 N, AD = 1·0 N.

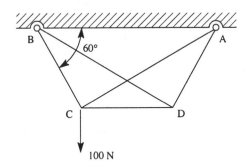

Fig. P.16.6

P.16.7 Calculate the forces in the members of the truss shown in Fig. P.16.7. The cross-sectional area of all horizontal members is 200 mm², that of the vertical members is 100 mm² while that of the diagonals is 300 mm²; E is constant throughout.

Ans. AB = FD = −29·2 kN, BC = CD = −29·2 kN, AG = GF = 20·8 kN, BG = DG = 41·3 kN, AC = FC = −29·4 kN, CG = 41·6 kN.

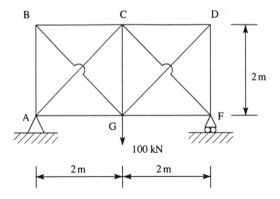

Fig. P.16.7

P.16.8   Calculate the forces in the members of the truss shown in Fig. P.16.8 and the vertical and horizontal components of the reactions at the supports; all members of the truss have the same cross-sectional properties.

Ans. $R_{A,V} = 67.52$ kN, $R_{A,H} = 70.04$ kN $= R_{F,H}$, $R_{F,V} = 32.48$ kN,
AB $= -32.49$ kN,   AD $= -78.31$ kN,   BC $= -64.98$ kN,
BD $= 72.65$ kN,   CD $= -100.0$ kN,   CE $= -64.98$ kN,   DE $= 72.65$ kN,
DF $= -70.04$ kN,   EF $= -32.49$ kN.

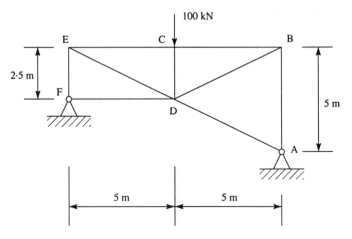

Fig. P.16.8

P.16.9   The plane truss shown in Fig. P.16.9(a) has one member (24) which is loosely attached at joint 2 so that relative movement between the end of the member and the joint may occur when the framework is loaded. This movement is a maximum of 0.25 mm and takes place only in the direction 24. Figure P.16.9(b) shows joint 2 in detail when the framework is unloaded. Find the value of P at which the member 24 just becomes an effective part of the truss and also the loads in all the members when $P = 10$ kN. All members have a cross-sectional area of 300 mm^2 and a Young's modulus of 70 000 N/mm^2.

Ans. $P = 2.95$ kN, $12 = 2.48$ kN, $23 = 1.86$ kN, $34 = 2.48$ kN,
  $41 = -5.64$ kN, $13 = 9.4$ kN, $42 = -3.1$ kN.

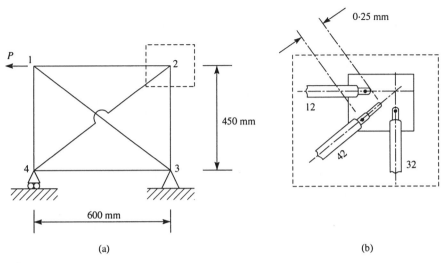

(a)   (b)

Fig. P.16.9

P.16.10   Figure P.16.10 shows a plane truss pinned to a rigid foundation. All members have the same Young's modulus of $70\,000$ N/mm^2 and the same cross-sectional area, A, except the member 12 whose cross-sectional area is $1.414A$.

Under some systems of loading, member 14 carries a tensile stress of 0.7 N/mm^2. Calculate the change in temperature which, if applied to member 14 only, would reduce the stress in that member to zero. The coefficient of linear expansion $\alpha = 2 \times 10^{-6}$/°C.

Ans. $5.5°$.

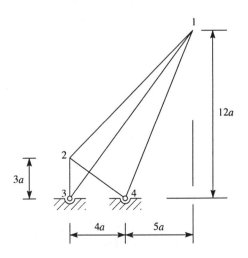

Fig. P.16.10

P.16.11   The truss shown in Fig. P.16.11 is pinned to a foundation at the points A and B and is supported on rollers at G; all members of the truss have the same axial rigidity $EA = 2 \times 10^9$ N.

Calculate the forces in all the members of the truss produced by a settlement of 15 mm at the support at G.

Ans.   FG = 1073·9 kN,   GH = −536·9 kN,   NF = −1073·9 kN,
DF = 1073·9 kN,   JH = −1610·8 kN,   DH = 1073·9 kN,
DC = 2147·7 kN,   CJ = 1073·9 kN,   AJ = −2684·6 kN,
AC = −1073·9 kN,   DJ = −1073·9 kN,   BC = 3221·6 kN.

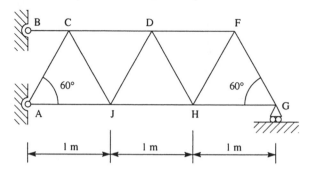

Fig. P.16.11

P.16.12   The cross-sectional area of the braced beam shown in Fig. P.16.12 is 4A and its second moment of area for bending is $Aa^2/16$. All other members have the same cross-sectional area, A, and Young's modulus is E for all members. Find, in terms of w, A, a and E, the vertical displacement of the point D under the loading shown.

Ans.   30 232 $wa^2/3AE$.

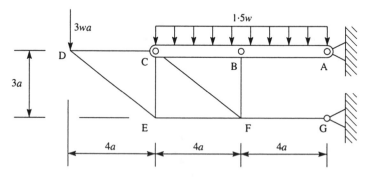

Fig. P.16.12

P.16.13   Determine the force in the vertical member BD (the king post) in the trussed beam ABC shown in Fig. P.16.13. The cross-sectional area of the king post is 2000 mm², that of the beam is 5000 mm² while that of the members AD and DC of the truss is 200 mm²; the second moment of area of the beam is 4.2×10^6 mm⁴ and Young's modulus, E, is the same for all members.

Ans.   91·6 kN.

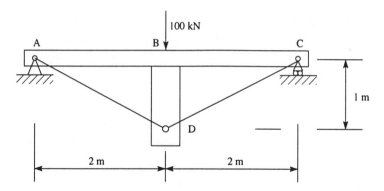

Fig. P.16.13

P.16.14 Determine the distribution of bending moment in the frame shown in P.16.14.

Ans. $M_B = 7wL^2/45$, $M_C = 8wL^2/45$. Parabolic distribution on AB, linear on BC and CD.

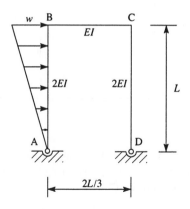

Fig. P.16.14

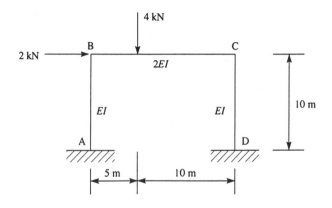

Fig. P.16.15

P.16.15   Use the flexibility method to determine the end moments in the members of the portal frame shown in Fig. P.16.15. The flexural rigidity of the horizontal member BC is $2EI$ while that of the vertical members AB and CD is EI.

Ans. $M_{AB} = -3.8$ kN m, $M_{BA} = -M_{BC} = -0.2$ kN m,

$M_{CB} = -M_{CD} = 8.2$ kN m, $M_{DC} = -7.8$ kN m.

P.16.16   Calculate the end moments in the members of the frame shown in Fig. P.16.16 using the flexibility method; all members have the same flexural rigidity, EI.

Ans. $M_{AB} = 14.7$ kN m, $M_{BA} = -M_{BC} = 84.8$ kN m,

$M_{CB} = -M_{CD} = 7.0$ kN m, $M_{DC} = 0$.

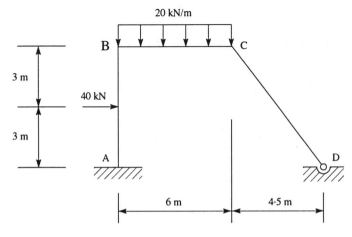

Fig. P.16.16

P.16.17   The two-pinned circular arch shown in Fig. P.16.17 carries a uniformly distributed load of 15 kN/m over the half-span AC. Calculate the support reactions and the bending moment at the crown C.

Ans. $R_{A,V} = 34.1$ kN, $R_{B,V} = 11.4$ kN, $R_{A,H} = R_{B,H} = 18.7$ kN,

$M_C = 1.76$ kN m.

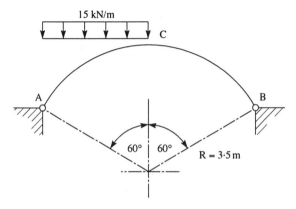

Fig. P.16.17

P.16.18 The two-pinned parabolic arch shown in Fig. P.16.18 has a second moment of area, I, that varies such that $I = I_0 \sec \theta$ where I_0 is the second moment of area at the crown of the arch and θ is the slope of the tangent at any point. Calculate the horizontal thrust at the arch supports and determine the bending moment in the arch at the loading points and at the crown.

Ans. $R_{A,H} = R_{B,H} = 168 \cdot 7$ kN, $M_D = 49 \cdot 6$ kN m, $M_C = -6 \cdot 1$ kN m.

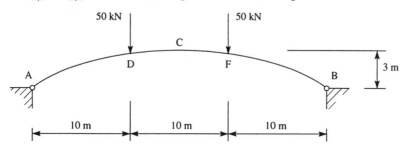

Fig. P.16.18

P.16.19 Show that, for a two-pinned parabolic arch carrying a uniformly distributed load over its complete span and in which the second moment of area of the cross-section varies as the secant assumption, the bending moment is everywhere zero.

P.16.20 The arch shown in Fig. P.16.20 is parabolic, the equation of its profile being $y = 0 \cdot 05 x (40 - x)$. If the second moment of area of the cross-section of the arch varies directly as the secant of its slope, calculate the reactions at the support points and the bending moment at the crown C.

Ans. $R_{A,V} = 10 \cdot 3$ kN, $R_{B,V} = 4 \cdot 7$ kN, $R_{A,H} = R_{B,H} = 0 \cdot 6$ kN,
$M_C = -44 \cdot 0$ kN m.

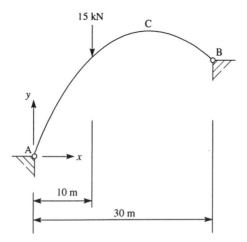

Fig. P.16.20

P.16.21 Use the slope–deflection method to solve P.16.3, P.16.4, P.16.5 and Ex. 16.20.

P.16.22   Use the slope–deflection method to determine the member end moments in the portal frame of Ex. 16.22.

P.16.23   Calculate the support reactions in the continuous beam shown in Fig. P.16.23 using the moment distribution method; the flexural rigidity, EI, of the beam is constant throughout.

Ans. $R_A = 2 \cdot 7$ kN, $R_B = 10 \cdot 6$ kN, $R_C = 3 \cdot 7$ kN, $M_A = -1 \cdot 7$ kN m.

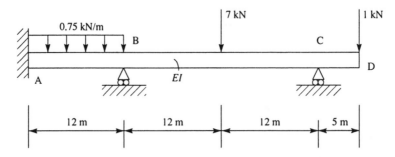

Fig. P.16.23

P.16.24   Calculate the support reactions in the beam shown in Fig P.16.24 using the moment distribution method; the flexural rigidity, EI, of the beam is constant throughout.

Ans. $R_C = 28 \cdot 2$ kN, $R_D = 17 \cdot 0$ kN, $R_E = 4 \cdot 8$ kN, $M_E = 1 \cdot 6$ kN m.

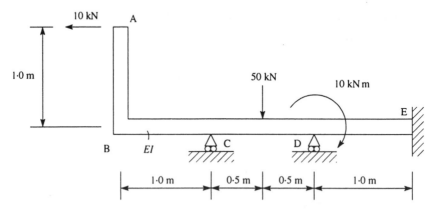

Fig. P.16.24

P.16.25   In the beam ABC shown in Fig. P.16.25 the support at B settles by 10 mm when the loads are applied. If the second moment of area of the spans AB and BC are $83 \cdot 4 \times 10^6$ mm^4 and $125 \cdot 1 \times 10^6$ mm^4, respectively, and Young's modulus, E, of the material of the beam is 207 000 N/mm^2, calculate the support reactions using the moment distribution method.

Ans. $R_A = 28 \cdot 6$ kN, $R_B = 15 \cdot 9$ kN, $R_C = 30 \cdot 5$ kN, $M_C = 53 \cdot 9$ kN m.

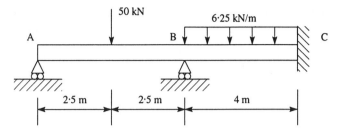

Fig. P.16.25

P.16.26 Calculate the end moments in the members of the frame shown in Fig. P.16.26 using the moment distribution method. The flexural rigidity of the members AB, BC and BD are $2EI$, $3EI$ and EI, respectively, and the support system is such that sway is prevented.

Ans. $M_{AB} = M_{CB} = 0$, $\quad M_{BA} = 30$ kN m, $\quad M_{BC} = -36$ kN m,
$M_{BD} = 6$ kN m, $M_{DB} = 3$ kN m.

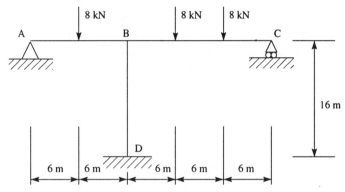

Fig. P.16.26

P.16.27 The frame shown in Fig. P.16.27 is pinned to the foundation of A and D and has members whose flexural rigidity is EI. Use the moment distribution method to calculate the moments in the members and draw the bending moment diagram.

Ans. $M_A = M_D = 0$, $\quad M_B = 11 \cdot 9$ kN m, $\quad M_C = 63 \cdot 2$ kN m.

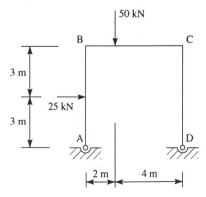

Fig. P.16.27

P.16.28   Use the moment distribution method to calculate the bending moments at the joints in the frame shown in Fig. P.16.28 and draw the bending moment diagram.

Ans. $M_{AB} = M_{DC} = 0$, $M_{BA} = 12\cdot7$ kN m $= -M_{BC}$, $M_{CB} = -13\cdot9$ kN m $= -M_{CD}$.

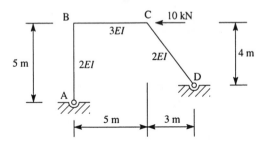

Fig. P.16.28

P.16.29   The frame shown in Fig. P.16.29 has rigid joints at B, C and D and is pinned to its foundation at A and G. The joint D is prevented from moving horizontally by the member DF which is pinned to a support at F. The flexural rigidity of the members AB and BC is $2EI$ while that of all other members is EI.

Use the moment distribution method to calculate the end moments in the members.

Ans. $M_{BA} = -M_{BC} = 2\cdot6$ kN m, $M_{CB} = -M_{CD} = 67\cdot7$ kN m,

$M_{DC} = -53\cdot5$ kN m, $M_{DF} = 26\cdot7$ kN m, $M_{DG} = 26\cdot7$ kN m.

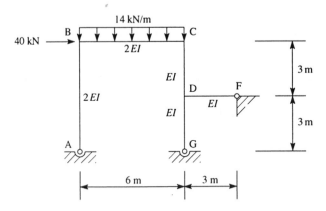

Fig. P.16.29

P.16.30   Figure P.16.30 shows a square symmetrical truss pinned to rigid supports at 2 and 4 and loaded with a vertical load, P, at 1. The axial rigidity EA is the same for all members.

Use the stiffness method to find the displacements at nodes 1 and 3 and hence solve for the member forces and support reactions.

Ans. $v_1 = 0\cdot707PL/AE,$ $v_3 = 0\cdot293PL/AE,$ $12 = 14 = 0\cdot5P,$

        $23 = 43 = -0\cdot207P,$ $13 = 0\cdot293P,$ $F_{z,2} = -F_{z,4} = 0\cdot207P,$

        $F_{y,2} = F_{y,4} = 0\cdot5P.$

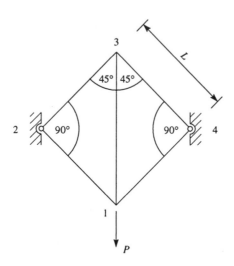

Fig. P.16.30

P.16.31 Form the matrices required to solve completely the plane truss shown in Fig. P.16.31 and determine the force in the member DE; all members have equal axial rigidity.

Ans. DE = 0.

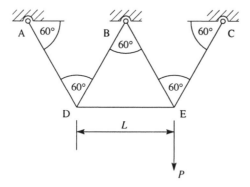

Fig. P.16.31

P.16.32 Use the stiffness method to find the ratio H/P for which the displacement of node 4 in the truss shown in Fig. P.16.32 is zero, and for that case find the displacements of nodes 2 and 3. All members have equal axial rigidity, *EA*.

Ans. $H/P = 0\cdot448,$ $v_2 = 0\cdot321PL/AE,$ $v_3 = 0\cdot481PL/AE.$

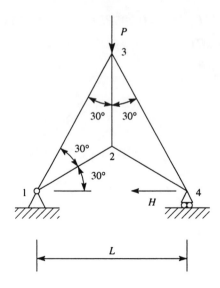

Fig. P.16.32

CHAPTER 17

Influence Lines

The structures we have considered so far have been subjected to loading systems that were stationary, i.e. the loads remained in a fixed position in relation to the structure. In many practical situations, however, structures carry loads that vary continuously. Thus a building supports a system of stationary loads which consist of its self-weight, the weight of any permanent fixtures such as partitions, machinery, etc., and also a system of imposed or 'live' loads which comprise snow loads, wind loads or any movable equipment. The structural elements of the building must then be designed to withstand the worst combination of these fixed and movable loads.

Other forms of movable load consist of vehicles and trains that cross bridges and viaducts. Again, these structures must be designed to support their self-weight, the weight of any permanent fixtures such as a road deck or railway track and also the forces produced by the passage of vehicles or trains. It is then necessary to determine the critical positions of the vehicles or trains in relation to the bridge or viaduct. Although these loads are moving loads, they are assumed to be moving or changing at such a slow rate that dynamic effects such as vibrations and oscillating stresses are absent.

The effects of loads that occupy different positions on a structure can be studied by means of *influence lines*. Influence lines give the value at a *particular* point in a structure of functions such as shear force, bending moment and displacement for *all* positions of a travelling unit load; they may also be constructed to show the variation of support reaction with the unit load position. From these influence lines the value of a function at a point can be calculated for a system of loads traversing the structure. For this we use the principle of superposition so that the structural systems we consider must be linearly elastic.

17.1 Influence lines for beams in contact with the load

We shall now investigate the construction of influence lines for support reactions and for the shear force and bending moment at a section of a beam when the travelling load is in continuous contact with the beam.

Consider the simply supported beam AB shown in Fig. 17.1(a) and suppose that we wish to construct the influence lines for the support reactions, R_A and R_B, and also for the shear force, S_K, and bending moment, M_K, at a given section K; all the influence lines are constructed by considering the passage of a unit load across the beam.

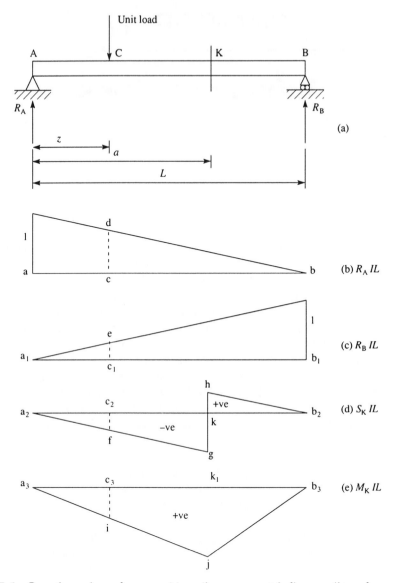

Fig. 17.1   Reaction, shear force and bending moment influence lines for a simply supported beam

R_A influence line

Suppose that the unit load has reached a position C, a distance z from A, as it travels across the beam. Then, considering the moment equilibrium of the beam about B we have

$$R_A L - 1(L - z) = 0$$

which gives

$$R_A = (L - z)L \qquad (17.1)$$

Hence R_A is a linear function of z and when $z = 0$, $R_A = 1$ and when $z = L$, $R_A = 0$; both these results are obvious from inspection. The influence line (IL) for R_A $(R_A IL)$ is then as shown in Fig. 17.1(b). Note that when the unit load is at C, the value of R_A is given by the ordinate cd in the R_A influence line.

R_B influence line

The influence line for the reaction R_B is constructed in an identical manner. Thus, taking moments about A

$$R_B L - 1z = 0$$

so that
$$R_B = z/L \tag{17.2}$$

Equation (17.2) shows that R_B is a linear function of z. Further, when $z = 0$, $R_B = 0$ and when $z = L$, $R_B = 1$, giving the influence line shown in Fig. 17.1(c). Again, with the unit load at C the value of R_B is equal to the ordinate $c_1 e$ in Fig. 17.1(c).

S_K influence line

The value of the shear force at the section K depends upon the position of the unit load, i.e. whether it is between A and K or between K and B. Suppose initially that the unit load is at the point C between A and K. Then the shear force at K is given by

$$S_K = -R_B$$

so that from Eq. (17.2)

$$S_K = -\frac{z}{L} \quad (0 \leqslant z \leqslant a) \tag{17.3}$$

The sign convention for shear force is that adopted in Section 3.2. We could have established Eq. (17.3) by expressing S_K in terms of R_A. Thus

$$S_K = R_A - 1$$

Substituting for R_A from Eq. (17.1) we obtain

$$S_K = \frac{L - z}{L} - 1 = -\frac{z}{L}$$

as before. Clearly, however, expressing S_K in the terms of R_B is the most direct approach.

We see from Eq. (17.3) that S_K varies linearly with the position of the load. Therefore, when $z = 0$, $S_K = 0$ and when $z = a$, $S_K = -a/L$, the ordinate kg in Fig. 17.1(d), and is the value of S_K with the unit load immediately to the left of K. Thus, with the load between A and K the S_K influence line is the line a_2g in Fig. 17.1(d) so that, when the unit load is at C, the value of S_K is equal to the ordinate $c_2 f$.

With the unit load between K and B the shear force at K is given by

$$S_K = +R_A \ (\text{or } S_K = 1 - R_B)$$

Substituting for R_A from Eq. (17.1) we have

$$S_K = (L - z)/L \qquad (a \leqslant z \leqslant L) \tag{17.4}$$

Again S_K is a linear function of load position. Therefore when $z = L$, $S_K = 0$ and when $z = a$, i.e. the unit load is immediately to the right of K, $S_K = (L - a)/L$ which is the ordinate kh in Fig. 17.1(d).

From Fig. 17.1(d) we see that the gradient of the line a_2g is equal to $[(-a/L) - 0]/a = -1/L$ and that the gradient of the line hb_2 is equal to $[0 - (L - a)/L]/(L - a) = -1/L$. Thus the gradient of the S_K influence line is the same on both sides of K. Furthermore, $gh = kh + kg$ or $gh = (L - a)/L + a/L = 1$.

M_K influence line

The value of the bending moment at K also depends upon whether the unit load is to the left or right of K. With the unit load at C

$$M_K = R_B(L - a) \qquad (\text{or } M_K = R_A a - 1(a - z))$$

which, when substituting for R_B from Eq. (17.2) becomes

$$M_K = (L - a)z/L \qquad (0 \leqslant z \leqslant a) \tag{17.5}$$

From Eq. (17.5) we see that M_K varies linearly with z. Therefore, when $z = 0$, $M_K = 0$ and when $z = a$, $M_K = (L - a)a/L$, which is the ordinate k_1j in Fig. 17.1(e).

Now with the unit load between K and B

$$M_K = R_A a$$

which becomes, from Eq. (17.1)

$$M_K = [(L - z)/L]a \qquad (a \leqslant z \leqslant L) \tag{17.6}$$

Again M_K is a linear function of z so that when $z = a$, $M_K = (L - a)a/L$, the ordinate k_1j in Fig. 17.1 (e), and when $z = L$, $M_K = 0$. The complete influence line for the bending moment at K is then the line a_3jb_3 as shown in Fig. 17.1(e). Hence the bending moment at K with the unit load at C is the ordinate c_3i in Fig. 17.1(e).

In establishing the shear force and bending moment influence lines for the section K of the beam in Fig. 17.1(a) we have made use of the previously derived relationships for the support reactions, R_A and R_B. If only the influence lines for S_K and M_K had been required, the procedure would have been as follows.

With the unit load between A and K

$$S_K = -R_B$$

Now, taking moments about A

$$R_B L - 1z = 0$$

whence

$$R_B = \frac{z}{L}$$

Thus

$$S_K = -\frac{z}{L}$$

This, of course, amounts to the same procedure as before except that the calculation of R_B follows the writing down of the expression for S_K. The remaining equations for the influence lines for S_K and M_K are derived in a similar manner.

We note from Fig. 17.1 that all the influence lines are composed of straight-line segments. This is always the case for statically determinate structures. We shall therefore make use of this property when considering other beam arrangements.

Example 17.1 Draw influence lines for the shear force and bending moment at the section C of the beam shown in Fig. 17.2(a).

In this example we are not required to obtain the influence lines for the support reactions. However, the influence line for the reaction R_A has been included to illustrate the difference between this influence line and the influence line for R_A in Fig. 17.1(b); the reader should verify the R_A influence line in Fig. 17.2(b).

Since we have established that influence lines for statically determinate structures consist of linear segments, they may be constructed by placing the unit load at different positions, which will enable us to calculate the principal values.

S_C influence line

With the unit load at A

$$S_C = -R_B = 0 \qquad \text{(by inspection)}$$

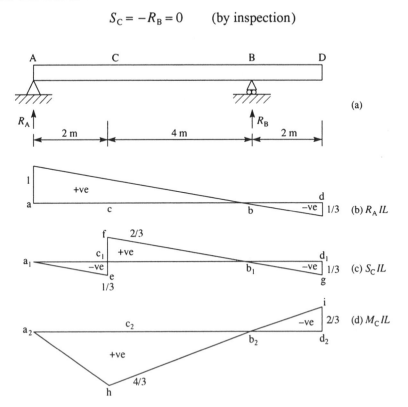

Fig. 17.2 Shear force and bending moment influence lines for the beam of Ex. 17.1

With the unit load immediately to the left of C

$$S_C = - R_B \qquad\qquad (i)$$

Now taking moments about A we have

$$R_B \times 6 - 1 \times 2 = 0$$

which gives

$$R_B = \frac{1}{3}$$

Therefore, from Eq. (i)

$$S_C = -\frac{1}{3} \qquad\qquad (ii)$$

Now with the unit load immediately to the right of C

$$S_C = +R_A \qquad\qquad (iii)$$

Taking moments about B gives

$$R_A \times 6 - 1 \times 4 = 0$$

whence

$$R_A = \frac{2}{3}$$

so that, from Eq. (iii)

$$S_C = +\frac{2}{3}$$

With the unit load at B

$$S_C = +R_A = 0 \qquad \text{(by inspection)} \qquad (v)$$

Placing the unit load at D we have

$$S_C = +R_A \qquad\qquad (vi)$$

Again taking moments about B

$$R_A \times 6 + 1 \times 2 = 0$$

from which $\qquad\qquad R_A = -1/3$

Hence $\qquad\qquad S_C = -1/3 \qquad\qquad (vii)$

The complete influence line for the shear force at C is then as shown in Fig. 17.2(c). Note that the gradient of each of the lines a_1e, fb_1, and b_1g is the same.

M_C influence line

With the unit load placed at A

$$M_C = +R_B \times 4 = 0 \qquad (R_B = 0 \text{ by inspection})$$

With the unit load at C

$$M_C = +R_A \times 2 = +\frac{4}{3}$$

in which $R_A = 2/3$ with the unit load at C (see above). With the unit load at B

$$M_C = +R_A \times 2 = 0 \ (R_A = 0 \text{ by inspection})$$

Finally, with the unit load at D

$$M_C = +R_A \times 2$$

but, again from the calculation of S_C, $R_A = -1/3$. Hence

$$M_C = -\frac{2}{3}$$

The complete influence line for the bending moment at C is shown in Fig. 17.2(d). Note that the line hb_2i is one continuous line.

17.2 Mueller–Breslau principle

A simple and convenient method of constructing influence lines is to employ the Mueller–Breslau principle which gives the shape of an influence line without the values of its ordinates; these, however, are easily calculated for statically determinate systems from geometry.

Consider the simply supported beam, AB, shown in Fig. 17.3(a) and suppose that a unit load is crossing the beam and has reached the point C a distance z from A. Suppose also that we wish to determine the influence line for the moment at the section K, a distance a from A. We now impose a virtual displacement, v_C, at C such that internal work is done only by the moment at K, i.e. we allow a change in gradient, θ_K, at K so that the lengths AK and KB rotate as rigid links as shown in Fig. 17.3(b). Therefore, from the principle of virtual work (Chapter 15), the external virtual work done by the unit load is equal to the internal virtual work done by the moment, M_K, at K. Thus

$$1v_C = M_K\theta_K$$

If we choose v_C so that θ_K is equal to unity

$$M_K = v_C \tag{17.7}$$

i.e. the moment at the section K due to a unit load at the point C, an arbitrary distance z from A, is equal to the magnitude of the virtual displacement at C. But, as we have seen in Section 17.1, the moment at a section K due to a unit load at a point C is the influence line for the moment at K. Therefore the M_K influence line may be constructed by introducing a hinge at K and imposing a unit change in angle at K; the displaced shape is then the influence line.

The argument may be extended to the construction of the influence line for the shear force, S_K, at the section K. Suppose now that the virtual displacement, v_C,

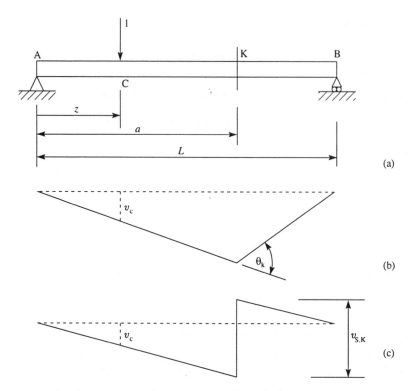

Fig. 17.3   Verification of the Mueller–Breslau principle

produces a shear displacement, $v_{S,K}$, at K as shown in Fig. 17.3(c). Again, from the principle of virtual work

$$1v_C = S_K v_{S,K}$$

If we choose v_C so that $v_{S,K} = 1$

$$S_K = v_C \qquad (17.8)$$

Hence, since the shear force at the section K due to a unit load at any point C is the influence line for the shear force at K, we see that the displaced shape in Fig. 17.3(c) is the influence line for S_K when the displacement at K produced by the virtual displacement at C is unity. A similar argument may be used to establish reaction influence lines.

The Mueller–Breslau principle demonstrated above may be stated in general terms as follows:

The shape of an influence line for a particular function (support reaction, shear force, bending moment, etc.) can be obtained by removing the resistance of the structure to that function at the section for which the influence line is required and applying an internal force corresponding to that function so that a unit displacement is produced at the section. The resulting displaced shape of the structure then represents the shape of the influence line.

Example 17.2   Use the Mueller–Breslau principle to determine the shape of the shear force and bending moment influence lines for the section C in the beam in Ex. 17.1 (Fig. 17.2(a)) and calculate the values of the principal ordinates.

In Fig. 17.4(b) we impose a unit shear displacement at the section C. In effect we are removing the resistance to shear of the beam at C by cutting the beam at C. We then apply positive shear forces to the two faces of the cut section in accordance with the sign convention of Section 3.2. Thus the beam to the right of C is displaced upwards while the beam to the left of C is displaced downwards. Since the slope of the influence line is the same on each side of C we can determine the ordinates of the influence line by geometry. Hence, in Fig. 17.4(b).

$$\frac{c_1 e}{c_1 a_1} = \frac{c_1 f}{c_1 b_1}$$

Therefore
$$c_1 e = \frac{c_1 a_1}{c_1 b_1} \, c_1 f = \frac{1}{2} c_1 f$$

Further, since
$$c_1 e + c_1 f = 1$$

$$c_1 e = \frac{1}{3}, \quad c_1 f = \frac{2}{3}$$

as before. The ordinate $d_1 g \, (= \tfrac{1}{3})$ follows.

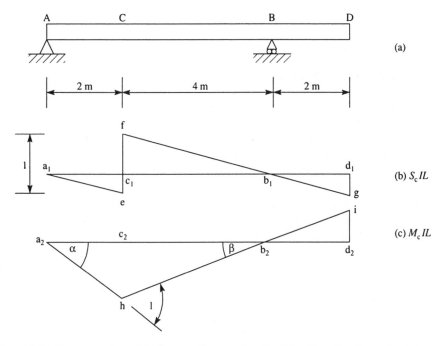

Fig. 17.4   Construction of influence lines using the Mueller–Breslau principle

In Fig. 17.4(c) we have, from the geometry of a triangle,

$$\alpha + \beta = 1 \qquad \text{(external angle = sum of opposite internal angles)}$$

Then, assuming that the angles α and β are small so that their tangents are equal to the angles in radians

$$\frac{c_2 h}{c_2 a_2} + \frac{c_2 h}{c_2 b_2} = 1$$

or

$$c_2 h \left(\frac{1}{2} + \frac{1}{4} \right) = 1$$

whence

$$c_2 h = \frac{4}{3}$$

as in Fig. 17.2(d). The ordinate $d_2 i \left(= \frac{2}{3} \right)$ follows from similar triangles.

17.3 Systems of travelling loads

Influence lines for beams are constructed, as we have seen, by considering the passage of a unit load across a beam or by employing the Mueller–Breslau principle. Once constructed, an influence line may be used to determine the value of the particular function for shear force, bending moment, etc., at a section of a beam produced by any system of travelling loads. These may be concentrated loads, distributed loads or combinations of both. Generally we require the maximum values of a function as the loads cross the beam.

Concentrated loads

By definition the ordinate of an influence line at a point gives the value of the function at a specified section of a beam due to a unit load positioned at the point. Thus, in the beam shown in Fig. 17.1(a) the shear force at K due to a unit load at C is equal to the ordinate $c_2 f$ in Fig. 17.1(d). Since we are assuming that the system is linear it follows that the shear force at K produced by a load, W, at C is $Wc_2 f$.

The argument may be extended to any number of travelling loads whose positions are fixed in relation to each other. In Fig. 17.5(a), for example, three concentrated loads, W_1, W_2 and W_3 are crossing the beam AB and are at fixed distances c and d apart. Suppose that they have reached the positions C, D and E, respectively. Let us also suppose that we require values of shear force and bending moment at the section K; the S_K and M_K influence lines are then constructed using either of the methods described in Sections 17.1 and 17.2.

Since the system is linear we can use the principle of superposition to determine the combined effects of the loads. Therefore, with the loads in the positions shown, and referring to Fig. 17.5(b),

$$S_K = -W_1 s_1 - W_2 s_2 - W_3 s_3 \tag{17.9}$$

in which s_1, s_2 and s_3 are the ordinates under the loads in the S_K influence line.

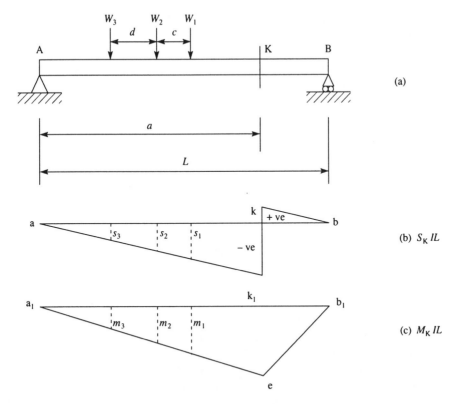

Fig. 17.5   Number of concentrated travelling loads

Similarly, from Fig. 17.5(c)

$$M_K = W_1 m_1 + W_2 m_2 + W_3 m_3 \qquad (17.10)$$

where m_1, m_2, and m_3 are the ordinates under the loads in the M_K influence line.

Maximum shear force at K

It can be seen from Fig. 17.5(b) that, as the loads W_1, W_2 and W_3 move to the right, the ordinates s_1, s_2 and s_3 increase in magnitude so that the shear force at K increases negatively to a peak value with W_1 just to the left of K. When W_1 passes to the right of K, the ordinate, s_1, becomes positive, then

$$S_K = +W_1 s_1 - W_2 s_2 - W_3 s_3$$

and the magnitude of S_K suddenly drops. As the loads move further to the right the now positive ordinate s_1 decreases in magnitude while the ordinates s_2 and s_3 increase negatively. Therefore a second peak value of S_K occurs with W_2 just to the left of K. When W_2 passes to the right of K the ordinate s_2 becomes positive and

$$S_K = +W_1 s_1 + W_2 s_2 - W_3 s_3$$

so that again there is a sudden fall in the negative value of S_K. A third peak value is reached with W_3 just to the left of K and then, as W_3 passes to the right of K, S_K becomes completely positive. The same arguments apply for positive values of S_K as the loads travel from right to left.

Thus we see that maximum positive and negative values of shear force at a section of a beam occur when one of the loads is at that section. In some cases it is obvious which load will give the greatest value, in other cases a trial-and-error method is used.

Maximum bending moment at K

A similar situation arises when determining the position of a set of loads to give the maximum bending moment at a section of a beam although, as we shall see, a more methodical approach than trial and error may be used when the critical load position is not obvious.

With the loads W_1, W_2 and W_3 positioned as shown in Fig. 17.5(a) the bending moment, M_K, at K is given by Eq. (17.10), i.e.

$$M_K = W_1 m_1 + W_2 m_2 + W_3 m_3$$

As the loads move to the right the ordinates m_1, m_2 and m_3 increase in magnitude until W_1 passes K and m_1 begins to decrease. Thus M_K reaches a peak value with W_1 at K. Further movement of the loads to the right causes m_2 and m_3 to increase, while m_1 decreases so that a second peak value occurs with W_2 at K; similarly, a third peak value is reached with W_3 at K. Thus the maximum bending moment at K will occur with a load at K. In some cases this critical load is obvious, or it may be found by trial and error as for the maximum shear force at K. However, alternatively, the critical load may be found as follows.

Suppose that the beam in Fig. 17.5(a) carries a system of concentrated loads, W_1, W_2, ..., W_j, ..., W_n, and that they are in any position on the beam. Then, from Eq. (17.10)

$$M_K = \sum_{j=1}^{n} W_j m_j \qquad (17.11)$$

Suppose now that the loads are given a small displacement δz. The bending moment at K then becomes $M_K + \delta M_K$ and each ordinate m becomes $m + \delta m$. Therefore, from Eq. (17.11)

$$M_K + \delta M_K = \sum_{j=1}^{n} W_j (m_j + \delta m_j)$$

or

$$M_K + \delta M_K = \sum_{j=1}^{n} W_j m_j + \sum_{j=1}^{n} W_j \delta m_j$$

whence

$$\delta M_K = \sum_{j=1}^{n} W_j \delta m_j$$

Therefore, in the limit as $\delta z \to 0$

$$\frac{\mathrm{d}M_\mathrm{K}}{\mathrm{d}z} = \sum_{j=1}^{n} W_j \frac{\mathrm{d}m_j}{\mathrm{d}z}$$

in which $\mathrm{d}m_j/\mathrm{d}z$ is the gradient of the M_K influence line. Therefore, if

$$\sum_{j=1}^{n} W_{j,\mathrm{L}}$$

is the sum of the loads to the left of K and

$$\sum_{j=1}^{n} W_{j,\mathrm{R}}$$

is the sum of the loads to the right of K, we have, from Eqs (17.5) and (17.6)

$$\frac{\mathrm{d}M_\mathrm{K}}{\mathrm{d}z} = \sum_{j=1}^{n} W_{j,\mathrm{L}} \left(\frac{L-a}{L}\right) + \sum_{j=1}^{n} W_{j,\mathrm{R}} \left(-\frac{a}{L}\right)$$

For a maximum value of M_K, $\mathrm{d}M_\mathrm{K}/\mathrm{d}z = 0$ so that

$$\sum_{j=1}^{n} W_{j,\mathrm{L}} \left(\frac{L-a}{L}\right) = \sum_{j=1}^{n} W_{j,\mathrm{R}} \frac{a}{L}$$

or

$$\frac{\sum_{j=1}^{n} W_{j,\mathrm{L}}}{a} = \frac{\sum_{j=1}^{n} W_{j,\mathrm{R}}}{L-a} \tag{17.12}$$

From Eq. (17.12) we see that the bending moment at K will be a maximum with one of the loads at K (from the previous argument) and when the load per unit length of beam to the left of K is equal to the load per unit length of beam to the right of K. Part of the load at K may be allocated to AK and part to KB as required to fulfil this condition.

Equation (17.12) may be extended as follows. Since

$$\sum_{j=1}^{n} W_j = \sum_{j=1}^{n} W_{j,\mathrm{L}} + \sum_{j=1}^{n} W_{j,\mathrm{R}}$$

then

$$\sum_{j=1}^{n} W_{j,\mathrm{R}} = \sum_{j=1}^{n} W_j - \sum_{j=1}^{n} W_{j,\mathrm{L}}$$

Substituting for

$$\sum_{j=1}^{n} W_{j,\mathrm{R}}$$

in Eq. (17.12) we obtain

$$\frac{1}{a}\sum_{j=1}^{n} W_{j,\mathrm{L}} = \left(\frac{1}{L-a}\right)\left(\sum_{j=1}^{n} W_j - \sum_{j=1}^{n} W_{j,\mathrm{L}}\right)$$

Rearranging we have

$$\frac{L-a}{a} = \frac{\displaystyle\sum_{j=1}^{n} W_j - \sum_{j=1}^{n} W_{j,\mathrm{L}}}{\displaystyle\sum_{j=1}^{n} W_{j,\mathrm{L}}}$$

whence

$$\frac{\displaystyle\sum_{j=1}^{n} W_j}{L} = \frac{\displaystyle\sum_{j=1}^{n} W_{j,\mathrm{L}}}{a} \tag{17.13}$$

Combining Eqs (17.12) and (17.13) we have

$$\frac{\displaystyle\sum_{j=1}^{n} W_j}{L} = \frac{\displaystyle\sum_{j=1}^{n} W_{j,\mathrm{L}}}{a} = \frac{\displaystyle\sum_{j=1}^{n} W_{j,\mathrm{R}}}{L-a} \tag{17.14}$$

Therefore, for M_K to be a maximum, there must be a load at K such that the load per unit length over the complete span is equal to the load per unit length of beam to the left of K and the load per unit length of beam to the right of K.

Example 17.3   Determine the maximum positive and negative values of shear force and the maximum value of bending moment at the section K in the simply supported beam AB shown in Fig. 17.6(a) when it is crossed by the system of loads shown in Fig. 17.6(b).

The influence lines for the shear force and bending moment at K are constructed using either of the methods described in Sections 17.1 and 17.2 as shown in Fig. 17.6(c) and (d).

Maximum negative shear force at K

It is clear from inspection that S_K will be a maximum with the 5 kN load just to the left of K, in which case the 3 kN load is off the beam and the ordinate under the 4 kN load in the S_K influence line is, from similar triangles, $-0 \cdot 1$. Then

$$S_\mathrm{K}(\mathrm{max}) = 5 \times (-0 \cdot 3) + 4 \times (-0 \cdot 1) = -1 \cdot 9 \text{ kN}$$

Maximum positive shear force at K

There are two possible load positions which could give the maximum positive value of shear force at K; neither can be eliminated by inspection. First we shall place the 3 kN load just to the right of K. The ordinates under the 4 kN load and 5 kN load are calculated from similar triangles and are $+0 \cdot 5$ and $+0 \cdot 3$, respectively. Then

$$S_\mathrm{K} = 3 \times 0 \cdot 7 + 4 \times 0 \cdot 5 + 5 \times 0 \cdot 3 = 5 \cdot 6 \text{ kN}$$

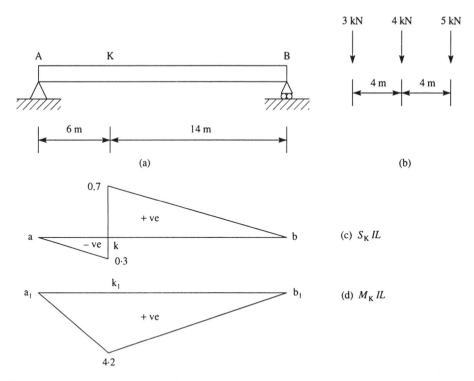

Fig. 17.6 Determination of the maximum shear force and bending moment at a section of a beam

Now with the 4 kN load just to the right of K, the ordinates under the 3 kN load and 5 kN load are $-0\cdot1$ and $+0.5$, respectively. Then

$$S_K = 3 \times (-0\cdot1) + 4 \times 0\cdot7 + 5 \times 0\cdot5 = 5\cdot0 \text{ kN}$$

Therefore the maximum positive value of S_K is $5\cdot6$ kN and occurs with the 3 kN load immediately to the right of K.

Maximum bending moment at K

We position the loads in accordance with the criterion of Eq. (17.14). The load per unit length of the complete beam is $(3+4+5)/20 = 0\cdot6$ kN/m. Therefore if we position the 4 kN load at K and allocate $0\cdot6$ kN of the load to AK the load per unit length on AK is $(3+0\cdot6)/6 = 0\cdot6$ kN/m and the load per unit length on KB is $(3\cdot4+5)/14 = 0\cdot6$ kN/m. The maximum bending moment at K therefore occurs with the 4 kN load at K; in this example the critical load position could have been deduced by inspection.

With the loads in this position the ordinates under the 3 kN and 5 kN loads in the M_K influence line are $1\cdot4$ and $3\cdot0$, respectively. Then

$$M_K(\text{max}) = 3 \times 1\cdot4 + 4 \times 4\cdot2 + 5 \times 3\cdot0 = 36\cdot0 \text{ kNm}$$

Distributed loads

Figure 17.7(a) shows a simply supported beam AB on which a uniformly distributed

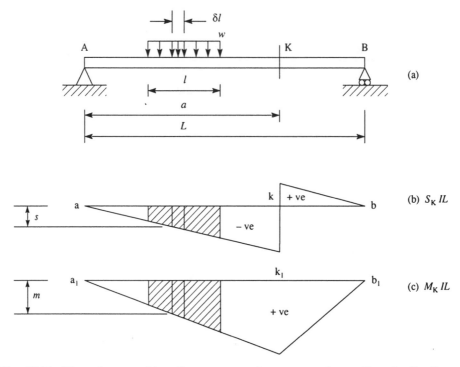

Fig. 17.7 Shear force and bending moment due to a moving uniformly distributed load

load of intensity w and length l is crossing from left to right. Suppose we wish to obtain values of shear force and bending moment at the section K of the beam. Again we construct the S_K and M_K influence lines using either of the methods described in Sections 17.1 and 17.2.

If we consider an elemental length δl of the load, we may regard this as a concentrated load of magnitude $w\delta l$. The shear force, δS_K, at K produced by this elemental length of load is then, from Fig. 17.7(b),

$$\delta S_K = w\delta ls$$

The total shear force, S_K, at K due to the complete length of load is then

$$S_K = \int_0^l ws\,dl$$

or, since the load is uniformly distributed

$$S_K = w \int_0^l s\,dl \qquad (17.15)$$

Hence $S_K = w \times$ *area under the projection of the load in the S_K influence line.*

Similarly $\qquad\qquad\qquad\qquad M_K = w \int_0^l m\,dl \qquad (17.16)$

so that $M_K = w \times$ *area under the projection of the load in the M_K influence line.*

Maximum shear force at K

It is clear from Fig. 17.7(b) that the maximum negative shear force at K occurs with the head of the load at K while the maximum positive shear force at K occurs with the tail of the load at K. Note that the shear force at K would be zero if the load straddled K such that the negative area under the load in the S_K influence line was equal to the positive area under the load.

Maximum bending moment at K

If we regard the distributed load as comprising an infinite number of concentrated loads, we can apply the criterion of Eq. (17.14) to obtain the maximum value of bending moment at K. Thus the load per unit length of the complete beam is equal to the load per unit length of beam to the left of K and the load per unit length of beam to the right of K. Therefore, in Fig. 17.8, we position the load such that

$$\frac{w\,ck_1}{a_1k_1} = \frac{w\,dk_1}{k_1b_1}$$

or
$$\frac{ck_1}{a_1k_1} = \frac{dk_1}{k_1b_1} \tag{17.17}$$

From Fig. 17.8
$$\frac{fc}{hk_1} = \frac{a_1c}{a_1k_1}$$

so that
$$fc = \frac{a_1c}{a_1k_1}\,hk_1 = \left(\frac{a_1k_1 - ck_1}{a_1k_1}\right)hk_1 = \left(1 - \frac{ck_1}{a_1k_1}\right)hk_1$$

Similarly
$$dg = \left(1 - \frac{dk_1}{b_1k_1}\right)hk_1$$

Therefore, from Eq. (17.17) we see that

$$fc = dg$$

and the ordinates under the extremities of the load in the M_K influence line are equal. It may also be shown that the area under the load in the M_K influence line is a maximum when fc = dg. This is an alternative method of deducing the position of

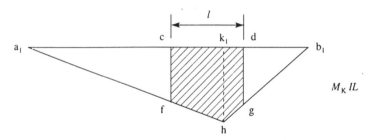

Fig. 17.8 Load position for maximum bending moment at K.

the load for maximum bending moment at K. Note that, from Eq. (17.17), K divides the load in the same ratio as it divides the span.

Example 17.4 A load of length 2 m and intensity 2 kN/m crosses the simply supported beam AB shown in Fig. 17.9(a). Calculate the maximum positive and negative values of shear force and the maximum value of bending moment at the quarter-span point.

The shear force and bending moment influence lines for the quarter-span point K are constructed in the same way as before and are shown in Fig. 17.9(b) and (c).

Maximum shear force at K

The maximum negative shear force at K occurs with the head of the load at K. In this position the ordinate under the tail of the load is −0·05. Hence

$$S_K(\text{max.} -\text{ve}) = -2 \times \tfrac{1}{2}(0·05 + 0·25) \times 2 = -0·6 \text{ kN}$$

The maximum positive shear force at K occurs with the tail of the load at K. With the load in this position the ordinate under the head of the load is 0·55. Thus

$$S_K(\text{max.} +\text{ve}) = 2 \times \tfrac{1}{2}(0·75 + 0·55) \times 2 = +2·6 \text{ kN}$$

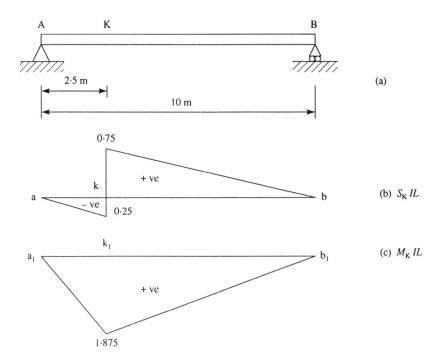

Fig. 17.9 Maximum shear force and bending moment at the quarter-span point in the beam of Ex. 17.4

Maximum bending moment at K

We position the load so that K divides the load in the same ratio that it divides the span. Therefore 0·5 m of the load is to the left of K and 1·5 m to the right of K. The ordinate in the M_K influence line under the tail of the load is then 1·5 as is the ordinate under the head of the load. The maximum value of M_K is thus given by

$$M_K(\text{max}) = 2[\tfrac{1}{2}(1\cdot5 + 1\cdot875) \times 0\cdot5 + \tfrac{1}{2}(1\cdot875 + 1\cdot5) \times 1\cdot5]$$

which gives

$$M_K(\text{max}) = 6\cdot75 \text{ kN m}$$

Diagram of maximum shear force

Consider the simply supported beam shown in Fig. 17.10 (a) and suppose that a uniformly distributed load of intensity w and length $L/5$ (any fraction of L may be chosen) is crossing the beam. We can draw a series of influence lines for the sections, A, K_1, K_2, K_3, K_4 and B as shown in Fig. 17.10(b) and then determine the maximum negative and positive values of shear force at each of the sections K_1, K_2, etc., by considering first the head of the load at K_1, K_2, etc., and then the tail of the load at A, K_1, K_2, etc. These values are then plotted as shown in Fig. 17.10(c).

With the head of the load at K_1, K_2, K_3, K_4 and B the maximum negative shear force is given by $w(ak_1)s_1$, $w(k_1 k_2)s_2$, and so on, where s_1, s_2, etc., are the mid-ordinates of the areas ak_1, k_1k_2, etc. Since s_1, s_2, etc. increase linearly, the maximum negative shear force also increases linearly at all sections of the beam between K_1 and B. At a section between A and K_1, the complete length of load will not be on the beam so that the maximum value of negative shear force at this section will not

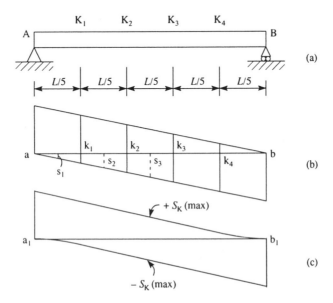

Fig. 17.10 Diagram of maximum shear force

lie on the straight line and the diagram of maximum negative shear force between A and K_1, will be curved; the maximum negative shear force should be calculated for at least one section between A and K_1.

An identical argument applies to the calculation of the maximum positive shear force which occurs with the tail of the load at a beam section. Thus, in this case, the non-linearity will occur as the load begins to leave the beam between K_4 and B.

Reversal of shear force

In some structures it is beneficial to know in which parts of the structure, if any, the maximum shear force changes sign. In Section 4.5, for example, we saw that the diagonals of a truss resist the shear forces and therefore could be in tension or compression depending upon their orientation and the sign of the shear force. If, therefore, we knew that the sign of the shear force would remain the same under the design loading in a particular part of a truss we could arrange the inclination of the diagonals so that they would always be in tension and would not be subject to instability produced by compressive forces. If, at the same time, we knew in which parts of the truss the shear force could change sign we could introduce counterbracing (see Section 17.5).

Consider the simply supported beam AB shown in Fig. 17.11(a) and suppose that it carries a uniformly distributed dead load (self-weight, etc.) of intensity w_{DL}. The shear force due to this dead load (the dead load shear (DLS)) varies linearly from $+w_{DL}L/2$ at A to $-w_{DL}L/2$ at B as shown in Fig. 17.11(b). Suppose now that a uniformly distributed live load of length less than the span AB crosses the beam. As for the beam in Fig. 17.10, we can plot diagrams of maximum positive and negative shear force produced by the live load; these are also shown in Fig. 17.11(b). Then, at any section of the beam, the maximum shear force is equal

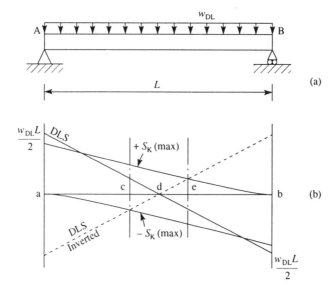

Fig. 17.11   Reversal of shear force in a beam

to the sum of the maximum positive shear force due to the live load and the dead load shear force, or the sum of the maximum negative shear force due to the live load and the dead load shear force. The variation in this maximum shear force along the length of the beam will be more easily understood if we invert the dead load shear force diagram.

Referring to Fig. 17.11(b) we see that the sum of the maximum positive shear force due to the live load and the dead load shear force is always positive between a and c. Furthermore, between a and c, the sum of the maximum negative shear force due to the live load and the dead load shear force is always positive. Similarly, between e and b the maximum shear force is always negative. However, between c and e the summation of the maximum positive shear force produced by the live load and the dead load shear force is positive, while the summation of the maximum negative shear force due to the live load and the dead load shear force is negative. Therefore the maximum shear force between c and e may be positive or negative, i.e. there is a possible *reversal* of maximum shear force in this length of the beam.

Example 17.5 A simply supported beam AB has a span of 5 m and carries a uniformly distributed dead load of 0·6 kN/m (Fig. 17.12(a)). A similarly distributed live load of length greater than 5 m and intensity 1·5 kN/m travels across the beam. Calculate the length of beam over which reversal of shear force occurs and sketch the diagram of maximum shear force for the beam.

The shear force at a section of the beam will be a maximum with the head or tail of the load at that section. Initially, before writing down an expression for shear force, we require the support reaction at A, R_A. Thus, with the head of the load at a

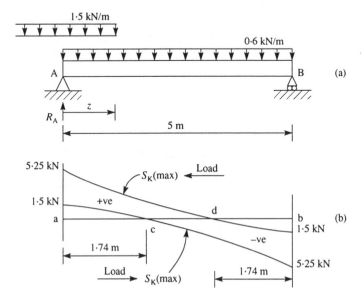

Fig. 17.12 Reversal of shear force in the beam of Ex. 17.5

section a distance z from A, the reaction, R_A, is found by taking moments about B. Thus

$$R_A \times 5 - 0 \cdot 6 \times 5 \times 2 \cdot 5 - 1 \cdot 5z(5 - z/2) = 0$$

whence $\qquad R_A = 1 \cdot 5 + 1 \cdot 5z - 0 \cdot 15z^2 \qquad\qquad$ (i)

The maximum shear force at the section is then

$$S(\text{max}) = R_A - 0 \cdot 6z - 1 \cdot 5z \qquad\qquad \text{(ii)}$$

or, substituting in Eq. (ii) for R_A from Eq. (i)

$$S(\text{max}) = 1 \cdot 5 - 0 \cdot 6z - 0 \cdot 15z^2 \qquad\qquad \text{(iii)}$$

Equation (iii) gives the maximum shear force at any section of the beam with the load moving from left to right. Then, when $z = 0$, $S(\text{max}) = 1 \cdot 5$ kN and when $z = 5$ m, $S(\text{max}) = -5 \cdot 25$ kN. Furthermore, from Eq. (iii) $S(\text{max}) = 0$ when $z = 1 \cdot 74$ m.

The maximum shear force for the load travelling from right to left is found in a similar manner. The final diagram of maximum shear force is shown in Fig. 17.12(b) where we see that reversal of shear force may take place within the length cd of the beam; cd is sometimes called the *focal length*.

Determination of the point of maximum bending moment in a beam

Previously we have been concerned with determining the position of a set of loads on a beam that would produce the maximum bending moment at a given section of the beam. We shall now determine the section and the position of the loads for the bending moment to be the absolute maximum.

Consider a section K a distance z_1 from the mid-span of the beam in Fig. 17.13 and suppose that a set of loads having a total magnitude W_T is crossing the beam. The bending moment at K will be a maximum when one of the loads is at K; let this

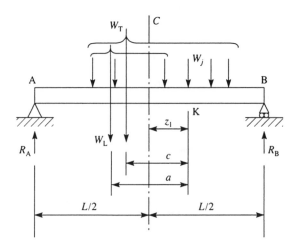

Fig. 17.13   Determination of the absolute maximum bending moment in a beam

load be W_j. Also, suppose that the centre of gravity of the complete set of loads is a distance c from the load W_j and that the total weight of all the loads to the left of W_j is W_L, acting at a distance a from W_j; a and c are fixed values for a given set of loads.

Initially we find R_A by taking moments about B.

Hence

$$R_A L - W_T \left(\frac{L}{2} - z_1 + c \right) = 0$$

which gives

$$R_A = \frac{W_T}{L} \left(\frac{L}{2} - z_1 + c \right)$$

The bending moment, M_K, at K is then given by

$$M_K = R_A \left(\frac{L}{2} + z_1 \right) - W_L a$$

or, substituting for R_A,

$$M_K = \frac{W_T}{L} \left(\frac{L}{2} - z_1 + c \right) \left(\frac{L}{2} + z_1 \right) - W_L a$$

Differentiating M_K with respect to z_1 we have

$$\frac{\mathrm{d}M_K}{\mathrm{d}z_1} = \frac{W_T}{L} \left[-1 \left(\frac{L}{2} + z_1 \right) + 1 \left(\frac{L}{2} - z_1 + c \right) \right]$$

or

$$\frac{\mathrm{d}M_K}{\mathrm{d}z_1} = \frac{W_T}{L} (-2z_1 + c)$$

For a maximum value of M_K, $\mathrm{d}M_K/\mathrm{d}z_1 = 0$ so that

$$z_1 = \frac{c}{2} \tag{17.18}$$

Therefore the maximum bending moment occurs at a section K under a load W_j such that the section K and the centre of gravity of the complete set of loads are positioned at equal distances either side of the mid-span of the beam.

To apply this rule we select one of the larger central loads and position it over a section K such that K and the centre of gravity of the set of loads are placed at equal distances on either side of the mid-span of the beam. We then check to determine whether the load per unit length to the left of K is equal to the load per unit length to the right of K. If this condition is not satisfied, another load and another section K must be selected.

Example 17.6 The set of loads shown in Fig. 17.14(b) crosses the simply supported beam AB shown in Fig. 17.14(a). Calculate the position and magnitude of the maximum bending moment in the beam.

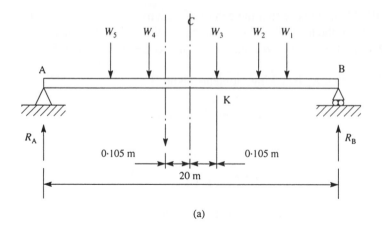

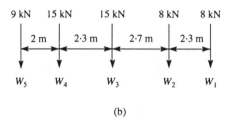

Fig. 17.14   Determination of absolute maximum bending moment in the beam of Ex. 17.6

The first step is to find the position of the centre of gravity of the set of loads. Thus, taking moments about the load W_5 we have

$$(9 + 15 + 15 + 8 + 8)\bar{z} = 15 \times 2 + 15 \times 4 \cdot 3 + 8 \times 7 \cdot 0 + 8 \times 9 \cdot 3$$

whence

$$\bar{z} = 4 \cdot 09 \text{ m}$$

Therefore the centre of gravity of the loads is $0 \cdot 21$ m to the left of the load W_3.

By inspection of Fig. 17.14(b) we see that it is probable that the maximum bending moment will occur under the load W_3. We therefore position W_3 and the centre of gravity of the set of loads at equal distances either side of the mid-span of the beam as shown in Fig. 17.14(a). We now check to determine whether this position of the loads satisfies the load per unit length condition. The load per unit length on $AB = 55/20 = 2 \cdot 75$ kN/m. Therefore the total load required on $AK = 2 \cdot 75 \times 10 \cdot 105 = 27 \cdot 79$ kN. This is satisfied by W_5, W_4 and part ($3 \cdot 79$ kN) of W_3.

Having found the load position, the bending moment at K is most easily found by direct calculation. Thus taking moments about B we have

$$R_A \times 20 - 55 \times 10 \cdot 105 = 0$$

whence

$$R_A = 27 \cdot 8 \text{ kN}$$

Hence

$$M_K = 27 \cdot 8 \times 10 \cdot 105 - 9 \times 4 \cdot 3 - 15 \times 2 \cdot 3 = 207 \cdot 7 \text{ kN m}$$

It is possible that in some load systems there may be more than one load position which satisfies both criteria for maximum bending moment but the corresponding bending moments have different values. Generally the absolute maximum bending moment will occur under one of the loads between which the centre of gravity of the system lies. If the larger of these two loads is closer to the centre of gravity than the other, then this load will be the critical load; if not then both cases must be analysed.

17.4 Influence lines for beams not in contact with the load

In many practical situations, such a bridge construction for example, the moving loads are not in direct contact with the main beam or girder. Fig. 17.15 shows a typical bridge construction in which the deck is supported by stringers that are mounted on cross beams which, in turn, are carried by the main beams or girders. The deck loads are therefore transmitted via the stringers and cross beams to the main beams. Generally, in the analysis, we assume that the segments of the stringers are simply supported at each of the cross beams. In Fig. 17.15 the portion of the main beam between the cross beams, for example FG, is called a *panel* and the points F and G are called *panel points*.

Figure 17.16 shows a simply supported main beam AB which supports a bridge deck via an arrangement of cross beams and stringers. Let us suppose that we wish to construct shear force and bending moment influence lines for the section K of the main beam within the panel CD. As before we consider the passage of a unit load; in this case, however, it crosses the bridge deck.

S_K influence line

With the unit load outside and to the left of the panel CD (position 1) the shear force, S_K, at K is given by

$$S_K = -R_B = -\frac{z_1}{L} \tag{17.19}$$

S_K therefore varies linearly as the load moves from A to C. Thus, from Eq. (17.19), when $z_1 = 0$, $S_K = 0$ and when $z_1 = a$, $S_K = -a/L$, the ordinate cf in the S_K influence line shown in Fig. 17.16(b). Furthermore, from Fig. 17.16(a) we see that

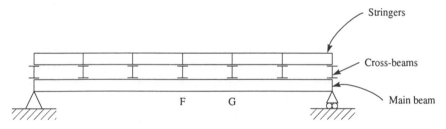

Fig. 17.15 Typical bridge construction

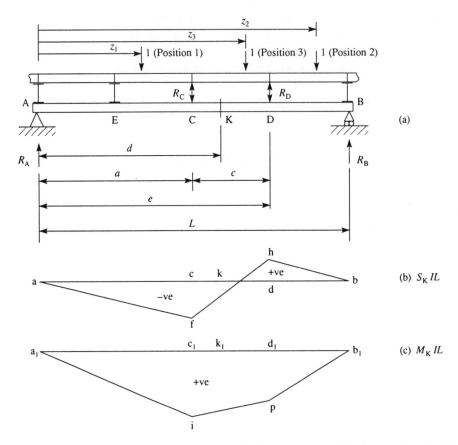

Fig. 17.16   Influence lines for a beam not in direct contact with the moving load

$S_K = S_C = S_D$ with the load between A and C, so that for a given position of the load the shear force in the panel CD has the same value at all sections.

Suppose now that the unit load is to the right of D between D and B (position 2). Then

$$S_K = +R_A = \frac{L - z_2}{L}$$ (17.20)

and is linear. Therefore when $z_2 = L$, $S_K = 0$ and when $z_2 = e$, $S_K = (L - e)/L$, the ordinate dh in the S_K influence line. Also, with the unit load between D and B, $S_K = S_C = S_D (= +R_A)$ so that for a given position of the load, the shear force in the panel CD has the same value at all sections.

Now consider the unit load at some point between C and D (position 3). There will now be reaction forces, R_C and R_D, as shown in Fig. 17.16(a) acting on the stringer and the beam where, by considering the portion of the stringer immediately above the panel CD as a simply supported beam, we see that $R_C = (e - z_3)/c$ and $R_D = (z_3 - a)/c$. Therefore the shear force at K is given by

$$S_K = -R_B + R_D \text{ (or } S_K = +R_A - R_C\text{)}$$

so that
$$S_K = -\frac{z_3}{L} + \frac{(z_3 - a)}{c} \qquad (17.21)$$

S_K therefore varies linearly as the load moves between C and D. Furthermore, when $z_3 = a$, $S_K = -a/L$, the ordinate cf in the S_K influence line, and when $z_3 = e$, $S_K = (L - e)/L$ the ordinate dh in the S_K influence line. Note that in the calculation of the latter value, $e - a = c$.

Note also that for all positions of the unit load between C and D, $S_K = -R_B + R_D$ which is independent of the position of K. Therefore, for a given load position between C and D, the shear force is the same at all sections of the panel.

M_K influence line

With the unit load in position 1 between A and C, the bending moment, M_K, at K is given by
$$M_K = R_B(L - d) = \frac{z_1}{L}(L - d) \qquad (17.22)$$

M_K therefore varies linearly with the load position between A and C. Also, when $z_1 = 0$, $M_K = 0$ and when $z_1 = a$, $M_K = a(L - d)/L$, the ordinate $c_1 i$ in the M_K influence line in Fig. 17.16(c).

With the unit load in position 2 between D and B
$$M_K = R_A d = \frac{(L - z_2)}{L} d \qquad (17.23)$$

Again, M_K varies linearly with load position so that when $z_2 = e$, $M_K = (L - e)d/L$, the ordinate $d_1 p$ in the M_K influence line. Furthermore, when $z_2 = L$, $M_K = 0$.

When the unit load is between C and D (position 3)
$$M_K = R_B(L - d) - R_D(e - d)$$

As before we consider the stringer over the panel CD as a simply supported beam so that $R_D = (z_3 - a)/c$. Then since
$$R_B = z_3/L$$

$$M_K = \frac{z_3}{L}(L - d) - \left(\frac{z_3 - a}{c}\right)(e - d) \qquad (17.24)$$

Equation (17.24) shows that M_K varies linearly with load position between C and D. Therefore, when $z_3 = a$, $M_K = a(L - d)/L$, the ordinate $c_1 i$ in the M_K influence line, and when $z_3 = e$, $M_K = d(L - e)/L$, the ordinate $d_1 p$ in the M_K influence line. Note, that in the latter calculation, $e - a = c$.

Maximum values of S_K and M_K

In determining maximum values of shear force and bending moment at a section of a beam that is not in direct contact with the load, certain points are worthy of note.

1. When the section K coincides with a panel point (C or D, say) the S_K and M_K influence lines are identical in geometry to those for a beam that is in direct contact with the moving load; the same rules governing maximum and minimum values therefore apply.

2. The absolute maximum value of shear force will occur in an end panel, AE or DB, when the S_K influence line will be identical in form to the bending moment influence line for a section in a simply supported beam that is in direct contact with the moving load. Therefore the same criteria for load positioning may be used for determining the maximum shear force, i.e. the load per unit length of beam is equal to the load per unit length to the left of E or D and the load per unit length to the right of E or D.

3. To obtain maximum values of shear force and bending moment in a panel, a trial-and-error method is the simplest approach remembering that, for concentrated loads, a load must be placed at the point where the influence line changes slope.

17.5 Forces in the members of a truss

In some instances the main beams in a bridge are trusses, in which case the cross beams are positioned at the joints of the truss. The shear force and bending moment influence lines for a panel of the truss may then be used to determine the variation in the truss member forces as moving loads cross the bridge.

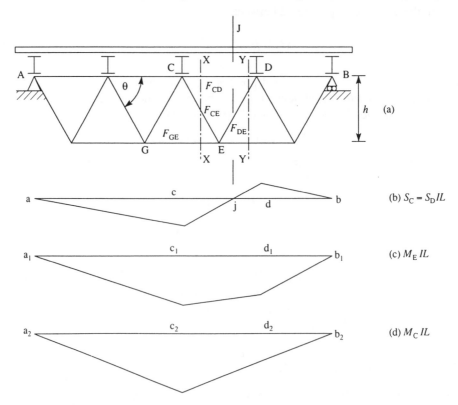

Fig. 17.17 Determination of forces in the members of a truss

Consider the simply supported Warren truss shown in Fig. 17.17(a) and suppose that it carries cross beams at its upper chord joints which, in turn, support the bridge deck. Alternatively, the truss could be inverted and the cross beams supported by the lower chord joints; the bridge deck is then the *through* type. Suppose also that we wish to determine the forces in the members CD, CE, DE and GE of the truss.

We have seen in Section 4.5 the mechanism by which a truss resists shear forces and bending moments. Thus shear forces are resisted by diagonal members, while bending moments are generally resisted by a combination of both diagonal and horizontal members. Therefore, referring to Fig. 17.17(a), we see that the forces in the members CE and DE may be determined from the shear force in the panel CD, while the forces in the members CD and GE may be found from the bending moments at E and C, respectively. We therefore construct the influence lines for the shear force in the panel CD and for the bending moment at E and C, as shown in Fig. 17.17(b), (c) and (d).

In Section 17.4 we saw that, for a given load position, the shear force in a panel such as CD is constant at all sections in the panel; we will call this shear force S_{CD}. Then, considering a section XX through CE, CD and GE, we have

$$F_{CE} \sin \theta = S_{CD}$$

so that

$$F_{CE} = \frac{S_{CD}}{\sin \theta} \qquad (17.25)$$

Similarly

$$F_{DE} = \frac{S_{CD}}{\sin \theta} \qquad (17.26)$$

From Fig. 17.17(b) we see that for a load position between A and J, S_{CD} is negative. Therefore, referring to Fig. 17.17(a), F_{CE} is compressive while F_{DE} is tensile. For a load position between J and B, S_{CD} is positive so that F_{CE} is tensile and F_{DE} is compressive. Thus F_{CE} and F_{DE} will always be of opposite sign; this may also be deduced from a consideration of the vertical equilibrium of joint E.

If we now consider the moment equilibrium of the truss at a vertical section through joint E we have

$$F_{CD}h = M_E$$

or

$$F_{CD} = \frac{M_E}{h} \qquad (17.27)$$

Since M_E is positive for all load positions (Fig. 17.17(c)), F_{CD} is compressive.

The force in the member GE is obtained from the M_C influence line in Fig. 17.17(d). Thus

$$F_{GE}h = M_C$$

which gives

$$F_{GE} = \frac{M_C}{h} \qquad (17.28)$$

F_{GE} will be tensile since M_C is positive for all load positions.

It is clear from Eqs (17.25)–(17.28) that the influence lines for the forces in the members could be constructed from the appropriate shear force and bending moment influence lines. Thus, for example, the influence line for F_{CE} would be identical in shape to the shear force influence line in Fig. 17.17(b) but would have the ordinates factored by $1/\sin\theta$. The influence line for F_{DE} would also have the S_{CD} influence line ordinates factored by $1/\sin\theta$ but, in addition, would have the signs reversed.

Example 17.17   Determine the maximum tensile and compressive forces in the member EC in the Pratt truss shown in Fig. 17.18(a) when it is crossed by a uniformly distributed load of intensity 2·5 kN/m and length 4 m; the load is applied on the bottom chord of the truss.

The vertical component of the force in the member EC resists the shear force in the panel DC. We therefore construct the shear force influence line for the panel DC as shown in Fig. 17.18(b). From Eq. (17.19) the ordinate $df = 2 \times 1\cdot4/(8 \times 1\cdot4) = 0\cdot25$ while from Eq. (17.20) the ordinate $cg = (8 \times 1\cdot4 - 3 \times 1\cdot4)/(8 \times 1\cdot4) = 0\cdot625$. Furthermore, we see that S_{DC} changes sign at the point j (Fig. 17.18(b)) where jd, from similar triangles, is 0·4.

The member EC will be in compression when the shear force in the panel DC is negative and its maximum value will occur when the head of the load is at j, thereby completely covering the length aj in the S_{DC} influence line. Therefore

$$F_{EC}\sin 45° = S_{DC} = 2\cdot5 \times \tfrac{1}{2} \times 3\cdot2 \times 0\cdot25$$

whence

$$F_{EC} = 1\cdot41 \text{ kN} \qquad\qquad \text{(compression)}$$

The force in the member EC will be tensile when the shear force in the panel DC is positive. Therefore to find the maximum tensile value of F_{EC} we must position the load within the part jb of the S_{DC} influence line such that the maximum value of S_{DC} occurs. Since the positive portion of the S_{DC} influence line is triangular, we may use the criterion previously established for maximum bending moment. Thus the load per unit length over jb must be equal to the load per unit length over jc and the load per unit length over cb. In other words, c divides the load in the same ratio that it

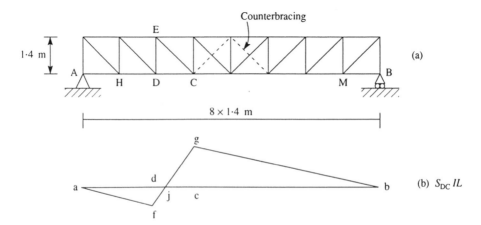

Fig. 17.18   Determination of the force in a member of the Pratt girder of Ex. 17.7

divides jb, i.e. 1:7. Therefore 0·5 m of the load is to the left of c, 3·5 m to the right. The ordinates under the extremities of the load in the S_{DC} influence line are then both 0·3125 m. Hence the maximum positive shear force in the panel CD is

$$S_{CD}(\text{max.} +\text{ve}) = 2·5[\tfrac{1}{2}(0·3125 + 0·625) \times 0·5 + \tfrac{1}{2}(0·625 + 0·3125) \times 3·5]$$

which gives $\qquad\qquad S_{CD}(\text{max.} +\text{ve}) = 4·69 \text{ kN}$

Then, since $\qquad\qquad\qquad F_{EC} \sin 45° = S_{CD}$

$$F_{EC} = 6·63 \text{ kN}$$

which is the maximum tensile force in the member EC.

Counterbracing

A diagonal member of a Pratt truss will, as we saw for the member EC in Ex. 17.7, be in tension or compression depending on the sign of the shear force in the particular panel in which the member is placed. The exceptions are the diagonals in the end panels where, in the Pratt truss of Fig. 17.18(a), construction of the shear force influence lines for the panels AH and MB shows that the shear force in the panel AH is always positive and that the shear force in the panel MB is always negative; the diagonals in these panels are therefore always in tension.

In some situations the diagonal members are unsuitable for compressive forces so that *counterbracing* is required. This consists of diagonals inclined in the opposite direction to the original diagonals as shown in Fig. 17.18(a) for the two centre panels. The original diagonals are then assumed to be carrying zero force while the counterbracing is in tension.

It is clear from Ex. 17.17 that the shear force in all the panels, except the two outer ones, of a Pratt truss can be positive or negative so that all the diagonals in these panels could experience compression. Therefore it would appear that all the interior panels of a Pratt truss require counterbracing. However, as we saw in Section 17.3, the dead load acting on a beam has a beneficial effect in that it reduces the length of the beam subjected to shear reversal. This, in turn, will reduce the number of panels requiring counterbracing.

Example 17.8 The Pratt truss shown in Fig. 17.19(a) carries a dead load of 1·0 kN/m applied at its upper chord joints. A uniformly distributed live load, which exceeds 9 m in length, has an intensity of 1·5 kN/m and is also carried at the upper chord joints. If the diagonal members are designed to resist tension only find which panels require counterbracing.

A family of influence lines may be drawn as shown in Fig. 17.19(b) for the shear force in each of the ten panels. We begin the analysis at the centre of the truss where the dead load shear force has its least effect; initially, therefore, we consider panel 5. The shear force, S_5, in panel 5 with the head of the live load at n_5 is given by

$$S_5 = 1·0(\text{area } n_5gb - \text{area } n_5qa) - 1·5(\text{area } n_5qa)$$

i.e. $\qquad\qquad S_5 = 1·0 \times \text{area } n_5gb - 2·5 \times \text{area } n_5qa \qquad\qquad\qquad (i)$

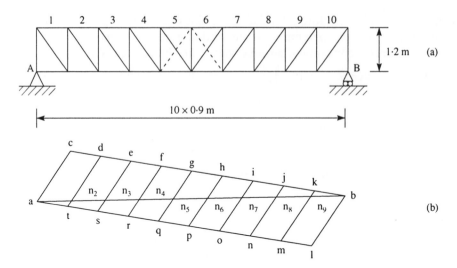

Fig. 17.19 Counterbracing in a Pratt truss

The ordinates in the S_5 influence line at g and q are found from similar triangles and are 0.5 and 0.4, respectively. Also, from similar triangles, n_5 divides the horizontal distance between q and g in the ratio $0.4:0.5$. Therefore, from Eq. (i)

$$S_5 = 1.0 \times \tfrac{1}{2} \times 5.0 \times 0.5 - 2.5 \times \tfrac{1}{2} \times 4.0 \times 0.4$$

which gives $S_5 = -0.75 \text{ kN}$

Therefore, since S_5 is negative, the diagonal in panel 5 will be in compression so that panel 5, and from symmetry panel 6, requires counterbracing.

Now with the head of the live load at n_4, $S_4 = 1.0(\text{area } n_4\text{fb} - \text{area } n_4\text{ra}) - 1.5(\text{area } n_4\text{ra})$.

The ordinates and base lengths in the triangles $n_4$fb and $n_4$ra are determined as before. Then

$$S_4 = 1.0 \times \tfrac{1}{2} \times 6.0 \times 0.6 - 2.5 \times \tfrac{1}{2} \times 3.0 \times 0.3$$

from which $S_4 = 0.675 \text{ kN}$

Therefore, since S_4 is positive, panel 4, and therefore panel 7, do not require counterbracing.

Clearly the remaining panels will not require counterbracing.

Note that for a Pratt truss having an odd number of panels the net value of the dead load shear force in the central panel is zero, so that this panel will always require counterbracing.

17.6   Influence lines for continuous beams

The structures we have investigated so far in this chapter have been statically determinate so that the influence lines for the different functions have comprised straight line segments. A different situation arises for statically indeterminate structures such as continuous beams.

Consider the two-span continuous beam ABC shown in Fig. 17.20(a) and let us suppose that we wish to construct influence lines for the reaction at B, the shear force at the section D in AB and the bending moment at the section F in BC.

The shape of the influence lines may be obtained by employing the Mueller–Breslau principle described in Section 17.2. Thus, in Fig. 17.20(b) we remove the support at B and apply a unit displacement in the direction of the support reaction, R_B. The beam will bend into the shape shown since it remains pinned to the supports at A and C. This would not have been the case, of course, if the span BC did not exist for then the beam would rotate about A as a rigid link and the R_B influence line would have been straight as in Fig. 17.1(c).

To obtain the shear force influence line for the section D we 'cut' the beam at D and apply a unit shear displacement as shown in Fig. 17.20(c). Again, since the beam is attached to the support at C, the resulting displaced shape is curved. Furthermore, the gradient of the influence line must be the same on each side of D because, otherwise, it would imply the presence of a moment causing a relative rotation. This is not possible since the displacement we have specified is due solely to shear. It follows that the influence line between A and D must also be curved.

The influence line for the bending moment at F is found by inserting a hinge at F and applying a relative unit rotation as shown in Fig. 17.20(d). Again the portion

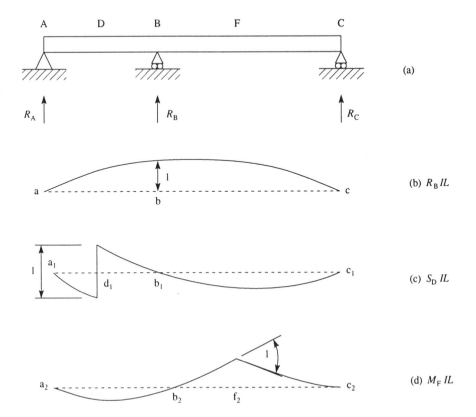

Fig. 17.20 Influence lines for a continuous beam using the Mueller–Breslau principle

ABF of the beam will be curved, as will the portion FC, since this part of the beam must rotate so that the sum of the rotations of the two portions of the beam at F is equal to unity.

Example 17.9 Construct influence lines for the reaction at B and for the shear force and bending moment at D in the two-span continuous beam shown in Fig. 17.21(a).

The shape of each influence line may be drawn using the Mueller–Breslau principle as shown in Fig. 17.21(b), (c) and (d). However, before they can be of direct use in determining maximum values, say, of the various functions due to the passage of loading systems, the ordinates must be calculated; for this, since the influence lines are comprised of curved segments, we need to derive their equations.

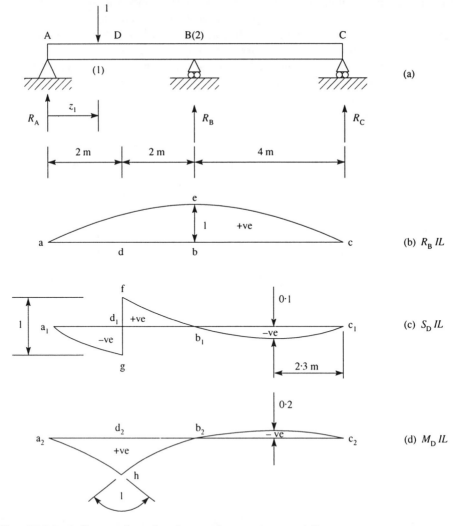

Fig. 17.21 Influence lines for the continuous beam of Ex. 17.9

However, once the influence line for a support reaction, R_B in this case, has been established, the remaining influence lines follow from statical equilibrium.

R_B influence line

Suppose initially that a unit load is a distance z_1 from A, between A and B. To determine R_B we may use the flexibility method described in Section 16.4. Thus we remove the support at B (point 2) and calculate the displacement, a_{21}, at B due to the unit load at z_1 (point 1). We then calculate the displacement, a_{22}, at B due to a vertically downward unit load at B. The total displacement at B due to the unit load at z_1 and the reaction R_B is then

$$a_{21} - a_{22}R_B = 0 \tag{i}$$

since the support at B is not displaced. In Eq. (i) the term $a_{22}R_B$ is negative since R_B is in the opposite direction to the applied unit load at B.

Both the flexibility coefficients in Eq. (i) may be obtained from a single unit load application since, from the reciprocal theorem (Section 15.4), the displacement at B due to a unit load at z_1 is equal to the displacement at z_1 due to a unit load at B. We therefore apply a vertically downward unit load at B.

The equation for the displaced shape of the beam is that for a simply supported beam carrying a central concentrated load. Therefore, from Eq. (iv) of Ex. 13.5

$$v = -\frac{1}{48EI}(4z^3 - 3L^2z) \tag{ii}$$

or, for the beam of Fig. 17.21(a)

$$v = -\frac{z}{12EI}(z^2 - 48) \tag{iii}$$

At B, when $z = 4$ m

$$v_B = \frac{32}{3EI} = a_{22} \tag{iv}$$

Furthermore, the displacement at B due to the unit load at z_1 (=displacement at z_1 due to a unit load at B) is, from Eq. (iii),

$$v_{z_1} = -\frac{z_1}{12EI}(z_1^2 - 48) = a_{21} \tag{v}$$

Substituting for a_{22} and a_{21} in Eq. (i) we have

$$-\frac{z_1}{12EI}(z_1^2 - 48) - \frac{32}{3EI}R_B = 0$$

from which

$$R_B = -\frac{z_1}{128}(z_1^2 - 48) \quad (0 \leqslant z_1 \leqslant 4 \cdot 0 \text{ m}) \tag{vi}$$

Equation (vi) gives the influence line for R_B with the unit load between A and B; the remainder of the influence line follows from symmetry. Eq. (vi) may be checked since we know the value of R_B with the unit load at A and B. Thus from Eq. (vi), when $z_1 = 0$, $R_B = 0$ and when $z_1 = 4 \cdot 0$ m, $R_B = 1$ as expected.

If the support at B were not symmetrically positioned, the above procedure would be repeated for the unit load on the span BC. In this case the equations for the deflected shape of AB and BC would be Eqs (xiv) and (xv) in Ex. 13.6.

In this example we require the S_D influence line so that we shall, in fact, need to consider the value of R_B with the unit load on the span BC. Therefore from Eq. (xv) in Ex. 13.6

$$v_{z_1} = \frac{1}{12EI} (z_1^3 - 24z_1^2 + 144z_1 - 128) \quad (4 \cdot 0 \text{ m} \leq z_1 \leq 8 \cdot 0 \text{ m}) \tag{vii}$$

Hence from Eq. (i)

$$R_B = \frac{1}{128} (z_1^3 - 24z_1^2 + 144z_1 - 128) \quad (4 \cdot 0 \text{ m} \leq z_1 \leq 8 \cdot 0 \text{ m}) \tag{viii}$$

A check on Eq. (viii) shows that when $z_1 = 4 \cdot 0$ m, $R_B = 1$ and when $z_1 = 8 \cdot 0$ m, $R_B = 0$.

S_D influence line

With the unit load to the left of D, the shear force, S_D, at D is most simply given by

$$S_D = +R_A - 1 \tag{ix}$$

where, by taking moments about C, we have

$$R_A \times 8 - 1(8 - z_1) + R_B \times 4 = 0 \tag{x}$$

Substituting in Eq. (x) for R_B from Eq. (vi) and rearranging gives

$$R_A = \frac{1}{256} (z_1^3 - 80z_1 + 256) \tag{xi}$$

whence, from Eq. (ix)

$$S_D = \frac{1}{256} (z_1^3 - 80z_1) \quad (0 \leq z_1 \leq 2 \cdot 0 \text{ m}) \tag{xii}$$

Therefore, when $z_1 = 0$, $S_D = 0$ and when $z_1 = 2 \cdot 0$ m, $S_D = -0 \cdot 59$, the ordinate $d_1 g$ in the S_D influence line in Fig. 17.21 (c).

With the unit load between D and B

$$S_D = +R_A$$

so that, substituting for R_A from Eq. (xi)

$$S_D = \frac{1}{256} (z_1^3 - 80z_1 + 256) \quad (2 \cdot 0 \text{ m} \leq z_1 \leq 4 \cdot 0 \text{ m}) \tag{xiii}$$

Thus, when $z_1 = 2 \cdot 0$ m, $S_D = +0 \cdot 41$, the ordinate $d_1 f$ in Fig. 17.21 (c) and when $z_1 = 4 \cdot 0$ m, $S_D = 0$.

Now consider the unit load between B and C. Again

$$S_D = +R_A$$

but in this case, R_B in Eq. (x) is given by Eq. (viii). Substituting for R_B from Eq. (viii) in Eq. (x) we obtain

$$R_A = S_D = -\frac{1}{256}(z_1^3 - 24z_1^2 + 176z_1 - 384) \quad (4 \cdot 0 \text{ m} \leqslant z_1 \leqslant 8 \cdot 0 \text{ m}) \qquad \text{(xiv)}$$

Therefore the S_D influence line consists of three segments, a_1g, fb_1 and b_1c_1.

M_D influence line

With the unit load between A and D

$$M_D = R_A \times 2 - 1(2 - z_1) \qquad \text{(xv)}$$

Substituting for R_A from Eq. (xi) in Eq. (xv) and simplifying, we obtain

$$M_D = \frac{1}{128}(z_1^3 + 48z_1) \quad (0 \leqslant z_1 \leqslant 2 \cdot 0 \text{ m}) \qquad \text{(xvi)}$$

When $z_1 = 0$, $M_D = 0$ and when $z_1 = 2 \cdot 0$ m, $M_D = 0 \cdot 81$, the ordinate d_2h in the M_D influence line in Fig. 17.21 (d).

Now with the unit load between D and B

$$M_D = R_A \times 2 \qquad \text{(xvii)}$$

Therefore, substituting for R_A from Eq. (xi) we have

$$M_D = \frac{1}{128}(z_1^3 - 80z_1 + 256) \quad (2 \cdot 0 \text{ m} \leqslant z_1 \leqslant 4 \cdot 0 \text{ m}) \qquad \text{(xviii)}$$

From Eq. (xviii) we see that when $z_1 = 2 \cdot 0$ m, $M_D = 0 \cdot 81$, again the ordinate d_2h in Fig. 17.21 (d). Also, when $z_1 = 4 \cdot 0$ m, $M_D = 0$.

Finally, with the unit load between B and C, M_D is again given by Eq. (xvii) but in which R_A is given by Eq. (xiv). Hence

$$M_D = -\frac{1}{128}(z_1^3 - 24z_1^2 + 176z_1 - 384) \quad (4 \cdot 0 \text{ m} \leqslant z_1 \leqslant 8 \cdot 0 \text{ m}) \qquad \text{(xix)}$$

The maximum ordinates in the S_D and M_D influence lines for the span BC may be found by differentiating Eqs (xiv) and (xix) with respect to z_1, equating to zero and then substituting the resulting values of z_1 back in the equations. Thus, for example, from Eq. (xiv),

$$\frac{dS_D}{dz_1} = -\frac{1}{256}(3z_1^2 - 48z_1 + 176) = 0$$

from which $z_1 = 5 \cdot 7$ m. Hence

$$S_D(\text{max}) = -0 \cdot 1$$

Similarly $M_D(\text{max}) = -0 \cdot 2$ at $z_1 = 5 \cdot 7$ m.

In this chapter we have constructed influence lines for beams, trusses and continuous beams. Clearly influence lines can be drawn for a wide variety of structures that carry moving loads. Their construction, whatever the structure, is based on considering the passage of a unit load across the structure.

Problems

P.17.1   Construct influence lines for the support reaction at A in the beams shown in Fig. P.17.1(a),(b) and (c).

Ans. (a) Unit load at C, $R_A = 1 \cdot 25$.
(b) Unit load at C, $R_A = 1 \cdot 25$; at D, $R_A = -0 \cdot 25$.
(c) Unit load between A and B, $R_A = 1$; at C, $R_A = 0$.

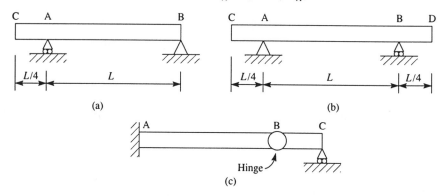

Fig. P.17.1

P.17.2   Draw influence lines for the shear force at C in the beams shown in Figs P.17.2(a) and (b).

Ans. Influence line ordinates
(a) $D = 0 \cdot 25$, $A = 0$, $C = \pm 0 \cdot 5$, $B = 0$.
(b) $D = 0 \cdot 25$, $A = B = 0$, $C = \pm 0 \cdot 5$, $E = -0 \cdot 25$.

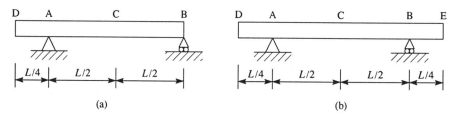

Fig. P.17.2

P.17.3   Draw influence lines for the bending moment at C in the beams shown in Fig. P.17.2(a) and (b).

Ans. Influence line ordinates
(a) $D = -0 \cdot 125L$, $A = B = 0$, $C = 0 \cdot 25L$.
(b) $D = E = -0 \cdot 125L$, $A = B = 0$, $C = 0 \cdot 25L$.

P.17.4   The simply supported beam shown in Fig. P.17.4 carries a uniformly distributed travelling load of length 10 m and intensity 20 kN/m. Calculate the maximum positive and negative values of shear force and bending moment at the section C of the beam.

Ans. $S_C = +37.5$ kN, -40.0 kN, $M_C = +550$ kN m, -80 kN m.

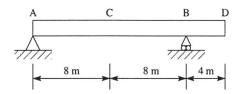

Fig. P.17.4

P.17.5   The beam shown in Fig. P.17.5 (a) is crossed by the train of four loads shown in Fig. P.17.5 (b). For a section at mid-span, determine the maximum sagging and hogging bending moments.

Ans. $+161.3$ kN m, -77.5 kN m.

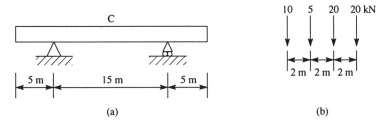

(a)

(b)

Fig. P.17.5

P.17.6   A simply supported beam AB of span 20 m is crossed by the train of loads shown in Fig. P.17.6. Determine the position and magnitude of the absolute maximum bending moment on the beam and also the maximum values of positive and negative shear force anywhere on the beam.

Ans. M(max) = 466.7 kN m under a central load 10.5 m from A.
    S(max +ve) = 104 kN at A, S(max −ve) = −97.5 kN at B.

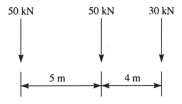

Fig. P.17.6

P.17.7   The three-span beam shown in Fig. P.17.7 has hinges at C and E in its central span. Construct influence lines for the reaction at B and for the shear force and bending moment at the sections K and D.

Ans. Influence line ordinates

R_B; A = 0, B = 1, C = 1·25, E = F = G = 0.
S_K; A = 0, K = ±0·5, B = 0, C = −0·25, E = 0.
S_D; A = B = 0, D = 1·0, C = 1·0, E = F = G = 0.
M_K; A = B = 0, K = 1·0, C = −0·5, E = F = G = 0.
M_D; A = B = D = 0, C = −0·5, E = F = G = 0.

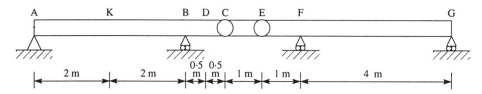

Fig. P.17.7

P.17.8   Draw influence lines for the reactions at A and C and for the bending moment at E in the beam system shown in Fig. P.17.8. Note that the beam AB is supported on the lower beam at D by a roller.

If two 10 kN loads, 5 m apart, cross the upper beam AB, determine the maximum values of the reactions at A and C and the bending moment at E.

Ans. R_A(max) = 16·7 kN, R_C(max) = 17·5 kN, M_E(max) = 58·3 kN m.

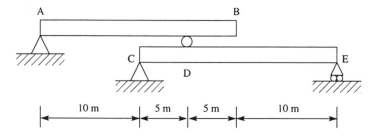

Fig. P.17.8

P.17.9   A simply supported beam having a span of 5 m has a self-weight of 0·5 kN/m and carries a travelling uniformly distributed load of intensity 1·2 kN/m and length 1 m. Calculate the length of beam over which shear reversal occurs.

Ans. The central 1·3 m (graphical solution).

P.17.10   Construct an influence line for the force in the member CD of the truss shown in Fig. P.17.10 and calculate the force in the member produced by the loads positioned at C, D and E.

Ans. 28·1 kN (compression).

P.17.11   The truss shown in Fig. P.17.11 carries a train of loads consisting of, left to right, 40 kN, 70 kN, 70 kN and 60 kN spaced at 2 m, 3 m and 3 m, respectively. If the self-weight of the truss is 15 kN/m, calculate the maximum force in each of the members CG, HD and FE.

Ans. CG = 763 kN, HD = −724 kN, FE = −326 kN.

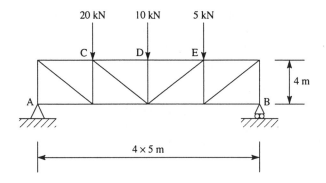

Fig. P.17.10

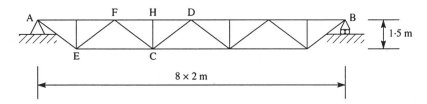

Fig. P.17.11

P.17.12 One of the main girders of a bridge is the truss shown in Fig. P.17.12. Loads are transmitted to the truss through cross beams attached at the lower panel points. The self-weight of the truss is 30 kN/m and it carries a live load of intensity 15 kN/m and of length greater than the span. Draw influence lines for the force in each of the members CE and DE and determine their maximum values.

Ans. CE = +37·3 kN, −65·3 kN, DE = +96·1 kN.

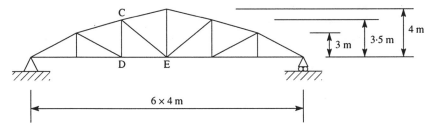

Fig. P.17.12

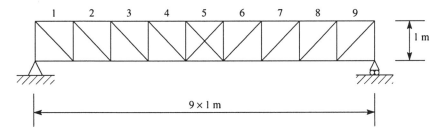

Fig. P.17.13

P.17.13   The Pratt truss shown in Fig. P.17.13 has a self-weight of 1·2 kN/m and carries a uniformly distributed live load longer than the span of intensity 2·8 kN/m, both being applied at the upper chord joints. If the diagonal members are designed to resist tension only, determine which panels require counterbracing.

Ans. Panels 4, 5 and 6.

P.17.14   Using the Mueller–Breslau principle sketch the shape of the influence lines for the support reactions at A and B, and the shear force and bending moment at E in the continuous beam shown in Fig. P.17.14.

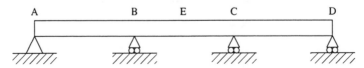

Fig. P.17.14

P.17.15   Construct influence lines for the reaction at A, the shear force at D and the bending moment at B in the continuous beam shown in Fig. P.17.15.

Ans.

$$R_A = (z_1{}^3 - 20z_1 + 32)/32 \qquad (0 \leqslant z_1 \leqslant 2\cdot0 \text{ m})$$
$$R_A = (-z_1{}^3 + 12z_1{}^2 - 44z_1 + 48)/32 \qquad (2\cdot0 \text{ m} \leqslant z_1 \leqslant 4\cdot0 \text{ m})$$
$$S_D = (z_1{}^3 - 20z_1)/32 \qquad (0 \leqslant z_1 \leqslant 1\cdot5 \text{ m})$$
$$S_D = (z_1{}^3 - 20z_1 + 32)/32 \qquad (1\cdot5 \text{ m} \leqslant z_1 \leqslant 2\cdot0 \text{ m})$$
$$M_B = (z_1{}^3 - 4z_1)/16 \qquad (0 \leqslant z_1 \leqslant 2\cdot0 \text{ m})$$
$$M_B = (-z_1{}^3 + 12z_1{}^2 - 44z_1 + 48)/16 \qquad (2\cdot0 \text{ m} \leqslant z_1 \leqslant 4\cdot0 \text{ m})$$

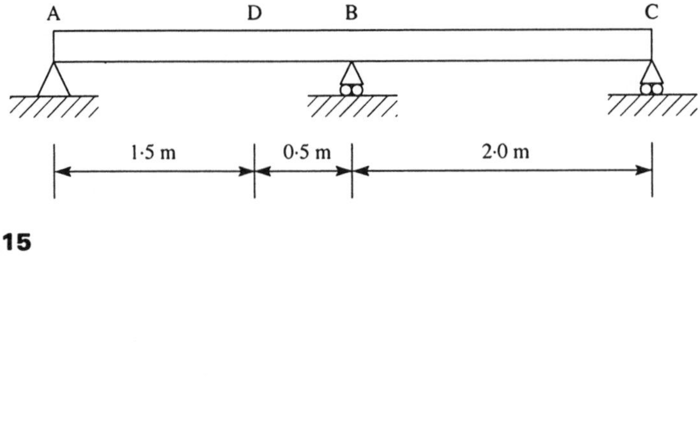

A   D   B   C

|← 1·5 m →|← 0·5 m →|← 2·0 m →|

Fig. P.17.15

CHAPTER 18

Structural Instability

So far, in considering the behaviour of structural members under load, we have been concerned with their ability to withstand different forms of stress. Their strength, therefore, has depended upon the strength properties of the material from which they are fabricated. However, structural members subjected to axial compressive loads may fail in a manner that depends upon their geometrical properties rather than their material properties. It is common experience, for example, that a long slender structural member such as that shown in Fig. 18.1(a) will suddenly bow with large lateral displacements when subjected to an axial compressive load (Fig. 18.1(b)). This phenomenon is known as *instability* and the member is said to *buckle*. If the member is exceptionally long and slender it may regain its initial straight shape when the load is removed.

Structural members subjected to axial compressive loads are known as *columns* or *struts*, although the former term is usually applied to the relatively heavy vertical members that are used to support beams and slabs; struts are compression members in frames and trusses.

It is clear from the above discussion that the design of compression members must take into account not only the material strength of the member but also its stability against buckling. Obviously the shorter a member is in relation to its cross-sectional dimensions, the more likely it is that failure will be a failure in compression of the material rather than one due to instability. It follows that in some intermediate range a failure will be a combination of both.

We shall investigate the buckling of long slender columns and derive expressions for the *buckling* or *critical load*; the discussion will then be extended to the design

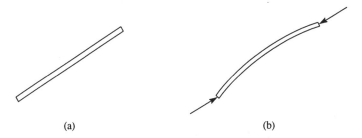

(a)   (b)

Fig. 18.1 Buckling of slender column

of columns of any length and to a consideration of beams subjected to axial load and bending moment.

18.1 Euler theory for slender columns

The first significant contribution to the theory of the buckling of columns was made in the eighteenth century by Euler. His classical approach is still valid for long slender columns possessing a variety of end restraints. Before presenting the theory, however, we shall investigate the nature of buckling and the difference between theory and practice.

We have seen that if an increasing axial compressive load is applied to a long slender column there is a value of load at which the column will suddenly bow or buckle in some unpredetermined direction. This load is patently the buckling load of the column or something very close to the buckling load. The fact that the column buckles in a particular direction implies a degree of asymmetry in the plane of the buckle caused by geometrical and/or material imperfections of the column and its load. Theoretically, however, in our analysis we stipulate a perfectly straight, homogeneous column in which the load is applied precisely along the perfectly straight centroidal axis. Theoretically, therefore, there can be no sudden bowing or buckling, only axial compression. Thus we require a precise definition of buckling load which may be used in the analysis of the perfect column.

If the perfect column of Fig. 18.2 is subjected to a compressive load P, only shortening of the column occurs no matter what the value of P. Clearly if P were to produce a stress greater than the yield stress of the material of the column, then material failure would occur. However, if the column is displaced a small amount by a lateral load, F, then, at values of P below the critical or buckling load, P_{CR}, removal of F results in a return of the column to its undisturbed position, indicating a state of stable equilibrium. When $P = P_{CR}$ the displacement does not disappear and the column will, in fact, remain in *any* displaced position so long as the displacement is small. Thus the buckling load, P_{CR}, is associated with a state of *neutral equilibrium*. For $P > P_{CR}$ enforced lateral displacements increase and the column is unstable.

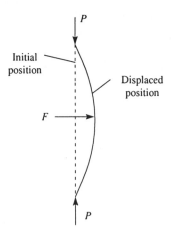

Fig. 18.2   Definition of the buckling load of a column

Buckling load for a pin-ended column

Consider the pin-ended column shown in Fig. 18.3. We shall assume that it is in the displaced state of neutral equilibrium associated with buckling so that the compressive axial load has reached the value P_{CR}. We also assume that the column has deflected so that its displacements, v, referred to the axes $0zy$ are positive. The bending moment, M, at any section Z is then given by

$$M = P_{CR}v$$

so that substituting for M from Eq. (13.3) we obtain

$$\frac{d^2v}{dz^2} = -\frac{P_{CR}}{EI}v \qquad (18.1)$$

Rearranging we obtain

$$\frac{d^2v}{dz^2} + \frac{P_{CR}}{EI}v = 0 \qquad (18.2)$$

The solution of Eq. (18.2) is of standard form and is

$$v = C_1 \cos \mu z + C_2 \sin \mu z \qquad (18.3)$$

in which C_1 and C_2 are arbitrary constants and $\mu^2 = P_{CR}/EI$. The boundary conditions for this particular case are $v = 0$ at $z = 0$ and $z = L$. The first of these gives $C_1 = 0$ while from the second we have

$$0 = C_2 \sin \mu L$$

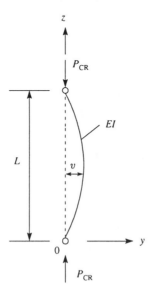

Fig. 18.3 Determination of buckling load for a pin-ended column

For a non-trivial solution (i.e. $v \neq 0$ and $C_2 \neq 0$) then

$$\sin \mu L = 0$$

so that $\mu L = n\pi$ where $n = 1, 2, 3, \ldots$

Hence
$$\frac{P_{CR}}{EI} L^2 = n^2 \pi^2$$

from which
$$P_{CR} = \frac{n^2 \pi^2 EI}{L^2} \qquad (18.4)$$

Note that C_2 is indeterminate and that the displacement of the column cannot therefore be found. This is to be expected since the column is in neutral equilibrium in its buckled state.

The smallest value of buckling load corresponds to a value of $n = 1$ in Eq. (18.4), i.e.

$$P_{CR} = \frac{\pi^2 EI}{L^2} \qquad (18.5)$$

The column then has the displaced shape $v = C_2 \sin \mu z$ and buckles into the longitudinal half sine-wave shown in Fig. 18.4(a). Other values of P_{CR} corresponding to $n = 2, 3, \ldots$ are

$$P_{CR} = \frac{4\pi^2 EI}{L^2}, \quad P_{CR} = \frac{9\pi^2 EI}{L^2}, \ldots$$

These higher values of buckling load correspond to more complex buckling modes as shown in Figs 18.4(b) and (c). Theoretically these different modes could be

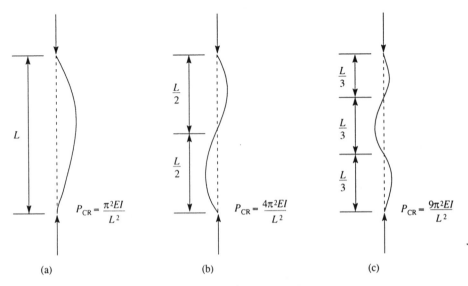

Fig. 18.4 Buckling modes of a pin-ended column

produced by applying external restraints to a slender column at the points of contraflexure to prevent lateral movement. However, in practice, the lowest value is never exceeded since high stresses develop at this load and failure of the column ensues. We are not therefore concerned with buckling loads higher than this.

Buckling load for a column with fixed ends

In practice, columns usually have their ends restrained against rotation so that they are, in effect, fixed. Figure 18.5 shows a column having its ends fixed and subjected to an axial compressive load that has reached the critical value, P_{CR}, so that the column is in a state of neutral equilibrium. In this case the ends of the column are subjected to fixing moments, M_F, in addition to axial load. Thus at any section Z the bending moment, M, is given by

$$M = P_{CR}v - M_F$$

Substituting for M from Eq. (13.3) we have

$$\frac{d^2v}{dz^2} = -\frac{P_{CR}}{EI}v + \frac{M_F}{EI} \tag{18.6}$$

Rearranging we obtain

$$\frac{d^2v}{dz^2} + \frac{P_{CR}}{EI}v = \frac{M_F}{EI} \tag{18.7}$$

the solution of which is

$$v = C_1 \cos \mu z + C_2 \sin \mu z + M_F/P_{CR} \tag{18.8}$$

where

$$\mu^2 = P_{CR}/EI$$

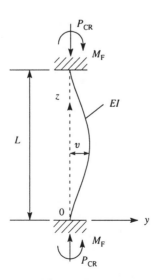

Fig. 18.5 Buckling of a slender column with fixed ends

When $z = 0$, $v = 0$ so that $C_1 = -M_F/P_{CR}$. Further $v = 0$ at $z = L$, hence

$$0 = -\frac{M_F}{P_{CR}} \cos \mu L + C_2 \sin \mu L + \frac{M_F}{P_{CR}}$$

which gives

$$C_2 = -\frac{M_F}{P_{CR}} \frac{(1 - \cos \mu L)}{\sin \mu L}$$

Hence Eq. (18.8) becomes

$$v = -\frac{M_F}{P_{CR}} \left[\cos \mu z + \frac{(1 - \cos \mu L)}{\sin \mu L} \sin \mu z - 1 \right] \tag{18.9}$$

Note that again v is indeterminate since M_F cannot be found. Also since $dv/dz = 0$ at $z = L$ we have from Eq. (18.9)

$$0 = 1 - \cos \mu L$$

whence

$$\cos \mu L = 1$$

and

$$\mu L = n\pi \text{ where } n = 0, 2, 4, \dots$$

For a non-trivial solution, i.e. $n \neq 0$, and taking the smallest value of buckling load ($n = 2$) we have

$$P_{CR} = \frac{4\pi^2 EI}{L^2} \tag{18.10}$$

Buckling load for a column with one end fixed and one end free

In this configuration the upper end of the column is free to move laterally and also to rotate as shown in Fig. 18.6. At any section Z the bending moment M is given by

$$M = -P_{CR}(\delta - v) \text{ or } M = P_{CR}v - M_F$$

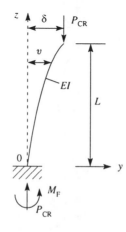

Fig. 18.6   Determination of buckling load for a column with one end fixed and one end free

Substituting for M in the first of these expressions from Eq. (13.3) (equally we could use the second) we obtain

$$\frac{d^2v}{dz^2} = \frac{P_{CR}}{EI}(\delta - v) \tag{18.11}$$

which, on rearranging, becomes

$$\frac{d^2v}{dz^2} + \frac{P_{CR}}{EI}v = \frac{P_{CR}}{EI}\delta \tag{18.12}$$

The solution of Eq. (18.12) is

$$v = C_1 \cos \mu z + C_2 \sin \mu z + \delta \tag{18.13}$$

where $\mu^2 = P_{CR}/EI$. When $z = 0$, $v = 0$ so that $C_1 = -\delta$. Also when $z = L$, $v = \delta$ so that from Eq. (18.13) we have

$$\delta = -\delta \cos \mu L + C_2 \sin \mu L + \delta$$

which gives

$$C_2 = \delta \frac{\cos \mu L}{\sin \mu L}$$

Hence

$$v = -\delta \left(\cos \mu z - \frac{\cos \mu L}{\sin \mu L} \sin \mu z - 1 \right) \tag{18.14}$$

Again v is indeterminate since δ cannot be determined. Finally we have $dv/dz = 0$ at $z = 0$. Hence from Eq. (18.14)

$$\cos \mu L = 0$$

whence

$$\mu L = n \frac{\pi}{2} \quad \text{where } n = 1, 3, 5, \ldots$$

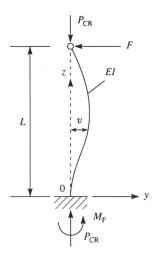

Fig. 18.7 Determination of buckling load for a column with one end fixed and the other end pinned

Thus taking the smallest value of buckling load (corresponding to $n = 1$) we obtain

$$P_{CR} = \frac{\pi^2 EI}{4L^2} \tag{18.15}$$

Buckling of a column with one end fixed, the other pinned

The column in this case is allowed to rotate at one end but requires a lateral force, F, to maintain its position (Fig. 18.7).

At any section Z the bending moment M is given by

$$M = P_{CR}v + F(L - z)$$

Substituting for M from Eq. (13.3) we have

$$\frac{d^2v}{dz^2} = -\frac{P_{CR}}{EI}v - \frac{F}{EI}(L - z) \tag{18.16}$$

which, on rearranging, becomes

$$\frac{d^2v}{dz^2} + \frac{P_{CR}}{EI}v = -\frac{F}{EI}(L - z) \tag{18.17}$$

The solution of Eq. (18.17) is

$$v = C_1 \cos \mu z + C_2 \sin \mu z - \frac{F}{P_{CR}}(L - z) \tag{18.18}$$

Now $dv/dz = 0$ at $z = 0$, thus

$$0 = \mu C_2 + \frac{F}{P_{CR}}$$

whence

$$C_2 = -\frac{F}{\mu P_{CR}}$$

When $z = L$, $v = 0$, hence

$$0 = C_1 \cos \mu L + C_2 \sin \mu L$$

which gives

$$C_1 = \frac{F}{\mu P_{CR}} \tan \mu L$$

Thus Eq. (18.18) becomes

$$v = \frac{F}{\mu P_{CR}} [\tan \mu L \cos \mu z - \sin \mu z - \mu(L - z)] \tag{18.19}$$

Also $v = 0$ at $z = 0$. Thus

$$0 = \tan \mu L - \mu L$$

or

$$\mu L = \tan \mu L \tag{18.20}$$

Equation (18.20) is a transcendental equation which may be solved graphically as shown in Fig. 18.8. The smallest non-zero value satisfying Eq. (18.20) is approximately 4·49.

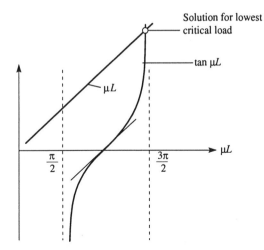

Fig. 18.8 Solution of a transcendental equation

This gives

$$P_{CR} = \frac{20 \cdot 2 \, EI}{L^2}$$

which may be written approximately as

$$P_{CR} = \frac{2 \cdot 05 \, \pi^2 EI}{L^2} \tag{18.21}$$

It can be seen from Eqs (18.5), (18.10), (18.15) and (18.21) that the buckling load in all cases has the form

$$P_{CR} = \frac{K^2 \pi^2 EI}{L^2} \tag{18.22}$$

in which K is some constant. Equation (18.22) may be written in the form

$$P_{CR} = \frac{\pi^2 EI}{L_e^2} \tag{18.23}$$

in which $L_e (=L/K)$ is the *equivalent length* of the column, i.e. (by comparison of Eqs (18.23) and (18.5)) the length of a pin-ended column that has the same buckling load as the actual column. Clearly the buckling load of any column may be expressed in this form so long as its equivalent length is known. By inspection of Eqs (18.5), (18.10), (18.15) and (18.21) we see that the equivalent lengths of the various types of column are:

Both ends pinned	$L_e = 1 \cdot 0L$
Both ends fixed	$L_e = 0 \cdot 5L$
One end fixed, one free	$L_e = 2 \cdot 0L$
One end fixed, one pinned	$L_e = 0 \cdot 7L$

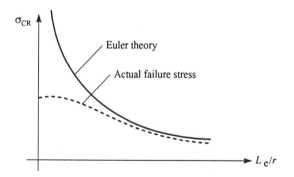

Fig. 18.9   Variation of critical stress with slenderness ratio

18.2   Limitations of the Euler theory

For a column of cross-sectional area A the critical stress, σ_{CR}, is, from Eq. (18.23)

$$\sigma_{CR} = \frac{P_{CR}}{A} = \frac{\pi^2 EI}{AL_e^2} \qquad (18.24)$$

The second moment of area, I, of the cross-section is equal to Ar^2 where r is the *radius of gyration* of the cross-section. Thus we may write Eq. (18.24) as

$$\sigma_{CR} = \frac{\pi^2 E}{(L_e/r)^2} \qquad (18.25)$$

Therefore for a column of a given material, the critical or buckling stress is inversely proportional to the parameter $(L_e/r)^2$. L_e/r is an expression of the proportions of the length and cross-sectional dimensions of the column and is known as its *slenderness ratio*. Clearly if the column is long and slender L_e/r is large and σ_{CR} is small; conversely, for a short column having a comparatively large area of cross-section, L_e/r is small and σ_{CR} is high. A graph of σ_{CR} against L_e/r for a particular material has the form shown in Fig. 18.9. For values of L_e/r less than some particular value, which depends upon the material, a column will fail in compression rather than by buckling so that σ_{CR} as predicted by the Euler theory is no longer valid. Thus in Fig. 18.9 the actual failure stress follows the dotted curve rather than the full line.

18.3   Failure of columns of any length

Empirical or semi-empirical methods are generally used to predict the failure of a column of any length: these then form the basis for safe load or safe stress tables given in Codes of Practice. One such method which gives good agreement with experiment is that due to Rankine.

Rankine theory

Suppose that P is the failure load of a column of a given material and of any length. Suppose also that P_S is the failure load in compression of a short column of the same material and that P_{CR} is the buckling load of a long slender column, again of the same material. The Rankine theory proposes that

$$\frac{1}{P} = \frac{1}{P_S} + \frac{1}{P_{CR}} \qquad (18.26)$$

Equation (18.26) is valid for a very short column since $1/P_{CR} \to 0$ and P then $\to P_S$; the equation is also valid for a long slender column since $1/P_S$ is small compared with $1/P_{CR}$; thus $P \to P_{CR}$. Equation (18.26) is therefore seen to hold for extremes in column length.

Now let σ_S be the yield stress in compression of the material of the column and A its cross-sectional area. Then

$$P_S = \sigma_S A$$

Also from Eq. (18.23)

$$P_{CR} = \frac{\pi^2 EI}{L_e^2}$$

Substituting for P_S and P_{CR} in Eq. (18.26) we have

$$\frac{1}{P} = \frac{1}{\sigma_S A} + \frac{1}{\pi^2 EI/L_e^2}$$

Thus

$$\frac{1}{P} = \frac{\pi^2 EI/L_e^2 + \sigma_S A}{\sigma_S A \pi^2 EI/L_e^2}$$

so that

$$P = \frac{\sigma_S A \pi^2 EI/L_e^2}{\pi^2 EI/L_e^2 + \sigma_S A}$$

Dividing top and bottom of the right-hand side of this equation by $\pi^2 EI/L_e^2$ we have

$$P = \frac{\sigma_S A}{1 + \dfrac{\sigma_S A L_e^2}{\pi^2 EI}}$$

But $I = Ar^2$ so that

$$P = \frac{\sigma_S A}{1 + \dfrac{\sigma_S}{\pi^2 E}\left(\dfrac{L_e}{r}\right)^2}$$

which may be written

$$P = \frac{\sigma_S A}{1 + k(L_e/r)^2} \qquad (18.27)$$

in which k is a constant that depends upon the material of the column. The failure stress in compression, σ_C, of a column of any length is then, from Eq. (18.27)

$$\sigma_C = \frac{P}{A} = \frac{\sigma_S}{1 + k(L_e/r)^2}$$ (18.28)

Note that for a column of a given material σ_C is a function of the slenderness ratio, L_e/r.

Initially curved column

An alternative approach to the Rankine theory bases a design formula on the failure of a column possessing a small initial curvature, the argument being that in practice columns are never perfectly straight.

Consider the pin-ended column shown in Fig. 18.10. In its unloaded configuration the column has a small initial curvature such that the lateral displacement at any value of z is v_o. Let us assume that

$$v_o = a \sin \pi \frac{z}{L}$$ (18.29)

in which a is the initial displacement at the centre of the column. Equation (18.29) satisfies the boundary conditions of $v_o = 0$ at $z = 0$ and $z = L$ and also $dv_o/dz = 0$ at $z = L/2$; the assumed deflected shape is therefore reasonable, particularly since we note that the buckled shape of a pin-ended column is also a half sine-wave.

Since the column is initially curved, an axial load, P, immediately produces bending and therefore further lateral displacements, v, measured from the initial displaced position. The bending moment, M, at any section Z is then

$$M = P(v + v_o)$$ (18.30)

If the column is initially unstressed, the bending moment at any section is proportional to the *change* in curvature at that section from its initial configuration and not its absolute value. Thus, from Eq. (13.3)

$$M = -EI \frac{d^2v}{dz^2}$$

so that

$$\frac{d^2v}{dz^2} = -\frac{P}{EI}(v + v_o)$$ (18.31)

Rearranging Eq. (18.31) we have

$$\frac{d^2v}{dz^2} + \frac{P}{EI} v = -\frac{P}{EI} v_o$$ (18.32)

Note that P is not, in this case, the buckling load for the column. Substituting for v_o from Eq. (18.29) we obtain

$$\frac{d^2v}{dz^2} + \frac{P}{EI} v = -\frac{P}{EI} a \sin \pi \frac{z}{L}$$ (18.33)

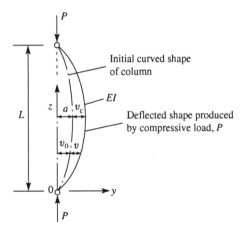

Fig. 18.10 Failure of an initially curved column

The solution of Eq. (18.33) is

$$v = C_1 \cos \mu z + C_2 \sin \mu z + \frac{\mu^2 a}{\pi^2/L^2 - \mu^2} \sin \pi \frac{z}{L} \qquad (18.34)$$

in which $\mu^2 = P/EI$. If the ends of the column are pinned, $v = 0$ at $z = 0$ and $z = L$. The first of these boundary conditions gives $C_1 = 0$ while from the second we have

$$0 = C_2 \sin \mu L$$

Although this equation is identical to that derived from the boundary conditions of an initially straight, buckled, pin-ended column, the circumstances are now different. If $\sin \mu L = 0$ then $\mu L = \pi$ so that $\mu^2 = \pi^2/L^2$. This would then make the third term in Eq. (18.34) infinite which is clearly impossible for a column in stable equilibrium ($P < P_{CR}$). We conclude, therefore, that $C_2 = 0$ and hence Eq. (18.34) becomes

$$v = \frac{\mu^2 a}{\pi^2/L^2 - \mu^2} \sin \pi \frac{z}{L} \qquad (18.35)$$

Dividing the top and bottom of Eq. (18.35) by μ^2 we obtain

$$v = \frac{a \sin \pi z/L}{\pi^2/\mu^2 L^2 - 1}$$

But $\mu^2 = P/EI$ and $a \sin \pi z/L = v_0$. Thus

$$v = \frac{v_0}{\dfrac{\pi^2 EI}{PL^2} - 1} \qquad (18.36)$$

From Eq. (18.5) we see that $\pi^2 EI/L^2 = P_{CR}$, the buckling load for a perfectly straight pin-ended column. Hence Eq. (18.36) becomes

$$v = \frac{v_0}{\dfrac{P_{CR}}{P} - 1} \tag{18.37}$$

It can be seen from Eq. (18.37) that the effect of the compressive load, P, is to increase the initial deflection, v_0, by a factor $1/(P_{CR}/P - 1)$. Clearly as P approaches P_{CR}, v tends to infinity. In practice this is impossible since material breakdown would occur before P_{CR} is reached.

If we consider displacements at the mid-height of the column we have, from Eq. (18.37),

$$v_c = \frac{a}{\dfrac{P_{CR}}{P} - 1}$$

Rearranging we obtain

$$v_c = P_{CR} \frac{v_c}{P} - a \tag{18.38}$$

Equation (18.38) represents a linear relationship between v_c and v_c/P. Thus in an actual test on an initially curved column a graph of v_c against v_c/P will be a straight line as the critical condition is approached. The gradient of the line is P_{CR} and its intercept on the v_c axis is equal to a, the initial displacement at the mid-height of the column. The graph (Fig. 18.11) is known as a Southwell plot and gives a convenient, non-destructive, method of determining the buckling load of columns.

The maximum bending moment in the column of Fig. 18.10 occurs at mid-height and is

$$M_{max} = P(a + v_c)$$

Substituting for v_c from Eq. (18.38) we have

$$M_{max} = Pa \left[1 + \frac{1}{\dfrac{P_{CR}}{P} - 1} \right]$$

or

$$M_{max} = Pa \left(\frac{P_{CR}}{P_{CR} - P} \right) \tag{18.39}$$

The maximum compressive stress in the column occurs in an extreme fibre and is, from Eq. (9.15)

$$\sigma_{max} = \frac{P}{A} + Pa \left(\frac{P_{CR}}{P_{CR} - P} \right) \left(\frac{c}{I} \right)$$

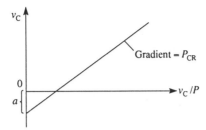

Fig. 18.11 Experimental determination of the buckling load of a column from a Southwell plot

in which A is the cross-sectional area, c is the distance from the centroidal axis to the extreme fibre and I is the second moment of area of the column's cross-section. Since $I = Ar^2$ (r = radius of gyration), we may rewrite the above equation as

$$\sigma_{max} = \frac{P}{A}\left[1 + \frac{P_{CR}}{P_{CR} - P}\left(\frac{ac}{r^2}\right)\right] \tag{18.40}$$

Now P/A is the average stress, σ, on the cross-section of the column. Thus, writing Eq. (18.40) in terms of stress we have

$$\sigma_{max} = \sigma\left[1 + \frac{\sigma_{CR}}{\sigma_{CR} - \sigma}\left(\frac{ac}{r^2}\right)\right] \tag{18.41}$$

in which $\sigma_{CR} = P_{CR}/A = \pi^2 E(r/L)^2$, (see Eq. (18.25)). The term ac/r^2 is an expression of the geometrical configuration of the column and is a constant for a given column having a given initial curvature. Therefore, writing $ac/r^2 = \eta$, Eq. (18.41) becomes

$$\sigma_{max} = \sigma\left[1 + \frac{\eta\sigma_{CR}}{\sigma_{CR} - \sigma}\right] \tag{18.42}$$

Expanding Eq. (18.42) we have

$$\sigma_{max}(\sigma_{CR} - \sigma) = \sigma[(1 + \eta)\sigma_{CR} - \sigma]$$

which, on rearranging, becomes

$$\sigma^2 - \sigma[\sigma_{max} + (1 + \eta)\sigma_{CR}] + \sigma_{max}\sigma_{CR} = 0 \tag{18.43}$$

the solution of which is

$$\sigma = \tfrac{1}{2}[\sigma_{max} + (1 + \eta)\sigma_{CR}] - \sqrt{\tfrac{1}{4}[\sigma_{max} + (1 + \eta)\sigma_{CR}]^2 - \sigma_{max}\sigma_{CR}} \tag{18.44}$$

The positive square root in the solution of Eq. (18.43) is ignored since we are only interested in the smallest value of σ. Equation (18.44) then gives the average stress, σ, in the column at which the maximum compressive stress would be reached for any value of η. Thus if we specify the maximum stress to be equal to σ_Y, the yield stress of the material of the column, then Eq. (18.44) may be written

$$\sigma = \tfrac{1}{2}[\sigma_Y + (1 + \eta)\sigma_{CR}] - \sqrt{\tfrac{1}{4}[\sigma_Y + (1 + \eta)\sigma_{CR}]^2 - \sigma_Y\sigma_{CR}} \tag{18.45}$$

It has been found from tests on mild steel pin-ended columns that failure of an initially curved column occurs when the maximum stress in an extreme fibre reaches the yield stress, σ_Y. Also, from a wide range of tests on mild steel columns, Robertson concluded that

$$\eta = 0{\cdot}003\left(\frac{L}{r}\right)$$

Substituting this value of η in Eq. (18.45) we obtain

$$\sigma = \tfrac{1}{2}[\sigma_Y + (1 + 0{\cdot}003\tfrac{L}{r})\sigma_{CR}] - \sqrt{\tfrac{1}{4}[\sigma_Y + (1 + 0{\cdot}003\tfrac{L}{r})\sigma_{CR}]^2 - \sigma_Y\sigma_{CR}} \quad (18.46)$$

In Eq. (18.46) σ_Y is a material property while σ_{CR} (from Eq. (18.25)) depends upon Young's modulus, E, and the slenderness ratio of the column. Thus Eq. (18.46) may be used to determine safe axial loads or stresses (σ) for columns of a given material in terms of the slenderness ratio. Codes of Practice tabulate maximum allowable values of average compressive stress against a range of slenderness ratios.

18.4   Effect of cross-section on the buckling of columns

The columns we have considered so far have had doubly symmetrical cross-sections with equal second moments of area about both centroidal axes. In practice, where columns frequently consist of I-section beams, this is not the case. Thus, for example, a column having the I-section of Fig. 18.12 would buckle about the centroidal axis about which the flexural rigidity, EI, is least, i.e. Gy. In fact, the most efficient cross-section from the viewpoint of instability would be a hollow circular section that has the same second moment of area about any centroidal axis and has as small an amount of material placed near the axis as possible. However, a disadvantage with this type of section is that connections are difficult to make.

In designing columns having only one cross-sectional axis of symmetry (e.g. a channel section) or none at all (i.e. an angle section having unequal legs) the least radius of gyration is taken in calculating the slenderness ratio. In the latter case the radius of gyration would be that about one of the principal axes.

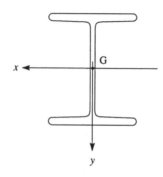

Fig. 18.12   Effect of cross-section on the buckling of columns

Another significant factor in determining the buckling load of a column is the method of end support. We saw in Section 18.1 that considerable changes in buckling load result from changes in end conditions. Thus a column with fixed ends has a higher value of buckling load than if the ends are pinned (cf. Eqs (18.5) and (18.10)). However, we have seen that by introducing the concept of equivalent length, the buckling loads of all columns may be referred to that of a pin-ended column no matter what the end conditions. It follows that Eq. (18.46) may be used for all types of end condition, provided that the equivalent length, L_e, of the column is used. Codes of Practice list equivalent or 'effective' lengths of columns for a wide variety of end conditions. Furthermore, although a column buckles naturally in a direction perpendicular to the axis about which EI is least, it is possible that the column may be restrained by external means in this direction so that buckling can only take place about the other axis.

18.5 Stability of beams under transverse and axial loads

Stresses and deflections in a linearly elastic beam subjected to transverse loads as predicted by simple beam theory are directly proportional to the applied loads. This relationship is valid if the deflections are small such that the slight change in geometry produced in the loaded beam has an insignificant effect on the loads themselves. This situation changes drastically when axial loads act simultaneously with the transverse loads. The internal moments, shear forces, stresses and deflections then become dependent upon the magnitude of the deflections as well as the magnitude of the external loads. They are also sensitive, as we observed in Section 18.3, to beam imperfections such as initial curvature and eccentricity of axial loads. Beams supporting both axial and transverse loads are sometimes known as *beam-columns* or simply as *transversely loaded columns*.

We consider first the case of a pin-ended beam carrying a uniformly distributed load of intensity w and an axial load, P, as shown in Fig. 18.13. The bending moment at any section of the beam is

$$M = Pv + \frac{wLz}{2} - \frac{wz^2}{2} = -EI\frac{d^2v}{dz^2} \quad \text{(from Eq. 13.3)}$$

giving

$$\frac{d^2v}{dz^2} + \frac{P}{EI}v = \frac{w}{2EI}(z^2 - Lz) \quad (18.47)$$

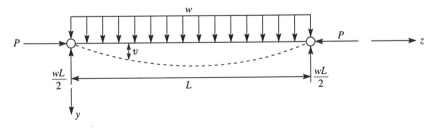

Fig. 18.13 Bending of a uniformly loaded beam-column

The standard solution of Eq. (18.47) is

$$v = C_1 \cos \mu z + C_2 \sin \mu z + \frac{w}{2P}\left(z^2 - Lz - \frac{2}{\mu^2}\right)$$

where C_1 and C_2 are unknown constants and $\mu^2 = P/EI$. Substituting the boundary conditions $v = 0$ at $z = 0$ and L gives

$$C_1 = \frac{w}{\mu^2 P}, \quad C_2 = \frac{w}{\mu^2 P \sin \mu L}(1 - \cos \mu L)$$

so that the deflection is determinate for any value of w and P and is given by

$$v = \frac{w}{\mu^2 P}\left[\cos \mu z + \left(\frac{1 - \cos \mu L}{\sin \mu L}\right)\sin \mu z\right] + \frac{w}{2P}\left(z^2 - Lz - \frac{2}{\mu^2}\right) \tag{18.48}$$

In beam-columns, as in beams, we are primarily interested in maximum values of stress and deflection. For this particular case the maximum deflection occurs at the centre of the beam and is, after some transformation of Eq. (18.48)

$$v_{\text{max}} = \frac{w}{\mu^2 P}\left(\sec \frac{\mu L}{2} - 1\right) - \frac{wL^2}{8P} \tag{18.49}$$

The corresponding maximum bending moment is

$$M_{\text{max}} = -Pv_{\text{max}} - \frac{wL^2}{8}$$

or, from Eq. (18.49) $\qquad M_{\text{max}} = \frac{w}{\mu^2}\left(1 - \sec \frac{\mu L}{2}\right) \tag{18.50}$

We may rewrite Eq. (18.50) in terms of the Euler buckling load, $P_{\text{CR}} = \pi^2 EI/L^2$, for a pin-ended column. Hence

$$M_{\text{max}} = \frac{wL^2}{\pi^2}\frac{P_{\text{CR}}}{P}\left(1 - \sec \frac{\pi}{2}\sqrt{\frac{P}{P_{\text{CR}}}}\right) \tag{18.51}$$

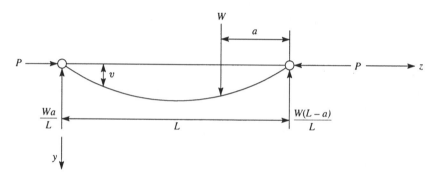

Fig. 18.14 Beam-column supporting a point load

As P approaches P_{CR} the bending moment (and deflection) becomes infinite. However, the above theory is based on the assumption of small deflections (otherwise d^2v/dz^2 would not be a close approximation for curvature) so that such a deduction is invalid. The indication is, though, that large deflections will be produced by the presence of a compressive axial load no matter how small the transverse load might be.

Let us consider now the beam-column of Fig. 18.14 with pinned ends carrying a concentrated load W at a distance a from the right-hand support.

For $z \leqslant L - a$,

$$EI \frac{d^2v}{dz^2} = -M = -Pv - \frac{Waz}{L} \tag{18.52}$$

and for $z \geqslant L - a$,

$$EI \frac{d^2v}{dz^2} = -M = -Pv - \frac{W}{L}(L - a)(L - z) \tag{18.53}$$

Writing

$$\mu^2 = P/EI$$

Equation (18.52) becomes

$$\frac{d^2v}{dz^2} + \mu^2 v = -\frac{Wa}{EIL} z$$

the general solution of which is

$$v = C_1 \cos \mu z + C_2 \sin \mu z - \frac{Wa}{PL} z \tag{18.54}$$

Similarly the general solution of Eq. (18.53) is

$$v = C_3 \cos \mu z + C_4 \sin \mu z - \frac{W}{PL}(L - a)(L - z) \tag{18.55}$$

where C_1, C_2, C_3 and C_4 are constants which are found from the boundary conditions as follows.

When $z = 0$, $v = 0$, therefore from Eq. (18.54) $C_1 = 0$. At $z = L$, $v = 0$ giving, from Eq. (18.55), $C_3 = -C_4 \tan \mu L$. At the point of application of the load the deflection and slope of the beam given by Eqs (18.54) and (18.55) must be the same. Hence, equating deflections,

$$C_2 \sin \mu(L - a) - \frac{Wa}{PL}(L - a) = C_4 [\sin \mu(L - a) - \tan \mu L \cos \mu(L - a)] - \frac{Wa}{PL}(L - a)$$

and equating slopes

$$C_2\mu \cos \mu(L - a) - \frac{Wa}{PL} = C_4\mu[\cos \mu(L - a) + \tan \mu L \sin \mu(L - a)] + \frac{W}{PL}(L - a)$$

Solving the above equations for C_2 and C_4 and substituting for C_1, C_2, C_3 and C_4 in Eqs (18.54) and (18.55) we have

$$v = \frac{W \sin \mu a}{P\mu \sin \mu L} \sin \mu z - \frac{Wa}{PL} z \quad \text{for} \quad z \le L - a \tag{18.56}$$

$$v = \frac{W \sin \mu(L-a)}{P\mu \sin \mu L} \sin \mu(L-z) - \frac{W}{PL}(L-a)(L-z) \quad \text{for} \quad z \ge L - a \tag{18.57}$$

These equations for the beam-column deflection enable the bending moment and resulting bending stresses to be found at all sections.

A particular case arises when the load is applied at the centre of the span. The deflection curve is then symmetrical with a maximum deflection under the load of

$$v_{max} = \frac{W}{2P\mu} \tan \frac{\mu L}{2} - \frac{WL}{4P}$$

Finally we consider a beam-column subjected to end moments, M_A and M_B, in addition to an axial load, P (Fig. 18.15). The deflected form of the beam-column may be found by using the principle of superposition and the results of the previous case. First we imagine that M_B acts alone with the axial load, P. If we assume that the point load, W, moves towards B and simultaneously increases so that the product $Wa = \text{constant} = M_B$ then, in the limit as a tends to zero, we have the moment M_B applied at B. The deflection curve is then obtained from Eq. (18.56) by substituting μa for $\sin \mu a$ (since μa is now very small) and M_B for Wa. Thus

$$v = \frac{M_B}{P} \left(\frac{\sin \mu z}{\sin \mu L} - \frac{z}{L} \right) \tag{18.58}$$

We find the deflection curve corresponding to M_A acting alone in a similar way. Suppose that W moves towards A such that the product $W(L-a) = \text{constant} = M_A$. Then as $(L-a)$ tends to zero we have $\sin \mu(L-a) = \mu(L-a)$ and Eq. (18.57) becomes

$$v = \frac{M_A}{P} \left[\frac{\sin \mu(L-z)}{\sin \mu L} - \frac{(L-z)}{L} \right] \tag{18.59}$$

The effect of the two moments acting simultaneously is obtained by superposition of the results of Eqs (18.58) and (18.59). Hence, for the beam-column of Fig. 18.15

$$v = \frac{M_B}{P} \left(\frac{\sin \mu z}{\sin \mu L} - \frac{z}{L} \right) + \frac{M_A}{P} \left[\frac{\sin \mu(L-z)}{\sin \mu L} - \frac{(L-z)}{L} \right] \tag{18.60}$$

Equation (18.60) is also the deflected form of a beam-column supporting eccentrically applied end loads at A and B. For example, if e_A and e_B are the eccentricities of P at the ends A and B, respectively, then $M_A = Pe_A$, $M_B = Pe_B$, giving a deflected form of

$$v = e_B \left(\frac{\sin \mu z}{\sin \mu L} - \frac{z}{L} \right) + e_A \left[\frac{\sin \mu(L-z)}{\sin \mu L} - \frac{(L-z)}{L} \right] \tag{18.61}$$

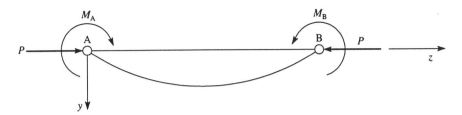

Fig. 18.15 Beam-column supporting end moments

Other beam-column configurations featuring a variety of end conditions and loading regimes may be analysed by a similar procedure.

18.6 Energy method for the calculation of buckling loads in columns (Rayleigh–Ritz method)

The fact that the total potential energy of an elastic body possesses a stationary value in an equilibrium state (see Section 15.3) may be used to investigate the neutral equilibrium of a buckled column. In particular the energy method is extremely useful when the deflected form of the buckled column is unknown and has to be 'guessed'.

First we shall consider the pin-ended column shown in its buckled position in Fig. 18.16. The internal or strain energy, U, of the column is assumed to be produced by bending action alone and is given by Eq. (9.21), i.e.

$$U = \int_0^L \frac{M^2}{2EI}\, dz \tag{18.62}$$

or alternatively, since $EI\, d^2v/dz^2 = -M$ (Eq. (13.3)),

$$U = \frac{EI}{2} \int_0^L \left(\frac{d^2v}{dz^2} \right)^2 dz \tag{18.63}$$

The potential energy, V, of the buckling load, P_{CR}, referred to the straight position of the column as datum, is then

$$V = -P_{CR}\delta$$

where δ is the axial movement of P_{CR} caused by the bending of the column from its

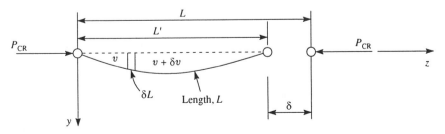

Fig. 18.16 Shortening of a column due to buckling

initially straight position. From Fig. 18.16 the length δL of the buckled column is

$$\delta L = (\delta z^2 + \delta v^2)^{1/2}$$

and since dv/dz is small then

$$\delta L \simeq \delta z \left[1 + \frac{1}{2}\left(\frac{dv}{dz}\right)^2 \right]$$

Hence

$$L = \int_0^{L'} \left[1 + \frac{1}{2}\left(\frac{dv}{dz}\right)^2 \right] dz$$

giving

$$L = L' + \int_0^{L'} \frac{1}{2}\left(\frac{dv}{dz}\right)^2 dz$$

Therefore

$$\delta = L - L' = \int_0^{L'} \frac{1}{2}\left(\frac{dv}{dz}\right)^2 dz$$

Since

$$\int_0^{L'} \frac{1}{2}\left(\frac{dv}{dz}\right)^2 dz$$

only differs from

$$\int_0^{L} \frac{1}{2}\left(\frac{dv}{dz}\right)^2 dz$$

by a term of negligible order, we write

$$\delta = \int_0^{L} \frac{1}{2}\left(\frac{dv}{dz}\right)^2 dz$$

giving

$$V = -\frac{P_{CR}}{2} \int_0^{L} \left(\frac{dv}{dz}\right)^2 dz \tag{18.64}$$

The total potential energy of the column in the neutral equilibrium of its buckled state is therefore

$$U + V = \int_0^{L} \frac{M^2}{2EI}\, dz - \frac{P_{CR}}{2} \int_0^{L} \left(\frac{dv}{dz}\right)^2 dz \tag{18.65}$$

or, using the alternative form of U from Eq. (18.63),

$$U + V = \frac{EI}{2} \int_0^{L} \left(\frac{d^2 v}{dz^2}\right)^2 dz - \frac{P_{CR}}{2} \int_0^{L} \left(\frac{dv}{dz}\right)^2 dz \tag{18.66}$$

We shall now assume a deflected shape having the equation

$$v = \sum_{n=1}^{\infty} A_n \sin \frac{n\pi z}{L} \tag{18.67}$$

This satisfies the boundary conditions of

$$(v)_{z=0} = (v)_{z=L} = 0, \quad \left(\frac{d^2v}{dz^2}\right)_{z=0} = \left(\frac{d^2v}{dz^2}\right)_{z=L} = 0$$

and is capable, within the limits for which it is valid and if suitable values for the constant coefficients, A_n, are chosen, of representing any continuous curve. We are therefore in a position to find P_{CR} exactly. Substituting Eq. (18.67) into Eq. (18.66) gives

$$U + V = \frac{EI}{2} \int_0^L \left(\frac{\pi}{L}\right)^4 \left(\sum_{n=1}^{\infty} n^2 A_n \sin \frac{n\pi z}{L}\right)^2 dz - \frac{P_{CR}}{2} \int_0^L \left(\frac{\pi}{L}\right)^2 \left(\sum_{n=1}^{\infty} n A_n \cos \frac{n\pi z}{L}\right)^2 dz \quad (18.68)$$

The product terms in both integrals of Eq. (18.68) disappear on integration leaving only integrated values of the squared terms. Thus

$$U + V = \frac{\pi^4 EI}{4L^3} \sum_{n=1}^{\infty} n^4 A_n^2 - \frac{\pi^2 P_{CR}}{4L} \sum_{n=1}^{\infty} n^2 A_n^2 \quad (18.69)$$

Assigning a stationary value to the total potential energy of Eq. (18.69) with respect to each coefficient, A_n, in turn, then taking A_n as being typical, we have

$$\frac{\partial(U + V)}{\partial A_n} = \frac{\pi^4 EI n^4 A_n}{2L^3} - \frac{\pi^2 P_{CR} n^2 A_n}{2L} = 0$$

from which

$$P_{CR} = \frac{\pi^2 EI n^2}{L^2}$$

as before.

We see that each term in Eq. (18.67) represents a particular deflected shape with a corresponding critical load. Hence the first term represents the deflection of the column shown in Fig. 18.16 with $P_{CR} = \pi^2 EI/L^2$. The second and third terms correspond to the shapes shown in Fig. 18.4(b) and (c) having critical loads of $4\pi^2 EI/L^2$ and $9\pi^2 EI/L^2$ and so on. Clearly the column must be constrained to buckle into these more complex forms. In other words, the column is being forced into an unnatural shape, is consequently stiffer and offers greater resistance to buckling, as we observe from the higher values of critical load.

If the deflected shape of the column is known, it is immaterial which of Eqs (18.65) or (18.66) is used for the total potential energy. However, when only an approximate solution is possible, Eq. (18.65) is preferable since the integral involving bending moment depends upon the accuracy of the assumed form of v, whereas the corresponding term in Eq. (18.66) depends upone the accuracy of d^2v/dz^2. Generally, for an assumed deflection curve v is obtained much more accurately than d^2v/dz^2.

Suppose that the deflection curve of a particular column is unknown or extremely complicated. We then assume a reasonable shape which satisfies as far as possible the end conditions of the column and the pattern of the deflected shape (Rayleigh–Ritz method). Generally the assumed shape is in the form of a finite series involving a series of unknown constants and assumed functions of z. Let us suppose that v is given by

$$v = A_1 f_1(z) + A_2 f_2(z) + A_3 f_3(z)$$

Substitution in Eq. (18.65) results in an expression for total potential energy in terms of the critical load and the coefficients A_1, A_2 and A_3 as the unknowns. Assigning stationary values to the total potential energy with respect to A_1, A_2 and A_3 in turn produces three simultaneous equations from which the ratios A_1/A_2, A_1/A_3 and the critical load are determined. Absolute values of the coefficients are unobtainable since the displacements of the column in its buckled state of neutral equilibrium are indeterminate.

As a simple illustration consider the column shown in its buckled state in Fig. 18.17. An approximate shape may be deduced from the deflected shape of a cantilever loaded at its free end. Thus, from Eq. (iv) of Ex. 13.1

$$v = \frac{v_0 z^2}{2L^3} (3L - z)$$

This expression satisfies the end conditions of deflection, viz. $v = 0$ at $z = 0$ and $v = v_0$ at $z = L$. In addition, it satisfies the conditions that the slope of the column is zero at the built-in end and that the bending moment, i.e. $d^2 v/dz^2$, is zero at the free end. The bending moment at any section is $M = P_{CR}(v_0 - v)$ so that substitution for M and v in Eq. (18.65) gives

$$U + V = \frac{P_{CR}^2 v_0^2}{2EI} \int_0^L \left(1 - \frac{3z^2}{2L^2} + \frac{z^3}{2L^3}\right)^2 dz - \frac{P_{CR}}{2} \int_0^L \left(\frac{3v_0}{2L^3}\right)^2 z^2 (2L - z)^2 \, dz$$

Integrating and substituting the limits we have

$$U + V = \frac{17}{35} \frac{P_{CR}^2 v_0^2 L}{2EI} - \frac{3}{5} P_{CR} \frac{v_0^2}{L}$$

Hence

$$\frac{\partial(U + V)}{\partial v_0} = \frac{17}{35} \frac{P_{CR}^2 v_0 L}{EI} - \frac{6 P_{CR} v_0}{5L} = 0$$

from which

$$P_{CR} = \frac{42EI}{17L^2} = 2 \cdot 471 \frac{EI}{L^2}$$

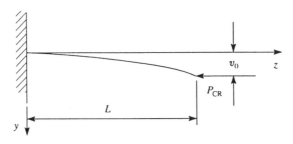

Fig. 18.17 Buckling load for a built-in column by the energy method

This value of critical load compares with the exact value (see Eq. (18.15)) of $\pi^2 EI/4L^2 = 2\cdot467 EI/L^2$; the error, in this case, is seen to be extremely small. Approximate values of critical load obtained by the energy method are always greater than the correct values. The explanation lies in the fact that an assumed deflected shape implies the application of constraints in order to force the column to take up an artificial shape. This, as we have seen, has the effect of stiffening the column with a consequent increase in critical load.

It will be observed that the solution for the above example may be obtained by simply equating the increase in internal energy (U) to the work done by the external critical load $(-V)$. This is always the case when the assumed deflected shape contains a single unknown coefficient such as v_0 in the above example.

In this chapter we have investigated structural instability with reference to the overall buckling or failure of columns subjected to axial load and also to bending. The reader should also be aware that other forms of instability occur. Thus the compression flange in an I-section plate girder can buckle laterally when the girder is subjected to bending moments unless it is restrained. Furthermore, thin-walled open section beams that are weak in torsion can exhibit torsional instability when subjected to axial load. These forms of instability are considered in more advanced texts.

Problems

P.18.1 A uniform column of length L and flexural rigidity EI is built-in at one end and is free at the other. It is designed so that its lowest buckling load is P. (Fig. P.18.1(a)). Subsequently it is required to carry an increased load and for that it is provided with a lateral spring at the free end (Fig. P.18.1(b)). Determine the necessary spring stiffness, k, so that the buckling load is $4P$.

Ans. $k = 4P\mu/(\mu L - \tan \mu L)$ where $\mu^2 = P/EI$.

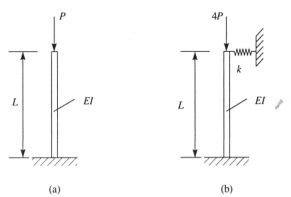

(a)   (b)

Fig. P.18.1

P.18.2 A pin-ended column of length L and flexural rigidity EI is reinforced to give a flexural rigidity $4EI$ over its central half. Determine its lowest buckling load.

Ans. $24\cdot2EI/L^2$.

P.18.3 A uniform pin-ended column of length L and flexural rigidity EI has an initial curvature such that the lateral displacement at any point between the column and the straight line joining its ends is given by

$$v_0 = a \frac{4z}{L^2}(L - z)$$

where a is the initial displacement at the mid-length of the column and the origin for z is at one end.

Show that the maximum bending moment due to a compressive axial load, P, is given by

$$M_{max} = \frac{8aP}{(\mu L)^2}\left(\sec\frac{\mu L}{2} - 1\right) \quad \text{where} \quad \mu^2 = \frac{P}{EI}$$

P.18.4 A compression member is made of circular-section tube having a diameter d and thickness t and is curved initially so that its initial deflected shape may be represented by the expression

$$v_0 = \delta \sin\left(\frac{\pi z}{L}\right)$$

in which δ is the displacement at its mid-length and the origin for z is at one end.

Show that if the ends are pinned, a compressive load, P, induces a maximum direct stress, σ_{max}, given by

$$\sigma_{max} = \frac{P}{\pi dt}\left[1 + \frac{1}{1 - \alpha}\frac{4\delta}{d}\right]$$

where $\alpha = P/P_{CR}$ and $P_{CR} = \pi^2 EI/L^2$. Assume that t is small compared with d so that the cross-sectional area of the tube is πdt and its second moment of area is $\pi d^3 t/8$.

P.18.5 In the experimental determination of the buckling loads for 12·5 mm diameter, mild steel, pin-ended columns, two of the values obtained were:

(i) length 500 mm, load 9800 N,
(ii) length 200 mm, load 26 400 N.

(a) Determine whether either of these values conforms to the Euler theory for buckling load.

(b) Assuming that both values are in agreement with the Rankine formula, find the constants σ_S and k. Take $E = 200\ 000$ N/mm^2.

Ans. (a) (i) conforms with Euler theory.
(b) $\sigma_S = 317$ N/mm^2, $k = 1 \cdot 16 \times 10^{-4}$.

P.18.6 A tubular column has an effective length of 2·5 m and is to be designed to carry a safe load of 300 kN. Assuming an approximate ratio of thickness to external diameter of 1/16, determine a practical diameter and thickness using the Rankine formula with $\sigma_S = 330$ N/mm^2 and $k = 1/7500$. Use a safety factor of 3.

Ans. Diameter = 128 mm, thickness = 8 mm.

P.18.7 A mild steel pin-ended column is 2·5 m long and has the cross-section shown in Fig. P.18.7. If the yield stress in compression of mild steel is 300 N/mm², determine the maximum load the column can withstand using the Robertson formula. Compare this value with that predicted by the Euler theory.

Ans. 576 kN, $P(\text{Rob.})/P(\text{Euler}) = 0·62$.

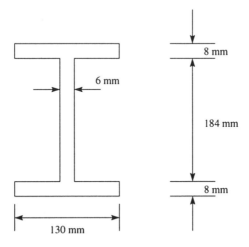

Fig. P.18.7

P.18.8 A compression member in a framework is subjected to an axial load of 20·2 tonnes and a bending moment of 0·1 t m; its effective length is 1·5 m. For practical purposes a channel section is the most suitable; design the section.

Ans. 203 mm × 76 mm × 23·8 kg is a suitable section.

P.18.9 A pin-ended column of length L has its central portion reinforced, the second moment of its area being I_2 while that of the end portions, each of length a, is I_1. Use the Rayleigh–Ritz method to determine the critical load of the column assuming that its centreline deflects into the parabola $v = kz(L - z)$ and taking the more accurate of the two expressions for bending moment.

In the case where $I_2 = 1·6I_1$ and $a = 0·2L$ find the percentage increase in strength due to the reinforcement and compare it with the percentage increase in weight on the basis that the radius of gyration of the section is not altered.

Ans. $P_{\text{CR}} = 15·2EI_1/L^2$, 52%, 36%.

P.18.10 A tubular column of length L is tapered in wall thickness so that the area and the second moment of area of its cross-section decrease uniformly from A_1 and I_1 at its centre to $0·2A_1$ and $0·2I_1$ at its ends, respectively.

Assuming a deflected centreline of parabolic form and taking the more correct form for the bending moment, use the Rayleigh–Ritz method to estimate its critical load; the ends of the column may be taken as pinned. Hence show that the saving in weight by using such a column instead of one having the same radius of gyration and constant thickness is about 15%.

Ans. $7EI_1/L^2$.

Index